INORGANIC SYNTHESES

Volume 26

Editor-in-Chief

HERBERT D. KAESZ

Department of Chemistry and Biochemistry
University of California, Los Angeles

•••

INORGANIC
SYNTHESES

Volume 26

WILEY

A Wiley-Interscience Publication

JOHN WILEY & SONS

New York   Chichester   Brisbane   Toronto   Singapore

546
I58m

Published by John Wiley & Sons, Inc.

Copyright © 1989 Inorganic Syntheses, Inc.

Library of Congress Catalog Number: 39-23015
ISBN 0-471-50485-8

Printed in the United States of America

10 9 8 7 6 5 4 3 2 1

238964

This volume is dedicated to
S. Young Tyree, Jr. (1920–1987),
who played a very important role in
the development of *Inorganic Syntheses*
as Secretary-Treasurer for several years
and as Editor-in-Chief of Volume IX

PREFACE

Jean'ne M. Shreeve stated in the Preface to Volume 24 of this series that, "preparative chemistry is the heart of chemistry." However, chemistry plays an important role in many scientific disciplines. Focusing on inorganic and metalorganic compounds for example, we note that metal complexes are playing an important role in autoradiography and cancer chemotherapy in the field of medicine. Mixed metal oxides are leading to new superconductors. Metal organic compounds are precursors to new materials, and volatile metal complexes are used in chemical vapor deposition of thin films in surface science, physics, and electronics. The same is true for the role of new and useful substances in organic chemistry, polymer chemistry, biochemistry, molecular biology, and so on. Thus, I am prompted to generalize and to suggest that *synthesis is a central discipline in science.*

Synthesis can be a difficult practice, as any novice to the field will attest. *Inorganic Syntheses* is an attempt to identify compounds of general interest, to select the best procedures for their preparation from generally available starting materials, and to verify that the procedures can be carried out successfully as written. I hope the scientific community finds this volume to be a useful collection, and continues to submit procedures and accept manuscripts for checking in continuing volumes of this series.

Contents. Submissions have been grouped into chemically related categories represented by the chapter headings. Because many of the submissions also incorporate syntheses of starting materials, some procedures are located outside of the category in which they would separately belong. For example, there are syntheses of compounds of the main group elements that appear beyond Chapter 1, such as procedures numbered 10.A, 10.C, 26, 27, 48.A, 50.A, 62.A, and 70.A. Similarly, procedures for some mononuclear complexes are located in Chapter 5, see 44.A, 59.A, 62.B, and 62.C.1, while that of a tetrarhenium derivative is located in Chapter 3, see 19.D.

Previous volumes of *Inorganic Syntheses* are available. Many of the volumes originally published by McGraw Hill, Inc. are available from R. E. Krieger Publishing Co., Inc., P.O. Box 9542, Melbourne, FL 32901. Please write this publisher for a current list. Volumes out of print with John Wiley & Sons, Inc. are also available from Krieger Publishing. Recent back volumes can be obtained from John Wiley & Sons, Inc., 605 Third Avenue, New York, NY 10158. Please write the publisher for a current list of available volumes.

Within each chapter the submissions are ordered, as far as possible, according to the columns of the periodic table. In Chapter 5, homopolynuclear complexes are presented first, followed by heteropolynuclear compounds. Many submissions, however, contain several parts. As a consequence, the location of some of the procedures can not adhere to this general sequence.

Acknowledgments.   Like editors of previous volumes of this series, I am indebted to many colleagues for contributions to this work. First there are Wolfgang Beck in Munich, Ekkehard Lindner in Tübingen, and John R. Shapley in Urbana, who assisted me in the selections and solicitations for three special collections: transition metal complexes containing weakly bonded anions (Chapter 3), metalocyclic complexes (Chapter 4), and polynuclear transition metal complexes (Chapter 5), respectively. Each has contributed a preface for the chapter he helped to form, following the pattern set in Volume XII when Alan G. MacDiarmid invited E. C. Ashby and myself to form a chapter on metal hydrides. As a consequence of such efforts, more than two-thirds of this volume consists of invited preparations.

The above-mentioned three valued colleagues and friends are also thanked for further participating as submitters and checkers. Indeed *all* the individuals so listed deserve thanks for the contributions that are the contents of this volume.

Editorial work was assisted by evaluations and comments on submitted manuscripts by members of Inorganic Syntheses. Foremost and unfailing in this group were J. C. Bailar, Jr., T. Moeller, D. F. Shriver, R. J. Angelici, and the late W. C. Fernelius. Advice on nomenclature was routinely provided by T. E. Sloan and W. H. Powell of Chemical Abstracts Service.

To all of the above-mentioned colleagues, and to my family from whom I have had to withdraw for many hours of solitary work, I extend my wholehearted thanks.

HERBERT D. KAESZ

Los Angeles, California
August 1988

NOTICE TO CONTRIBUTORS
AND CHECKERS

The *Inorganic Syntheses* series is published to provide all users of inorganic substances with detailed and foolproof procedures for the preparation of important and timely compounds. Thus the series is the concern of the entire scientific community. The Editorial Board hopes that all chemists will share in the responsibility of producing *Inorganic Syntheses* by offering their advice and assistance in both the formulation of and the laboratory evaluation of outstanding syntheses. Help of this kind will be invaluable in achieving excellence and pertinence to current scientific interests.

There is no rigid definition of what constitutes a suitable synthesis. The major criterion by which syntheses are judged is the potential value to the scientific community. For example, starting materials or intermediates that are useful for synthetic chemistry are appropriate. The synthesis also should represent the best available procedure, and new or improved syntheses are particularly appropriate. Syntheses of compounds that are available commercially at reasonable prices are not acceptable. We do not encourage the submission of compounds that are unreasonably hazardous, and in this connection, less dangerous anions generally should be employed in place of perchlorate.

The Editorial Board lists the following criteria of content for submitted manuscripts. Style should conform with that of previous volumes of *Inorganic Syntheses*. The introductory section should include a concise and critical summary of the available procedures for synthesis of the product in question. It should also include an estimate of the time required for the synthesis, an indication of the importance and utility of the product, and an admonition if any potential hazards are associated with the procedure. The Procedure should present detailed and unambiguous laboratory directions and be written so that it anticipates possible mistakes and misunderstandings on the part of the person who attempts to duplicate the procedure. Any unusual equipment or procedure should be clearly described. Line drawings should be included when they can be helpful. All safety measures should be stated clearly. Sources of unusual starting materials must be given, and, if possible, minimal standards of purity of reagents and solvents should be stated. The scale should be reasonable for normal laboratory operation, and any problems involved in scaling the procedure either up or down should be discussed. The criteria for judging the purity of the final product should be delineated clearly. The section on Properties should supply and discuss those physical and chemical characteristics that are relevant to judging the purity of the product and to permitting its handling and use in an intelligent manner. Under References, all pertinent literature citations should be listed in order. A style sheet is available from the Secretary of the Editorial Board. Authors are requested to avoid procedures involving perchlorate salts due to the high risk of explosion in combination with organic or organometallic substances. It is also suggested if at all possible to avoid the use of benzene as a solvent owing to its carcinogenic properties.

The Editorial Board determines whether submitted syntheses meet the general specifications outlined above, and the Editor-in-Chief sends the manuscript to an independent laboratory where the procedure must be satisfactorily reproduced.

Each manuscript should be submitted in duplicate to the Secretary of the Editorial Board, Professor Jay H. Worrell, Department of Chemistry, University of South Florida, Tampa, FL 33620. The manuscript should be typewritten in English. Nomenclature should be consistent and should follow the recommendations presented in *Nomenclature of Inorganic Chemistry*, 2nd ed., Butterworths & Co., London, 1970, and in *Pure and Applied Chemistry*, Volume 28, No. 1 (1971). Abbreviations should conform to those used in publications of the American Chemical Society, particularly *Inorganic Chemistry*.

Chemists willing to check syntheses should contact the editor of a future volume or make this information known to Professor Worrell.

TOXIC SUBSTANCES AND
LABORATORY HAZARDS

Chemicals and chemistry are by their very nature hazardous. Chemical reactivity implies that reagents have the ability to combine. This process can be sufficiently vigorous as to cause flame, an explosion, or, often less immediately obvious, a toxic reaction.

The obvious hazards in the syntheses reported in this volume are delineated, where appropriate, in the experimental procedure. It is impossible, however, to foresee every eventuality, such as a new biological effect of a common laboratory reagent. As a consequence, *all* chemicals used and *all* reactions described in this volume should be viewed as potentially hazardous. Care should be taken to avoid inhalation or other physical contact with all reagents and solvents used in procedures described in this volume. In addition, particular attention should be paid to avoiding sparks, open flames, or other potential sources that could set fire to combustible vapors or gases.

A list of 400 toxic substances may be found in the *Federal Register*, Vol. 40, No. 23072, May 28, 1975. An abbreviated list may be obtained from *Inorganic Syntheses*, Volume 18, p. xv, 1978. A current assessment of the hazards associated with a particular chemical is available in the most recent edition of *Threshold Limit Values for Chemical Substances and Physical Agents in the Workroom Environment* published by the American Conference of Governmental Industrial Hygienists.

The drying of impure ethers can produce a violent explosion. Further information about this hazard may be found in *Inorganic Syntheses*, Volume 12, p. 317. A hazard associated with the synthesis of tetramethyldiphosphine disulfide [*Inorg. Synth.*, **15**, 186 (1974)] is cited in *Inorganic Syntheses*, Volume 23, p. 199.

CONTENTS

Chapter Six   SOLID STATE

INORGANIC SYNTHESES

Volume 26

Chapter One

COMPOUNDS OF THE MAIN GROUP ELEMENTS AND THE LANTHANIDES

1. POTASSIUM TETRADECAHYDRONONABORATE(1 –)

$$B_{10}H_{14} + [OH]^- \longrightarrow [B_{10}H_{13}]^- + H_2O$$
$$[B_{10}H_{13}]^- + [OH]^- \longrightarrow [B_{10}H_{13}OH]^{2-}$$
$$[B_{10}H_{13}OH]^{2-} + [H_3O]^+ + H_2O \longrightarrow [B_9H_{14}]^- + B(OH)_3 + H_2$$

Submitted by PATRICK J. DOLAN,* DONALD F. GAINES,† DAVID C. MOODY,* CATERINA K. NELSON,† and RILEY SCHAEFFER*
Checked by SHELDON G. SHORE‡ and STEVEN H. LAWRENCE

There are two synthetically useful routes to salts of the tetradecahydrononaborate(1 –) ion, $[B_9H_{14}]^-$. The first, which involves the reaction of NaH and pentaborane(9), B_5H_9, in tetrahydrofuran (THF) in a 1:2 molar ratio,[1] suffers from two drawbacks:

(1) If a tetraalkylammonium salt is not present, fairly severe contamination by $[B_{11}H_{14}]^-$ salts occurs,[2] so the procedure is most useful for preparing tetraalkylammonium salts. The potassium salt is often more desirable as it is the specified precursor in the syntheses of iso-B_9H_{15},[3] B_8H_{12},[4] and several metallaboranes.[5]

*Chemistry Department, Indiana University, Bloomington, IN 47401.
†Chemistry Department, University of Wisconsin, Madison, WI 53706.
‡Department of Chemistry, The Ohio State University, Columbus, OH 43210.

(2) The synthesis employs B_5H_9, a pyrophoric liquid, and requires special handling procedures.

The second route to salts of the $[B_9H_{14}]^-$ anion, described below, is accomplished by base degradation of $B_{10}H_{14}$ (which is a moderately air-stable solid) followed by acid hydrolysis, yielding an aqueous solution of the $[B_9H_{14}]^-$ anion.[6] The potassium salt is precipitated by addition of K_2CO_3, and the $K[B_9H_{14}]$ is then purified by extraction with diethyl ether.

- **Caution.** *The materials used in this synthesis are toxic and should be handled only in a hood by persons wearing gloves. The diethyl ether extraction must be done with peroxide-free diethyl ether to minimize the possibility of an explosion during the final evacuation.*

Procedure

The $B_{10}H_{14}$ used in the following procedure should be freshly sublimed at a temperature below 80 °C. Nitrogen should be admitted to the sublimation apparatus only after it has been cooled to room temperature.

A solution of 17 g (0.30 mol) KOH in 100 mL H_2O in a 1-L, three-necked flask is stirred in an ice bath under N_2 until the temperature of the solution reaches 0 °C. Then, 22 g (0.18 mol) of sublimed, freshly powdered $B_{10}H_{14}$ is added, whereupon the solution rapidly turns yellow. This solution is stirred (at 0 °C) for ~ 3 h, by which time most ($\sim 90\%$) of the $B_{10}H_{14}$ has dissolved.

The solution is slowly acidified to pH 2 with $12\,M$ HCl, while the temperature is maintained at 0 °C. Much frothing accompanies the acidification. The pH may be monitored using pH paper. The aqueous solution is filtered and then extracted twice with 30-mL portions of hexane to remove any residual decaborane. Then 50 mL of *peroxide-free* diethyl ether is added to the aqueous solution, followed by 50 g of anhydrous potassium carbonate, K_2CO_3. The diethyl ether layer is separated, and the aqueous layer is extracted three times with 30-mL portions of ether. The extracts are combined in a 400-mL beaker, which is covered with a watch glass and placed in a vacuum desiccator. The desiccator is evacuated by means of a water aspirator for 4 h, or nearly to dryness. The desiccator is then evacuated for an additional 24 h using a liquid nitrogen trapped two-stage vacuum pump. The desiccator is filled with dry nitrogen and opened. The beaker is transferred to a glove bag for product removal. Yield: 23–26 g (80–90%). The salt may be further purified by dissolving in peroxide-free diethyl ether, whereupon residual salts such as K_2CO_3, KCl, or KOH remain undissolved. After filtration or decantation, evaporation gives a highly purified product. The $K[B_9H_{14}]$ is shown to be boron pure by comparison of its ^{11}B NMR spectrum with that previously reported.[7]

This reaction may also be run on a 10% scale, but the volumes of the hexane and diethyl ether extractions are *not* reduced. The yield is 90%.

If an alkylammonium salt is desired, it can be obtained by a metathesis reaction. In a typical reaction, 3 g of $K[B_9B_{14}]$ is dissolved in 30 mL of deoxygenated distilled water, and a solution of 8.8 g of $[(C_2H_5)_4N]Br$ in 20 mL of deoxygenated distilled water is added. A precipitate forms immediately. The solution is filtered and the solid is dried by evacuation using a liquid nitrogen trapped two-stage vacuum pump. The product is obtained in 90% yield.

Properties

Potassium tetradecahydrononaborate(1 −) is a white solid that is soluble in ethers, water, and acetonitrile. The 86.6 MHz ^{11}B NMR spectrum in CD_3CN consists of three doublets of equal area: a low-field doublet at − 8.0 ppm ($J = 142$ Hz) and two high-field doublets at − 20.5 ppm ($J = 132$ Hz) and − 23.6 ppm ($J = 142$ Hz). The compound $K[B_9B_{14}]$ exhibits no detectable decomposition (via ^{11}B NMR) following exposure to air for 1 day. In diethyl ether solution, however, there is significant decomposition after 1 day's exposure to air.

Salts of the $[B_9H_{14}]^-$ anion have been used as the primary source for salts of *nido*-$B_9H_{12}^-$, which in turn are readily oxidized to *n*-$B_{18}H_{22}$.[5a] In metallaborane chemistry salts of the $[B_9H_{14}]^-$ anion have proved very versatile cluster fragment precursors. Metals ranging from manganese[5b,c] to platinum[8] have been inserted into $[B_9H_{14}]^-$ to produce a number of new classes of metallaboranes.[9]

References

1. J. B. Leach, M. A. Toft, F. L. Himpsl, and S. G. Shore, *J. Am. Chem. Soc.*, **103**, 988–989 (1981).

2. P. C. Keller, private communication.

3. J. Dobson, P. C. Keller, and R. Schaeffer, *J. Am. Chem. Soc.*, **87**, 3522–3523 (1965).

4. J. Dobson, P. C. Keller, and R. Schaeffer, *Inorg. Chem.*, **7**, 399–402 (1968).

5. (a) J. W. Lott and D. F. Gaines, *Inorg. Chem.*, **13**, 2261–2267 (1974). (b) J. C. Calabrese, M. B. Fischer, D. F. Gaines, and J. W. Lott, *J. Am. Chem. Soc.*, **96**, 6318–6323 (1974); (c) D. F. Gaines, C. K. Nelson, and G. A. Steehler, *J. Am. Chem. Soc.*, **106**, 7266–7267 (1984).

6. L. E. Benjamin, S. F. Stafiej, and E. A. Takacs, *J. Am. Chem. Soc.*, **85**, 2674 (1963).

7. P. C. Keller, *Inorg. Chem.*, **9**, 75 (1970).

8. A. R. Kane, L. J. Guggenberger, and E. L. Muetterties, *J. Am. Chem. Soc.*, **92**, 2571–2572 (1970).

9. N. N. Greenwood and J. D. Kennedy, in *Metal Interactions with Boron Clusters*, R. N. Grimes (ed.), Plenum Press, New York, 1982, pp. 43–118.

2. BROMOTRIMETHYLSILANE

$$(C_6H_5)_3P + Br_2 + (CH_3)_3SiOSi(CH_3)_3 \xrightarrow{Zn} 2BrSi(CH_3)_3 + (C_6H_5)_3PO$$

Submitted by CLAUDIO PALOMO* and JESUS M. AIZPURUA*
Checked by JOYCE Y. COREY[†] and JANET BRADDOCK[†]

The methods for the preparation of bromotrimethylsilane include the treatment of hexamethyldisiloxane with bromine reagents such as phosphorus tribromide,[1,2] 2, 2, 2-tribromo-1, 3, 2 λ^5-benzodioxaphosphole (catechyl phosphorus tribromide),[2] and aluminium tribromide,[3] the yields being in the range from 73 to 87%. A further method[4] shows that bromotrimethylsilane can be obtained from aminosilanes and hydrobromic acid, but the yield (55%) is lower than in the former methods. Other methods involve the use of expensive reagents, such as hexamethyldisilane, 1, 4-bis(trimethylsilyl)-2, 5-cyclohexadiene, and 1, 4-dihydro-1, 4-bis(trimethylsilyl)naphthalene.[5] Finally, some alternative procedures have been developed for its *in situ* preparation.[5]

The following procedure is an operatively simple route for the synthesis of bromotrimethylsilane on a preparative laboratory scale from reagents that are readily accessible and inexpensive. This could be a method of choice in some laboratories despite the fact that bromotrimethylsilane is now commercially available (Petrach Systems, Aldrich, or Alpha). Moreover, the procedure also serves as a suitable method for the synthesis of azidotrimethylsilane and isocyanatotrimethylsilane, and is specially useful for the preparation of cyanotrimethylsilane. Thus a mixture of triphenylphosphine dibromide, hexamethyldisiloxane, and a catalytic amount of powdered metal zinc in 1, 2-dichlorobenzene is heated under reflux to produce bromotrimethylsilane in nearly quantitative yield, which is simultaneously distilled over a suspension of the corresponding pseudohalogenoacid salt in N, N-dimethylformamide as solvent.[6]

Procedure

■ **Caution.** *Owing to the lachrymatory properties of bromine reagent and halosilane product, this reaction should be carried out in a well-ventilated hood.*

*Department to Organic Chemistry, University of Euskal Herria, Donostia, Apdo 1072, 20080 Spain.
[†]Department of Chemistry, University of Missouri, St. Louis, MO 63121.

Source of Reagents

Triphenylphosphine was obtained from BASF, and the freshly opened material was used without further purification or drying. 1, 2-Dichlorobenzene from Panreac, was used without purification or drying. These materials were obtained from Aldrich by the Checkers. Hexamethyldisiloxane was obtained from hydrolysis of chorotrimethylsilane and distilled before use. Thus, a mixture of chlorotrimethylsilane (253 mL, 2.00 mol), ice (150 g), and water (200 mL) was vigorously stirred for 2 h, and then, the inorganic layer was separated using a separatory funnel. The organic layer was successively washed with water (50 mL), 2 N sodium hydroxide (100 mL), and water again (50 mL), after which, the crude siloxane was dried over $MgSO_4$ and purified by distillation to afford pure hexamethyldisiloxane (314.8 g, 97%)(bp 101–102 °C) and (bp 102 °C).[7] The chlorotrimethylsilane was obtained from Wacker-Chemie in Germany, or may be obtained from Petrach Systems, Aldrich, or Alpha in the United States.

In a dry 250-mL round-bottomed flask fitted with a magnetic stirring bar and a funnel equipped with a drying tube are placed (39.2 g, 0.15 mol) of triphenylphosphine and 75 mL of 1, 2-dichlorobenzene. After this, 16.8 g (22 mL, 0.104 mol) of hexamethyldisiloxane and a catalytic amount (0.5 g) of powdered zinc metal are added to this stirred mixture and the flask is cooled in an ice bath. Bromine 27.2 g (8.71 mL, 0.17 mol), dissolved in 25 mL of 1, 2-dichlorobenzene is placed in the addition funnel and added dropwise to the mixture over a 20 to 30-min period. During the addition, the colorless or yellow suspension becomes bright yellow. It is very important to achieve yellow coloration of the mixture. An excess of bromine may be used for this purpose. This fact is critical for a good yield and purity of the product. After the addition is completed, the cooling bath is removed and the reaction flask is placed in a silicone oil bath.

■ **Caution.** *Due to temperatures reaching 200–215 °C, a silicone oil bath is to be used. Ordinary hydrocarbon oil baths will ignite at these temperatures.*

Under an atmosphere of dry nitrogen gas, the dropping funnel is replaced by a fractional distillation apparatus provided with a 10-cm Vigreux column and a 5-cm water-cooled condenser. The system is heat-dried under a flow of nitrogen gas with an electric heat gun or may be assembled from oven-dried glassware. The mixture is then heated over a period of 20–30 min until the silicone oil bath reaches 130–140 °C. During this time, the initial yellow precipitate is partially dissolved and the mixture refluxes smoothly. The silicone oil bath is maintained at this temperature for an additional 30 min. Heating is then increased over the period of 1 h to achieve distillation of the product; the final oil temperature reaches 200 to 215 °C.

The yield of bromotrimethylsilane is 30.60 g (96%, based on hexamethyldisiloxane used) (bp 80–81 °C). Other characteristics,[5] (bp 79.9 °C/754 torr, d_4^{20}: 1.188. Bromotrimethylsilane hydrolyzes more rapidly than chlorotrimethylsilane. The purity of the product is 98% as established by NMR in CCl_4 solution. A single peak at 0.5 ppm downfield from tetramethylsilane is observed, the only impurities being siloxane hydrolysis products.

Properties

Bromotrimethylsilane has proven to be useful for a wide variety of applications most of them being reviewed.[5,8] Other recent applications are mild cleavage of oxiranes,[9] the synthesis of glycosyl bromides,[10] the selective cleavage of tetrahydro-2,5-dimethoxyfuran and tetrahydro-2,6-dimethoxypyran,[11] the cleavage of esters and ethers,[12] and the synthesis of benzyl bromides.[13]

References

1. W. F. Gilliam, R. N. Meals, and R. O. Sauer, *J. Am. Chem. Soc.*, **68**, 1161 (1948).

2. H. Gross, C. Böck, B. Costisella, and J. J. Gloede, *Prakt. Chem.*, **320**, 344 (1978).

3. M. G. Voronkov, B. N. Dolgov, and N. A. Dimitrieva, *Proc. Acad. Sci. USSR*, **84**, 959 (1952).

4. D. L. Bailey, L. H. Sommer, and F. C. Withmore, *J. Am. Chem. Soc.*, **70**, 435 (1948).

5. A. H. Schmidt, *Aldrichimica Acta*, **14**, 31 (1981) and references cited therein.

6. J. M. Aizpurua and C. Palomo, *Nouv. J. Chim.*, **8**, 51 (1984).

7. A. E. Pierce, *Silylation of Organic Compounds*, Pierce Chemical Co., Rockford, III (1968).

8. E. Colvin, *Silicon in Organic Synthesis*, Butterworths, London, 1981, p. 288.

9. H. R. Kricheldorf, G. Morber, and W. Regel, *Synthesis*, 383 (1981).

10. J. W. Gillard and M. Israel, *Tetrahedron Lett.*, **24**, 513 (1981).

11. T. H. Chan and S. D. Lee, *Tetrahedron Lett.*, **24**, 1225 (1983).

12. E. C. Friedrich and G. Delucca, *J. Org. Chem.*, **48**, 1678 (1983).

13. C. Palomo and J. M. Aizpurua, *Tetrahedron Lett*, **25**, 1103 (1984).

3. TRIMETHYLPHOSPHINE

$$P(OC_6H_5)_3 + 3CH_3MgBr \xrightarrow{Bu_2O} P(CH_3)_3 + 3Mg(OC_6H_5)Br$$

Submitted by M. L. LUETKENS, Jr.,* A. P. SATTELBERGER,*,† H. H.
MURRAY,‡ J. D. BASIL,‡ and J. P. FACKLER, Jr.‡
Checked by R. A. JONES§ and D. E. HEATON§

Trimethylphosphine is a ligand of proven utility in organometallic chemistry. However, due to its expense when purchased from commercial sources [Strem Chemicals] or its poor[1] to moderate[2] yields when isolated as the silver iodide adduct $[AgI(PMe_3)]_4$, its potential is not fully realized. Frequently, large quantities of the phosphine are required, for example, as a reactive solvent[3] or in the field of lanthanide and actinide chemistry, wherein the lability of the phosphine ligand may require crystallization from neat trimethylphosphine.[4] Similar considerations apply in exploratory synthetic early transition metal chemistry.[5] In transition metal ylide chemistry,[6] access to quantities of PMe_3 is also very desirable, particularly when excess ylide is required. The volatility of PMe_3 facilitates work-up of reaction mixtures and the ligand provides convenient proton NMR signals for product characterization. Trimethylphosphine derivatives usually have good solubility properties and show enhanced crystallizability relative to analogs containing other tertiary phosphines.

The following procedure is a hybrid of several trimethylphosphine syntheses.[1a,7,8] The key features of the new procedure are the use of triphenyl phosphite[8] in place of phosphorus(III) halides[1a,7] and dibutyl ether in place of diethyl ether.[1a,9] The present procedure has the following advantages over the preparations reported previously in *Inorganic Syntheses*:[2,7] (a) large quantities ($\sim$ 1 mol) of PMe_3 can be made in 1 day in a "one-pot synthesis" from an easily handled and fairly innocuous and inexpensive phosphorus compound, $P(OC_6H_5)_3$, and (b) the lower boiling PMe_3 (bp 39–40 °C) is easily separated from the reaction mixture (dibutyl ether, bp 142 °C) by distillation at atmospheric pressure giving the phosphine in high yield and purity.

*Department of Chemistry, The University of Michigan, Ann Arbor, MI 48109.
†Present address: Los Alamos National Laboratory, Los Alamos, NM 87545.
‡Department of Chemistry, Texas A & M University, College Station, TX 77843.
§Department of Chemistry, University of Texas, Austin, TX 78712.

Starting Materials

Dibutyl ether (Aldrich, 99%) is dried and freed from dissolved molecular oxygen by distillation under nitrogen from a solution of the solvent and sodium benzophenone ketyl.

■ **Caution.** *Dibutyl ether should be stored under nitrogen or argon, over sodium wire to prevent the formation of peroxides.*

Triphenyl phosphite (Aldrich, 97%) is purified by fractional vacuum distillation (bp 181–183 °C, 1 torr) and stored in the dark under nitrogen. A $2 M$ solution of CH_3MgBr in dibutyl ether may be purchased from Alfa Inorganics, or prepared according to the following procedure. If the CH_3MgBr solution is prepared in the laboratory, bromomethane (Air Products) is used without further purification.

Procedure

■ **Caution.** *Bromomethane and trimethylphosphine are volatile, toxic materials. This reaction must be carried out in an efficient fume hood. Trimethylphosphine may ignite spontaneously in air[2] and must be handled under an inert atmosphere at all times.*

All operations are performed under an atmosphere of prepurified nitrogen. An oven-dried, 3-L, three-necked round-bottomed flask equipped with an N_2 gas inlet, a precision mechanical stirrer, and a jacketed 250-mL pressure-equalizing dropping funnel (Fig. 1) vented through an oil bubbler, is charged with Grignard grade magnesium turnings (101 g, 4.15 mol) and ~1.5 L of purified dibutyl ether under a countercurrent of N_2. The mixture is then cooled to 0 °C in a large tub of ice water. The jacket of the dropping funnel is filled with a mixture of powdered Dry Ice–acetone, and bromomethane (200 mL, 365.6 g, 3.85 mol) is carefully poured into the dropping funnel. If a metal gas cylinder equipped with a manual control (needle) valve is the source of bromomethane, liquid can be withdrawn by *inverting* the cylinder. Tygon tubing is a suitable material for liquid transfer to the dropping funnel. The cold bromomethane is added dropwise to the stirred suspension. The Grignard reaction usually starts after the first 5 to 10 mL of bromomethane is added. If it does not, a few small crystals of iodine should be added before resuming the addition of bromomethane. The remaining bromomethane is added after the Grignard reaction has been initiated, as evidenced by a shiny surface appearing on the magnesium turnings. The addition of bromomethane should be carried out over a 2-h period.

After the addition of bromomethane is complete, the dropping funnel is replaced with a stopper, the ice tub is removed, and the Grignard solution is stirred at room temperature for 4 h (or overnight if this is more convenient).

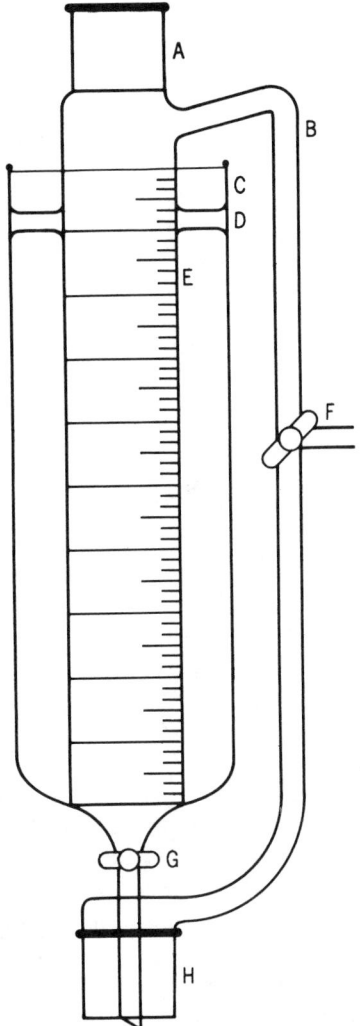

Fig. 1. Jacketed 250-mL pressure-equalizing dropping funnel: (A) 24/40 female ground joint, (B) 10-mm glass tubing, (C) 80-mm tubing, (D) solid glass braces, (E) 250-mL graduated cylinder, (F) three-way, 2-mm glass stopcock, (G) 2-mm glass stopcock, (H) 24/40 male ground joint.

The stopper is then replaced with a 500-mL pressure-equalizing dropping funnel and the white Grignard suspension is cooled to $\sim -5\,°C$ with a large ice–salt bath. A solution of triphenyl phosphite (300 mL, 355.2 g, 1.145 mol) and dibutyl ether (300 mL) is added dropwise to the Grignard solution, with

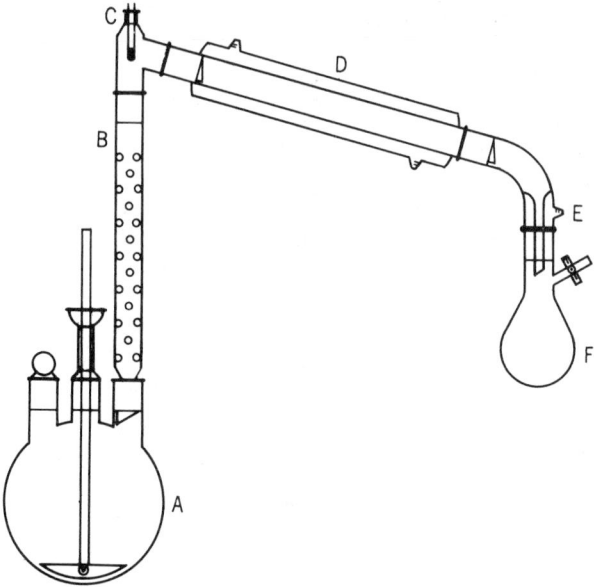

Fig. 2. First distillation assembly: (A) 3-L reaction flask, (B) 15-in Vigreaux column, (C) 24/40 distillation head with 150 °C thermometer, (D) water-cooled condenser, (E) to glass "tee," which is also connected to an N_2 source and a mineral oil bubbler, (F) 200-mL Schlenk receiver flask.

vigorous stirring, over the course of 2.5 h. The ice–salt bath should be replenished periodically to maintain the temperature of this exothermic reaction below ~ 10 °C. After the phosphite addition is complete, the cooling bath is removed and the dropping funnel is replaced with an N_2 purged distillation assembly (Fig. 2) under N_2 flow. The latter is connected via a glass "tee" to the nitrogen source and a mineral oil bubbler. A slow flow of N_2 is maintained during the following operation. The 200-mL receiving flask is cooled to − 78 °C (Dry Ice–acetone) and the contents of the reaction flask are brought to a *gentle* reflux using a 3-L heating mantle. Ideally, at the early stages of the distillation, the butyl ether condenses about one-quarter of the way up the Vigreaux column. A "cloud" of trimethylphosphine vapor appears on the upper part of the column, condenses, and collects in the receiver.

Product is collected until the still head temperature reaches 110 °C. At this point the heating mantle is removed, the nitrogen flow is increased slightly, and the distillation flask is allowed to cool. When the latter is cool enough to touch, the receiving flask is carefully disconnected under a positive N_2 flow, and capped. The *warm* residue in the distillation flask (which solidifies on cooling to room temperature) may be poured into waste solvent bottles for disposal or slowly flushed down the fume hood drain with copious amounts of water. The Schlenk flask, which contains ~ 150 mL of liquid (PMe$_3$ and Bu$_2$O), is fitted with a magnetic stirring bar and a clean N$_2$ purged distillation assembly, which consists of a 6-in Vigreaux column, a distillation head with a thermometer [a Kontes Vigreaux Distillation Head (K-287450) works well here], a water-cooled condenser, and a 100-mL Schlenk receiving flask. As before, the distillation assembly is connected via a "tee" to the nitrogen source and a mineral oil bubbler. The flask is placed in an oil bath (atop a stirrer hot plate) and the receiver is cooled to $-78\,°C$. The oil bath is heated to $60\,°C$ and pure trimethylphosphine distills at 39 to $40\,°C$ from the stirred solution. The temperature of the oil bath is slowly raised to $\sim 110\,°C$ during the latter stages of the distillation. When the distillation is complete, the oil bath is removed, the nitrogen flow is increased slightly, and the distillation flask is allowed to cool to room temperature. At this point, the receiver is carefully disconnected under a positive N$_2$ flow and tightly capped. Any residual trimethylphosphine in the distillation flask is destroyed by adding Clorox bleach (dilute aqueous sodium hypochlorite). The yield of PMe$_3$ is typically 70 to 74 g or 80 to 85% based on P(OC$_6$H$_5$)$_3$. The ^{1}H NMR (C$_6$D$_6$, 360 MHz): δ 0.79 (d, $^2J_{HP} =$ 2.75 Hz).

Properties

Trimethylphosphine is a colorless, pyrophoric[2] liquid with a very unpleasant odor. It forms air-stable phosphonium salts with a variety of alkyl halides, and the corresponding oxide and sulfide upon reaction with O$_2$ and sulfur.

Trimethylphosphine may be stored as the air-stable silver iodide complex [AgI(PMe$_3$)]$_4$.[7] Other physicochemical properties are described in previous volumes of this series.[2,7]

References

1. (a) A. B. Burg and R. I. Wagner, *J. Am. Chem. Soc.*, **75**, 3872 (1953); (b) J. G. Evans, P. L. Goggin, R. J. Goodfellow, and J. G. Smith, *J. Chem. Soc. A*, **1968**, 464.

2. R. T. Markham, E. A. Dietz, Jr., and D. R. Martin, *Inorg. Synth.*, **16**, 153 (1974).

3. V. C. Gibson, C. E. Graimann, P. M. Hare, M. L. H. Green, J. A. Bandy, P. D. Grebenik, and K. Prout, *J. Chem. Soc. Dalton Trans.*, **1985**, 2025.

4. P. G. Edwards, R. A. Andersen, and A. Zalkin, *J. Am. Chem. Soc.*, **103**, 7792 (1981).

5. (a) G. A. Rupprecht, L. W. Messerle, J. D. Fellmann, and R. R. Schrock, *J. Am. Chem. Soc.*, **102**, 6236 (1980); (b) J. D. Fellmann, R. R. Schrock, and G. A. Rupprecht, *J. Am. Chem. Soc.*, **103**, 5752 (1981); (c) A. P. Sattelberger, R. B. Wilson, Jr., and J. C. Huffman, *J. Am. Chem. Soc.*, **102**, 7111 (1980); (d) M. L. Luetkens, Jr., J. C. Huffman, and A. P. Sattelberger, *J. Am. Chem. Soc.*, **105**, 4474 (1983); (e) K. W. Chiu, G. G. Howard, H. S. Rzepa, R. N. Sheppard, G. Wilkinson,

A. M. R. Galas, and M. B. Hursthouse, *Polyhedron*, **1**, 441 (1982); (f) G. S. Girolami, V. V. Mainz, R. A. Andersen, S. H. Vollmer, and V. W. Day, *J. Am. Chem. Soc.*, **103**, 3953 (1981); (g) P. R. Sharp and R. R. Schrock, *J. Am. Chem. Soc.*, **102**, 1430 (1980).

6. H. Schmidbaur and R. Franke, *Inorg. Chim. Acta*, **13**, 79 (1975).
7. R. Thomas and K. Eriks, *Inorg. Synth.*, **9**, 59 (1967).
8. W. Wolfsberger and H. Schmidbaur, *Synth. Inorg. Met. Org. Chem.*, **4**, 149 (1974).
9. P. R. Sharp, Ph.D. thesis, Massachusetts Institute of Technology, 1980.

4. PHOSPHORUS TRIFLUORIDE

$$SbF_3 + PCl_3 \xrightarrow{\text{SbCl}_5} PF_3 + SbCl_3$$

Submitted by RONALD J. CLARK* and HELEN BELEFANT*
Checked by STANLEY M. WILLIAMSON[†]

Phosphorus trifluoride is a ligand that is used extensively in coordination chemistry. It substitutes readily into various metal carbonyl complexes using either thermal or photochemical techniques. As a ligand, it is unique in its similarity to carbon monoxide in lower-valent organometallic compounds. In its role as a model for CO, a number of studies are possible that cannot be done on the carbonyls themselves.[1] The name normally used for PF_3 in complexes is trifluorophosphine.

Although PF_3 is inherently an inexpensive material, the fact that it currently has no industrial uses results in a high cost from speciality suppliers. It can be prepared in a number of ways almost all of which are based upon the interaction of phosphorus trichloride with some mild fluorinating agent.[2] A convenient procedure is the interaction of PCl_3 with solid ZnF_2.[3] However, the yield of PF_3 is strongly dependent on the reactivity of the ZnF_2, which can vary greatly depending on the process of manufacture. The use of AsF_3 as fluorinating agent has obvious disadvantages particularly when sizable quantities of PF_3 are needed.[4] The use of NaF as the fluorinating agent has many attractive features but the reaction does not progress with consistent rates, yields varying from 20 to 80% in 48 h. Another method found in the literature is to bubble HF gas through a tower of liquid PCl_3.[5] However, one not only has to scrub 3 mol of HCl from every mole of the PF_3, but the use of HF is not desirable except in the most skilled of hands. (Hydrogen chloride

*Department of Chemistry, Florida State University, Tallahassee, FL 32306.
[†]Department of Chemistry, University of California, Santa Cruz, CA 95064.

and PF$_3$ have almost identical boiling points so they cannot be separated by vacuum techniques.)

Our resolution to this dilemma has been to work out a compact procedure based upon the fluorinating agent SbF$_3$ catalyzed by SbCl$_5$ using acetonitrile as a solvent (i.e., the Swarts reaction).[6]

Procedure

■ **Caution.** *Owing to the scale of the reaction, care must be exercised regarding free flow of gas through the apparatus, especially the collection of product in the cold trap. Mineral oil bubblers are used to monitor gas flow and ensure that the product collecting in the cold trap has not plugged the flow.*

Owing to the toxicity of PF$_3$, this reaction should be carried out in a well-ventilated hood.

The heart of the apparatus is a 1-L, three-necked flask equipped with a dropping funnel, gas purge tube, a low temperature reflux cold finger, and a strong magnetic stirrer (see Fig. 1). Alternatively, a motor driven stirrer and a gas tight gland can be used. The dropping funnel is arranged so that the PCl$_3$

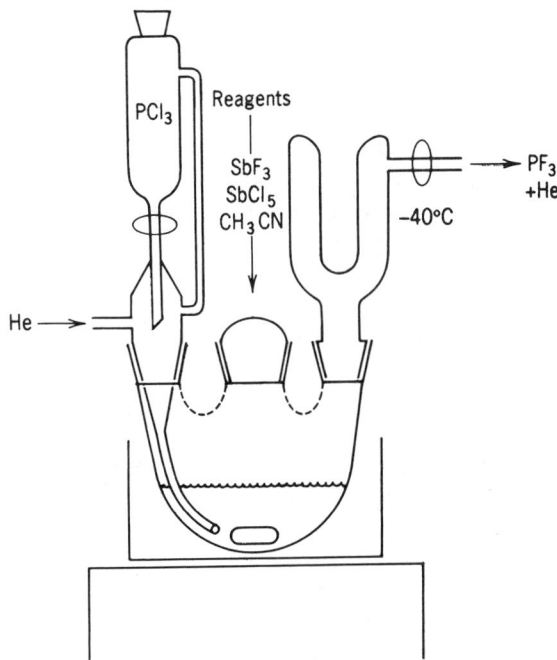

Fig. 1. Apparatus for the fluorination of PCl$_3$.

and gas purge enter below the surface of the solvent. Following the cold finger are two traps cooled to $-78\,°C$, a 4A molecular sieve trap $\sim 40\,cm$ in length, and a final $-196\,°C$ product trap.

Oil-filled bubblers are placed at the start of the gas train and after the PF_3 condenser to monitor the gas flow. About 300 mL of acetonitrile, from a fresh bottle with a water content of no more than 0.05%, is admitted to the flask. The acetonitrile can be dried over 3A molecular sieves. As purge gas, helium is preferred but nitrogen can also be used. Under a gas purge, 150 g (0.84 mol) of SbF_3 is dropped in over a period of a few minutes with vigorous stirring. If the order of addition is reversed, the SbF_3 tends to form clumps and resists stirring. Three milliters of (0.02 mol) $SbCl_5$ is added. The cold finger is cooled to $\sim -40\,°C$ by the judicious addition of Dry Ice to acetone. A suitable thermometer or thermocouple device should be used to monitor this bath temperature. Finally, 60 mL (95 g, 0.69 mol) of PCl_3 is placed in the dropping funnel and is added to the flask over the course of $\sim 1\,h$. If excess SbF_3 is used and the PCl_3 is introduced below the surface of the solvent, the tendency to form $PClF_2$ and PCl_2F is kept quite low. These by-products are monitored by chloride test of the product,[7] which should be negative (i.e., no visible chloride in the test).

The reaction is mildly exothermic; a room temperature water bath should be placed around the reaction flask to control the reaction temperature and particularly the vapor pressure of CH_3CN to reduce the problems of product clean-up. The PF_3 collects as a fluffy solid in the $-196\,°C$ trap and occupies a volume of about five times its liquid volume. It is essential that this trap be fairly large. Typical dimensions for the large trap are $27 \times 5\,cm$; the inner tube reaches no closer than 10 cm of the bottom. The acetonitrile traps are $13 \times 2.5\,cm$. In addition, it is extremely important that gases come in the sidearms of all the traps rather than the central tube in order to avoid plugging. This can occur either with CH_3CN transported into the $-78\,°C$ trap, or with PF_3 collecting in the $-196\,°C$ trap. Liquid nitrogen should be added to the $-196\,°C$ trap frequently and in small increments to avoid large pressure surges.

If nitrogen is being used as a purge gas, a moderate amount becomes trapped in the lattice of the frozen PF_3 at $-196\,°C$ and its removal necessitates a series of subsequent freeze–pump–thaw cycles. After completion of the PCl_3 addition, a gas purge of 20 min is used to sweep the PF_3 out of the reaction flask into the $-196\,°C$ trap. The PF_3 is then isolated from the rest of the system without the removal of the liquid nitrogen. Then it is connected to a vaccum system and the helium or nitrogen pumped off. Once it is under vaccum, the liquid nitrogen can be removed and the snowy PF_3 allowed to become warm enough to melt. It is advisable to monitor the pressure during the thaw cycle, however, PF_3 has wide liquid range (-152 to $-102\,°C$), and will not build too much pressure as long as some solid PF_3 is in evidence, or shortly after the disappearance of the last solid PF_3. The liquid should be refrozen by liquid nitrogen. If the system manometer returns to zero pressure, the product is free of purge gas. If not, additional freeze–pump–thaw cycles are needed. Yield: ~ 50 g or 83%.

If the PF_3 is not to be used immediately, it can be condensed into a metal–pressure vessel at liquid nitrogen temperature using vaccum techniques.

■ **Caution.** *Only stainless steel (i.e., 316 SS) or aluminum pressure vessels approved for cryogenic service should be used. The use of lecture bottles of mild steel must be avoided, since such lecture bottles lose their strength at cryogenic temperatures.*

The $-78\,°C$ trap and molecular sieve trap are designed to remove CH_3CN. Since traces of moisture are inevitably present, this means that HCl is also likely to be produced; the molecular sieve trap will also remove it. An alternate

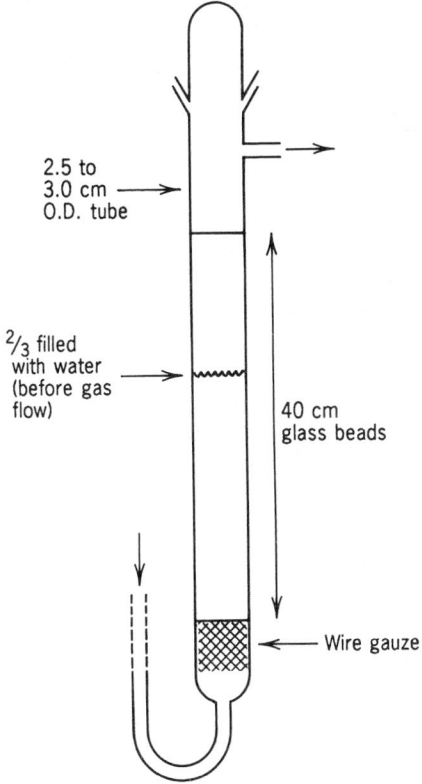

Fig. 2. Apparatus for alternate method of purification of PF_3.

method of purification is to pass the product through a water tower packed with glass beads as shown in Fig. 2.

The HCl dissolves quickly in the water but the PF_3 hydrolysis is slow. The product is then given a final purification from water and other materials by a vacuum distillation from a methylcyclohexane slush bath ($-128\,°C$). However, the molecular sieve trap works better and avoids the 10 to 20% loss that occurs from PF_3 hydrolysis. The product should give a negative chloride test[7] indicating absence of HCl and/or by-products $PClF_2$ or PCl_2F.

Properties

The gas-phase infrared (IR) spectrum contains two strong vibrational stretching bands in the region from 800 to $1000\,cm^{-1}$, which are indicative of PF_3. However, the gas-phase IR absorptivity of the P—F bands is so much stronger than the bands of impurities such as HCl and CH_3CN that IR spectroscopy is not a good method to check purity unless one is careful. To detect the presence of HCl or CH_3CN at approximately the 1% level by IR requires gas pressures of ~ 700 torr in the 10-cm cell. At that pressure, the P—F stretching bands are totally off scale and various broad combination and overtone bands are seen as strong absorptions in the regions of 1700 and $1200\,cm^{-1}$.

The gas has a boiling point of $-102\,°C$ and a freezing point of $-152\,°C$. In high-pressure cylinders, the deviation from PVT ideality is quite significant. At higher pressures, room temperature gauge readings are about one half to two thirds of what one would calculate based upon the mass of PF_3 present. Thus pressure is not a good measure of yield unless the data are carefully corrected.

Phosphorus trifluoride does not have a particularly sharp odor that one would expect of an acid gas until the concentration becomes higher than is safe. It combines with hemoglobin[8] in a manner comparable to carbon monoxide. It should also be assumed that the subsequent hydrolysis products are highly undesirable in the body. Thus, one is strongly advised to work with PF_3 only with a good hood system or a well-designed vacuum system. Pressure storage vessels should be examined for slow leaks before use in storage of PF_3.

References

1. R. J. Clark and M. A. Busch, *Acct. Chem. Res.*, **6**, 246 (1973).
2. R. Schmutzler, *Advances in Fluorine Chemistry*, Vol. 5, M. Stacey, J. C. Tatlow, and A. G. Sharpe (eds.), Butterworths, Washington, 1965, p. 37.
3. A. A. Williams, *Inorg. Synth.*, **5**, 95 (1957).

4. C. J. Hoffman, *Inorg. Synth.*, **4**, 149 (1953).

5. G. Brauer, *Handbook of Preparative Inorganic Chemistry*, Vol. 1, 2nd ed., Academic Press, New York, 1963, p. 189.

6. F. Swarts, *Bull. Acad. R. Belg.*, **24**, 309 (1892).

7. The chloride test is performed by bubbling a sample of the product into a $AgNO_3/HNO_3$ solution. Alternately, vacuum techniques can be used to transfer a sample of the product into a flask followed by the admission of the silver nitrate solution. T. L. Brown and H. E. LeMay, *Qualitative Inorganic Chemistry*, Prentice Hall, New Jersey, 1984.

8. J. W. Irvine and G. Wilkinson, *Science*, **113**, 742 (1951).

5. (η⁵-CYCLOPENTADIENYL)LANTHANIDE COMPLEXES FROM THE METALLIC ELEMENTS

Submitted by GLEN B. DEACON,* GEOFF N. PAIN,* and TRAN D. TUONG*
Checked by WILLIAM J. EVANS,† KEITH R. LEVAN,† and RAUL DOMINGUEZ†

Tris(η⁵-cyclopentadienyl)lanthanides were the first authentic organolanthanides to be prepared[1] and bis(η⁵-cyclopentadienyl)lanthanide(II) compounds have played a germinal part in the development of lower oxidation state organolanthanide chemistry.[2] These cyclopentadienyls are sources of coordination compounds of structural interest and are reagents for the synthesis of other organolanthanides, for example, bis- and mono(η⁵-cyclopentadienyl)lanthanide(III) derivatives.[2]

Early syntheses of tris(η⁵-cyclopentadienyl)lanthanides employed reactions of anhydrous lanthanide trihalides with alkali metal, magnesium, or beryllium cyclopentadienides, generally followed by sublimation of the crude product.[1,2] Bis(η⁵-cyclopentadienyl)europium(II) and bis(η⁵-cyclopentadienyl)ytterbium(II) have been obtained by reactions of the free metals with cyclopentadiene in liquid ammonia, again with sublimation work-up.[3-5] Recently, reductive transmetalation reactions of thallium(I)[6,7] and mercury(II)[8,9] cyclopentadienides with lanthanide elements have been developed as a route to trivalent and divalent (η⁵-cyclopentadienyl)lanthanide complexes. These methods avoid the need for anhydrous lanthanide halides, highly air-sensitive metal cyclopentadienide reagents, and use of liquid ammonia. For transmetallation, thallium(I) cyclopentadienide is the more attractive reagent, since it is more readily prepared, is less heat and light sensitive, and has better storage characteristics than bis(cyclopentadienyl)mercury.[10,11]

*Department of Chemistry, Monash University, Clayton, Victoria, Australia, 3168.
†Department of Chemistry, University of California, Irvine, CA 92717.

In the following sections we detail transmetalation syntheses from $(\eta^5\text{-}C_5H_5)Tl$ (see Ref. 12) of three $(\eta^5\text{-cyclopentadienyl})$lanthanide complexes, which can be isolated analytically pure in nearly quantitative yields simply by evaporation of the filtered reaction solvent. These simplify and optimize our reported syntheses.[6,7]

General Comments

The organolanthanides described are very sensitive to oxygen and water both in solution and as solids, but they can be handled and stored under purified nitrogen or argon (BASF R3/11 oxygen removal catalyst and molecular sieves), and in dry solvents. For reliable results, "grease-free" Schlenk apparatus incorporating polytetrafluoroethylene O-rings and "Rotaflo," "Young," or similar taps should be used. (See Refs. 13 and 14 for general techniques in organometallic chemistry.)

A small amount of mercury metal aids initiation of transmetalation, but is not essential. The apparatus used in all three reactions is shown in Fig. 1. This consists of a 100-mL Schlenk flask containing a magnetic stirring bar, on top of which are attached (in order) a condenser, a Schlenk filter [covered with a Whatman Microfibre filter paper, a thin (~ 3 mm) layer of dried diatomaceous earth (Sigma, grade 1) and a second glass fiber filter paper], and a further Schlenk flask (100 mL, preweighed). Because of the use of poisonous materials (see Cautionary Note in procedure in Section A), the apparatus should be placed in a well-ventilated fume hood. Owing to the air-sensitive nature of the products, the apparatus should be thoroughly flushed with purified nitrogen or argon.

The powdered lanthanide element, thallium(I) cyclopentadienide and a trace of mercury metal are placed into the first flask. An inert atmosphere is established by evacuation of the apparatus and backfilling with N_2 or Ar at least three times. Dried and degassed solvent is then added to the reagents by syringe and the reaction is carried out as described in each of the following procedures. After reaction and cooling, the apparatus is inverted, and the reaction mixture is filtered into the second flask by gravity or by reduced pressure. The filter is washed with solvent (2×5 mL) added by syringe, and the filtrate is evaporated slowly to dryness under vacuum at room temperature yielding the pure cyclopentadienyllanthanide (see comments immediately following). All operations including handling of solids must be carried out under purified argon or nitrogen.

Lanthanide metals were obtained from Research Chemicals as powders. The metals can be weighed and stored in air, but for long term storage, especially of neodymium, use of a dry box is recommended.

■ **Caution.** *Finely divided residues after reaction can be pyrophoric.*

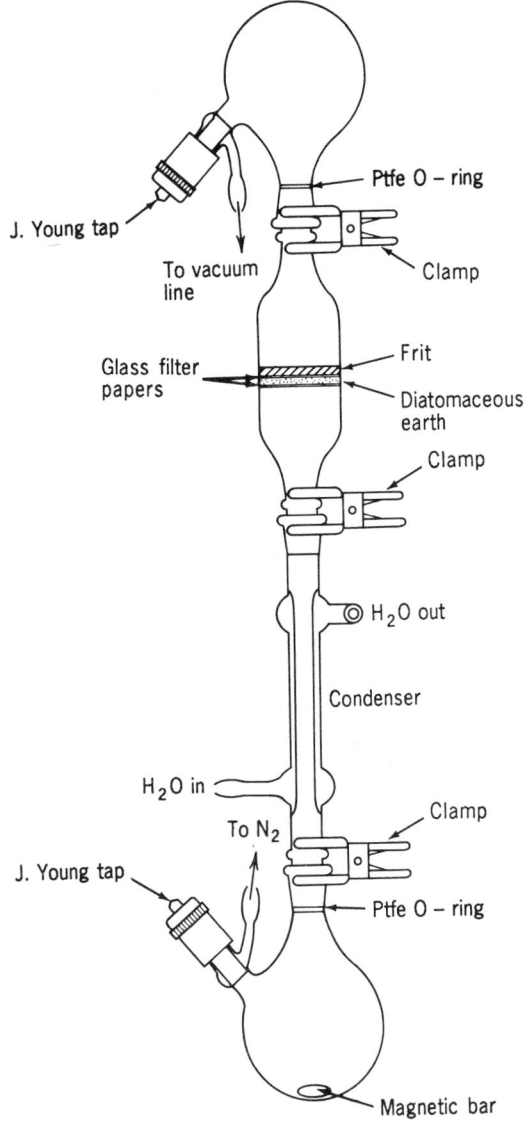

Fig. 1. Apparatus for the preparation of (η^5-cyclopentadienyl)lanthanide complexes from the metallic elements.

Peroxide-free tetrahydrofuran (THF) (predried first over KOH pellets followed by sodium wire) and 1,2-dimethoxyethane (dme) (predried over sodium wire) are distilled from sodium benzophenone ketyl and stored under purified nitrogen in Schlenk vessels equipped with Teflon Rotaflo or Young taps.

A. TRIS(η^5-CYCLOPENTADIENYL)(TETRAHYDROFURAN)-NEODYMIUM(III)

$$Nd + 3Tl(C_5H_5) + C_4H_8O \xrightarrow{\text{THF}} Nd(C_5H_5)_3(C_4H_8O) + 3Tl\downarrow$$

■ **Caution.** *Compounds of thallium and thallium metal are extremely poisonous. Protective rubber gloves should be used in handling thallium compounds and their solutions, and all operations should be carried out in a well-ventilated fume hood. Disposal of thallium and mercury-containing residues must be handled according to procedures established for poisonous metals.*

Procedure

The Schlenk apparatus is charged with Nd (1.45 g, 10.05 mmol), Hg (0.170 g, 0.85 mmol), Tl(C_5H_5) (1.090 g, 4.04 mmol), and THF (25 mL). The mixture is heated at reflux point for 20 h with magnetic stirring. After filtration, washing of filter, and evaporation of solvent, the pure complex is obtained as pink-blue crystals (pink in daylight, blue in artificial light). [The color appears dependent on crystal size, since the checkers report that, isolated as a microcrystalline solid, the compound is light blue in natural light, but lavender under a light bulb, whereas larger crystals (from THF) appear pink in sunlight.] Yield: 0.500 g (90%). (The checkers report 73% when the product was scraped from the flask in a dry box.)

Anal. Calcd. for $C_{19}H_{23}NdO$: C, 55.4; H, 5.6. Found: C, 55.6; H, 5.6%.

Properties

The complex $Nd(C_5H_5)_3(C_4H_8O)$ is unstable to air both in solution or as the solid.[7] It must be handled using techniques for air-sensitive compounds.[13,14] The IR spectrum of a Nujol mull prepared under inert atmosphere shows absorptions at 1335 (m), 1254 (m), 1009 (vs), 925 (m), 858 (m) (characteristic of coordinated THF), 795 (vs), 768 (s), 740 (s, sh), and 670 (m) cm^{-1}. The Vis–near IR absorptions (in THF solution) are λ_{max} (ε): 323 (75), 528 (8), 531 (7), 576 (21),

582 (22), 588 (123), 591 (94), 594 (133), 597 (92), 600 (23), 605 (15), 618 (5), 622 (4), 733 (6), 737 (8), 740 (12), 743 (5), 755 (8), 792 (9), 800 (6), 802 (4)nm.

The mass spectrum is identical with that previously reported for the unsolvated compound.[15]

B.   TRIS(η^5-CYCLOPENTADIENYL)(TETRAHYDROFURAN)-SAMARIUM(III)

$$Sm + 3Tl(C_5H_5) + C_4H_8O \xrightarrow{\text{THF}} Sm(C_5H_5)_3(C_4H_8O) + 3Tl\downarrow$$

■ *See Cautionary note, Section A.*

Procedure

The Schlenk apparatus is charged with Sm (2.52 g, 16.75 mmol), Hg (0.10 g, 0.5 mmol), Tl(C_5H_5) (1.980 g, 7.35 mmol), and THF (50 mL). The mixture is heated at reflux point for 4 h with magnetic stirring. After work-up, as described previously, the pure complex is obtained as bright yellow crystals. Crystal size may be increased by slow evaporation of the solvent, and this can give crystals suitable for X-ray characterization.[7] Yield: 0.979 g (96%). [The checkers' isolation method (see Section A) gave 72%].

Anal. Calcd. for $C_{19}H_{23}OSm$: C, 54.6; H, 5.5. Found: C, 54.9; H, 5.5%.

Properties

The complex $Sm(C_5H_5)_3(C_4H_8O)$ is unstable to air both in solution or as the solid.[7] Crystallographic data[7] show the complex to be isostructural with $Gd(\eta^5-C_5H_5)_3$(thf), hence samarium has pseudotetrahedral stereochemistry and formal 10 coordination. The IR spectrum of a Nujol mull prepared under inert atmosphere shows absorptions at 1335 (m), 1257 (m), 1009 (vs), 858 (m) (characteristic of coordinated THF), 790 (vs), 773 (s, br), 750 (s, sh), and 667 (m) cm^{-1}. The Vis–near IR absorptions (in THF solution) are $\lambda_{max}(\varepsilon)$: 296 (1584), 424 (134), 429 (134), 1050 (4), 1075 (5), 1104 (4), 1206 (15), 1227 (19), 1268 (8), 1320 (14) nm. The mass spectrum agrees with that previously reported for the unsolvated compound.[15] Heating at 140 to 150 °C under vacuum ($\times 10^{-3}$ torr), gives orange tris(η^5-cyclopentadienyl)samarium(III) with IR absorption as previously reported.[16]

C. BIS(η^5-CYCLOPENTADIENYL)(1, 2-DIMETHOXYETHANE)-YTTERBIUM(II)

$$\text{Yb} + 2\text{Tl}(C_5H_5) + C_4H_{10}O_2 \xrightarrow{\text{dme}} \text{Yb}(C_5H_5)_2(C_4H_{10}O_2) + 2\text{Tl}\downarrow$$

■ *See cautionary note, Section A.*

Procedure

The Schlenk apparatus is charged with Yb (3.003 g, 17.35 mmol), Hg (0.1 g, 0.5 mmol), Tl(C_5H_5) (2.000 g, 7.42 mmol), and 1,2-dimethoxyethane (dme, 50 mL). The mixture is heated for 2 h during which time the color of the reaction mixture changes from an olive green to the final blue-green. Monitoring the reaction by Vis–near IR spectroscopy reveals tris(η^5-cyclopentadienyl)ytterbium(III) (olive green) to be the first detectable organometallic product. This is then reduced to bis(η^5-cyclopentadienyl)ytterbium(II) (blue-green) by the excess metallic ytterbium.[7]

Work-up, as above, gives the pure complex as green crystals. Yield: 1.40 g (96%). (The checkers' isolation method gave 68%).

Anal. Calcd. for $C_{14}H_{20}O_2$Yb: C, 42.8; H, 5.1. Found: C, 42.9; H, 4.9%.

Properties

The complex Yb(C_5H_5)$_2$($C_4H_{10}O_2$) is very air sensitive both in solution and in the solid state.[7] The IR spectrum as a Nujol mull prepared under inert atmosphere shows absorptions at 3070 (m), 1276 (m), 1240 (m), 1190 (m), 1108 (m), 1090 (m), 1058 (vs), 1002 (vs), 854 (s), 825 (m), 772 (s, sh), 750 (vs), and 738 (vs) cm^{-1}. Visible–near IR spectrum (in 1, 2-dimethoxyethane), $\lambda_{max}(\varepsilon)$: 382 (582), 624 (226) nm. The complex dissolves in THF to form a red-purple solution, and attempted dissolution in benzene results in precipitation of unsolvated Yb(C_5H_5)$_2$, and formation of a green solution with a dme: Yb(C_5H_5)$_2$ ratio (by ^{1}H NMR) of > 1:1. (The checkers were able to dissolve a sample completely in benzene.) In a sealed capillary under nitrogen, the compound turns yellow at 140 °C and brick red at 150 °C, the green color being restored on cooling. On being heated under vacuum at 140 to 150 °C, unsolvated bis(η^5-cyclopentadienyl)ytterbium is obtained. The ^{1}H NMR spectrum of Yb(C_5H_5)$_2$ in C_4D_8O solution (acetone-d_6 lock tube) shows a cyclopentadienyl resonance at δ 5.64 (s) downfield from external Me$_4$Si. The crystal and molecular structure of Yb(η^5-C_5H_5)$_2$($C_4H_{10}O_2$) has been determined.[17]

References

1. G. Wilkinson and J. M. Birmingham, *J. Am. Chem. Soc.*, **76**, 6210 (1954); **78**, 42 (1956).
2. T. J. Marks and R. D. Ernst, "Scandium, Yttrium, and the Lanthanides and Actinides," in *Comprehensive Organometallic Chemistry*, Vol. 3 G. Wilkinson, F. G. A. Stone, and E. W. Abel (eds.) Pergamon, Oxford, 1982, Chapter 21; H. Schumann, *Angew. Chem. Int. Ed. Engl.*, **23**, 474 (1984).
3. E. O. Fischer and H. Fischer, *J. Organomet. Chem.*, **3**, 181 (1965).
4. R. G. Hayes and J. L. Thomas, *Inorg. Chem.*, **8**, 2521 (1969).
5. F. Calderazzo, R. Pappalardo, and S. Losi, *J. Inorg. Nucl. Chem.*, **28**, 987 (1966).
6. G. B. Deacon, A. J. Koplick, and T. D. Tuong, *Polyhedron,* **1**, 423 (1982).
7. G. B. Deacon, A. J. Koplick, and T. D. Tuong, *Aust. J. Chem.*, **37**, 517 (1984).
8. G. Z. Suleimanov, L. F. Rybakova, Ya. A. Nuriev, T. Kh. Kurbanov, and I. P. Beletskaya, *J. Organomet. Chem.*, **235**, C19 (1982).
9. G. Z. Suleimanov, T. Kh. Kurbanov, Ya. A. Nuriev, L. F. Rybakova, and I. P. Beletskaya, *Dokl. Chem.*, **265**, 254 (1982).
10. F. A. Cotton and G. Wilkinson, *Advanced Inorganic Chemistry*, 4th ed., Wiley, New York, 1980, p. 1163.
11. G. Wilkinson and T. S. Piper, *J. Inorg. Nucl. Chem.*, **2**, 32 (1955).
12. A. J. Nielson, C. E. F. Rickard, and J. M. Smith, *Inorg. Synth.*, **24**, 97 (1986).
13. D. F. Shriver, *The Manipulation of Air-Sensitive Compounds*, McGraw-Hill, New York, 1969.
14. R. B. King, "Transition-Metal Compounds," in *Organometallic Syntheses*, Vol. I, J. J. Eisch and R. B. King (eds.), 1965; Vol. 2, J. J. Eisch, "Nontransition-Metal Compounds," 1981; Academic Press, New York.
15. J. L. Thomas and R. G. Hayes, *J. Organomet. Chem.*, **23**, 487 (1970).
16. G. W. Watt and E. W. Gillow, *J. Am. Chem. Soc.*, **91**, 775 (1969).
17. G. B. Deacon, P. I. MacKinnon, T. W. Hambley, and J. C. Taylor, *J. Organomet. Chem.*, **259**, 91 (1983).

Chapter Two

MONONUCLEAR TRANSITION METAL COMPLEXES Part I

6. AMMONIUM [(1*R*)-(*ENDO, ANTI*)]-3-BROMO-1,7-DIMETHYL-2-OXOBICYCLO[2.2.1]HEPTANE-7-METHANESULFONATE,* AND SYNTHESIS AND RESOLUTION OF *cis*-DICHLOROBIS(1,2-ETHANEDIAMINE)CHROMIUM(III) CHLORIDE

Submitted by A. HAMMERSHØI,[†] E. HANSSON,[†] and J. SPRINGBORG[†]
Checked by C. J. HAWKINS[‡]

The chiral $(+)_{589}$-α-bromocamphor-π-sulfonate ion ([BCS]$^-$) has been widely used as a resolving agent for resolutions of cationic species.[1] Here an improved and simplified method for the preparation of its ammonium salt is presented as well as a method for the synthesis and the resolution of the *cis*-[Cr(en)$_2$Cl$_2$]$^+$ ion.

---

*Throughout the text, this anion may also be termed $(+)_{589}$-α-bromocamphor-π-sulfonate, or [BCS]$^-$.

†Chemistry Department, Royal Veterinary and Agricultural University, Thorvaldsensvej 40, DK-1871 Frederiksberg C, Denmark.

‡Department of Chemistry, University of Queensland, St. Lucia, Brisbane 4067, Australia.

A.   AMMONIUM [(1*R*)-(*ENDO, ANTI*)]-3-BROMO-1,7-DIMETHYL-2-OXOBICYCLO[2.2.1]HEPTANE-7-METHANESULFONATE [AMMONIUM (+)$_{589}$-α-BROMOCAMPHOR-π-SULFONATE]

$$Cl-SO_3H + H_2O \longrightarrow HCl + H_2SO_4$$

$$HCl + H_2SO_4 + 3NH_3 \longrightarrow NH_4Cl + (NH_4)_2SO_4$$

$$H[BCS] + NH_3 \longrightarrow NH_4[BCS]$$

In the past, the [BCS]$^-$ ion has been synthesized by sulfonation of bromocamphor, (1*R*)-3-bromo-1,7,7-trimethylbicyclo[2.2.1]heptan-2-one and was isolated usually as the water-soluble ammonium salt. An early method by Kipping and Pope[2] utilized chloroform as the solvent and chlorosulfonic acid as the sulfonating agent. A later method by Kaufmann[3] employed fuming sulfuric acid as both solvent and sulfonating agent. In either case, tedious and lengthy work-up procedures were used. In the procedure by Kipping and Pope,[2] acid neutralization and removal of the sulfate ion were achieved by addition of calcium carbonate. Excess calcium ions were subsequently precipitated with ammonium carbonate. In general, this procedure involved manipulating large quantities of insoluble material and gave rise to tedious filtrations. Furthermore, the overall yield was hardly impressive.

For the Kaufmann method,[3] even larger amounts of sulfate salt had to be removed before the [BCS]$^-$ salt was isolated. This was accomplished by the addition of excess ammonia and coprecipitation of ammonium sulfate and ammonium α-bromocamphor-π-sulfonate, followed by extraction of the latter compound with ethanol. This method is clearly superior to previous methods, but from our experience the yields are variable, presumably because optimal sulfonation yields are critically dependent on reaction conditions, which consequently need to be carefully controlled. The following method combines the easily reproducible sulfonation step of the older method with an easy isolation procedure, which takes advantage of the fact that $NH_4[BCS]$ is only sparingly soluble in concentrated ammonium chloride solution. The ammonium salt may also be available commercially (Aldrich).

Procedure

■ **Caution.** *Chlorosulfonic acid, $ClSO_3H$, is a very strong acid and must be handled with care. Accidental contact with water releases gaseous HCl with spattering due to overheating. It should be handled in a well-ventilated hood.*

In a 1-L three-necked flask fitted with a condenser and a mechanical stirrer is placed a solution of (1R)-3-bromo-1,7,7-trimethylbicyclo[2.2.1]heptan-2-one [$(+)_{589}$-α-bromo-π-camphor, 200 g, 0.87 mol; Merck] in chloroform (275 mL). With continuous stirring, chlorosulfonic acid (115 mL, 1.74 mol) is added from a dropping funnel over a period of 20 min and the resulting solution is heated to reflux for 6 h. After cooling to room temperature, the reaction mixture is cautiously poured onto crushed ice (400 g) in a 1-L beaker. By means of a separatory funnel, the chloroform phase in removed. The aqueous phase is cooled in ice to 10 °C before 12 M ammonia (170 mL, 2.0 mol) is slowly added with stirring and cooling. After cooling to 5 °C, the deposited product, $NH_4[BCS]$, is collected on a filter, and the solid is sucked almost dry before washing with ice cold 5 M NH_4Cl (100 mL), followed by washing with an acetone–ethanol mixture (1:1, v/v; 75 mL) and finally twice with acetone (100 mL). Yield of the first batch of crude product is 86 g. The mother liquor and washings are combined and concentrated in a rotary evaporator (40 °C) to 500 mL. A second batch of crystalline product separates, and after cooling to 5 °C this is collected as above, yielding a further 43 g of product.

A pure product is obtained by reprecipitation with NH_4Cl as follows: The crude product (129 g) is dissolved in boiling water (265 mL), and the hot solution is filtered. The filtrate is allowed to cool to room temperature with crystal formation. When the temperature has reached 25 to 30 °C, portions of 5 M NH_4Cl (13 × 5 mL) are added over a period of 15 min with intermittent cautious stirring with a glass rod. The resulting mixture is cooled in ice to below 5 °C. The crystalline product is collected on a filter and the mother liquor is thoroughly removed by suction. With continued suction the crystals are washed with ice-cold acetone–ethanol (1:1, v/v; 2 × 50 mL). Suction is then interrupted, and the crystals are thoroughly washed with acetone (2 × 100 mL) before drying in air. Yield: 96 g (34%).

Anal. Calcd. for $C_{10}H_{18}BrNO_4S$: C, 36.59; H, 5.53; Br, 24.34; N, 4.27; S, 9.77. Found: C, 36.2; H, 5.5; Br, 24.4; N, 4.4; S, 9.9.

Properties

The product is soluble in water, slightly soluble in ethanol, and essentially insoluble in acetone and diethyl ether. The specific rotation is $[\alpha]_{589}^{25} = 84°$ (1.0% aqueous solution). The absolute configuration of the $[BCS]^-$ ion has been established in several X-ray studies.[4]

B. *cis*-DICHLOROBIS(1, 2-ETHANEDIAMINE)CHROMIUM(III) CHLORIDE MONOHYDRATE

$$\textit{trans-}[Cr(H_2O)_4Cl_2]Cl \cdot 2H_2O + 4DMF$$

$$\longrightarrow \textit{cis-}[Cr(DMF)_4Cl_2]Cl + 6H_2O$$

$$\textit{cis-}[Cr(DMF)_4Cl_2]Cl + 2NH_2CH_2CH_2NH_2$$

$$\longrightarrow \textit{cis-}[Cr(en)_2Cl_2]Cl + 4DMF$$

$$DMF = N, N\text{-dimethylformamide} = HCON(CH_3)_2$$
$$en = 1, 2\text{-ethanediamine}$$

Several methods are described in the literature for the preparation of this complex.[5] The best method is that described by Pedersen[5] and a modification is given in the following procedure.

Procedure

■ **Caution.** *Avoid skin contact and breathing the vapors (DMF, 1, 2-ethanediamine, and hydrogen chloride). The synthesis must be performed in a well-ventilated fume hood.*

In a 1-L one-necked round-bottomed flask, $[Cr(H_2O)_4Cl_2]Cl \cdot 2H_2O$ (200 g, 0.75 mol) is added to DMF (600 mL). The mixture is stirred with a glass rod until all or most of the chromium(III) salt is dissolved. The flask is placed in an electric heating mantle equipped with a condenser for distillation and a thermometer, and both the flask and the condenser are insulated (aluminum foil and cotton wool). Then, 300 mL of solvent is distilled off (normal pressure, 760 torr). During distillation the vapor temperature reaches 148 °C (bp of DMF is 153 °C).

A 2-L beaker is placed on an electric heating plate, clamped to a rack, and equipped with a mechanical stirrer (rod and wing in stainless steel). This reaction mixture is placed in the beaker and allowed to cool to 100 °C. From a dropping funnel, en (100 mL, 1.50 mol) is added dropwise ($3-6\,mL\,min^{-1}$) with stirring. The heat of reaction will increase the temperature, which should not exceed 120 °C during the addition of en. If the temperature increases above 120 °C, the addition of en is interrupted until the temperature has fallen to ~ 115 °C.

The mixture is then heated with continued stirring to 150 °C and kept at this temperature for 1 h. The reaction mixture sometimes becomes very viscous towards the end of addition, and at that time it may be difficult to achieve an effective stirring. This does not affect the yield.

The mixture is then cooled slowly to room temperature. The crystals are collected on a filter, washed three times with 300-mL portions of 96% ethanol and dried in the air. Yield: 162–183 g (73–82%) of crude anhydrous product containing variable amounts of DMF of crystallization. The product is ground in a mortar. A 20-g quantity is extracted as fast as possible on the filter with water (60–65 °C) in portions of ~ 30 mL (total exctration volume 90–100 mL). Each successive portion of solution is immediately filtered into a single ice-cooled 500-mL suction flask containing 100 mL of ice-cold 12 M hydrochloric acid and equipped with a magnetic stirrer. After the mixture is cooled with continued stirring for a further 20 min, the product is isolated by filtration and is washed with 96% ethanol (three 10 mL portions) and diethyl ether. Drying in air gives 11.0 g (55%) of pure *cis*-$[Cr(en)_2Cl_2]Cl \cdot H_2O$.

Anal. Calcd. for $CrC_4H_{18}N_4Cl_3O$: C, 16.20; H, 6.12; N, 18.89; Cl, 35.86. Found: C, 15.9; H, 6.2; N, 18.9; Cl, 35.9.

Properties

cis-Dichlorobis(1, 2-ethanediamine)chromium(III) chloride is a purple crystalline solid that is quite soluble in water as in 12 M HCl at 25 °C. It hydrolyzes in aqueous solution yielding *cis*-$[Cr(en)_2Cl(H_2O)]^{2+}$ with a half-life of about 0.5 h at 25 °C.[6] It is a very useful starting material for the synthesis of other cis complexes, for example, *cis*-$[Cr(en)_2(H_2O)(OH)][S_2O_6]$ and *cis*-$[Cr(en)_2(OSO_2CF_3)_2](CF_3SO_3)$.[7,8] Resolution of the Λ enantiomer is described in Section C.

C. $(+)_{589}$-Λ-*cis*-DICHLOROBIS(1, 2-ETHANEDIAMINE)- CHROMIUM(III) CHLORIDE MONOHYDRATE

$$2cis\text{-}[Cr(en)_2Cl_2]Cl \cdot H_2O + NH_4[BCS]$$

$$\longrightarrow \Delta\text{-}cis\text{-}[Cr(en)_2Cl_2]Cl$$

$$+ \Lambda\text{-}cis\text{-}[Cr(en)_2Cl_2][BCS] + NH_4Cl + 2H_2O$$

$$\Lambda\text{-}cis\text{-}[Cr(en)_2Cl_2][BCS] + HCl + H_2O$$

$$\longrightarrow \Lambda\text{-}cis\text{-}[Cr(en)_2Cl_2]Cl\cdot H_2O + H[BCS]$$

$$H[BCS] + NH_3 \longrightarrow NH_4[BCS]$$

$(+)_{589}$-Λ-*cis*-Dichlorobis(1, 2-ethanediamine)chromium(III)   chloride monohydrate was originally isolated by Werner[9] from racemic *cis*-$[Cr(en)_2Cl_2]Cl\cdot H_2O$ through the $[BCS]^-$ salt. This method has since been modified by Selbin and Bailar[10] and LeMay and Bailar[11] and later by Springborg and Schäffer.[12] The following procedure is based on the latter modification, and it has the advantage that the yield based on $NH_4[BCS]$ is increased more than twofold while the yield based on Cr(III) is only slightly decreased. Racemic *cis*-$[Cr(en)_2Cl_2]Cl\cdot H_2O$, reprecipitated from water, is used here (see procedure in Section B). Finally, an easy method for recovery of the $NH_4[BCS]$ salt is included.

Procedure

Finely ground *cis*-$[Cr(en)_2Cl_2]Cl\cdot H_2O$ (60.0 g, 0.202 mol) is quickly (within 2 min) dissolved in water (1000 mL) in a 2-L beaker at 20 °C, and $NH_4[BCS]$ (40.0 g, 0.122 mol) is added portionwise within 5 min with vigorous stirring. Precipitation of the diastereoisomer commences during the addition of the $NH_4[BCS]$ salt, and after another 5 min the precipitate is collected, washed with 96% ethanol (3 × 100 mL), and once with diethyl ether; it is then dried in air. The yield is 32 g of $(+)_{589}$-Λ-$[Cr(en)_2Cl_2][BCS]$. This is dissolved in ice-cold 12 M hydrochloric acid (130 mL). Immediately after dissolution, ice-cold 96% ethanol (260 mL) is added to the solution while the mixture is cooled in ice and stirred vigorously. After 5 min the precipitate is collected and washed with 96% ethanol (2 × 100 mL) once with diethyl ether and then air dried. The product $NH_4[BCS]$ is recovered from the filtrate as described below.

The yield of optically pure $(+)_{589}$-Λ-*cis*-$[Cr(en)_2Cl_2]Cl\cdot H_2O$ is 16.5 g {46% based on $NH_4[BCS]$ or 55% based on Cr(III)}. The specific rotation is $[\alpha]_{589}^{25} = 268(3)°$ (0.075% aqueous solution, extrapolated to time of dissolution), and no change in rotation is observed, either upon further reprecipitation with 96% ethanol from 12 M hydrochloric acid ($[\alpha]_{589}^{25} = 271°$), or by repeated reprecipitation with $NH_4[BCS]$ $[\alpha]_{589}^{25} = 268°$).

The filtrate from the conversion of the $[BCS]^-$ salt to the chloride salt contains H[BCS], which is readily recovered as follows: The filtrate (~ 400 mL) is concentrated to a volume of 80 mL in a rotatory evaporator, whereby most of the ethanol is removed. Concentrated ammonia (12 M, ~ 34 mL) is added to the aqueous and strongly acidic solution to give pH ~ 8. Precipitation of $NH_4[BCS]$ commences almost immediately and the mixture is cooled in ice for $\frac{1}{2}$ h before the precipitate is isolated and recrystallized as described above. Yield: ~ 10 g.

Anal. Calcd. for $CrC_4H_{18}N_4Cl_3O$: C, 16.20; H, 6.12; N, 18.89; Cl, 35.86. Found: C, 16.0; H, 6.3; N, 18.7; Cl, 35.8.

Properties

The chiral chloride salt is stable for years when kept dry and in the dark, preferentially in a refrigerator at $-15\,°C$. When heated, the solid salt dehydrates and undergoes racemization.[11] In acid solution the Λ-*cis*-$[Cr(en)_2Cl_2]^+$ cation aquates to Λ-*cis*-$[Cr(en)_2(H_2O)Cl]^{2+}$ followed by racemization of this complex. In acid solution, Ag^+-assisted hydrolysis leads to the Λ-*cis*-$[Cr(en)_2(H_2O)_2]^{3+}$ complex.[12]

References

1. P. Newman, *Optical Resolution Procedures for Chemical Compounds*, Vol. I, Optical Resolution Information Center, Manhattan College, Riverdale, New York, 1979.
2. F. S. Kipping and W. J. Pope, *J. Chem. Soc.*, **67**, 356 (1895).
3. G. B. Kaufmann, *J. Prakt. Chem.*, **33**, 295 (1966).
4. A. Hammershøi, A. M. Sargeson, and W. L. Steffen, *J. Am. Chem. Soc.*, **106**, 2819 (1984), and references cited therein.
5. E. Pedersen, *Acta Chem. Scand.*, **24**, 3362 (1970).
6. C. S. Garner and D. A. House, in *Transition Metal Chemistry*, Vol. VI, Carlin, R. L. (ed.), Marcel Dekker, New York, 1970, p. 187.
7. J. Springborg and C. E. Schäffer, *Inorg. Synth.*, **18**, 75 (1978).
8. P. A. Lay and A. M. Sargeson, *Inorg. Synth.*, **24**, 283 (1986).
9. A. Werner, *Ber. Dtsch. Chem. Ges.*, **44**, 3132 (1911).
10. J. Selbin and J. C. Bailar, *J. Am. Chem. Soc.*, **79**, 4285 (1957).
11. H. E. Le May and J. C. Bailar, *J. Am. Chem. Soc.*, **90**, 1729 (1968).
12. J. Springborg and C. E. Schäffer, *Acta Chem. Scand.*, **A30**, 787 (1976).

7. ORGANOMETALLIC COMPLEXES OF ISOMERIC ACYL ISOCYANIDES: CHROMIUM CARBONYL (ACYL ISOCYANIDE) AND (ACYL CYANIDE) COMPLEXES

Submitted by ASHRAF A. ISMAIL,* IAN S. BUTLER,* and GERARD JAOUEN[†]
Checked by ROBERT J. ANGELICI and GEORGE N. GLAVEE[‡]

Acyl isocyanides ($:CNCOR$, $R = CH_3$, C_6H_5, etc.)[1] are highly reactive but they can be stabilized by coordination to metals[2] like other reactive species such as carbenes.[3] We present here a convenient, general synthetic route to acyl isocyanide complexes via electrophilic attack of acyl halides on anionic metal cyanide complexes. The procedure is applicable to a wide variety of complexes with arene rings, containing either electron-donating [e.g., CH_3, $CH(CH_3)_2$, $C(CH_3)_3$, OCH_3] or electron-withdrawing substituents (e.g., CO_2CH_3, $COCH_3Cl$, NO_2). The functional group on the acyl moiety may be C_6H_5, $N(CH_3)_2$, OC_2H_5, SC_2H_5, and so on. The complexes with the $CNCOCH_3$ ligand are apparently too unstable to be isolated.

The compound selected to illustrate this procedure is $Cr(\eta^6\text{-}C_6H_5CO_2CH_3)(CO)_2(CNCOC_6H_5)$. Compounds of this type have already been shown to be active catalytically but sometimes with a different activity from that of the well-known parent tricarbonyl derivatives.[4] For example, reaction of *cis*-cyclooctene with $Cr(\eta^6\text{-}1\text{-}CO_2CH_3\text{-}2\text{-}CH_3C_6H_4)(CO)_2$-$(CNCOC_6H_5)$ for 15 h at 60 °C in CCl_4–tetrahydrofuran (THF) solution affords 60% conversion into 1-chloro-2-(trichloromethyl)cyclooctane.[2]

The stronger electron-withdrawing capacity of the CNCOR ligands compared to CO is demonstrated by the facile displacement of the arene ring in the $Cr(\eta^6\text{-arene})(CO)_2(CNCOR)$ derivatives by CO under pressure (~ 100 atm) to yield $Cr(CO)_5(CNCOR)$. An alternative route to certain other pentacarbonyl acyl isocyanide complexes [e.g., $Cr(CO)_5(CNCOC_6H_4Cl)$, $Cr(CO)_5(CNCOCH_2Cl)$] involves the thermolysis of arenediazonium pentacarbonylcyanochromates in various solvents.[5] The displacement of arene rings is extremely difficult for the analogous tricarbonyl complexes under the same conditions.

The synthesis of $Cr(CO)_5(CNCOC_6H_5)$ from $Cr(\eta^6\text{-}C_6H_5CO_2CH_3)(CO)_2$-$(CNCOC_6H_5)$ is also described here, together with that of the isomeric

*Department of Chemistry, McGill University, 801 Sherbrooke St. West, Montreal, Quebec, Canada H3A 2K6.
[†]Ecole Nationale Supérieure de Chimie de Paris, 11 rue Pierre et Marie Curie, 75231 Paris Cédex 05, France.
[‡]Department of Chemistry, Iowa State University, Ames, IA 50011.

ligand complex containing the $Cr(CO)_5(NCCOC_6H_5)$. There are very few examples of complexes of such ligand isomers in organometallic chemistry, and these two are the first complexes known for the CNCOR and NCCOR ligands. The $Cr(CO)_5(CNCOC_6H_5)$ derivative is expected to be active catalytically.[6] It should also be pointed out that performing similar chemistry on the $Cr(\eta^6\text{-arene})(CO)_2(CNCOR)$ derivatives to that for the closely related thiocarbonyl complexes[7] leads to the formation of the chiral species $Cr(\eta^6\text{-arene})(CO)(CNCOR)L$, which may in the future prove to have some utility as catalysts in asymmetric organic synthesis.

■ **Caution.** *Carbon monoxide and metal carbonyl complexes are highly toxic and must be handled at all times with care in an efficient hood.*

All the reactions and transfers of materials described in the following syntheses should be performed under a nitrogen atmosphere.

A. (BENZOYL ISOCYANIDE)DICARBONYL(η^6-METHYL BENZOATE)CHROMIUM (0)

$$Cr(\eta^6\text{-}C_6H_5CO_2CH_3)(CO)_3 + KCN$$

$$\xrightarrow{hv} K[Cr(\eta^6\text{-}C_6H_5CO_2CH_3)(CO)_2(CN)] + CO$$

$$K[(\eta^6\text{-}C_6H_5CO_2CH_3)(CO)_2(CN)] + C_6H_5COCl$$

$$\longrightarrow Cr(\eta^6\text{-}C_6H_5CO_2CH_3)(CO)_2(CNCOC_6H_5) + KCl$$

Procedure

■ **Caution.** *KCN is highly toxic and should be handled with suitable protection toward contact with the skin.*

The starting material $Cr(\eta^6\text{-}C_6H_5CO_2CH_3)(CO)_3$ can be easily prepared as described previously in *Inorganic Syntheses*[8] or may be purchased from Strem.

A mixture of $Cr(\eta^6\text{-}C_6H_5CO_2CH_3)(CO)_3$ (2.0 g, 7.4 mmol) and KCN (1.0 g, 15.4 mmol) is placed in a Pyrex ultraviolet (UV) irradiation vessel (capacity 350 mL)[9] equipped with a magnetic stirring bar and fitted with a water-cooled quartz finger containing a 150-W high-pressure mercury lamp. Deoxygenated CH_3OH (200 mL) is transferred to the vessel under a slight nitrogen pressure via stainless steel cannula.[10] The vessel is wrapped in aluminum foil, and the orange solution is irradiated with UV light for 3 h at room temperature with constant stirring.

■ **Caution.** *Exposure of the eyes to UV light is extremely hazardous and must be avoided at all times.*

After turning off the lamp, the reaction mixture is transferred to a 500-mL, round-bottomed flask and the solvent is stripped off under reduced pressure on a rotary evaporator to afford solid $K[Cr(\eta^6\text{-}C_6H_5CO_2CH_3)(CO_2(CN))]$. Infrared (IR) spectrum (CH_2Cl_2): $\nu_{(CO)}$ 1910(s)(a'), 1830(s)(a'')cm^{-1}; ^1H NMR spectrum: $\delta_{C_6H_5Cr)}$ 5.82(m, 2), 5.12(m, 1), 4.84(m, 2), and $\delta_{(OCH_3)}$ 3.47(s, 3) ppm. The residue is treated immediately with C_6H_5COCl (1.0 g, 7.1 mmol) and CH_2Cl_2 (20 mL). The resulting solution is stirred for 15 min, and then the volume of the solvent is reduced under vacuum. Reaction completion is checked by the absence of the $\nu_{(CO)}$ peaks due to the anion. Note that if there are electron-withdrawing groups better than $-CO_2CH_3$ attached to the ring, it is necessary to warm the reaction mixture slightly (35 °C) and to allow it to stand for ~ 20 min.

The pure benzoyl isocyanide derivative is finally obtained as red crystals (1.40–2.23 g, 50–80% yield, mp 78 °C) following thin-layer chromatography (TLC) on silica gel plates (40 × 20 cm) in a large developing tank (43 × 23 × 29 cm) using 700 mL of $(C_2H_5)_2O$–petroleum ether (1:4) as eluent. The desired product is contained in the red band on the TLC plate.

Anal. Calcd. for $C_{18}H_{13}NO_5Cr$: C, 57.6; H, 3.46; N, 3.73; MW, 375.0199. Found: C, 57.3; H, 3.52; N, 3.86; MW (high-resolution mass spectrum) 375.0200.

Properties

(Benzoyl isocyanide)dicarbonyl(η^6-methyl benzoate)chromium (0) is an air-stable, red crystalline solid that is soluble in most common organic solvents. Its IR spectrum in CCl_4 solution exhibits two strong $\nu_{(CO)}$ bands at 1987 (a') and 1922 (a'') cm^{-1}, but no peak attributable to $\nu_{(CN)}$ is observed. A recent vibrational study, including ^{13}C and ^{15}N labeling of the CN group, has shown that the $\nu_{(CN)}$ mode contributes to the 1987 cm^{-1} peak and to a shoulder on the 1922 cm^{-1} peak (at $\sim$ 1908 cm^{-1}) in the unlabeled species.[11] In the ^1H NMR spectrum in CDCl$_3$, the product displays resonances at $\delta_{(C_6H_5)}$ 8.45 (m, 2), 7.93 (m, 3), $\delta_{(C_6H_5)Cr}$ 6.45 (m, 2), 5.65 (m, 3), $\delta_{(OCH_3)}$ 4.05 (s, 3) ppm. The ^{13}C NMR data [downfield relative to $(CH_3)_4Si$, $\delta = 0.00$ ppm] in CH_2Cl_2 are 231.6 (s, $\delta_{(CO)_2}$), 223.8 (s, $\delta_{(CN)}$), and 157.0 (s, $\delta_{(CO)R}$) ppm. An X-ray analysis of the crystal and molecular structure has confirmed the formulation of $(\eta^6\text{-}C_6H_5CO_2CH_3)Cr(CO)_2(CNCOC_6H_5)$.[2] Crystals of two different morphologies (triclinic, $P1$, $Z = 2$; monoclinic $P2_1/c$, $Z = 4$) are obtained upon slow recrystallization from pentane. The stereochemistry is essentially the same in both crystal forms; the bent, two-electron donating, N-α-functionalized isocyanide ligand [C—$\hat{N}$—C = 168(1)°] is coordinated linearly to the chromium atom [Cr—$\hat{C}$—N = 178.8(9)°]. The complex reacts with aryl phosphines (PR_3) under the influence of UV light to give (η^6-$C_6H_5CO_2CH_3)Cr(CO)(CNCOC_6H_5)PR_3$. Attempts to prepare the corresponding isomer with $NCCOC_6H_5$ in place of $CNCOC_6H_5$ have been unsuccessful.

B.   (BENZOYL ISOCYANIDE)PENTACARBONYLCHROMIUM(0)

$$Cr(\eta^6\text{-}C_6H_5CO_2CH_3)(CO)_2(CNCOC_6H_5) + 3CO$$
$$\longrightarrow Cr(CO)_5(CNCOC_6H_5) + C_6H_5CO_2CH_3$$

- *See cautionary note in Section A.*

Procedure

A solution of $Cr(\eta^6\text{-}C_6H_5CO_2CH_3)(CO)_2(CNCOC_6H_5)$ (0.5 g, 1.3 mmol) in THF (50 mL freshly distilled from sodium benzophenone ketyl) is placed in a stainless steel autoclave (capacity 75 mL), and the solution is degassed by repeated freeze–thaw cycles. Carbon monoxide gas is introduced into the autoclave until the internal pressure at room temperature is ~ 100 atm. The reaction vessel is then heated at 100 °C for 5 h. After the vessel has cooled to room temperature, the excess CO gas is vented off in an efficient hood. The THF solvent is reduced in volume to ~ 7 mL on a rotary evaporator. Product purification is achieved by chromatography on silica gel (60 G, Merck) on a 75 × 3-cm column using deoxygenated $(C_2H_5)_2O$ as eluent. Following subsequent solvent removal under reduced pressure, orange-yellow crystals are obtained. Yield: 0.14–0.25 g (40–70%), mp 85 °C.

Anal. Calcd. for $C_{13}H_5NO_6Cr$: C, 48.3; H, 1.55; N, 4.65; MW, 322.9522. Found: C, 48.0; H, 1.60; N, 4.61; MW (high-resolution mass spectrum), 322.9521.

Properties

(Benzoyl isocyanide)pentacarbonylchromium(0) is an orange-yellow crystalline solid that is soluble in most common organic solvents. In its IR spectrum in cyclohexane, the characteristic peaks are $\nu_{(CN)}$ 2130 (w) (a_1); $\nu_{(CO)}$ 2030 (w) (a_1^{eq}), 1980 (m) (a_1^{ax}), 1950 (s) (e); $\nu_{(C=O)R}$ 1720 cm^{-1}. The corresponding Raman vibrations for the solid are $\nu_{(CN)}$ 2108; $\nu_{(CO)}$ 2010 (w), 1993 (s), 1986 (s), 1941 (w); $\nu_{(C=O)R}$ 1684 (w) cm^{-1}. The ^{13}C NMR resonances in CH_2Cl_2 appear at 214.5 $\delta_{(CO)_{trans}}$, 213.4 $\delta_{(CO)_{cis}}$, 186.5 $\delta_{(CN)}$, 154.0 $\delta_{(CO)R}$ ppm; for the phenyl ring the signals are at 130.8 $\delta_{C(4)}$, 130.5 $\delta_{C(2,6)}$, and 129.4 $\delta_{C(1,3,5)}$. The octahedral coordination and $CNCOC_6H_5$ arrangement have been confirmed by X-ray diffraction. The crystals are monoclinic (space group, $P2_1/c$, $Z = 4$) with

Cr—C (isocyanide) = 1.928(3) Å,   Cr—C (carbonyls) = 1.900(4) Å,   and Cr—$\hat{C}$—N = 173.9(3).[12]

C. (BENZOYL CYANIDE)PENTACARBONYLCHROMIUM(0)

$$Cr(CO)_6 + C_4H_8O \xrightarrow{hv} Cr(CO)_5(C_4H_8O) + CO$$

$$Cr(CO)_5(C_4H_8O) + C_6H_5COCN \longrightarrow Cr(CO)_5(NCCOC_6H_5) + C_4H_8O$$

Procedure

■ **Caution.** *Benzoyl cyanide is extremely toxic and should be handled with suitable precautions to prevent human contact.*

■ **Caution.** *Exposure of the eyes to UV light is extremely hazardous and must be avoided at all times.*

Hexacarbonylchromium(0) (1.0 g, 4.5 mmol) is placed in the photochemical cell described earlier, and THF (C_4H_8O, 200 mL; dried over sodium wire) is transferred to the vessel. Irradiation of the solution under a nitrogen atmosphere using a 100-W high pressure mercury lamp for 2 h at ice-water temperature affords a bright yellow solution indicative of the formation of $Cr(CO)_5(C_4H_8O)$ (IR in C_4H_8O: $v_{(CO)}$ 1938 cm^{-1}). Benzoyl cyanide (available commercially from Aldrich) (1.0 g, 7.5 mmol) is added to the reaction mixture, which immediately becomes bright red. After stirring the mixture for 1 h and subsequent solvent removal under reduced pressure, a red solid is isolated. This $Cr(CO)_5(NCCOC_6H_5)$ product is purified by subliming (25 °C/0.01 torr) out any unreacted $Cr(CO)_6$ and C_6H_5COCN. Yield: 1.3 g (88%).

Anal. Calcd. for $C_{13}H_5NO_6Cr$: C, 48.3; H, 1.55; N, 4.65; MW 323. Found: C, 48.0; H, 1.77; N, 4.42; MW (mass spectrum) 323.

Properties

(Benzoyl cyanide)pentacarbonylchromium(0) is an air-stable, red solid that is soluble in most organic solvents. Its IR spectrum in CCl_4 solution displays the following characteristic peaks: $v_{(CN)}$ 2200(w)(a_1); $v_{(CO)}$ 2070(2)a_1^{eq}), 1957(s)(e), 1940(m) (a_1^{ax}); $v_{(C=O)R}$ 1672 cm^{-1}. The relevant ^{13}C NMR resonances in CD_2Cl_2 are 218.3 $\delta_{(CO)trans}$, 213.3 $\delta_{(CO)cis}$, 164.2 $\delta_{(CN)}$, 137.2 $\delta_{(CO)R}$; the phenyl resonances are at 132.9 $\delta_{C(4)}$, 130.2 $\delta_{C(1,3,5)}$ ppm. For the free ligand: $v_{(CN)}$ 2222(w), $v_{(C=O)R}$ 1683 cm^{-1}; ^{13}C NMR (CD_2Cl_2), $\delta_{(CN)}$ 168.0, $\delta_{(CO)R}$ 136.9, $\delta_{(C4)}$ 133.3, $\delta_{C(2,6)}$ 130.4, $\delta_{C(1,3,5)}$ 129.5 ppm. There is no evidence for the interconversion of the two ligand isomers in solution. They exhibit quite distinct spectral properties and so can be easily distinguished from one another. Especially useful in this respect are the $\nu_{(CN)}$ modes in the IR and $\delta_{(CN)}$ resonances in the ^{13}C NMR spectra.

References

1. G. Höfle and B. Lange, *Angew. Chem.*, **89**, 272 (1977) and references therein.

2. P. Le Maux, G. Simonneaux, G. Jaouen, L. Ouahab, and P. Batail, *J. Am. Chem. Soc.*, **100**, 432 (1978).

3. E. O. Fischer, *Angew. Chem.*, **86**, 651 (1974).

4. G. Jaouen, in *Transition Metals in Organic Synthesis*, Vol. 2, H. Alper (ed.), Academic Press, New York, 1978.

5. W. P. Fehlhammer and F. Degel, *Angew. Chem. Int. Ed. Engl.*, **18**, 75 (1979).

6. M. S. Wrighton, D. S. Gimley, M. A. Schroeder, and D. L. Morse, *Pure Appl. Chem.*, **41**, 671 (1975).

7. (a) G. Jaouen and R. Dabard, *J. Organometal. Chem.*, **72**, 377 (1974); (b) G. Jaouen and G. Simonneaux, *Inorg. Synth.*, **19**, 197 (1979).

8. C. A. L. Mahaffy and P. L. Pauson, *Inorg. Synth.*, **19**, 154 (1979).

9. I. S. Butler, D. Cozak, S. R. Stobart, and K. R. Plowman, *Inorg. Synth.*, **19**, 193 (1979).

10. D. F. Shriver and M. A. Drezdzon, *The Manipulation of Air-Sensitive Compounds*, 2nd ed., Wiley-Interscience, New York, 1986, p. 22.

11. P. Caillet and P. Le Maux, *J. Organometal. Chem.*, **243**, 51 (1983).

12. J. Y. Le Marouille and P. Caillet, *Acta Cryst.*, **38B**, 267 (1982).

8. DIAMMONIUM PENTACHLOROOXOMOLYBDATE(V)

$$2Na_2MoO_4 + 2N_2H_4 + 16HCl \longrightarrow 2H_2[MoOCl_5] + N_2 + 2NH_4Cl \\ + 4NaCl + 6H_2O$$

$$H_2[MoOCl_5] + 2NH_4OH \xrightarrow[\substack{(2)\,HCl\,gas \\ in\,the\,cold}]{(1)\,60\,^\circ C} (NH_4)_2[MoOCl_5] + 2H_2O$$

Submitted by A. SYAMAL* and M. R. MAURYA[†]
Checked by K. DEHNICKE[‡] and F. SCHMOCK[‡]

Diammonium pentachlorooxomolybdate(V) is commonly used as a starting material for the syntheses of oxomolybdenum(V) complexes. It was first

*Department of Applied Sciences and Humanities, Kurukshetra University, Kurukshetra 132119, Haryana, India.
[†]Department of Chemistry, Regional Engineering College, Kurukshetra 132119, Haryana, India.
[‡]Fachbereich Chemie, Philipps-Universität Marburg, Hans-Meerwein Str. D-3550 Marburg, Federal Republic of Germany.

prepared by the reduction of a concentrated hydrochloric acid solution of ammonium molybdate(VI) with hydriodic acid,[1] and later by the electrolytic reduction of a hydrochloric acid solution of molybdenum(VI) trioxide.[2] It is commonly prepared[3] by the reduction of a hydrochloric acid solution of ammonium molybdate(VI) with mercury, treating the solution with $(NH_4)_2CO_3$ and CO_2 gas, heating the mixture to form a precipitate of $MoO(OH)_3$, and finally allowing $MoO(OH)_3$ to react with $(NH_4)_2CO_3$ and HCl gas. This method, however, have several disadvantages: a large amount of hazardous mercury is required, the time required for the preparation is ~ 2 days, an intermediate compound, $MoO(OH)_3$, must be isolated and dried overnight, and the yield is low (45–54%).

A simple, inexpensive, and less time consuming (~ 4 h) synthesis of diammonium pentachlorooxomolybdate(V) giving a high yield (87.8–89.9%) is described below. Isolation of intermediate compounds is not necessary and only one gas (HCl) is required for the synthesis.

Procedure

■ **Caution.** *Since HCl gas is a lachrymatory substance this procedure should be carried out in a well-ventilated hood.*

To a mixture of hydrazine hydrate (3.0 g, 0.06 mol) and concentrated analytical reagent grade HCl (100 mL, specific gravity 1.18) (*the use of a lower grade of concentrated HCl leads to lower yield*) sodium molybdate(VI) dihydrate (7.2 g, 0.03 mol) is added in five portions with vigorous shaking. The mixture is then digested on a water bath for 2 h. During digestion, sodium molybdate(VI) is completely reduced to Mo(V), forming a green solution along with white precipitates of NaCl and NH_4Cl. The mixture is then cooled in an ice bath in a stoppered flask for 1 h and filtered. Ammonium hydroxide (10 mL, specific gravity 0.9) is added dropwise to the filtrate at 60 °C and the solution cooled in an ice bath.

Dry HCl gas is required in the next step of the procedure. It may be obtained from a commercial gas cylinder or generated immediately before use by addition of concentrated HCl (12 M) slowly through a separatory funnel onto concentrated sulfuric acid (H_2SO_4, 36 M), held on an Erlenmeyer flask. The apparatus for generating HCl gas is shown in Fig. 1.

Dry HCl gas is passed for $\frac{1}{2}$ h through the ice-cold solution prepared in the previous step, and an emerald green solid is precipitated. This is separated by filtration through a sintered glass Gooch crucible (No. 4 pore size 5–15 μm), washed with concentrated HCl, and dried in a vacuum desiccator over solid NaOH. Yield: 8.5–8.7 g (87.8–89.9%).

The nonrecrystallized product is quite pure and is suitable for synthetic use. For a purer product the drying step is skipped and the compound is recrystallized as follows: The product is suspended in 10 mL of concentrated

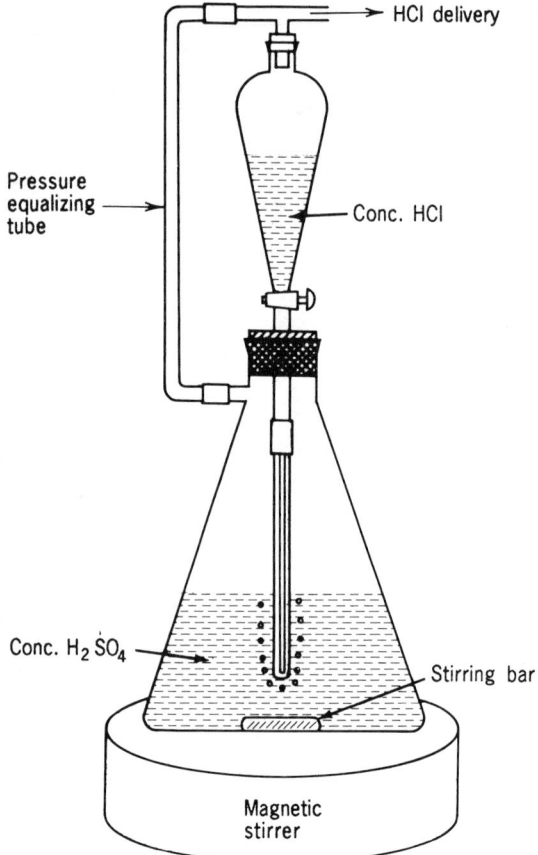

Fig. 1. Hydrogen chloride generator.

analytical reagent grade HCl, heated to 80 °C, and dissolved by adding a minimum amount of water. The solution is then cooled in an ice bath and the compound is reprecipitated by passing a stream of dry HCl gas through the ice-cold solution for $\frac{1}{2}$ h. The compound is separated by filtration, washed with concentrated HCl, and dried as before. It should be stored in a stoppered bottle in a desiccator. Yield after recrystallization: 7.5–7.65 g (77.5–79.0%).

Anal. Cacld. for $(NH_4)_2[MoOCl_5]$: Mo, 29.49; H, 2.46; N, 8.60; Cl, 54.53%. Found: Mo, 29.3; N, 8.7; Cl, 54.8% (authors); N, 8.48; H, 2.50; Cl, 54.92% (checkers).

Properties

Diamonium pentachlorooxomolybdate(V) is an emerald green solid, stable in air. It is hygroscopic and should be stored in a stoppered vial in a desiccator. In concentrated HCl, a solution ($\geqslant 10\,M$) of the compound is green. In dilute HCl ($< 10\,M$) the solution is greenish-brown or borwnish-red. The compound undergoes extensive ionic dissociation in aqueous solution. It is insoluble in benzene, chloroform, dichloromethane, and carbon tetrachloride. It is soluble (with decomposition) in ethanol, methanol, acetone, and pyridine; a white solid of ammonium chloride precipitates from all these solutions immediately. The compound dissolves in dimethyl sulfoxide without decomposition. The electronic spectrum in $10\,M$ HCl contains the following absorptions: $14{,}100(\varepsilon_{max} = 11)$, $22{,}500(\varepsilon_{max} = 10)$, $28{,}200(\varepsilon_{max} = 570)$, $32{,}200(\varepsilon_{max} = 5300)$, and $41{,}700\,\mathrm{cm}^{-1}$ ($\varepsilon_{max} = 3600$). Infrared absorptions for the compound (in a Nujol mull) are, $974\,(\mathrm{vs})$ ($v_{Mo=O}$, A_1), $385\,(\mathrm{sh})$ (v_s $MoCl_4$, A_1), $370\,(\mathrm{vs})$ (v_{as} $MoCl_4$, E), $235\,(\mathrm{m})$ ($v_{Mo-Cl_{trans}}$, A_1).

The compound exhibits the effective magnetic moment of 1.67 BM at 295 °K and possesses an octahedral structure. The compound is a good starting material[4] for $Mo_2O_3(N, N\text{-diethyldithiocarbamate})$ (Ref. 5), $Mo_2O_3(\text{acetylacetone})_2$ (Ref. 6), $MoOCl(8\text{-hydroxyquinoline})$ (Ref. 7), $MoO(OH)Cl_2$ (*o*-phenanthroline) (Ref. 7), $MoOCl_3(2,2'\text{-bipyridyl})$ (Ref. 8), $Mo_2O_3Cl_4$ $(2,2'\text{-bipyridyl})_2$ (Ref. 8), $Mo_2O_4Cl_2(2,2'\text{-bipyridyl})_2$ (Ref. 8), $Ba[Mo_2O_4(H_2O)_2(C_2O_4)_2]\cdot3H_2O$ (Ref. 9), $\{MoOCl[N\text{-(hydroxyethyl)-}$ salicylideneimine]\}_2$,[10] and $MoOCl(H_2O)$ [N-(hydroxyphenyl)salicylideneimine].[11]

References

1. P. Klason, *Ber.*, **34**, 148 (1901).

2. F. Foerster and E. Fricke, *Z. Angew. Chem.*, **36**, 458 (1923); R. G. James and W. Wardlaw, *J. Chem. Soc.*, **1927**, 2145.

3. W. G. Palmer, *Experimental Inorganic Chemistry*, Oxford University Press, London, 1954, p. 406.

4. F. A. Cotton and G. Wilkinson, *Advanced Inorganic Chemistry*, 3rd ed., Wiley, New York, 1972, p. 965.

5. F. W. Moore and M. L. Larson, *Inorg. Chem.*, **6**, 998 (1967).

6. F. W. Moore and M. L. Larson, *Inorg. Chem.*, **2**, 881 (1963).

7. P. C. H. Mitchell and R. J. P. Williams, *J. Chem. Soc.*, **1962**, 4570.

8. P. C. H. Mitchell, *J. Inorg. Nucl. Chem.*, **25**, 963 (1963).

9. F. A. Cotton and S. M. Moorehouse, *Inorg. Chem.*, **4**, 1377 (1965).

10. A. Syamal and M. A. Bari Niazi, *Trans. Metal Chem.*, **10**, 54 (1985).

11. K. Dey, R. K. Maiti, and J. K. Bhar, *Trans. Metal Chem.*, **6**, 346 (1981).

9. CARBYNE COMPLEXES OF TUNGSTEN

Submitted by E. O. FISCHER* and D. Wittmann*
Checked by A. Mayr[†] and A. McDermott[†]

Cationic carbyne (methylidyne) complexes[1] of the type $[M(CO)_5(CNEt_2)][BF_4]$ (M = Cr, Mo, W), are obtained from carbene (methylidene) complexes $(M(CO)_5[C(NEt_2)OEt]$ and boron trifluoride.[2–4] Their reaction with anionic nucleophiles affords access to a large variety of otherwise unavailable carbene and neutral carbyne complexes. For M = Cr, addition of X^- to the carbyne carbon leads to carbene complexes $Cr(CO)_5[C(NEt_2)X]$.[2,5–11] In case of M = Mo or W, substitution of the *trans*-CO group by X^- occurs predominantly to give neutral carbyne complexes *trans*-$[MX(CO)_4(CNEt_2)]$.[3,4] If strongly reducing nucleophiles are used, reductive dimerization of the cationic carbyne complexes yielding $[M(CO)_5(CNEt_2)]_2$ competes with both addition and substitution.[11,12] The synthesis of $[(CO)_5W(CNEt_2)][BF_4]$ and its reaction with $(NEt_4)(OCN)$ to form *trans*-$[W(OCN)(CO)_4(CNEt_2)]$ are described.

Procedure

■ **Caution.** *Boron trifluoride is very corrosive. Inhalation of the vapors and contact with the skin is very dangerous. It should be handled only in a well-ventilated hood.*

A. PENTACARBONYL[(DIETHYLAMINO)-METHYLIDYNE]TUNGSTEN TETRAFLUOROBORATE

$$W(CO)_5[C(NEt_2)OEt] + 2BF_3 \longrightarrow [(CO)_5W(CNEt_2)][BF_4] + \{BF_2OEt\}$$

All operations must be carried out under a nitrogen atmosphere, free from oxygen and moisture. The solvents must be dried by the usual methods and be saturated with nitrogen.[13]

The reaction apparatus is a Schlenk tube ($\sim$ 400-mL volume) with an N_2 inlet (Fig. 1). A glass tube reaches into the reaction mixture. It is tightly fixed to the cap of the Schlenk tube; the latter is linked to an escape valve filled with silicon oil. A solution of $W(CO)_5[C(NEt_2)OEt]^{14a,b}$ (4.54 g, 10 mmol) in

*Anorganisch-chemisches Institut der Technischen Universität, München, Lichtenbergstr. 4, D-8046 Garching, Federal Republic of Germany.
[†]Department of Chemistry, Princeton University, Princeton, NJ 08540.

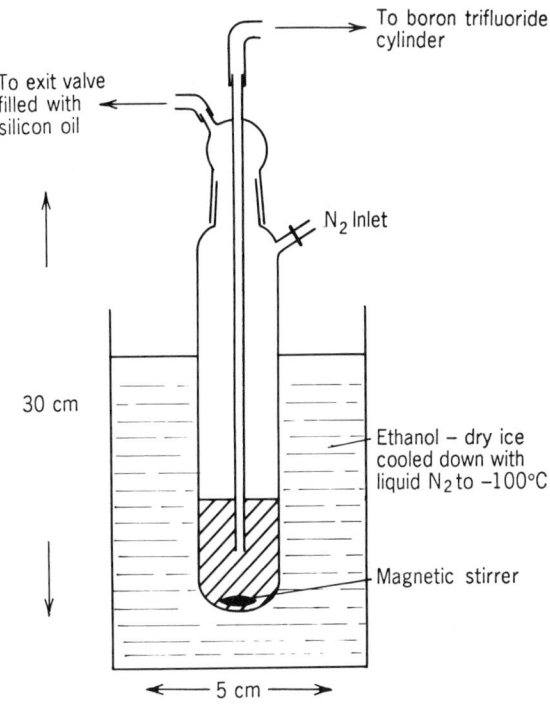

Fig. 1. Apparatus for the reaction of W(CO)$_5$[C(NEt$_2$)OEt] with BF$_3$.

40 mL of dichloromethane is placed into the Schlenk tube under nitrogen, and then cooled to $-100\,°C$ under stirring. Boron trifluoride gas is then slowly admitted to the reaction mixture over a period of 15 to 30 min to permit maintenance of the $-100\,°C$ temperature. The product is light sensitive and subdued lighting or light shielding is recommended for the next and subsequent steps of product work-up.

The light brown color turns to red whereupon the temperature is permitted to rise to $-78\,°C$ within 30 min. Excess boron trifluoride is permitted to evaporate and escape through the exit valve. The cap with the glass tube is removed and the Schlenk tube is stoppered. Residual boron trifluoride is removed *in vacuo* at $-60\,°C$ for 1 h. The product is precipitated at $-60\,°C$ with 80 mL of cold diethyl ether. The solvent is decanted and the red residue dried *in vacuo* at or below $-50\,°C$ for 1 h. At this temperature, it is reprecipitated from dichloromethane (40 mL) with diethyl ether (80 mL) precooled to $-50\,°C$. After removal of the solvent as above, 3.5 g (71%) of the title compound is obtained.

Anal. Calcd. for $C_{10}H_{10}BF_4NO_5W$: C, 24.27; H, 2.04; B, 2.18; F, 15.36; N, 2.83; W, 37.17. Found: C, 24.32; H, 2.15; B, 2.32; F, 15.10; N, 3.01; W, 37.55. Working on a smaller scale of 1.90 g (4.19 mmol) of starting material, Checkers report a 61% yield. Low temperatures of -50 or $-60\,^{\circ}$C must be maintained during work-up.

Properties

The product is a light red solid, soluble in dichloromethane, insoluble in diethyl ether and pentane. It can be stored under an inert gas at $-30\,^{\circ}$C for a few hours or at $-78\,^{\circ}$C for several months. It is less stable in solution, IR $(CH_2Cl_2, -40\,^{\circ}C, \gamma\ CO)$: 2143, 2100, 2065, and 2012 cm^{-1}; ^{1}H NMR $(CD_2Cl_2, -40\,^{\circ}C)$: $\delta = 1.49$ (t, 6H) and 3.69 (q, 4H). Upon warming, a fast rearrangement is observed accompanied by CO loss, to give *trans-* $[(BF_4)W(CO)_4(CNEt_2)]$ (2118, 2014, and 1978 cm^{-1}).

B.   *trans-* TETRACARBONYL[(DIETHYLAMINO)METHYLIDYNE]- (ISOCYANATO)TUNGSTEN

$$[W(CO)_5(CNEt_2)][BF_4] \xrightarrow{\ -CO\ }$$

$$\left. \begin{array}{l} \{trans\text{-}\ [W(BF_4)(CO)_4(CNEt_2)]\} \\ trans\text{-}\ [W(OCN)(CO)_4(CNEt_2)] \end{array} \right] + NEt_4(OCN)$$

Procedure

All operations are carried out under dry, oxygen-free nitrogen, using dry and nitrogen-saturated solvents. The silica gel (Merck) (0.06–0.20 mm) was held 5 h *in vacuo* and saturated with nitrogen.

In a 100-mL round-bottomed flask equipped with N_2 inlet, $[W(CO)_5(CNEt_2)][BF_4]$ (2.1 g, 4.2 mmol) is added to 30 mL of dichloromethane at $-40\,^{\circ}$C, followed by tetraethylammonium cyanate (1.5 g, 8.7 mmol) while stirring. After 30 min, the orange solution is taken directly to column chromatography[14b] (silica gel, $-40\,^{\circ}$C, column length 40 cm, width 3 cm). Elution with dichloromethane–diethyl ether (1:1) into a cooled ($-40\,^{\circ}$C) Schlenk tube with N_2 inlet yields a dark orange solution, which is reduced *in vacuo* to 15 mL. The product (0.95 g, 54%) is precipitated with 25 mL of pentane at $-40\,^{\circ}$C and washed twice with 20 mL of cold pentane.

Anal. Calcd. for $C_{10}H_{10}N_2O_5W$: C, 28.46; H, 2.39; N, 6.64; O, 18.95; W, 43.56. Found: C, 28.39; H, 2.43; N, 6.68; O, 18.93; W, 43.95.

Properties

The orange crystalline complex is soluble in dichloromethane, acetone, and diethyl ether, but it is insoluble in pentane. It should be stored under an inert gas at $-30\,^{\circ}$C. Infrared spectra (1, 1, 2-trichloroethane, $-20\,^{\circ}$C, v_{CO}): 2106, 2010, 1980, 2226 (v_{CN}) cm^{-1}. ^{1}H NMR (acetone-d$_6$, $-40\,^{\circ}$C): $\delta = 1.38$ (t, 6H) and 3.51 (q, 4H).

References

1. For a survey see: K. H. Dötz, H. Fischer, P. Hoffmann, F. R. Kreissl, U. Schubert and K. Weiss *Transition Metal Carbene Complexes*, Verlag Chemie, Weinheim, 1983, p. 176

2. E. O. Fischer, W. Kleine, and F. R. Kreissl, *Angew. Chem.*, **88**, (1976) 646; *Angew.Chem. Int. Ed. Engl.*, **15**, 616 (1976).

3. E. O. Fischer, D. Wittmann, D. Himmelreich, U. Schubert, and K. Ackermann, *Chem. Ber.*, **115**, 3141 (1982).

4. E. O. Fischer, D. Wittmann, D. Himmelreich, R. Cai, K. Ackermann, and D. Neugebauer, *Chem. Ber.*, **115**, 3152 (1982).

5. E. O. Fischer, W. Kleine, F. R. Kreissl, H. Fischer, P. Friedrich, and G. Huttner, *J. Organomet. Chem.*, **128**, C 49 (1977).

6. E. O. Fischer, D. Himmelreich, R. Cai, H. Fischer, U. Schubert, and B. Zimmer-Gasser, *Chem. Ber.* **114**, 3209 (1981).

7. H. Fischer, E. O. Fischer, R. Cai, and D. Himmelreich, *Chem. Ber.*, **116**, 1009 (1983).

8. U. Schubert, E. O. Fischer, and D. Wittmann, *Angew. Chem.*, **92**, 662 (1980); *Angew. Chem. Int. Ed. Engl.*, **19**, 643 (1980).

9. E. O. Fischer, R. B. A. Pardy, and U. Schubert, *J. Organomet. Chem.*, **181**, 37 (1979).

10. H. Fischer, E. O. Fischer, and R. Cai, *Chem. Ber.*, **115**, 2707 (1982).

11. E. O. Fischer and R. Reitmeier, *Z. Naturforsch.*, **38**, 582 (1983).

12. E. O. Fischer, D. Wittmann, D. Himmelreich, and D. Neugebauer, *Angew. Chem.*, **94**, 451 (1982); *Angew. Chem. Int. Ed. Engl.*, **21**, 444 (1982).

13. D. F. Shriver, *The Manipulation of Air-Sensitive Compounds*, McGraw-Hill, New York, 1969.

14. (a) E. O. Fischer, U. Schubert, W. Kleine, and H. Fischer, *Inorg. Synth.* **19**, 168 (1979); (b) This compound is prepared like the homologous chromium complex (p. 168); (c) A description of the chromatography column is found on p. 166.

10. 2,2-DIMETHYLPROPYLIDYNE TUNGSTEN(VI) COMPLEXES AND PRECURSORS FOR THEIR SYNTHESES*

Submitted by RICHARD R. SCHROCK,[†] JOSE SANCHO,[†]
and STEVEN F. PEDERSON[†]
Checked by SCOTT C. VIRGIL[‡] and ROBERT H. GRUBBS[‡]

In the last eight years since its first synthesis,[1] trineopentylneopentylidynetungsten(VI) has become a versatile intermediate in the synthesis of organometallic compounds of catalytic and structural interest. The compound was first synthesized in 25% yield by the reaction of tungsten hexachloride with six equivalents of neopentyllithium.[1] In 1982, a greatly improved synthesis of trineopentylneopentylidynetungsten(VI) was developed from trichlorotrimethoxytungsten, and several useful synthetic intermediates were prepared.[2] Some of these tungsten alkylidyne compounds catalyze the productive metathesis of internal acetylenes.[3] Well-characterized olefin metathesis catalysts[4] along with a variety of other tungsten complexes[5-9] also have been prepared from trineopentylneopentylidynetungsten(VI) and trichloro(1,2-dimethoxyethane)neopentylidynetungsten(VI).

The air and moisture sensitivities of the reagents and products involved in the preparations of tungsten compounds require the use of dry box or standard Schlenk[10] tube techniques using an atmosphere of dry, oxygen-free argon or nitrogen. Diethyl ether and predried 1,2-dimethoxyethane are vacuum transferred from sodium benzophenone ketyl and stored under argon. Pentane is vacuum transferred from sodium benzophenone ketyl to which tetraglyme has been added in order to solubilize the ketyl.

A. METHOXYTRIMETHYLSILANE

$$(CH_3)_3SiNHSi(CH_3)_3 + 2CH_3OH \xrightarrow{\ (CH_3)_3SiCl\ } 2CH_3OSi(CH_3)_3 + NH_3$$

Procedure

To a 500-mL three-necked round-bottomed flask equipped with a magnetic stirbar, addition funnel, and a Dewar condenser filled with an ice–salt slush

---

*Throughout this work these compounds also may be referred to as trineopentyl-neopentylidynetungsten(VI) and trichloro(1,2-dimethoxyethane)neopentylidynetungsten(VI), respectively. The author express thanks to S. C. Virgil for detailed procedures A, B, and C.

[†]Department of Chemistry, Massachusetts Institute of Technology, Cambridge, MA 02139.
[‡]Department of Chemistry, California Institute of Technology, Pasadena, CA 91125.

($-10\,°C$) and connected to an oil bubbler is added hexamethyldisilazane [Aldrich, 98%] (161 g, 1.0 mol), and 1 drop of chlorotrimethylsilane. The addition funnel is charged with absolute methanol (64 g, 2.0 mol) and addition of the methanol is begun at a rate of 2 drops per second. Ammonia is immediately evolved and detected from the bubbler and soon the mixture begins to reflux at a temperature of $\sim 30\,°C$. Use of the $-10\,°C$ Dewar condenser is necessary for keeping the volatile product from being carried through the condenser with the ammonia. After the 30 min addition of methanol, the addition funnel is replaced with a stopper and the solution is slowly warmed as ammonia is evolved over a period of 2 h. The condenser is replaced with a 40-cm Vigreaux column and reflux–distillation head protected with argon. The solution is refluxed 1 day under argon to drive off more ammonia then distilled slowly collecting only the fraction boiling at $57.0\,°C$. Yield: 187 g (90%).

B. TRICHLOROTRIMETHOXYTUNGSTEN

$$WCl_6 + 3CH_3OSi(CH_3)_3 \xrightarrow{15\,°C} (CH_3O)_3WCl_3 + 3(CH_3)_3SiCl$$

Although the preparation of $(CH_3O)_3WCl_3$ has been reported,[11] it is important to stress that the reactants be pure and to make modifications to accommodate a preparative scale. The reaction fails partially or completely if trace amounts of HCl are present during the reaction; it is believed that HCl catalyzes the decomposition of methoxytungsten chlorides to insoluble white tungsten oxides and either chloromethane or dimethyl ether. In this procedure, granular dri-Na® is used to scavenge trace HCl as it is generated in the reaction.

Procedure

To an oven-dried 250-mL Schlenk flask under argon and equipped with a magnetic stirbar are added 0.5 g of granular dri-Na® [Baker, 10% sodium and 90% lead as alloy] and, via syringe, methoxytrimethylsilane (30 mL, 22.6 g, 0.217 mol, 3.03 equiv). Sublimed, pulverized tungsten hexachloride* (28.4 g, 0.072 mol) is loaded into an oven-dried powder addition funnel (Kontes) and connected to the Schlenk flask. The flask is cooled in a 15 °C ice–water bath

---

*Tungsten hexachloride is sublimed *in vacuo*. It is desirable to remove relatively volatile $WOCl_4$ efficiently from WCl_6 before collecting the WCl_6 sublimate. On a large scale this becomes somewhat problematic, as the $WOCl_4$ migrates to the relatively cool interior of the sample that is being sublimed where it sublimes slowly along with WCl_6 from the hot exterior zone.

and vented to an oil bubbler. Tungsten hexachloride is added in small portions over a 1-h period to the efficiently stirred reaction mixture. Since each added quantity of WCl_6 reacts slowly, it is important to stir the reaction efficiently. After about one-third of the required WCl_6 has been added, yellow crystals of $(CH_3O)_3WCl_3$ begin to form in the yellow solution. After all of the WCl_6 has been added, the mixture of yellow crystals in $\sim 40\,mL$ of solution is stirred at room temperature for an additional 15 min. The addition funnel is replaced with a stopper, an auxilliary trap is connected, all volatile components are removed *in vacuo*, and the solid product is kept *in vacuo* for 3 h. Absolute diethyl ether (50 mL) is added and the solid dissolved by warming the solution slightly. The cloudy yellow solution is filtered through Celite® to remove $\sim 0.1\,g$ of white precipitate, which is rinsed with 5 mL of additional diethyl ether. The clear yellow solution is cooled to $-80\,°C$ and a total of 23.4 g (85% yield) of pure $W(CH_3O)_3Cl_3$ is collected by filtration. The yellow spars are stored in an inert atmosphere.

C.   (2,2-DIMETHYLPROPYL)MAGNESIUM CHLORIDE

$$(CH_3)_3CCH_2Cl + Mg \xrightarrow{\text{diethyl ether}} (CH_3)_3CCH_2MgCl$$

Procedure

A 250-mL three-necked round-bottomed flask equipped with a magnetic stirbar and an Allihn condenser is oven-dried, assembled hot with well-greased joints, and kept under vacuum as it is allowed to cool to room temperature. The apparatus is filled with argon and magnesium turnings [Fisher 99.8% Grignard turnings] (9 g, 0.37 mol) are added followed by 60 mL of absolute diethyl ether via syringe. An addition funnel containing neopentyl chloride (29 g, 0.27 mol)* is connected and 5 to 10% of the neopentyl chloride is added. The solution is heated until it refluxes gently and 0.1 mL of 1, 2-dibromoethane is added to initiate the reaction.

■ **Caution.** *1, 2-Dibromoethane is a suspected carcinogen. It should be used only in an efficient hood. Gloves should be worn.*

The reaction soon continues to reflux without external heating and the

---

*This Grignard reaction is particularly difficult to initiate and depends greatly on the purity of the neopentyl chloride. Commercial neopentyl chloride [Eastern] is stirred over four consecutive portions of concentrated sulfuric acid for a period of 6 h each followed by 1 h for the mixture to completely settle before draining off the darkened acid layer. The fourth portion of acid is only slightly colored. The clear neopentyl chloride is washed with water followed by aqueous sodium bicarbonate solution, dried over calcium chloride, then filtered from the drying agent before use. Pure neopentyl chloride distils at 83 to 84 °C and is inert to concentrated sulfuric acid at room temperature.

remainder of the neopentyl chloride is added dropwise during 1 h. After all the neopentyl chloride has been added, the reaction soon subsides and external heat is again applied in order to maintain a gentle reflux for 6 h. The mixture is filtered through Celite® in a Schlenk frit and the filter cake rinsed with diethyl ether.[†] Titration of a 1-mL aliquot against 2-butanol using 5-methyl-1, 10-phenanthroline as an indicator[12] suggests that the yield is usually 80 to 85% (0.22–0.23 mol), although higher yields are not uncommon.

D.   TRIS(2, 2-DIMETHYLPROPYL)(2, 2-DIMETHYLPROPYLIDY-NE)TUNGSTEN(VI)

$$(CH_3O)_3WCl_3 + 6(CH_3)_3CCH_2MgCl$$

$$\xrightarrow[-5-0\,°C]{\text{diethyl ether}} [(CH_3)_3CCH_2]_3W{\equiv}CC(CH_3)_3$$

Procedure

A double Schlenk tube (Ref. 10 pp. 153–154) equipped with two high-vacuum stopcocks and a medium porosity fritted disc separating two 600-mL capacity tubes is employed (B-1, B-2; Fig. 1). Neopentylmagnesium chloride in diethyl ether (2.8 M, 81 mL, 224 mmol) is transferred via cannula into one side of the double Schlenk tube (B-1) and diluted to $\sim 1\,M$ with 150 mL of absolute diethyl ether. The Grignard solution (under argon) is cooled in a $-5\,°C$ ice–salt bath. A solution of $W(CH_3O)_3Cl_3$ (14.3 g, 37.4 mmol) in 50 mL of absolute diethyl ether (Schlenk tube A; Fig. 1) is added in portions via cannula to the vigorously stirred Grignard solution over a period of 30 min. Soon after the addition of $W(CH_3O)_3Cl_3$ is begun, neopentane is evolved through the bubbler and the reaction mixture turns clear brown. Soon thereafter, magnesium salts precipitate abundantly, rendering efficient stirring more difficult. If the $W(CH_3O)_3Cl_3$ solution is not quickly dispersed through the reaction mixture, $W(CH_3O)_3Cl_3$ crystallizes as green-brown coated crystals that react relatively slowly. After the addition of the trichlorotrimethoxy-tungsten solution is complete, the mixture is diluted by adding 100 mL of diethyl ether and stirred as it warms to room temperature overnight.

The reaction mixture is filtered into B-2 (Fig. 1) and the magnesium salts extracted with solvent that has been condensed back into B-1.[10] Three such extractions yield a final wash that is essentially colorless. The solvent is then condensed back into B-1, and the solvent and suspended magnesium salts

---

[†]Under ideal conditions, very little suspended solid (presumably insoluble magnesium salts) is present in the Grignard solution, and filtration through Celite® is unnecessary.

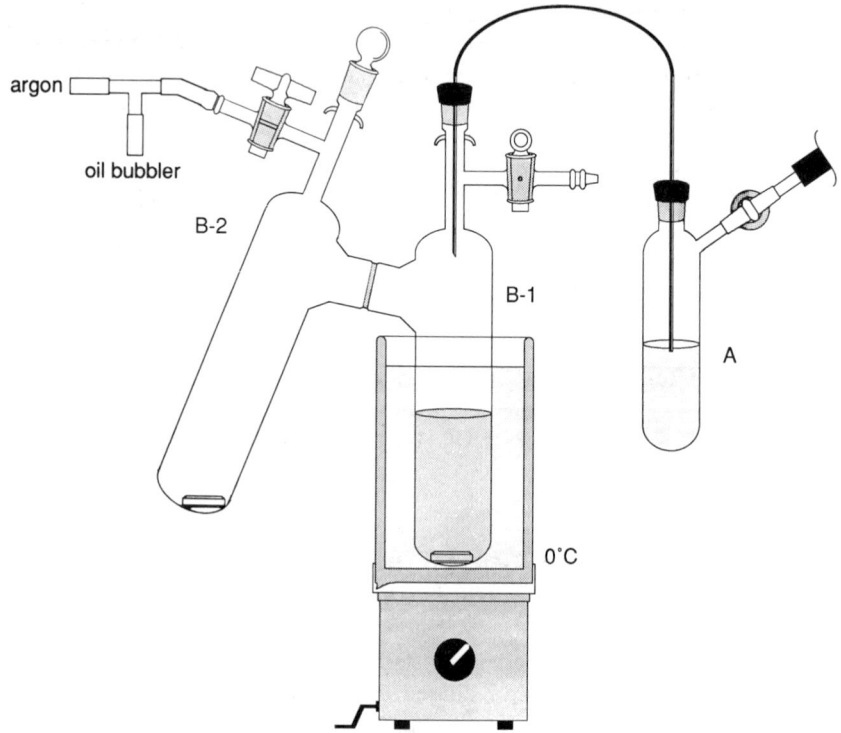

Fig. 1.   Preparation of trineopentylneopentylidynetungsten(VI).

removed via canula for quenching by pouring onto powdered Dry Ice. The septum in B-1 is replaced by a greased stopper and the apparatus evacuated at 1 μ for 4 h while heating B-2 to 40 °C in order to remove all diethyl ether from the brown oil. B-2 is then cooled to room temperature and 100 mL of pentane is added. The solution is then filtered back into B-1 and the pentane removed *in vacuo*. The brown oil is heated to 60 °C in an oil bath at high vacuum (1 μ) for 4 h, thereby removing volatile side products such as $(CH_3)_3CCH_2CH_2C(CH_3)_3$. (The vacuum must be applied to the same side of the apparatus as the sample in order to remove the side products efficiently.) About 20 mL of dark brown oil is obtained that is already rather pure product according to its ^1H NMR spectrum. The crude product is transferred into a short-path, large-diameter distillation apparatus (Fig. 2)* via cannula. A small amount of pentane is used to wash the last traces of the oil into the distillation

---

*The crude product can also be purified by sublimation at this scale,[1] but this technique often results in complications if it is attempted at a larger scale.

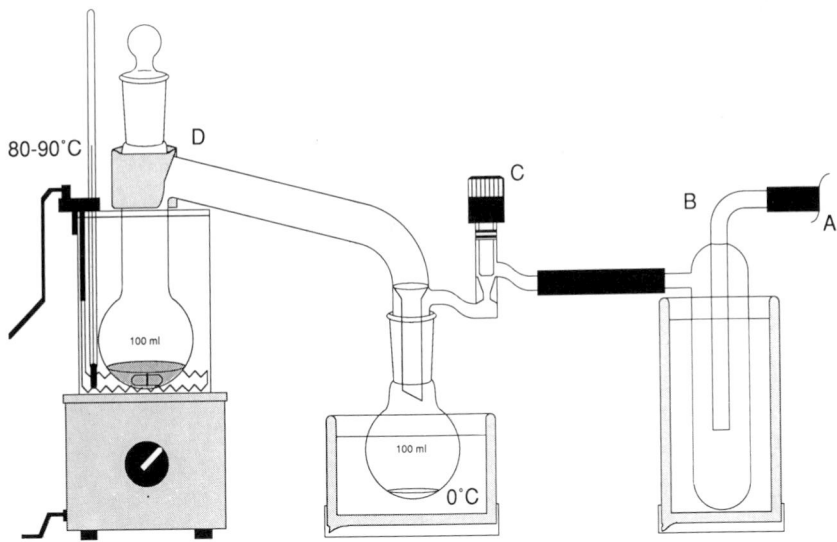

Fig. 2. High vacuum distillation of trineopentylneopentylidynetungsten(VI).

apparatus. A vacuum of 1×10^{-4} torr is applied at A through the auxilliary trap B that is cooled in liquid nitrogen (Fig. 2). The flask and its neck are heated slowly to $80\,°C$ at 1×10^{-4} torr as the pentane and traces of other volatile impurities are evolved. After the oil bath reaches $80\,°C$ and all volatile impurities have been trapped out (usually after 1 h at $80\,°C$), the yellow liquid begins to distil. During distillation it is necessary to warm the stillhead periodically with a heat gun in order to prevent crystallization of the liquid in the air-cooled condenser. Under no circumstances should the pot, stillhead, or condenser be heated above $110\,°C$, as the product decomposes rapidly near $140\,°C$. Distillation proceeds between 75 to $90\,°C$ at a moderate rate and is complete in 30 min. Yellow, crystalline $[(CH_3)_3CCH_2]_3W \equiv CC(CH_3)_3$ [10.5–11.5 g, 60–66% based on $(CH_3O)_3WCl_3$] is stored in a dry box.

Properties

Yellow, crystalline $[(CH_3)_3CCH_2]_3W \equiv CC(CH_3)_3$ melts at 47 to $48\,°C$ to an orange-yellow oil. At $140\,°C$, it decomposes rapidly to a black mass as gases are evolved. The compound is very air sensitive, reacting instantly with air to give a brown-black insoluble oil. The complex $[(CH_3)_3CCH_2]_3W \equiv CC(CH_3)_3$ is extremely soluble in ether and hydrocarbon solvents. It was thought to be a dimer in early studies,[1] although it is now believed that steric arguments militate against that possibility. ^{1}H NMR (C_6D_6) δ 1.56 [s, 9H, $CC(CH_3)_3$], δ 1.14 [s, 27H, $CH_2C(CH_3)_3$], δ 0.97 [s, 6H, $CH_2C(CH_3)_3$].

E. TRICHLORO(1, 2-DIMETHOXYETHANE)(2, 2-DIMETHYL-PROPYLIDYNE)TUNGSTEN(VI)

$$[(CH_3)_3CCH_2]_3W\equiv CC(CH_3)_3 + 3HCl + CH_3OCH_2CH_2OCH_3$$

$$\xrightarrow[0\,°C]{\text{pentane}} CH_3O-\overset{\displaystyle Cl \underset{\displaystyle Cl}{}}{\underset{\displaystyle O \quad Cl}{W}}\equiv CC(CH_3)_3 + 3(CH_3)_4C$$

Procedure

The double Schlenk tube (Fig. 1) equipped with a medium porosity fritted disc is a convenient apparatus for preparing $Cl_3(dme)W\equiv CC(CH_3)_3$. The complex $[(CH_3)_3CCH_2]_3W\equiv CC(CH_3)_3$ (10.0 g, 21.4 mmol) is dissolved in pentane (100 mL) containing 1, 2-dimethoxyethane (dme) (6.7 mL, 5.8 g, 64.4 mmol) and the solution is cooled to 0 °C using an ice–water bath. Anhydrous HCl in diethyl ether (Aldrich)* (1.0 M, 80 mL, 80 mmol, 3.75 equiv.) is added dropwise via syringe over a 5-min period while the reaction is stirred. During the first half of the addition, the solution becomes progressively darker brown with no sign of a purple product. During the final half of the addition, the reaction turns a deep purple color and blue-purple crystals of $Cl_3(dme)W\equiv CC(CH_3)_3$ precipitate abundantly.† The reaction is stirred 30 min at 0 °C and then cooled to − 30 °C. The blue-purple product is filtered off, washed twice with pentane, and dried *in vacuo* to give 8.65 g (90%) of pure $Cl_3(dme)W\equiv CC(CH_3)_3$. It can be stored inside a dry box at 25 °C.

Properties

The complex $Cl_3(dme)W\equiv CC(CH_3)_3$ is a blue-purple air-sensitive solid. It is essentially insoluble in pentane, but soluble in toluene, dichloromethane.

---

*Alternatively, HCl may be added as a gas to $[(CH_3)_3CCH_2]_3W\equiv CC(CH_3)_3$ in a mixture of pentane and ether.
†Deposits of insoluble reddish residue will be observed among the blue-purple precipitate if a typical source of technical grade HCl is used in the reaction. Extraction or recrystallization of the crude product with ether may then be necessary in order to obtain pure material.

and diethyl ether at 25 °C. Deep violet crystals can be obtained by cooling a diethyl ether solution to -60 °C. The ^{1}H NMR (C_6D_6) δ 3.60 and 3.25 (ps, 6H), $CH_3OCH_2CH_2OCH_3$), δ 2.83 and 2.92 (pdd, 4H, $CH_3OCH_2CH_2OCH_3$), δ 1.28 (s, 9H, C$C(CH_3)_3$). The complex Cl_3(dme)W$\equiv$ C$C(CH_3)_3$ is an important intermediate for preparing olefin and acetylene metathesis catalysts.[3,4] It is also a useful starting material for preparing certain tungstenacyclobutadiene, η^3-cyclopropenyl, and η^5-cyclopentadienyl complexes.[5,6]

References

1. D. N. Clark and R. R. Schrock, *J. Am. Chem. Soc.*, **100**, 6774–6776 (1978).

2. R. R. Schrock, D. N. Clark, J. Sancho, J. H. Wengrovius, S. M. Rocklage, and S. F. Pedersen, *Organometallics*, **1**, 1645–1651 (1982).

3. J. H. Wengrovius, J. Sancho, and R. R. Schrock, *J. Am. Chem. Soc.*, **103**, 3932–3934 (1981). M. R. Churchill, J. W. Ziller, J. H. Freudenberger, and R. R. Schrock, *Organometallics*, **3**, 1554–1562.

4. C. J. Schaverien, J. C. Dewan, and R. R. Schrock, *J. Am. Chem. Soc.*, **108**, 2171–2773 (1986).

5. S. F. Pedersen, R. R. Schrock, M. R. Churchill, and H. J. Wasserman, *J. Am. Chem. Soc.*, **104**, 6809–6811 (1982).

6. (a) M. R. Churchill, J. W. Ziller, J. H. Freudenberger, and R. R. Schrock, *Organometallics*, **3**, 1554 (1984); (b) J. H. Freudenberger, R. R. Schrock, M. R. Churchill, A. L. Rheingold, and J. W. Ziller, *Organometallics*, **3**, 1563 (1984); (c) R. R. Schrock, S. F. Pedersen, M. R. Churchill, and J. W. Ziller, *Organometallics*, **3**, 1574 (1984).

7. (a) R. R. Schrock, J. S. Murdzek, J. H. Freudenberger, M. R. Churchill, and J. W. Ziller, *Organometallics*, **5**, 25 (1986); (b) J. H. Freudenberger, and R. R. Schrock, *Organometallics*, **5**, 1411 (1986).

8. R. R. Schrock, *J. Organometal. Chem.*, **300**, 249 (1986).

9. R. R. Schrock, *Acc. Chem. Res.*, **19**, 342 (1986).

10. D. F. Shriver, *The Manipulation of Air-Sensitive Compounds*, McGraw-Hill, New York, 1978.

11. L. B. Handy, K. G. Sharp, and F. E. Brinckman, *Inorg. Chem.*, **11**, 523–531 (1972).

12. S. C. Watson and J. F. Eastham, *J. Organometal. Chem.*, **9**, 165–168 (1967).

11. ZEROVALENT IRON ISOCYANIDE COMPLEXES

Submitted by MICHEL O. ALBERS,* ERIC SINGLETON,* and NEIL J. COVILLE[†]
Checked by ARTHUR Y.-J. CHEN,[‡] RUSTY BLANSKI,[‡] and HERBERT D. KAESZ[‡]

A recent review has highlighted the extensive and interesting chemistry of metal isocyanide complexes.[1] Although synthetic procedures are varied, a vast number are based on substitution in metal carbonyl complexes by isocyanides. Such procedures are, however, not always successful. This is especially so in cases where multiple substitution of CO is required, as in the syntheses of homoleptic isocyanide complexes. Many of the inherent difficulties are illustrated by the reaction of iron pentacarbonyl with isocyanides.

Pentacarbonyl iron is fairly inert to substitution reactions, and attempts to prepare $Fe(CO)_{5-n}(CNR)_n$ $(n = 1-5)$ by the direct reaction of $Fe(CO)_5$ with isocyanides in Carius tubes has produced only the complexes $Fe(CO)_{5-n}(CNR)_n$ $(n = 1$ and $2)$.[2] The products were obtained as mixtures that required separation. Other syntheses, including photochemical[3] and trimethyl-amine N-oxide promoted[4a] displacement of carbonyl groups, or other means,[4b] give the same products in variable yield. Procedures based on diiron nonacarbonyl[5] and triiron dodecacarbonyl[6] have produced similar results. The only zerovalent iron complex $Fe(CO)_{5-n}(CNR)_n$ where $n > 2$ is the complex $Fe(CNR)_5$ prepared either by metal vapor synthesis techniques[7] or by sodium amalgam reduction of iron(II) bromide in the presence of isocyanide.[8]

In this contribution we describe facile, high-yield syntheses of the series of zerovalent iron isocyanide complexes $Fe(CO)_{5-n}(CNC_6H_3Me_2-1,3)_n$ $(n = 1-5)$. The starting material is iron pentacarbonyl, and cobalt(II) chloride is used as a catalyst to achieve the stepwise replacement of carbonyl groups by 2-isocyano-1,3-dimethylbenzene.[4,9]

Although the synthetic procedure is exemplified by the use of 2-isocyano-1,3-dimethylbenzene, the series of complexes $Fe(CO)_{5-n}(CNR)_n$ $(n = 1-5)$ can be obtained for other aromatic isocyanides, and for some alkyl isocyanides (e.g., $n = 1-4$, R = t-Bu; $n = 1-3$, R = Me, C_6H_{11}, $C_6H_5CH_2$) using a similar technique.[4]

*National Chemical Research Laboratory, Council for Scientific and Industrial Research, P.O. Box 395, Pretoria 0001, Republic of South Africa.
[†]Department of Chemistry, University of the Witwatersrand, 1 Jan Smuts Avenue, Johannesburg 2001, Republic of South Africa.
[‡]Department of Chemistry and Biochemistry, University of California, Los Angeles, CA 90024-1569.

General Procedure

■ **Caution.** *The use of the volatile, toxic iron pentacarbonyl necessitates that all manipulations be carried out in a well-ventilated hood. In addition, the carbon monoxide evolved in the reaction is an odorless, extremely toxic gas, and care should be exercised that the apparatus vents into the best ventilated region of the hood. The compound 2-isocyano-1,3-dimethylbenzene is a vile smelling, volatile solid that is best handled in a well-ventilated hood, using protective gloves.*

Unless otherwise stated, all manipulations are performed under nitrogen using standard Schlenk techniques.[10] Solvents are all of Analar Grade and were dried and distilled under nitrogen prior to use.[11] The 2-isocyano-1,3-dimethylbenzene was purchased from Fluka AG and is used without further purification. The cobalt(II) chloride catalyst is obtained by heating $CoCl_2 \cdot 6H_2O$ under vacuum at $\sim 50\,°C$ for 5 h. The material so obtained is blue-pink in color and analyzed approximately as $CoCl_2 \cdot 2H_2O$.[4] Column chromatography on 60 to 200 μm of silica gel is used throughout. Melting points are all recorded in air in a well-ventilated hood, and are uncorrected.

A. TETRACARBONYL(2-ISOCYANO-1, 3-DIMETHYLBENZENE)IRON(0)

■ *See cautionary note under General Procedure.*

1. $Fe(CO)_5 + 2, 6\text{-}Me_2C_6H_3NC \longrightarrow Fe(CO)_4(2\text{-}CNC_6H_3Me_2\text{-}1,3) + CO$

Procedure

A 250-mL two-necked reaction flask containing a magnetic stirrer bar is equipped with a reflux condenser attached to a check-valve oil bubbler to observe venting of gases. The flask is purged with dry, oxygen-free nitrogen and charged with pentacarbonyliron (11.76 g, 60.0 mmol), 2-isocyano-1,3-dimethylbenzene (3.93 g, 30.0 mmol), and toluene (80 mL). The yellow reaction solution is brought to 85 °C using an oil bath and a magnetic stirrer–heater. Heating is continued at 85 °C until CO evolution ceases, generally requiring between 15 and 30 min. It is important to stop heating the reaction mixture as soon as the reaction is complete. Otherwise, a darkening occurs that imparts an undesirable color to the product. Purification to remove this discoloration has proved to be difficult.[4]

The reaction mixture is cooled to room temperature and the solvent and excess $Fe(CO)_5$ are removed under vacuum. The product is obtained as a yellow, crystalline solid. Yield: 8.6 g (96%). This material is sufficiently pure for most purposes, but it may be recrystallized from pentane.

Anal. Calcd. for $C_{13}H_9NO_4Fe$: C, 52.2; H, 3.05; N, 4.7. Found: C, 52.45; H, 3.15; N, 4.7, mp 82–83 °C.

$$2. \quad Fe(CO)_5 + 1,3\text{-}Me_2C_6H_3NC$$

$$\xrightarrow{CoCl_2} Fe(CO)_4(2\text{-}CNC_6H_3Me_2\text{-}1,3) + CO$$

Procedure

- See cautionary note under *General Procedure.*

The same equipment is used as described in the preceding procedure; this method has the advantage that stoichiometric quantities of reagents are used. The reaction flask (250 mL) containing a magnetic stirrer bar is charged with pentacarbonyl iron (3.92 g, 20.0 mmol), $CoCl_2 \cdot 2H_2O$ (0.050 g, 0.3 mmol), and toluene (100 mL). The reaction mixture is heated to 85 °C, $Me_2C_6H_3$1,3-NC (2.62 g, 20.0 mmol) is then added and the reaction mixture is maintained there until CO evolution ceases ($\sim$ 5 min). The green reaction solution is cooled to 0 °C, filtered and passed down a short silica gel column (1 × 10 cm) in order to separate remaining traces of catalyst, indicated by a green coloration. Further small quantities of toluene are used as eluent. The solvent is then removed from the eluent under vacuum to give the product as a yellow, crystalline solid. Yield: 5.4 g (90%).

Anal. Calcd. for $C_{13}H_9N\,O_4Fe$: C, 52.2; H, 3.05; N, 4.7. Found: C, 52.32; H, 3.01; N, 4.69, mp 82–83 °C.

B.   TRICARBONYLBIS(2-ISOCYANO-1, 3-DIMETHYLBENZENE)-IRON(0)

- See cautionary note under *General Procedure.*

$$Fe(CO)_4(CNC_6H_3Me_2\text{-}1,3) + 1,3\text{-}Me_2C_6H_3NC$$

$$\xrightarrow{CoCl_2} Fe(CO)_3(CNC_6H_3Me_2\text{-}1,3)_2 + CO$$

Procedure

A 100-mL two-necked reaction flask containing a magnetic stirrer bar is equipped with a reflux condenser connected to a check-valve oil bubbler to observe evolution of gases. The flask is charged with $Fe(CO)_4(2\text{-}CNC_6H_3Me_2\text{-}1,3)$ (1.79 g, 6.0 mmol), $CoCl_2 \cdot 2H_2O$ (0.034 g, 0.2 mmol), and toluene (30 mL). The reaction mixture is brought to 85 °C using an oil bath and a magnetic

TABLE I. Selected Spectroscopic Data for the Complexes $Fe(CO)_{5-n}(2\text{-}CNC_6H_3Me_2\text{-}1,3)_n$ ($n = 1\text{-}5$) (see Ref. 4).

Complex	IR[a] Frequencies (cm^{-1})		NMR[b] τ Values
	$\nu_{(NC)}$	$\nu_{(CO)}$	CH_3
$Fe(CO)_4(2\text{-}CNC_6H_3Me_2\text{-}1,3)$	2151	2051, 1999, 1975	8.07
$Fe(CO)_3(2\text{-}CNC_6H_3Me_2\text{-}1,3)_2$	2108	2000, 1938	7.92
$Fe(CO)_2(2\text{-}CNC_6H_3Me_2\text{-}1,3)_3$	2065, 2045 (sh)	1940, 1906	7.83
$Fe(CO)(2\text{-}CNC_6H_3Me_2\text{-}1,3)_4$	2045, 1990	1903	7.69
$Fe(2\text{-}CNC_6H_3Me_2\text{-}1,3)_5$	2028, 1960, 1920 (sh)		7.61

[a] $n = 1$ recorded in hexane; $n = 2$ recorded in CHCl$_3$; $n = 3\text{-}5$ recorded in C$_6$H$_6$.
[b] $n = 1\text{-}5$ recorded in C$_6$D$_6$ relative to TMS.

stirrer–heater device. When the reaction mixture has reached 85 °C, 2-isocyano-1, 3-dimethylbenzene (0.786 g, 6.0 mmol) is added to give an immediate green coloration. Vigorous CO evolution occurs for a period of ~ 10 min. The completion of the reaction is indicated by the cessation of CO evolution, but is best confirmed by IR spectroscopy (Table I). The reaction mixture is cooled to 0 °C, filtered to remove the catalyst, and then passed down a short silica gel column (~ 20 g, wrapped in foil to minimize photochemical decomposition) in order to remove any remaining traces of catalyst. Portions of toluene may be used as eluent. The solvent is removed under reduced pressure to give the product as a yellow, crystalline solid. Yield: 2.12 g (88%). This material is sufficiently pure for most purposes, but it may be recrystallized from dichloromethane–pentane to give the analytically pure product.

Anal. Calcd. for $C_{21}H_{18}N_2O_3Fe$: C, 62.7; H, 4.50; N, 6.95. Found: C, 62.3; H, 4.60; N, 6.95, mp 132–134 °C (dec).

C. DICARBONYLTRIS(2-ISOCYANO-1, 3-DIMETHYLBEN-ZENE)IRON(0)

■ *See cautionary note under General Procedure.*

$$Fe(CO)_4(CNC_6H_3Me_2\text{-}1,3) + 2(1,3\text{-}Me_2C_6H_3NC)$$

$$\xrightarrow{\text{CoCl}_2} Fe(CO)_2(CNC_6H_3Me_2\text{-}1,3)_3 + 2CO$$

Procedure

The same equipment is used as described in the procedure in Section B. The flask is charged with $Fe(CO)_4(2\text{-}CNC_6H_3Me_2\text{-}1,3)$ (1.79 g, 6.0 mmol), $CoCl_2 \cdot 2H_2O$ (0.034 g, 0.2 mmol), and toluene (30 mL), and the mixture is heated to 85 °C with stirring. To the hot solution is added 2-isocyano-1, 3-dimethylbenzene (1.57 g, 12.0 mmol). It is crucial to add the isocyanide to the hot solution. Otherwise, catalyst deactivation (believed to be due to isocyanide polymerization) occurs. This results in sluggish and incomplete reaction.[4,12]

The reaction begins immediately as evidenced by a green coloration and vigorous evolution of CO. Continued heating at 85 °C (~ 5 min) gives an orange reaction solution. The end of the reaction is indicated when the evolution of CO ceases, but is best confirmed by IR spectroscopy (see Table I). Cooling to 0 °C, followed by filtration to remove the catalyst, gives a clear, orange solution. The volume of the solution is reduced under vacuum to 10 to 15 mL. Addition of pentane (15–30 mL) (with cooling to -20 or -78 °C if necessary) gives the product as a yellow, crystalline solid. Further cycles of solvent removal and addition of pentane followed by crystallization, may be necessary. Yield: 2.1 g (70%).

Anal. Calcd. for $C_{29}H_{27}N_3O_2Fe$: C, 68.91; H, 5.35; N, 8.32. Found: C, 68.82; H, 5.51; N, 8.40.

D. CARBONYLTETRAKIS(2-ISOCYANO-1, 3-DIMETHYLBEN-ZENE)IRON(0)

■ *See cautionary note under General Procedure.*

$$Fe(CO)_4(CNC_6H_3Me_2\text{-}1, 3) + 3(1, 3\text{-}Me_2C_6H_3NC)$$

$$\xrightarrow{CoCl_2} Fe(CO)(CNC_6H_3Me_2\text{-}1, 3)_4 + 3CO$$

Procedure

The same equipment is used as described in the procedure in Section B. The reaction flask is charged with $Fe(CO)_4(2\text{-}CNC_6H_3Me_2\text{-}1,3)(1.79$ g, 6.0 mmol), $CoCl_2 \cdot 2H_2O$ (0.034 g, 0.2 mmol), and toluene (30 mL), and these components are heated to 85 °C with magnetic stirring, using an oil bath. To the hot mixture is added 2-isocyano-1, 3-dimethylbenzene (2.36 g, 18.0 mmol). There is an immediate green coloration and vigorous evolution of CO. Completion of the reaction is indicated by the cessation of CO evolution, but is best confirmed by IR spectroscopy (see Table I). Filtration of the cold (0 °C) reaction solution gives a clear, orange solution. The volume is reduced under vacuum to ~ 10 to 15 mL. Addition of pentane (15–30 mL) gives the product as an orange, crystalline solid. If the product does not crystallize immediately, solvent should be removed under vacuum to reduce the volume again to ~ 10 to 15 mL, followed by addition of pentane (15–30 mL) and cooling to -20 or -78 °C. Repeat this cycle if necessary. Yield: 2.8 g (76%).

Anal. Calcd. for $C_{37}H_{36}N_4O$ Fe: C, 73.03; H, 5.92; N, 9.21. Found: C, 72.26; H, 5.99; N, 8.99.

E. PENTAKIS(2-ISOCYANO-1, 3-DIMETHYLBENZENE)IRON(0)

■ *See cautionary note under General Procedure.*

$$Fe(CO)_4(CNC_6H_3Me_2\text{-}1, 3) + 4(1, 3\text{-}Me_2C_6H_3NC)$$

$$\xrightarrow{CoCl_2} Fe(CNC_6H_3Me_2\text{-}1, 3)_5 + 4CO$$

Procedure

The same equipment is used as described in the procedure in Section B. The flask is charged with $Fe(CO)_4(2\text{-}CNC_6H_3Me_2\text{-}1,3)(1.79\,g, 6.0\,mmol)$, $CoCl_2\cdot 2H_2O$ (0.034 g, 0.2 mmol), and toluene (30 mL). These components are heated to 85 °C, and 2-isocyano-1, 3-dimethylbenzene (3.15 g, 24.0 mmol) is added to the hot solution. There is an immediate green coloration and vigorous evolution of CO. Completion of the reaction is indicated by the cessation of CO evolution, but is best confirmed by IR spectroscopy (see Table I). Cooling to 0 °C followed by filtration to remove the catalyst gives a clear, red solution of the product. The volume of the solution is reduced under vacuum to ~ 10 to 15 mL. Addition of pentane (15–30 mL) followed by cooling to -20 or -78 °C, gives the product as a red crystalline solid. Repeated toluene–pentane crystallization cycles at -20 or -78 °C may be necessary if the first obtained product is an oil. Yield: 2.9 g (68%).

Anal. Calcd. for $C_{45}H_{45}N_5Fe$: C, 75.95; H, 6.33; N, 9.85. Found: C, 75.33; H, 6.15; N, 9.05.

Properties

The complexes $Fe(CO)_4(CNC_6H_3Me_2\text{-}1,3)$ and $Fe(CO)_3(CNC_6H_3Me_2\text{-}1,3)_2$ are yellow, crystalline solids.[4] They are air stable and mildly light sensitive. The compounds are soluble in most common organic solvents. The complexes $Fe(CO)_{5-n}(CNC_6H_3Me_2\text{-}1,3)_n(n=3\text{-}5)$[4] are yellow $(n=3)$, orange $(n=4)$, and red $(n=5)$, air- and light-sensitive materials, susceptible to oxidation by O_2 and by solvents such as $CHCl_3$ and CH_2Cl_2. The compounds are soluble in, and best handled in hydrocarbon or ether solvents.

The compounds $Fe(CO)_{5-n}(CNC_6H_3Me_2\text{-}1,3)_n$ $(n=1\text{-}5)$ are characterized by IR vibrational spectroscopy $[\nu_{(CO)}$ and $\nu_{(NC)}$, 1800–2200-cm^{-1} region] and by NMR spectroscopy (aromatic and methyl protons). Selected spectroscopic data for these complexes are given in the Table I. Infrared spectroscopy may conveniently be used for the monitoring of the progress of the substitution reaction, and NMR spectroscopy for an estimate of product purity. The complexes may also be characterized by mass spectrometry (operating temperatures 25–200 °C) a molecular ion being observed in each case.[4] This series of complexes have been shown to be precursors to a varied and rich chemistry, particularly involving the isonitrile ligand and the ease of oxidation of the electron-rich iron(0) complexes.[4,13,14] In addition, the spectroscopic and structural details of these complexes have attracted attention within the context of five coordination at iron(0) centers.[9,13,15]

References

1. E. Singleton and H. E. Oosthuizen, *Adv. Organometal. Chem.*, **22**, 209 (1983).
2. W. Hieber and D. von Pigenot, *Chem. Ber.*, **89**, 193 (1956).
3. W. Strohmeier and F. J. Müller, *Chem. Ber.*, **102**, 3613 (1969).
4a. M. O. Albers, N. J. Coville, and E. Singleton, *J. Chem. Soc. Dalton Trans.*, **1982**, 1069.
4b. S. B. Butts and D. F. Shriver, *J. Organometal. Chem.*, **169**, 191 (1979).
5. A. Reckziegel and M. Bigorgne, *J. Organometal. Chem.*, **3**, 341 (1965).
6. S. Grant, J. Newman, and A. R. Manning, *J. Organometal. Chem.*, **96**, C11 (1975).
7. D. Gladkowski and F. R. Scholer, Abstracts of Papers from the Centennial American Chemical Society Meeting, New York, 1976, INOR 133.
8. J. M. Basset, D. E. Berry, G. K. Barker, M. Green, J. A. K. Howard, and F. G. A. Stone, *J. Chem. Soc. Dalton Trans.*, **1979**, 1003.
9. M. O. Albers, N. J. Coville, T. V. Ashworth, E. Singleton, and H. E. Swanepoel, *J. Chem. Soc. Chem. Commun.*, **1980**, 489.
10. D. F. Shriver and M. A. Drezdzon, *The Manipulation of Air-Sensitive Compounds*, 2nd ed., McGraw-Hill, New York, 1986.
11. D. D. Perrin, W. L. F. Armarego, and D. R. Perrin, *Purification of Laboratory Chemicals*, 2nd ed., Pergamon Press, Oxford, 1980.
12. M. O. Albers, N. J. Coville, T. V. Ashworth, E. Singleton, and H. E. Swanepoel, *J. Organometal. Chem.*, **199**, 55 (1980).
13. M. O. Albers, E. Singleton, and N. J. Coville, unpublished results.
14. J. M. Bassett, M. Green, J. A. K. Howard, and F. G. A. Stone, *J. Chem. Soc. Dalton Trans.*, **1980**, 1779; J. M. Bassett, L. J. Farrugia, and F. G. A. Stone, *J. Chem. Soc. Dalton Trans.*, **1980**, 1789.
15. G. W. Harris, J. C. A. Boeyens, and N. J. Coville, *Acta Crystallogr. Sect. C*, **39**, 1180 (1983).

12. TETRACARBONYLIRON(0) COMPLEXES CONTAINING GROUP V DONOR LIGANDS

Submitted by MICHEL O. ALBERS,* ERIC SINGLETON,* and NEIL J. COVILLE[‡]

Checked by CARLTON E. ASH,[‡] CHRISTINE C. KIM,[‡] and MARCETTA Y. DARENSBOURG[‡]

Since the first reported synthesis of $Fe(CO)_4(PPh_3)$ by Reppe and Schweckendiek in 1948[1] there have been numerous attempts to prepare this complex in

*National Chemical Research Laboratory, Council for Scientific and Industrial Research, P. O. Box 395, Pretoria 0001, Republic of South Africa.
[†]Department of Chemistry, University of the Witwatersrand, 1 Jan Smuts Avenue, Johannesburg 2001, Republic of South Africa.
[‡]Department of Chemistry, Texas A & M University, College Station, TX 77843-3255.

high yield and in particular, free from contamination with $Fe(CO)_3(PPh_3)_2$. Direct procedures have included the thermal[2] as well as the photochemical[3] reaction between triphenylphosphine and iron pentacarbonyl. Indirect synthetic methods have included the reaction between triphenylphosphine and triiron dodecacarbonyl,[4] and the reduction of iron(II) carbonyl halide complexes with phenyllithium in the presence of triphenylphosphine.[5] More recent synthetic procedures have included the use of a combination of high temperature and photochemical irradiation,[6] main group metal hydrides as promoters of carbonyl substitution on iron pentacarbonyl,[7] the use of Wilkinson's catalyst, $RhCl(PPh_3)_3$, as a stoichiometric decarbonylation reagent,[8] and iron carbonyl anions as CO substitution catalysts.[9] A similar checkered variety of synthetic methods also exists for the synthesis of other $Fe(CO)_4L$ complexes, where L is a large range of phosphine, phosphite, arsine, and stibine ligands.[10] In general, however, these methods all suffer from a number of disadvantages that include long reaction times, forcing conditions, the use of expensive reagents, and most significantly, the formation of mixtures of $Fe(CO)_4L$ and $Fe(CO)_3L_2$ with concomitant low yields of $Fe(CO)_4L$.

In this chapter, we describe the high-yield, selective synthesis of the complex $Fe(CO)_4(PPh_3)$ directly from iron pentacarbonyl and triphenylphosphine in boiling toluene using cobalt(II) chloride as a catalyst.[11] The method may also be generalized to a large variety of other Group V donor ligands and we provide a guide to a suitable choice of catalysts for such reactions.[11,12]

General Procedure

■ **Caution.** *The use of the volatile, toxic iron pentacarbonyl necessitates that all manipulations be carried out in a well-ventilated hood. In addition, carbon monoxide, evolved in the reaction, is an odorless, extremely toxic gas, and care must be exercised that the apparatus vents into the best ventilated region of the hood.*

All reactions are routinely carried out under an inert atmosphere and in a well-ventilated fume hood. Standard ground glass apparatus is used throughout.

All solvents are of Analar Grade and are freshly dried and distilled under nitrogen prior to use.[13] The catalysts $CoX_2 \cdot nH_2O$ (X = Cl, I) are dried under vacuum (0.1 torr, 50 °C, 5 h) to give materials that are pink, analyzing approximately as $CoCl_2 \cdot 2H_2O$ or $CoI_2 \cdot 4H_2O$.[11] These turn color when added to the reaction mixtures, turning blue (X = Cl) or brown (X = I). The catalyst $\{Fe(\eta^5\text{-}C_5Me_5)(CO)_2\}_2$ was purchased from Strem and is used without further purification.

The catalyst $\{Fe(\eta^5\text{-}C_5Me_5)(CO)_2\}_2$ is synthesized by the method of King and Bisnette.[14] Pentamethylcyclopentadiene was purchased from Strem.

Column chromatography utilized 63 to 200 μm of silica gel and 63 to 200 μm of neutral alumina activity grade 1. Melting points are all recorded in air in a well-ventilated hood, and are uncorrected. The reactions are all routinely monitored by infrared (IR) spectroscopy using a 0.05-mm pathlength cell.

A. TETRACARBONYL(TRIPHENYLPHOSPHINE)IRON(0)

■ *See cautionary note under general procedure.*

$$Fe(CO)_5 + PPh_3 \xrightarrow{CoCl_2} Fe(CO)_4(PPh_3) + CO$$

Procedure

A 50-mL two-necked flask containing a magnetic stirrer bar is equipped with a reflux condenser and connected to a check-valve oil bubbler to observe evolution of gases. The flask is charged with PPh_3 (2.26 g, 10.0 mmol), $CoCl_2 \cdot 2H_2O$ (0.050 g, 0.3 mmol) and toluene (30 mL). The stirred solution is brought to reflux using an oil bath temperature controlled at 120 °C, with a magnetic stirrer–heater device. When reflux is achieved, pentacarbonyliron (3.92 g, 20 mmol) is added. The course of the reaction is followed by monitoring the changes in the region from 1900 to 2100 cm^{-1} of the IR spectrum. Reflux is continued for 2 h or until the spectrum remains invariant with time. The reaction solution is cooled, and catalyst and traces of excess ligand are removed by eluting the reaction solution through a column (21 × 1,5 cm) containing three layers: $CoCl_2 \cdot 6H_2O$ (5 g) neutral alumina (20 g) and silica gel (20 g). Benzene is used as eluent.

■ **Caution.** *Vapors of benzene are harmful, and toluene should be substituted if possible. The procedure must be carried out in a well-ventilated hood.*

The solvent and excess pentacarbonyliron are removed under reduced pressure and the product is recrystallized from dichloromethane–hexane. Yield: 3.56 g (83% based on consumed pentacarbonyliron).

Anal. Calcd. for $C_{22}H_{15}O_4PFe$: C, 61.43, H, 3.52. Found: C, 61.31, H, 3.53.

B. OTHER TETRACARBONYL (GROUP V DONOR LIGAND) IRON(0) COMPLEXES

The general applicability of the reaction catalyzed by cobalt(II) halide is illustrated by the examples in Table I. An important feature of the catalyzed reaction is the ability to prepare $Fe(CO)_4L$ in high yield with little contamination by $Fe(CO)_3L_2$. Typically the final reaction product contains $<5\%$ $Fe(CO)_3L_2$ except for L = $P(OEt)_3$, where $\sim 10\%$ $Fe(CO)_3\{P(OEt)_3\}_2$ is

TABLE I. Reaction Conditions and Spectroscopic Data for the Complexes $Fe(CO)_4L$ [L = PPh_3, $AsPh_3$, $SbPh_3$, $PMePh_2$, PMe_2Ph, PCy_3, $P(n\text{-}Bu)_3$, $P(OPh)_3$, $P(OEt)_3$, $P(OMe)_3$] using $CoX_2 \cdot nH_2O$ (X = Cl, I) as Catalyst.[a]

	$CoCl_2$		CoI_2		IR Frequencies $\nu_{(CO)}$[b](cm^{-1})			Melting Point[c] (°C)	References
	Yield(%)	Time[d](h)	Yield(%)	Time[d](h)					
$Fe(CO)_4(PPh_3)$	76–83[e]	2	95	0.5	2052	1978	1940	204–205	1–9
$Fe(CO)_4(AsPh_3)$	86	2	99[e]	0.5	2048	1972	1942	178–179	4,6,7
$Fe(CO)_4(SbPh_3)$	90	5	97[e]	1	2045	1970	1938	133–135	4,6,7
$Fe(CO)_4(PMePh_2)$	98[e]	1	78	1.5	2058	1977	1947	Oil	16
$Fe(CO)_4(PMe_2Ph)$	91[e]	1.3	93	4	2055	1973	1937	49–50	16
$Fe(CO)_4(PCy_3)$	45–60[e]	3	58	5	2043	1964	1930	174–175	3
$Fe(CO)_4[P(n\text{-}Bu)_3]$	38	6	50	6	2045	1968	1932	Oil	3,6
$Fe(CO)_4[P(OPh)_3]$	79	4	75–95[e]	0.5	2067	1997	1960	68–69	7,17
$Fe(CO)_4[P(OEt)_3]$	15[f]	6	59	6	2062	1983	1950	Oil	7,17
$Fe(CO)_4[P(OMe)_3]$	nr[g]	6	nr	6	2053	1986	1949	44–45	6

[a] See Sections A and B and Ref. 11.
[b] Recorded in $CHCl_3$.
[c] Recorded in air, uncorrected.
[d] Monitored by IR spectroscopy until the spectrum remained invariant with time or for a period of 6 h.
[e] Verified by checkers.
[f] Maximum yield estimated by IR.
[g] nr = no reaction.

produced. Excess $Fe(CO)_5$ is used in the reactions to ensure that $Fe(CO)_4L$ is the major product. The excess iron pentacarbonyl is readily removed together with the solvent, and if stored under nitrogen, may be reused. Yields are thus based on consumed $Fe(CO)_5$ or alternatively, on added ligand. Purification of the product requires the removal of both unreacted ligand and catalyst. This is achieved by the use of a chromatographic column made up of three layers.[11] The top layer consists of $CoCl_2 \cdot 6H_2O$ and is used to remove the small amounts of unreacted ligand via the reaction $CoX_2 + 2L \rightarrow CoX_2L_2$.[15] This is followed by a layer of silica gel and a layer of alumina that remove the catalyst from the product.

Other catalysts may also be used for the syntheses of $Fe(CO)_4L$ complexes. Of particular use are $\{Fe(\eta^5\text{-}C_5H_5)(CO)_2\}_2$ and $\{Fe(\eta^5\text{-}C_5Me_5)(CO)_2\}_2$, which have proved in many ways to be complementary to the cobalt(II) halide catalysts.[12] Details are outlined in Table II. An identical synthetic procedure to that outlined above for $CoCl_2 \cdot 2H_2O$ as catalyst is used except that 0.1 mmol of catalyst is used and the reaction is carried out in 20 mL of toluene as solvent. Once the reaction is complete, the solvent is removed and the crude product is chromatographed on silica gel.[12] There is usually little, if any, unreacted ligand present, obviating the need for a $CoCl_2 \cdot 6H_2O$ layer. These

TABLE II. Reaction Conditions for the Synthesis of the Complexes $Fe(CO)_4L$ [$L = PPh_3$, $AsPh_3$, $SbPh_3$, $PMePh_2$, PMe_2Ph, PCy_3, $P(n\text{-}Bu)_3$, $P(OEt)_3$, $P(OMe)_3$] using $[Fe(\eta^5\text{-}C_5H_5)(CO)_2]_2$ and $[Fe(\eta^5\text{-}C_5Me_5)(CO)]_2$ as Catalysts[a].

	$[Fe(\eta^5\text{-}C_5H_5)(CO)_2]_2$		$[Fe(\eta^5\text{-}C_5Me_5)(CO)_2]_2$	
	Yield(%)	Time[b](h)	Yield(%)	Time[b](h)
$Fe(CO)_4(PPh_3)$	90	5	97	1.25
$Fe(CO)_4(AsPh_3)$	88	5	94	1.25
$Fe(CO)_4(SbPh_3)$	92	2.75	97	1.25
$Fe(CO)_4(PMePh_2)$	98	1.75	96	1.25
$Fe(CO)_4(PMe_2Ph)$	93	0.5	91	1.25
$Fe(CO)_4(PCy_3)$	69	5	90	1.5
$Fe(CO)_4[P(n\text{-}Bu)_3]$	98–100[c]	1	93	1.25
$Fe(CO)_4[P(OPh)_3]$	83–93	4	94	1.5
$Fe(CO)_4[P(OEt)_3]$	93[c]	1.5	90	1.25
$Fe(CO)_4[P(OMe)_3]$	56–83[d]	0.75	86	1.5

[a] See Section B and Ref. 12; cy = cyclohexyl.

[b] Monitored by IR spectroscopy until the reaction had gone to completion.

[c] Verified by checkers.

[d] The checkers were unable to repet these yields. Crude yields were 56% and the yield of isolated, purified product was $\sim 15\%$. The synthesis of the checkers' choice for $Fe(CO)_4[P(OMe)_3]$ is that of Ref. 9.

reactions may all be scaled up or down in size without affecting the yields of products noticeably.

Properties

The properties of the complexes $Fe(CO)_4L$ [$L = PPh_3$, $AsPh_3$, $SbPh_3$, $PMePh_2$, PMe_2Ph, PCy_3, $P(n\text{-}Bu)_3$, $P(OPh)_3$, $P(OEt)_3$, and $P(OMe)_3$] have been extensively studied and reviewed.[10,16,17] The complexes are all air stable.

References

1. W. Reppe and W. J. Schweckendiek, *Liebigs Ann. Chem.*, **560**, 104 (1948).
2. F. A. Cotton and R. V. Parish, *J. Chem. Soc.*, 1960, 1440.
3. W. Strohmeier and F.-J. Müller, *Chem. Ber.*, **102**, 3613 (1969); M. O. Albers and N. J. Coville, unpublished results.
4. A. F. Clifford and A. K. Mukherjee, *Inorg. Chem.*, **2**, 151 (1963); A. F. Clifford and A. K. Mukherjee, *Inorg. Synth.*, **8**, 185 (1966).
5. T. A. Manuel, *Inorg. Chem.*, **2**, 854 (1963).
6. H. L. Condor and M. Y. Darensbourg, *J. Organometal. Chem.*, **67**, 93 (1974).
7. W. O. Siegel, *J. Organometal. Chem.*, **92**, 321 (1975).
8. Yu. S. Varshavsky, E. P. Shestakova, N. V. Kiseleva, T. G. Cherkasova, N. A. Buzina, L. S. Bresler, and V. A. Kormer, *J. Organometal. Chem.*, **170**, 81 (1979).
9. S. B. Butts and D. F. Shriver, *J. Organometal. Chem.*, **169**, 191 (1979).
10. Gmelin, *Handbuch der Anorganische Chemie*, Eisen-Organische Verbindungen, Teil B, U. Kruerke (ed.), Springer Verlag, Berlin, 1978.
11. M. O. Albers, N. J. Coville, T. V. Ashworth, and E. Singleton, *J. Organometal. Chem.*, **217**, 385 (1981).
12. M. O. Albers, N. J. Coville, and E. Singleton, *J. Organometal. Chem.*, **232**, 261 (1982).
13. D. D. Perrin, W. L. F. Armarego, and D. R. Perrin, *Purification of Laboratory Chemicals*, 2nd ed., Pergamon, Oxford, 1980.
14. R. B. King and M. B. Bisnette, *J. Organometal. Chem.*, **8**, 287 (1967).
15. F. A. Cotton, O. D. Faut, D. M. L. Goodgame, and R. H. Holm, *J. Am. Chem. Soc.*, **83**, 1780 (1961).
16. P. M. Treichel, W. M. Douglas, and W. K. Dean, *Inorg. Chem.*, **11**, 1615 (1972).
17. J. D. Cotton and R. L. Heazelwood, *Aust. J. Chem.*, **22**, 2673 (1969).

13. *cis*-TETRAAMMINEDIHALORUTHENIUM(III) HALIDE COMPLEXES

$$[Ru(NH_3)_5Cl]Cl_2 + H_2dhn + 2OH^- \longrightarrow [Ru(NH_3)_4(dhn)]^+$$
$$+ NH_3 + 3Cl^- + 2H_2O$$

$$H_2dhn =$$

$$[Ru(NH_3)_4dhn]^+ + 3HX\,(X = Cl, Br) \longrightarrow \textit{cis-}[Ru(NH_3)_4X_2]X$$
$$+ H_2dhn + H^+$$

Submitted by S. D. PELL,* M. M. SHERBAN,* V. TRAMONTANO,*
and M. J. CLARKE*
Checked by K. HANSONGNERN† and R. A. KRAUSE†

While *cis*-tetraamminedichlororuthenium(III) is widely used as a starting material for obtaining a series of *cis*-tetraammineruthenium complexes for the metal in both the II and III oxidation states, little improvement has been made over the original synthesis reported by Gleu 50 years ago.[1,2] *cis*-Tetraammineruthenium complexes are also of interest since they exhibit both good antitumor activity[3] and mutagenicity.[4] Despite attempts at optimization[5] Gleu's method, which involves the formation of the air sensitive and slow to crystallize intermediate $[Ru^{II}(H_2O)(NH_3)_5]S_2O_6$ followed by ammine displacement with oxalate and its subsequent removal in strong acid, requires 2 days to complete and results in yields which are typically 30 to 40%.

1, 2-Benzenediolato (catecholato) ligands have such a strong affinity for Ru^{III} in basic media that they rapidly displace *cis*-ammines from $[Ru(NH_3)_6]^{3+}$ in a single step.[6] However, since they are easily removed in strong acid, their use retains the versatility of Gleu's method, while reducing the time required to $\sim 5\,h$ (elapsed time), and increasing yields to 50% for multigram batches.

*Department of Chemistry, Boston College, Chestnut Hill, MA 02167.
†Department of Chemistry, University of Connecticut, Storrs, CT 06268.

A. *cis*-TETRAAMMINEDICHLORORUTHENIUM(III) CHLORIDE

Procedure

A 5 g (17 mmol) sample of $[Ru(NH_3)_5Cl]Cl_2$[7] is placed in a 250-mL
Erlenmeyer flask and dispersed in 50 mL of water.[7,*] This is placed in a 55 °C
water bath and a twofold excess of 2,3-naphthalenediol (H_2dhn) (34 mmol,
5.45 g), which had been dissolved in 45 mL of 1:1 (v/v) water/ethanol at 55 °C,
is added with constant stirring. After a few minutes the reaction mixture is
slowly adjusted to a pH of 9.4 to 9.5 (with the aid of a pH meter) by the
dropwise addition of fresh 3 *M* NaOH. An intense blue-violet color rapidly
develops due to the formation of the catecholato chelate complex[6] and the
reaction is allowed to continue for ∼ 40 min.

The reaction mixture is then filtered through Whatman No. 1 filter paper
and evaporated to dryness on a rotary evaporator at 40 °C under vacuum in a
1000-mL round-bottomed flask. Foaming may occur. The blue residue is
scraped off the sides of the flask and transferred to a 250-mL Erlenmeyer flask.
The remaining residue is dissolved in a minimum of water and added to the
scrapings. Concentrated HCl (∼ 100 mL) is added and the mixture is stirred at
50 °C on a stirring hotplate for 10 to 15 min. During this time, the solution turns
a red-violet color. After cooling, the precipitated ligand and unreacted starting
material are eliminated by suction filtration. Most of the remaining ligand is
removed by transferring the solution to a large separatory funnel and washing 2
to 3 times with 15-mL portions of $CHCl_3$.

■ **Caution.** *Because of the toxicity of $CHCl_3$, this procedure must be
carried out in an efficient fume hood.*

The aqueous (violet) layer is then transferred to a 400-mL beaker and
an equal volume of absolute ethanol added. The desired product forms as a
flocculent yellow precipitate upon cooling and is filtered off after 1 to 3 h of
refrigeration followed by washings with $CHCl_3$, ethanol, and diethyl ether.
Yields are typically 50% (2.71 g) and vary with batch size. Preparations using
0.5 g of $[Ru(NH_3)_5Cl]Cl_2$ gave yields of ∼ 30%.

Anal. Calcd. for $[Ru(NH_3)_5Cl]Cl_2 \cdot H_2O$: H, 4.81; N, 19.09; Ru, 36.23. Found:
H, 4.76; N, 19.10; Ru, 36.21.

Properties

The product is an amorphous to microcrystalline yellow solid, which is stable

---

*Following the method of Allen, et al.,[7] $[Ru(NH_3)_5Cl]Cl_2$ was prepared from 5-g quantities of
$RuCl_3 \cdot 3H_2O$ and recrystallized from 0.1 *M* HCl. Highly crystalline samples should be ground to
facilitate their dissolution and reaction with the catechol.

in air. It is readily soluble in water, and weakly acidic solutions are stable for long periods. Dissolving in concentrated HCl followed by the addition of two volumes of ethanol and cooling affords needlelike crystals. The UV–VIS spectrum: λ, (ε): 260 nm $(4.8 \times 10^2 \, M^{-1} \, cm^{-1})$; 310 nm $(13.6 \times 10^3 \, M^{-1} \, cm^{-1})$; 352 nm $(1.6 \times 10^3 \, M^{-1} \, cm^{-1})$.[8,9] The compound undergoes an irreversible reduction $(E^\circ = -0.11 \, V)$ with loss of one or both halides. Following reduction, the synthesis of a number of complexes with a variety of ligands is easily accomplished by direct substitution for one or both halides.[9]

B. *cis*-TETRAAMMINEDIBROMORUTHENIUM(III) BROMIDE

cis-Tetraamminedibromoruthenium(III) bromide is similarly prepared except that following transfer of the 1, 2-naphthalenediolato complex to the 250-mL Erlenmeyer, 100 mL of *fresh* 48% HBr is added.* Yield: 50% (3.49 g). The optical spectrum is similar to that reported by Hartman and Buschbeck.[10]

Anal. Calcd. for $[H_{12}N_4Br_3Ru]$: H, 2.96; N, 13.70; Ru, 24.7. Found: H, 3.28; N, 13.90; Ru, 24.8.

Properties

The product is an amorphous or microcrystalline orange solid, which is stable in air. It is stable in water at low pH, but less soluble than the dichloro species. Reactivity is generally similar to the corresponding dichloro complex. The UV–visible spectrum: λ, (ε): 345 nm (inflection); 380 nm $(1.1 \times 10^3 \, M^{-1} \, cm^{-1})$; 435 nm $(1.3 \times 10^3 \, M^{-1} \, cm^{-1})$.[11]

References

1. K. Gleu and W. Breuel, *Z. Anorg. Allg. Chem.*, **237**, 350–355 (1938).
2. Gmelin, *Handbuch der Anorganischen Chemie*, **63**, pp. 54–55 (1974).
3. M. J. Clarke, *Metal Ions Biol. Syst.*, **11**, 231–283 (1980).
4. R. E. Yasbin, C. R. Matthews, and M. J. Clarke, *Chem. Biol. Interact.*, **30**, 355–365 (1980).
5. S. E. Diamond, "Functional Group Modifications Employing Ruthenium Ammines," Ph.D. thesis, Standford University, 1975.
6. S. D. Pell, R. B. Salmonsen, A. Abelliera, and M. J. Clarke, *Inorg. Chem.*, **23**, 387–392 (1984).
7. A. D. Allen, F. Bottomley, R. O. Harris, V. P. Reinsalu, and C. V. Senoff, *J. Am. Chem. Soc.*, **89**, 5595–5599 (1967) and *Inorg. Synth.*, **12**, 1–7 (1969).

*Checker employed concentrated HBr with 1 mg of Na_2SO_3 added per 200 mL followed by purging the solution with argon bubbling.

8. J. N. Armor, "Studies in the Reactivity of Ruthenium Ammines," Ph.D. thesis, Standford University, 1970.

9. D. A. Chaisson, R. E. Hintze, D. H. Stuermer, J. D. Petersen, D. P. McDonald, and P. C. Ford, *J. Am. Chem. Soc.*, **94**, 6555 (1972). J. A. Marchant, T. Matsubara, and P. C. Ford, *Inorg. Chem.*, **16**, 2160 (1977). T. Matsubara and P. C. Ford, *Inorg. Chem.*, **15**, 1107 (1976).

10. H. Hartmann and C. Buschbeck, *Physik. Chem.*, **11**, 120–135 (1957).

14. (η^4-1, 5-CYCLOOCTADIENE)RUTHENIUM(II) COMPLEXES

Submitted by MICHEL O. ALBERS,* TERENCE V. ASHWORTH,* HESTER
E. OOSTHUIZEN,* and ERIC SINGLETON*
Checked by JOSEPH S. MEROLA and RAYMOND T. KACMARCIK[†]

Di-μ-halo-(η^4-1, 5-cyclooctadiene)ruthenium(II)polymers, [RuX$_2$(η^4-C$_9$H$_{12}$)]$_x$(X = Cl, Br, I; $x > 2$) have been prepared in 30 to 40% yield from ruthenium halides and 1, 5-cyclooctadiene in boiling ethanol.[1] These polymers are useful starting materials for the preparation of a series of monomeric ruthenium(II) complexes by halogen-bridge cleavage reactions with nitrogen-donor molecules.[2] Of these reactions, only acetonitrile,[3,4] and hydrazines[5] have given facile reactions leading to high yields of a series of air-stable, neutral, and cationic ruthenium(II) complexes. Below is described the preparation of the polymer [RuCl$_2$(η^4-C$_8$H$_{12}$)]$_x$ ($x > 2$) in 90 to 96% yield, together with its conversion to the neutral complexes [RuX$_2$(η^4-C$_8$H$_{12}$) (NCR)$_2$] (X = Cl, Br; R = CH$_3$, C$_6$H$_5$), and the salts [Ru × (η^4-C$_8$H$_{12}$) (NCCH$_3$)$_3$] [PF$_6$] (X = Cl, Br), [Ru(η^4-C$_8$H$_{12}$) (NCCH$_3$)$_4$] [PF$_6$]$_2$ and [Ru(η^4-C$_8$H$_{12}$) (NH$_2$NHR)$_4$]Y$_2$ (R = H, Y = BPh$_4$; R = CH$_3$, Y = BPh$_4$, PF$_6$). The syntheses of [RuCl$_2$(η^4-C$_8$H$_{12}$) (NCCH$_3$)$_2$] (Ref. 4), [RuCl(η^4-C$_8$H$_{12}$) (NCMe)$_3$] [PF$_6$] (Ref. 3), [Ru(η^4-C$_8$H$_{12}$) (NCCH$_3$)$_4$] [PF$_6$]$_2$, (Ref. 3, 6) and [Ru(η^4-C$_8$H$_{12}$) (NH$_2$NHR)$_4$]Y$_2$ (Ref. 5) have been previously reported. The preparation of [RuCl$_2$(η^4-C$_8$H$_{12}$) (NCCH$_3$)$_2$] has been optimized to rationalize inconsistencies in the described procedure, whereas the remainder follow the procedures described in the literature.

GENERAL PROCEDURE

All reactions are performed in air. Solvents of analytical purity are used throughout, as further purification and drying are found to have no effect on

*National Chemical Research Laboratory, Council for Scientific and Industrial Research, P.O. Box 395, Pretoria 0001, Republic of South Africa.
[†]Exxon Research & Engineering Co., Corporate Research, Route 22 East, Annandale, NJ 08801.

product yields. All reagents are of analytical purity and used as purchased. Hydrazine was obtained from Baker and methylhydrazine from Fluka AG. Melting points are all recorded in air in a well-ventilated hood and are uncorrected.

A. Di-μ-CHLORO(η^4-1,5-CYCLOOCTADIENE)RUTHENIUM(II)

$$2RuCl_3 \cdot xH_2O + 2C_8H_{12} + CH_3CH_2OH \longrightarrow 2[RuCl_2(\eta^4\text{-}C_8H_{12})]_x$$
$$+ 2HCl + CH_3CHO$$

Procedure

A 750-mL Erlenmeyer flask containing a large Teflon-coated stirring bar and fitted with a reflux condenser is charged with a mixture of $RuCl_3 \cdot xH_2O$ (40.0 g, ~ 0.2 mol) and 1,5-cyclooctadiene (50 mL, 0.4 mol) in absolute ethanol (400 mL) and heated under reflux with stirring for 24 h. During this time the brown product precipitates from solution. On completion of the reaction, the solution is cooled to room temperature and filtered. Washing with diethyl ether (50 mL) and drying, gives the analytically pure product. Yield: 51–54 g (91–96%). The procedure was checked using 4 g of $RuCl_3 \cdot xH_2O$, whereupon the observed yield is 80%.

Anal. Calcd. for $C_8H_{12}Cl_2Ru$: C, 34.29; H, 4.29; Cl, 25.40. Found: C, 34.75; H, 4.40; Cl, 25.16.

B. BIS(ACETONITRILE)DICHLORO(η^4-1,5-CYCLOOCTADIENE) RUTHENIUM(II)

$$[RuCl_2(\eta^4\text{-}C_8H_{12})]_x + 2CH_3CN + xH_2O$$
$$\longrightarrow [RuCl_2(\eta^4\text{-}C_8H_{12})(NCCH_3)_2] \cdot xH_2O$$

Procedure

A 100-mL Erlenmeyer flask containing a Teflon-coated stirring bar and fitted with a reflux condenser is charged with $[RuCl_2(\eta^4\text{-}C_8H_{12})]_x$ (1.4 g, 5.0 mmol) and acetonitrile (50 mL). 1,5-cyclooctadiene (2 mL) is added to the mixture to suppress the formation of $[RuCl_2(NCCH_3)_4]$. The suspension is heated under reflux for 5 h. Longer reaction times increase the amount of $[RuCl_2(NCCH_3)_4]$ formed at the expense of the desired product. The reaction mixture is filtered while still hot to separate unchanged starting material and the volume of the filtrate is reduced under vacuum to ~ 25 to 30 mL. Cooling to $-20\,°C$ for 12 h gives a crop of orange crystals of the desired product. Filtration, followed by washing with diethyl ether (10 mL) and vacuum drying, gives the analytically pure product. Yield: 0.7–0.8 g (38–44%). This material contains varying amounts of water of crystallization.

Anal. Calcd. for 1 mol of water per mole of product, $C_{12}H_{20}N_2O\ Cl_2Ru$: C, 37.90; H, 5.30; N, 7.37; Cl, 18.65. Found: C, 37.84; H, 5.29; N, 7.38; Cl, 18.78, mp dec. at 170 °C without melting.

C.   BIS(BENZONITRILE)DICHLORO(η^4-1,5-CYCLOOCTADIENE)-RUTHENIUM(II)

$$[RuCl_2(\eta^4\text{-}C_8H_{12})(NCMe)_2] + 2C_6H_5CN$$
$$\longrightarrow [RuCl_2(\eta^4\text{-}C_8H_{12})(NCC_6H_5)_2] + 2CH_3CN$$

Procedure

A 100-mL Erlenmeyer flask containing a magnetic stirring bar and fitted with a reflux condenser is charged with $[RuCl_2(\eta^4\text{-}C_8H_{12})(NCCH_3)_2]$ (0.76 g, 2.0 mmol) and benzonitrile (10 mL). In order to complete dissolution, the reaction mixture may be gently warmed. Ethanol (40 mL) is added to the solution, which is then heated under reflux for 45 min. The volume of the solution is reduced to ~ 5 mL by evaporation under reduced pressure. Addition of hexane (20 mL) and cooling to $-20\,°C$ for 4 h gives the product as an orange precipitate. Yield: 0.40–0.78 g (40–78%).

Anal. Calcd. for $C_{22}H_{22}N_2Cl_2Ru$: C, 54.30; H, 4.55; N, 5.75; Cl, 14.60. Found: C, 53.95; H, 4.85; N, 5.70; Cl, 14.25, mp 132 °C.

D.   BIS(ACETONITRILE)DIBROMO(η^4-1,5-CYCLOOCTADIENE)-RUTHENIUM(II)

$$[RuCl_2(\eta^4\text{-}C_8H_{12})]_x + 2CH_3CN + 2LiBr$$
$$\longrightarrow [RuBr_2(\eta^4\text{-}C_8H_{12})(NCCH_3)_2] + 2LiCl$$

Procedure

A 100-mL Erlenmeyer flask containing a magnetic stirring bar and fitted with a reflux condenser is charged with $[RuCl_2(\eta^4\text{-}C_8H_{12})]_x$ (1.4 g, 5.0 mmol), lithium bromide (1.65 g, 20.0 mmol), and acetonitrile (50 mL). The suspension is heated under reflux for 12 h. The reaction mixture is filtered while still hot, and the residue (unreacted $[RuCl_2(\eta^4\text{-}C_8H_{12})]_x$, $\sim 0.5\,g$) is washed with acetonitrile (5 mL). The combined filtrate and washings are heated under reflux for a further 15 min. The volume of the reaction solution is then reduced to $\sim 2\,mL$ under reduced pressure (water aspirator). Dichloromethane (25 mL) is added, and the precipitated lithium halides are removed by filtration, and washed with portions of dichloromethane (5 × 5 mL). Ethanol (15 mL) is added to the combined filtrate and washings, and the resulting solution is concentrated under reduced pressure until crystals begin to form. The solution is cooled to $-20\,°C$ for 2 h to give orange microcrystals of the product. Yield: 0.73 g (50% based on consumed starting material, 32% overall yield). The material contains varying amounts of water of crystallization.

Anal. Calcd. for $C_{12}H_{18}N_2Br_2Ru$: C, 31.95; H, 4.00; N, 6.20; Br, 35.40. Found: C, 32.15; H, 3.95; N, 6.10; Br, 35.50, mp dec. without melting at 200 °C.

E.   BIS(BENZONITRILE)DIBROMO(η^4-1, 5-CYCLOOCTADIENE)-RUTHENIUM(II)

$$[RuBr_2(\eta^4\text{-}C_8H_{12})(NCCH_3)_2] + 2C_6H_5CN$$

$$\longrightarrow [RuBr_2(\eta^4\text{-}C_8H_{12})(NCC_6H_5)_2] + 2CH_3CN$$

Procedure

A 100-mL Erlenmeyer flask containing a magnetic stirring bar and fitted with a reflux condenser is charged with $[RuBr_2(\eta^4\text{-}C_8H_{12})(NCCH_3)_2]$ (0.9 g, 2.0 mmol) and benzonitrile (10 mL). The mixture may be gently heated in order to effect complete dissolution. Ethanol (40 mL) is added, and the reaction solution is heated under reflux for 45 min. The volume of the solvent is reduced under vacuum to $\sim 5\,mL$ and hexane (20 mL) is added. Cooling to $-20\,°C$ for 1 h gives the product as an orange, crystalline solid. Yield: 0.72 g (62%).

Anal. Calcd. for $C_{22}H_{22}N_2Br_2Ru$: C, 45.80; H, 3.85; N, 4.85; Br, 27.70. Found: C, 45.95; H, 3.80; N, 4.90; Br, 28.00, mp 194 °C.

F.   TRIS(ACETONITRILE)CHLORO(η^4-1, 5-CYCLOOCTADIENE)-RUTHENIUM(II) HEXAFLUOROPHOSPHATE(1 −)

$$[RuCl_2(\eta^4\text{-}C_8H_{12})(NCCH_3)_2] + CH_3CN + NH_4[PF_6]$$

$$\longrightarrow [RuCl(\eta^4\text{-}C_8H_{12})(NCCH_3)_3][PF_6] + NH_4Cl$$

Procedure

A 50-mL Erlenmeyer flask containing a magnetic stirring bar and fitted with a reflux condenser is charged with [RuCl$_2$(η^4-C$_8$H$_{12}$) (NCCH$_3$)$_2$] (0.76 g, 2.0 mmol) and acetonitrile (20 mL). A quantity of NH$_4$PF$_6$ (0.67 g, 4.0 mmol) in acetonitrile (10 mL) is added. The solution is heated under reflux for 15 min, cooled to room temperature, and filtered to remove precipitated NH$_4$Cl. The filtrate is concentrated under reduced pressure to a volume of $\sim$ 5 mL. Addition of ethanol (10 mL) and diethyl ether (20 mL), followed by cooling to $-$ 20 °C for 2 h gives the product as a microcrystalline, yellow solid. Yield: 0.8 g (80%).

Anal. Calcd. for C$_{14}$H$_{21}$N$_3$Cl PF$_6$Ru: C, 32.80; H, 4.15; N, 8.20; Cl, 6.90. Found: C, 32.70; H, 3.90; N, 7.80; Cl, 6.65, mp dec without melting above 170 °C.

G. TRIS(ACETONITRILE)BROMO(η^4-1, 5-CYCLOOCTADIENE)-RUTHENIUM(II) HEXAFLUOROPHOSPHATE(1 $-$)

$$[RuBr_2(\eta^4\text{-}C_8H_{12})(NCCH_3)_2] + CH_3CN + NH_4[PF_6]$$

$$\longrightarrow [RuBr(\eta^4\text{-}C_8H_{12})(NCCH_3)_3][PF_6] + NH_4Br$$

Procedure

A 50-mL Erlenmeyer flask containing a magnetic stirring bar and fitted with a reflux condenser is charged with [RuBr$_2$(η^4-C$_8$H$_{12}$) (NCCH$_3$)$_2$] (1.1 g, 2.6 mmol) and acetonitrile (20 mL). A quantity of NH$_4$PF$_6$ (0.84 g, 5.0 mmol) in acetonitrile (10 mL) is added. The solution is heated under reflux for 15 min, cooled to room temperature, and filtered to remove the precipitate NH$_4$Br. The filtrate is concentrated under reduced pressure to a volume of $\sim$ 5 mL. Addition of ethanol (10 mL) and cooling to $-$ 20 °C for 12 h gives the product as an orange, crystalline solid. Yield: 1.0 g (67%).

Anal. Calcd. for C$_{14}$H$_{21}$N$_3$Br PF$_6$Ru: C, 30.20; H, 3.80; N, 7.55; Br, 14.35. Found: C, 30.30; H, 3.70; N, 7.90; Br, 14.35, mp dec without melting above 180 °C.

H. TETRAKIS(ACETONITRILE) (η^4-1, 5-CYCLOOCTADIENE)-RUTHENIUM(II) BIS[HEXAFLUOROPHOSPHATE(1 $-$)]

$$[RuCl(\eta^4\text{-}C_8H_{12})(NCCH_3)_3][PF_6] + CH_3CN + Ag[PF_6]$$

$$\longrightarrow [Ru(\eta^4\text{-}C_8H_{12})(NCCH_3)_4][PF_6]_2 + AgCl$$

Procedure

A 50-mL Erlenmeyer flask containing a magnetic stirring bar and fitted with a reflux condenser is charged with $[RuCl(\eta^4\text{-}C_8H_{12})(NCCH_3)_3][PF_6]$ (1.26 g, 2.0 mmol) and acetonitrile (20 mL). A quantity of $Ag[PF_6]$ (0.53 g, 2.1 mmol) in acetonitrile (10 mL) is added. The solution is stirred at room temperature for 3 h and then filtered to remove the precipitated AgCl. The filtrate is concentrated under reduced pressure to a volume of ~ 10 mL, followed by the addition of ethanol (20 mL), and cooling to $-20\,°C$ for 2 h. The product is obtained as a white, crystalline solid. Yield: 1.36 g (90%).

Anal. Calcd. for $C_{16}H_{24}N_4F_{12}P_2Ru$: C, 29.00; H, 3.65; N, 8.45. Found: C, 28.60; H, 3.50; N, 8.65, mp dec without melting above 215 °C.

I. (η⁴-1,5-CYCLOOCTADIENE)TETRAKIS(HYDRAZINE)DICHLORO-RUTHENIUM(II) BIS[TETRAPHENYLBORATE(1 −)]

$$[RuCl_2(\eta^4\text{-}C_8H_{12})]_x + 4N_2H_4 + 2Na[BPh_4] + 2H_2O$$
$$\longrightarrow [RuCl_2(\eta^4\text{-}C_8H_{12})(N_2H_4)_4][BPh_4]_2 \cdot 2H_2O + 2NaCl$$

Procedure

■ **Caution.** *Anhydrous hydrazines and solutions containing hydrazines are corrosive and can cause burns. They are also suspected to be carcinogens. Therefore, protective clothing must be worn, and the reactions should be performed in a well-ventilated hood. Affected areas of the eyes or skin should be irrigated with large volumes of water. If swallowed, the mouth should be washed out with copious amounts of water, and water to drink should also be administered.*

A 100-mL Erlenmeyer flask is charged with $[RuCl_2(\eta^4\text{-}C_8H_{12})]_x$ (2.0 g, 7.0 mmol), methanol (50 mL), and anhydrous hydrazine (95%, 3 mL, 95 mmol). The mixture is heated on a steam-bath for ~ 5 min to give a yellow-red solution. Water (10 mL) is added and the solution is filtered. Addition of a filtered solution of $Na[BPh_4]$ (5.0 g, 15 mmol) in methanol (25 mL) followed by cooling (0 °C) for 1 h gives the product as a cream colored precipitate. The product is obtained by filtration followed by washing with chilled ethanol (2 mL) and diethyl ether (5 mL), and drying under vacuum. Yield: 4.0–4.2 g (56–61%). This material is sufficiently pure for purposes of synthesis.

Anal. Calcd. for $C_{56}H_{72}N_8O_2B_2Ru$: C, 66.47; H, 7.17; N, 11.07. Found: C, 66.27; H, 7.18; N, 12.15, mp dec without melting above 160 °C. The product may, however, be recrystallized from a hydrazine–ethanol mixture giving a cream precipitate containing ethanol of crystallization.

J. (η^4-1,5-CYCLOOCTADIENE)TETRAKIS(METHYL-HYDRAZINE)RUTHENIUM(II) BIS[HEXAFLUORO-PHOSPHATE(1 −)]

$$[RuCl_2(\eta^4\text{-}C_8H_{12})]_x + 4NH_2HNMe + 2NH_4[PF_6]$$
$$\longrightarrow [Ru(\eta^4\text{-}C_8H_{12})(NH_2NHMe)_4][PF_6]_2 + 2NH_4Cl$$

Procedure

■ **Caution.** *Anhydrous hydrazines and solutions containing hydrazines are corrosive and can cause burns. They are also suspected to be carcinogens. All handling must be done with protective clothing and gloves, and must be carried out in a well-ventilated hood.*

A 50-mL Erlenmeyer flask is charged with $[RuCl_2(\eta^4\text{-}C_8H_{12})]_x$ (1.0 g, 3.6 mmol), methanol (20 mL), and methylhydrazine (98%, 2 mL, 38 mmol). The mixture is heated on a steambath for ∼ 10 min forming a dark red solution. On cooling, the solution is filtered, and a filtered solution of $NH_4[PF_6]$ (2.5 g, 15 mmol) in water (15 mL) is added. The volume of this solution is reduced under vacuum to ∼ 10 to 15 mL. Cooling at 0 °C for 1 h gives a precipitate that is separated by filtration and washed with chilled ethanol (2 mL) and diethyl ether (5 mL). Drying under vacuum gives the product as a cream colored solid. Yield: 1.5–1.7 g (61–69%). This material is sufficiently pure for purposes of further synthesis.

Anal. Calcd. for $C_{12}H_{36}N_8F_{12}P_2Ru$: C, 21.09; H, 5.31; N, 16.40. Found: C, 21.26; H, 5.14; N, 16.34, mp dec without melting above 140 °C.

K. (η^4-1,5-CYCLOOCTADIENE)TETRAKIS(METHYLHYDRA-ZINE)RUTHENIUM(II) BIS[TETRAPHENYLBORATE(1 −)]

$$[RuCl_2(\eta^4\text{-}C_8H_{12})]_x + 4NH_2NHMe + 2Na[BPh_4] + H_2O$$
$$\longrightarrow [Ru(\eta^4\text{-}C_8H_{12})(NH_2NHMe)_4][BPh_4]_2 \cdot H_2O + 2NaCl$$

Procedure

■ **Caution.** *See cautionary note under the procedures in Sections I and J.*
The complex $[Ru(\eta^4\text{-}C_8H_{12})(NH_2NHMe)_4][BPh_4]_2$ is prepared in the same manner as the analogous hexfluorophosphate salt described previously, except that a filtered solution of $Na[BPh_4]$ (2.5 g, 7.3 mmol) in a mixture of methanol (15 mL) and water (15 mL) is added to achieve precipitation of the

dication. The isolated product contains 1 mol of water per mole of product. Yield: 2.3–2.4 g (61–65%).

Anal. Calcd. for $C_{60}H_{78}N_8OB_2Ru$; C, 68.63; H, 7.49; N, 10.67. Found: C, 68.61; H, 7.24; N, 10.55, mp 138–140 °C.

Properties

[RuCl$_2$(η^4-C$_8$H$_{12}$)]$_x$. The complex is a brown solid, highly insoluble in most organic solvents. It is thought to have a polymeric, chloride-bridged structure.[1] It reacts via halogen-bridged cleavage and is a synthetic precursor to a variety of ruthenium(II) complexes.[2]

$$[RuX_2(\eta^4\text{-}C_8H_{12})(NCR)_2] \quad (X = Cl, Br; R = CH_3, C_6H_5)$$
$$[RuX(\eta^4\text{-}C_8H_{12})(NCCH_3)_3][PF_6] \quad (X = Cl, Br) \quad \text{and}$$
$$[Ru(\eta^4\text{-}C_8H_{12})(NCCH_3)_4][PF_6]_2$$

Some properties of these compounds that facilitate identification and assessment of purity are given in the Table I. All the compounds are air stable in the solid state as well as in solution. They are soluble in dichloromethane, acetonitrile, and nitromethane; less soluble in chloroform and acetone, and insoluble in alcohols, ethers, and alkanes. The complex $[RuCl_2(\eta^4\text{-}C_8H_{12})(NCCH_3)_2]$ is also soluble in water. It may therefore have an extensive aqueous chemistry.

TABLE I.   Selected Spectroscopic Properties of the Complexes

$$[RuX_2(\eta^4\text{-}C_8H_{12})(NCR)_2] \quad (X = Cl, Br; R = CH_3, C_6H_5)$$
$$[RuX(\eta^4\text{-}C_8H_{12})(NCCH_3)_3][PF_6] \quad (X = Cl, Br) \quad \text{and}$$
$$[Ru(\eta^4\text{-}C_8H_{12})(NCCH_3)_4][PF_6]_2$$

	IR[a] Frequencies (cm^{-1}) $\nu_{(CN)}$	NMR[b] δ Values NCCH$_3$
$[RuCl_2(\eta^4\text{-}C_8H_{12})(NCCH_3)_2]$	2414 (w), 2305 (w)	2.68 (6H)
$[RuCl_2(\eta^4\text{-}C_8H_{12})(NCC_6H_5)_2]$	2300 (m)	
$[RuBr_2(\eta^4\text{-}C_8H_{12})(NCCH_3)_2]$	2400 (w), 2300 (w)	2.70 (6H)
$[RuBr_2(\eta^4\text{-}C_8H_{12})(NCC_6H_5)_2]$	2300 (m)	
$[RuCl(\eta^4\text{-}C_8H_{12})(NCCH_3)_3][PF_6]$	2320 (m)	2.62 (6H), 2.40 (3H)
$[RuBr(\eta^4\text{-}C_8H_{12})(NCCH_3)_3][PF_6]$	2400 (w), 2300 (w)	2.70 (6H), 2.47 (3H)
$[Ru(\eta^4\text{-}C_8H_{12})(NCCH_3)_4][PF_6]_2$	2320 (m)	2.68 (6H), 2.42 (6H)

[a]Recorded as Nujol mull.
[b]Recorded in CD$_2$Cl$_2$ relative to TMS.

In general the cyanide ligands are labile in all the compounds although complete substitution of these ligands does not occur readily.[3,6] The neutral complexes $[RuX_2(\eta^4\text{-}C_8H_{12})(NCCH_3)_2]$ (X = Cl and Br) form the dimeric species $[Ru_2X(\mu\text{-}X)_3(\eta^4\text{-}C_8H_{12})_2(NCCH_3)]$ when heated under reflux in acetone.[7] The cationic complexes $[RuX(\eta^4\text{-}C_8H_{12})(NCCH_3)_3][PF_6]$ (X = Cl, Br) possess a meridional configuration of acetonitrile ligands, and the acetonitrile ligand trans to diene undergoes exchange with its deuterated analog.[7]

$[Ru(\eta^4\text{-}C_8H_{12})(N_2H_4)_4][BPh_4]_2$. This material is air stable but mildly light sensitive. For prolonged storage it is best kept under an inert atmosphere in the dark. It can be handled with ease in air both in the solid state and in solution.

The compound is soluble in acetone, sparingly soluble in methanol and ethanol, but insoluble in dichloromethane, diethyl ether, and nonpolar solvents such as pentane. Care must be exercised when heating this compound in acetone, as boiling in this solvent over 10 to 15 min results in the deposition of $[1,3\text{-}\eta^3\text{-}5,6\text{-}\eta^2\text{-cyclooctadienyl}][\eta^6\text{-tetraphenylborato}(1^-)]$ruthenium(II) as an insoluble yellow powder in $\sim 35\%$ yield.[5]

The IR spectrum (Nujol mull) shows weak bands at 3360–3100, 1612–1606, 1149–1140, and 920 cm^{-1} assignable to $\nu_{(NH)}$, $\delta_{(NH)asymm}$, $\delta_{(NH)symm}$, and $\nu_{(NN)}$, respectively. Typical bands at 710, 740, and 755 cm^{-1} associated with the tetraphenylborate anion are also present. A weak absorption at 3530 cm^{-1} can be assigned to the water of crystallization. The NMR spectrum recorded in CD_3CN consists of the following resonances: 6.04 ppm (2H, NH$_2$); 4.72 (2H, NH$_2$); 3.3–4.2 (16H, CH, and NH$_2$); 2.1 (8H, CH$_2$); 6.5–7.5 (4OH, aromatic).

$[Ru(\eta^4\text{-}C_8H_{12})(NH_2NHMe)_4]X_2$ (X = $[PF_6]$ or $[BPh_4]$). The methyl-hydrazine complexes are similar in solubility and stability to the hydrazine complex. The IR spectrum (Nujol mull) shows weak bands at 3330–3200, 1620–1610, and 1135 cm^{-1}, assignable to $\nu_{(NH)}$, $\delta_{(NH)asymm}$, and $\delta_{(NH)symm}$, respectively. For the hexafluorophosphate complex, an intense band at 840 cm^{-1} is assignable to the vibrations of the $[PF_6]$ anion. For the tetraphenylborate complex, bands at 715, 745, and 760 cm^{-1} are assignable to the $[BPh_4]$ anion. A weak absorption at 3480–3520 cm^{-1} in the spectrum of the latter complex is assignable to the water of crystallization.

The NMR spectra [recorded in $(CD_3)_2CO$ relative to tetramethylsilane(TMS)] both show broad, ill-defined resonances in the region from 1.6 to 4.3 ppm of the spectrum, with the aromatic protons of the tetraphenylborate anion appearing between 6.8 and 7.6 ppm.

Some reactivity patterns of both the hydrazine and methylhydrazine complexes are given by Singleton and coworkers.[5]

References

1. E. W. Abel, M. A. Bennett, and G. Wilkinson, *Chem. Ind.*, 1959, 1516.

2. G. Wilkinson (ed.), *Comprehensive Organometallic Chemistry*, Vol. 4, Pergamon Press, Oxford, 1982, pp. 748–750.

3. T. V. Ashworth and E. Singleton, *J. Organometal. Chem.*, 77, C31 (1974).

4. B. F. G. Johnson, J. Lewis, and I. E. Ryder, *J. Chem. Soc. Dalton Trans*, 1977, 719.

5. T. V. Ashworth, E. Singleton, and J. J. Hough, *J. Chem. Soc. Dalton Trans.*, 1977, 1809.

6. R. R. Schrock, B. F. G. Johnson, and J. Lewis, *J. Chem. Soc. Dalton Trans.*, 1974, 951.

7. T. V. Ashworth and E. Singleton, unpublished results.

15. PENTACARBONYLHYDRIDORHENIUM

Submitted by MICHAEL A. URBANCIC and JOHN R. SHAPLEY*
Checked by NANCY N. SAUER and ROBERT J. ANGELICI[†]

$$Re(CO)_5 Br \xrightarrow{\text{Zn–H}^+} ReH(CO)_5$$

Hieber and coworkers first prepared the pentacarbonyl hydride complex of rhenium[1] by protonation of the corresponding pentacarbonyl anion, $[Re(CO)_5]^-$. Modified versions of the original syntheses have been reported,[2] but these procedures still require extensive vacuum-line manipulations. Furthermore, the yield of $ReH(CO)_5$ is only $\sim 30\%$ based on pure $Na[Re(CO)_5]$,[1,3] which means that the overall yield from $Re_2(CO)_{10}$ is much lower.

Pentacarbonylhydridorhenium can be prepared readily in solution in $\sim 85\%$ yield by the reaction of $Re(CO)_5 Br$ with zinc and acetic acid in methanol.[4] This procedure has been modified to allow the convenient isolation of pure $ReH(CO)_5$ in high yield.

■ **Caution.** *Volatile metal carbonyls are highly toxic and should be handled in a well-ventilated hood.*

Procedure

Into a 100-mL single-neck (14/20) round-bottomed Schlenk flask are placed a Teflon-coated magnetic stirring bar, 2.50 g (6.16 mmol) of powdered

*Department of Chemistry and the Materials Research Laboratory, University of Illinois, Urbana, IL 61801.
[†]Department of Chemistry, Iowa State University, Ames, IA 50011.

Re(CO)$_5$Br,[5] and 2.5 g (38 mmol) of zinc dust. The flask is then capped with a rubber serum stopper and evacuated. Into another Schlenk flask is placed 50 mL of 2,5,8,22,14-pentaoxapentadecane (tetraglyme, Aldrich) and 4 mL of 85% phopshoric acid. This flask is also capped with a serum stopper and the solution is degassed *in vacuo* for 2 h at room temperature. At this point the acid solution is transferred under nitrogen to the flask containing the solids via a double-tipped needle (cannula). The flask is wrapped with aluminum foil for protection from light, and the mixture is stirred under nitrogen until no solid Re(CO)$_5$Br remains. This should take ~ 24 h.

When the reaction is complete, 5 g of P$_4$O$_{10}$ is added to the flask and the mixture is stirred vigorously to remove the water in the solution. After 30 min, the product is ready for collection by vacuum distillation at room temperature. This is accomplished by using the apparatus shown in the Fig. 1, which is assembled inside a nitrogen-filled glove bag. The reaction flask is attached to a Schlenk filter tube by means of a U-shaped adapter. The adapter should be as small as possible, preferably not larger than 7 × 10 cm. A column of P$_4$O$_{10}$ (10 × 25 mm) on the frit serves to remove any last traces of water during the distillation. A layer of glass wool (~ 2 mm thick) on the frit beneath the P$_4$O$_{10}$ helps to keep the frit unclogged. A 5-mL pear-shaped flask is used as the receiver and is shielded from light by aluminum foil. The receiver is cooled at − 78 °C, and the apparatus is evacuated with a vacuum pump. The reaction mixture bubbles slowly as the hydride distills out.

If the top of the column of P$_4$O$_{10}$ becomes syrupy, it may prevent flow through the column during the distillation. In this case, the top portion of the column should be removed and placed into the reaction flask along with some additional P$_4$O$_{10}$ to react with the remaining water.

Normally, three fractions are collected over a 12-h period to complete the distillation. Yield: 1.67 g, 5.10 mmol 84%.

Anal. Calcd. for ReH(CO)$_5$: C, 18.35; 0.31. Found: C, 18.25; H, 0.21.

Properties

Pure ReH(CO)$_5$ is a colorless liquid with a density of 2.30 g mL^1 (determined at 24 °C) and a melting point of 12.5 °C.[1b] It is weakly acidic, more soluble in nonpolar than polar solvents, and practically insoluble in water.[1] The equilibrium vapor pressure (6–100 °C) is given by the equation.[1b]

$$\log_{10} P(\text{mm}) = 8.598 - 2353.6/T$$

Three major infrared absorptions occur in the carbonyl stretching region near 2014, 2005, and 1982 cm^{-1} (cyclohexane)[3,6] and the molar absorptivities

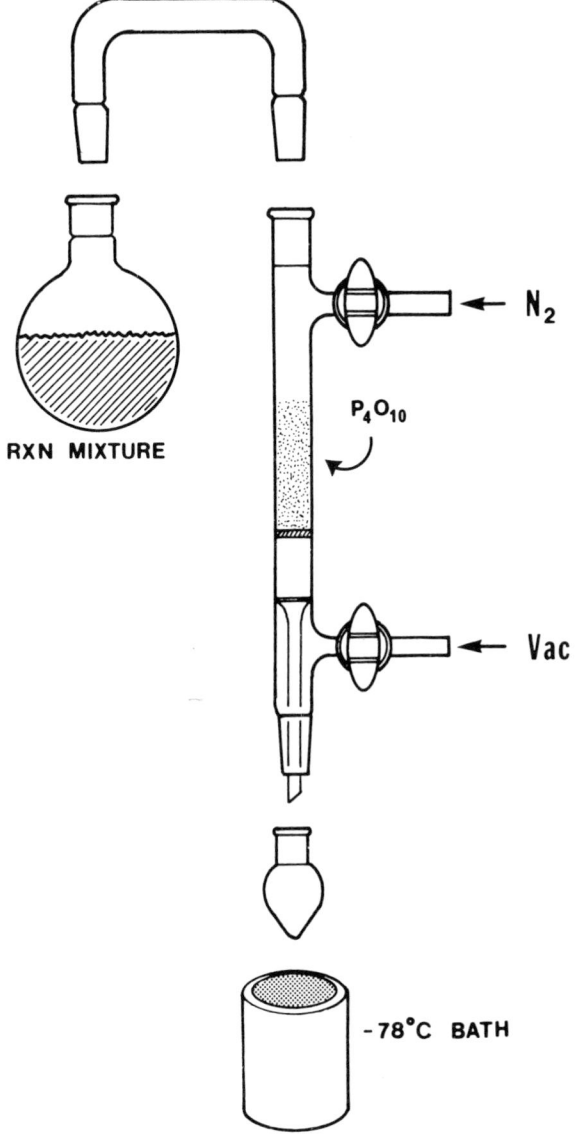

RXN MIXTURE

N$_2$

P$_4$O$_{10}$

Vac

-78°C BATH

Fig. 1. Apparatus for vacuum distillation of the reaction mixture containing HRe(CO)$_5$.

at these positions are 86, 22, and $2.7 \, L \, g \, cm^{-1}$, respectively.* The metal–hydrogen stretching absorptions are at $1882 \, cm^{-1}$ for the hydride and $1313 \, cm^{-1}$ for the deuteride (cyclohexane).[3] A singlet is observed in the ^1H NMR spectrum at $\delta - 5.88$ (hexane.[6] The gas-phase IR[1b] and Raman[7] spectra as well as the UV–Vis spectrum[8] have also been reported.

Contrary to early reports,[1b] pentacarbonylhydridorhenium is air stable.[6] The neat liquid, however, is moderately sensitive to light, turning yellow with formation of $Re_3H(CO)_{14}$.[9] Pentacarbonylhydridorhenium is stable up to $100 \, °C$, whereupon it decomposes to form $Re_2(CO)_{10}$ and H_2.[1] Substitution of carbonyl groups in $ReH(CO)_5$ by tertiary phosphines and other Lewis bases is well known and apparently occurs by a radical chain pathway.[6] More detailed surveys of the reactivity of $ReH(CO)_5$ have been summarized elsewhere.[10,11]

References

1. (a) W. Hieber and G. Braun, *Z. Naturforsch.*, **14b**, 132 (1959); (b) W. Beck, W. Hieber, and G. Braun, *Z. Anorg. Allg. Chem.*, **308**, 23 (1961).

2. (a) R. B. King and F. G. A. Stone, *Inorg. Synth.*, **7**, 198 (1963); (b) J. J. Eisch and R. B. King (eds.), *Organometallic Synthesis*, Vol. I, Academic Press, New York, 1965, p. 158; (c) W. F. Edgell and W. M. Risen, Jr., *J. Am. Chem. Soc.*, **88**, 5451 (1966).

3. P. S. Braterman, R. W. Harrill, and H. D. Kaesz, *J. Am. Chem. Soc.*, **89**, 2851 (1967).

4. J. R. Shapley, G. A. Pearson, M. Tachikawa, G. E. Schmidt, M. R. Churchill, and F. J. Hollander, *J. Am. Chem. Soc.*, **99**, 8064 (1977).

5. $Re(CO)_5Br$ was prepared by "titrating" a dichloromethane solution of $Re_2(CO)_{10}$ with liquid bromine at room temperature and recrystallizing from 1:2 acetone–methanol. See: S. P. Schmidt, W. C. Trogler, and F. Basolo, *Inorg. Chem.*, **23**, 41 (1985).

6. B. H. Byers and T. L. Brown, *J. Am. Chem. Soc.*, **99**, 2527 (1977).

7. A. Davison and J. W. Faller, *Inorg. Chem.*, **6**, 845 (1967).

8. G. B. Blakney and W. F. Allen, *Inorg. Chem.*, **10**, 2763 (1971).

9. Ultraviolet photolysis of neat $HRe(CO)_5$ yields $HRe_3(CO)_{14}$ in high yield; Mass spectrum, m/z 954 $(M^+, {}^{187}Re)$ and $(M^-, xCO)^+$, $x = 1–14$; IR (cyclohexane) ν_{CO} 2100(w), 2047(vs), 2014(m), 1996(s), 1979(m), 1969(w), 1933(m) cm^{-1}. Compare: W. Fellmann and H. D. Kaesz, *Inorg. Nucl. Chem. Lett.*, **2**, 63 (1966).

10. (a) A. P. Ginsberg, in *Transition Metal Chemistry*, Vol. I, R. L. Carlin (ed.), Marcel Dekker, New York, 1965, p. 111; (b) H. D. Kaesz and R. B. Saillant, *Chem. Rev.*, **72**, 231 (1972).

11. (a) D. Giusto, *Inorg. Chim. Acta Rev.*, **6**, 91 (1972); (b) N. M. Boag and H. D. Kaesz, in *Comprehensive Organometallic Chemistry*, Vol. IV, G. Wilkinson, F. G. A. Stone, E. W. Abel (eds.); Pergamon, New York, 1982, p. 172.

*Determined in cyclohexane solutions from 0.4 to $1 \, mg \, mL^{-1}$ at $2014 \, cm^{-1}$ and from 0.4 to $4 \, mg \, mL^{-1}$ at 2005 and $1982 \, cm^{-1}$.

16. ORGANOMETALLIC FLUORO COMPLEXES

Submitted by E. Horn,*F. L. WIMMER,* and M. R. SNOW*
Checked by PENNY PANGAGIOTIDOU[†] and R. COTTON[†]

■ **Caution.** *Skin contact with all fluorides should be avoided.*

Fluoride complexes of organometallics can be readily prepared by halide abstraction reactions. The general reaction scheme for a carbonyl species:

$$(CO)_x RMX + AgY \xrightarrow{\text{solvent}} (CO)_x RMY + AgX$$

can be made to work satisfactorily provided the reagents are suitably dry, the solvent is suitably inert, and that rigid "plastic" vessels are used. Silver salts tend to be solvated. Silver fluoride prepared from aqueous solution always contains some water even when extensively heated and dried over phosphorus pentoxide. Such silver fluoride will invariably yield the aqua complex.[1] ■ However, if silver (hydrogen difluoride), $Ag[F-H-F]$, is used in which the fluoride is solvated by HF, the desired fluoride can be obtained (procedure in Section B). The use of Pyrex vessels can lead to contamination by fluoroboric acids such as $HOBF_3$. This weak acid may give products that coordinate without deprotonation or simply hydrogen bond into the lattice.[1] When silver fluoride is employed (to prepare $[Re(CO)_3 F]_4 \cdot 4H_2 O$, procedure in Section A) no borates as might be derived from glass apparatus are detected in the product. Dichloromethane is a satisfactory solvent when the reactions are at room temperatures and complete in under an hour. Long reaction times especially at elevated temperatures lead to chloride formation.[2] A polar, but unreactive solvent is called for such as fluorobenzene (procedure in Section A).

Fluoroanions such as $[BF_4]^-$, $[PF_6]^-$, and $[AsF_6]^-$ can be considered as potential sources of fluoride ions. The silver salts tend to be wet and abstraction reactions lead to hydrolysis[3] (see the procedure in Section C) or yield either the aqua or fluoroanion complexes ($[BF_4]^-$ or $[AsF_6]^-$.[4] Using dry reagents such as $NO[PF_6]$ can lead to the oxidative addition of fluoride (see the procedure in Section D).[5]

*Department of Physical and Inorganic Chemistry, The University of Adelaide, G.P.O. Box 498, Adelaide, South Australia, Australia 5001.
[†]Department of Inorganic Chemistry, The University of Melbourne, Parkville, Victoria, Australia 3052.

A.   TETRAKIS[TRICARBONYLFLUORORHENIUM(I)]-TETRAHYDRATE

$$4Re(CO)_5Br + 4AgF\cdot(H_2O)_{0.5}(XS)$$
$$\longrightarrow [Re(CO)_3F]_4\cdot 4H_2O + 8CO + 4AgBr$$

Procedure[2]

Silver fluoride is prepared by slowly dissolving excess silver oxide in hydrofluoric acid (50%) over 24 h. The filtered solution is evaporated to dryness *in vacuo* at 35 °C and then heated to 60 °C for 1 h. The yellow solid as stored over P_2O_5 away from light and powdered before use.

A mixture of $Re(CO)_5Br$ dissolved in ~ 8 mL of fluorobenzene and finely divided and freshly prepared silver fluoride (30 mg, 0.24 mmol, two-fold excess) is stirred in a glass vessel at room temperature for 2 h. The solution is filtered to remove the precipitated silver bromide and excess silver fluoride. The filtrate is heated to reflux at 90 to 100 °C for 5 h, cooled to room temperature, and the solvent is removed under vacuum to give the yellow product. Yield: 44 mg, 0.036 mmol (75% yield). This can be recrystallized from a diethyl ether solution into which fluorobenzene is allowed to diffuse and dried *in vacuo* to give needle or prismatic crystals.

Anal. Calcd. for $[Re(CO)_3F]_4\cdot 4H_2O$: C, 11.7; H, 0.7; F, 6.2%. Found: C, 12.5; H, 0.5; F, 6.2%.

Properties

The crystals are tetragonal, space group $I\bar{4}$, with a 11.716(5), c 8.988(3) Å, and $Z = 2$. There is no parent peak in the mass spectrum, the highest is for $Re_4(CO)F_3C$. The compound exhibits infrared(IR) CO stretching frequencies at 2043 (ms) and 1932 (s) (fluorobenzene) and at 2092 (m), 2041 (s), 1932 (s, sh), and 1901 (s) (Nujol). Water bands at 3570 (sh), and 3530 (w) (sharp) appear in the Nujol spectrum.

B.   (2, 2'-BIPYRIDINE)TRICARBONYLFLUORORHENIUM(I)

$$Re(CO)_3(bipy)Br + AgHF_2 \longrightarrow Re(CO)_3(bipy)F + AgBr + HF$$

Procedure[1]

The bromotricarbonyl complex is prepared from $Re(CO)_5Br$ and 2, 2'-bipyridine (bipy) in light petroleum ether (30–40 °C) by reflux, and isolated by literature methods.[6,7] In a rigid plastic vessel a mixture of dichloromethane (20 mL), $Re(CO)_3(bipy)Br$ (90 mg, 0.18 mmol), and AgF_2H (31 mg, 0.21 mmol, 17% excess)* is stirred under a dry nitrogen atmosphere for 30 min. The solution is filtered, concentrated to ~ 5 mL and the yellow product (yield 70%) precipitated and washed with light petroleum ether and dried *in vacuo*.

Anal. Calcd. for $Re(CO)_3(bipy)F$: C, 35.1; H, 1.8; F, 4.3; N, 6.3%. Found: C, 35.0; H, 1.9; F, 4.2; N, 6.3%.

Properties

The fluorides are less soluble in halogenated solvents (e.g., $CHCl_3$, CH_2Cl_2, C_6H_5Cl, C_6H_5F, and C_6F_6) than the bromides. The solubility is dependent on the ligand and decreases in the following order $(CH_3)_2N—CH_2—CH_2—N(CH_3)_2 > Ph_2P—CH_2—CH_2—PPh_2 > bipy$. In these solutions the fluorides are stable to halide substitution (e.g., from $[Ph_4As]Cl$) for > 24 h at room temperature. The complexes give excellent mass spectra with partially decarbonylated species more prominent than the parent ones. The IR stretching frequencies in the CO region for the bipy complex are 2040 (s) and 1926 (s) (CH_2Cl_2) and 2032 (sm) and 1902 (s) (Nujol), well separated from the parent bromide.

C. (2, 2'-BIPYRIDINE)TRICARBONYL-(PHOSPHORODIFLUORIDATO)RHENIUM(I)

$$Re(CO)_3(bipy)Br + Ag[PF_6] + 2H_2O$$
$$\longrightarrow Re(CO)_3(bipy)(O_2PF_2) + AgBr + HF$$

Procedure[3]

Silver hexafluorophosphate (76 mg, 0.3 mmol) is added to a solution of $Re(CO)_3(bipy)Br$ (53.6 mg, 0.10 mmol) in dichloromethane (15 mL). The mixture is stirred for 30 min, filtered into a cold flask, and concentrated to a small volume by applying vacuum. Light petroleum ether at $0\,°C$ is added to precipitate a fine yellow powder. The solvent is decanted and the product dried on the vacuum line yields 36.3 mg, 0.07 mmol (69%).

Anal. Calcd. for $Re(CO)_3(bipy)PO_2F_2,(CH_2Cl_2)_{0.25}$: C, 29.0; H, 1.6; F, 6.9; N, 5.1; P, 5.6%, Found: C, 29.2; H, 1.7; F, 7.1; N, 5.2; P, 5.5%.

*Available from Ozark-Mahoning Co., 1870 South Boulder, Tulsa, OK 74119.

Properties

The compound is stable in air at room temperature. It is very soluble in dichloromethane from which it crystallizes with a small quantity of dichloromethane as shown by X-ray structure analysis.[3] The IR stretching frequencies in the CO region are at 2034 (s, sh), 2030 (vs), 1923 (s, sh), and 1901 (vs) (Nujol) and in dichloromethane at 2035 (vs), 1929 (vs), and 1918 (vs, sh). The ^{31}P NMR spectrum exhibits a triplet, $\delta - 12.7$ pp, from external $(C_2H_5O)_3PO$, $J = 969$ Hz.

D.   DICARBONYLBIS[1, 2-ETHANEDIYLBIS(DIPHENYLPHOS-PHINE)]FLUOROMOLYBDENUM(II) HEXAFLUORO-PHOSPHATE(1 −)

$$Mo(CO)_2(dppe)_2 + 2NOPF_6 \longrightarrow [MoF(CO)_2(dppe)_2]PF_6 + 2NO + PF_5$$

$$where \; dppe = PPh_2CH_2CH_2PPh_2$$

Procedure[5]

The complex *cis*-$Mo(CO)_2(dppe)_2$ [1,2-ethanediylbis(diphenylphosphine)] is prepared by the literature method.[8] In a dry box, a quantity (427 mg, 0.450 mmol) of this complex is mixed with NOPF$_6$* (162 mg, 0.923 mmol). The NO[PF$_6$] must be freshly prepared or from a freshly opened bottle. Dichloromethane (6 mL) is added and the mixture is rapidly stirred for 10 to 15 min until the solid has completely dissolved giving a green solution. This is eluted through a 5-cm column (diameter, ~7 mm) of chromatographic alumina† with dichloromethane. The bright yellow eluate is concentrated to a few milliliters by a water pump. Methanol (10 mL) is added and the mixture is cooled in an ice bath giving canary yellow crystals. These are filtered, washed with methanol, and dried *in vacuo*. Yield: 260 mg (52%).

Anal. Calcd. for [MoF(CO)$_2$(dppe)$_2$]PF$_6$: C, 58.3; H, 4.4; F, 12.0; P, 13.9%. Found: C, 58.2; H, 4.4; F, 12.1; P, 13.8%. Other materials result if solvents other than dichloromethane are used for the preparation.[5]

Properties

The complex is extremely soluble in dichloromethane and very soluble in

---

*NO[PF$_6$] and dppe can be obtained from Strem Chemicals, 7 Mulliken Way, Dexter Industrial Park, Newburyport MA 01950.
†Alumina, aluminium oxide (Type 507C, neutral), can be obtained from Fluka Chemie AG, Industriestrasse 25, CH-9470, Buchs, Switzerland.

acetone, tetrahydrofuran, and chloroform. The complex is air stable in dichloromethane for periods in excess of 25 h in the absence of acid. The CO stretching frequencies are 1945 (s) and 1889 (s) (CH_2Cl_2). The ^{19}F NMR spectrum of the cation exhibits a sole non-Gaussian resonance at $+62.4$ relative to CF_3CO_2H.

References

1. E. Horn and M. R. Snow, *Aust. J. Chem.*, **37**, 35–45 (1984).
2. E. Horn and M. R. Snow, *Aust. J. Chem.*, **34**, 737–743 (1981).
3. E. Horn and M. R. Snow, *Aust. J. Chem.*, **33**, 2369–76 (1980).
4. E. Horn and M. R. Snow, *Aust. J. Chem.*, **37**, 1375–93 (1984).
5. M. R. Snow and F. L. Wimmer, *Aust. J. Chem.*, **29**, 2349, 2361 (1976).
6. E. W. Abel and S. P. Tyfield, *Can. J. Chem.*, **47**, 4627 (1969).
7. E. W. Abel and G. Wilkinson, *J. Chem. Soc.*, 1959, 1501.
8. J. Chatt and H. R. Watson, *J. Chem. Soc.*, 1961, 4980. F. Zingales and F. Canziani, *Gazz. Chim. Ital.*, **92**, 343 (1962).

17. (TETRAHYDROTHIOPHENE)GOLD(I) OR GOLD(III) COMPLEXES

Submitted by RAFAEL USON,* ANTONIO LAGUNA,* and MARIANO LAGUNA*

Checked by D. A. BRIGGS,† H. H. MURRAY,† and J. P. FACKLER, JR.†

Gold(I) or gold(III) complexes containing dialkyl sulfides (SMe_2) or tetrahydrothiophene (SC_4H_8) have frequently been used as precursors for the synthesis of new gold complexes, since they are generally stable at room temperature and their sulfide ligand can readily be replaced by other neutral or anionic ligands[1-4]. In the following sections, the preparations of $AuCl(SC_4H_8)$, $Au(C_6F_5)(SC_4H_8)$ and $Au(C_6F_5)_3(SC_4H_8)$, and the syntheses of the gold(I) complexes (NBu_4) $[Au(C_6F_5)Br]$, $[Au(pdma)_2]$ $[Au(C_6F_5)_2]$ [pdma = 1,2-phenylenebis(dimethylarsine)], and $(C_6F_5)Au$ $(NCS)Au(PPh_3)$, obtained by the displacement of the tetrahydrothiophene ligand, are described.

■ **Caution.** *All the reactions must be run in a well-ventilated fume hood because tetrahydrothiophene may irritate the cornea of the eye and has a strongly disagreeable odor.*

*Department of Inorganic Chemistry, University of Zaragoza, Spain.
†Department of Chemistry, Texas A&M University, College Station, TX 77843.

A. CHLORO(TETRAHYDROTHIOPHENE)GOLD(I)

$$H[AuCl_4] + SC_4H_8 + H_2O \longrightarrow AuCl(SC_4H_8) + OSC_4H_8 + 3HCl$$

Procedure

A 200-mL flask fitted with a Teflon-coated magnetic stirring bar is charged with a solution of hydrogen tetrachloroaurate(III) (H[AuCl_4]·4H_2O,[5] 6.18 g, 15 mmol) in a mixture of 10 mL of water and 50 mL of ethanol. Dropwise addition of tetrahydrothiophene (2.8 mL, 31.75 mmol) gives rise to the formation of a bulky yellow precipitate of $AuCl_3(SC_4H_8)$, which is transformed into the white solid of $AuCl(SC_4H_8)$ while the addition is continued. The mixture is stirred for 15 min at room temperature and the white precipitate is filtered, washed with two 10-mL portions of ethanol and vacuum dried. Yield: 4.57 g (95%).

Anal. Calcd. for C_4H_8AuClS: C, 15.0; H, 2.5; Au, 61.45. Found: C, 15.1; H, 2.45; Au, 61.7.

Properties

Chloro(tetrahydrothiophene)gold(I) is a white solid, which decomposes very slowly at room temperature when dry. At 0 °C it can be stored over a long time. It is soluble in acetone, benzene, chloroform and dichloromethane, insoluble in diethyl ether, ethanol and hexane. Its IR spectrum shows a strong band at $332 \, cm^{-1}$, $(v_{(Au-Cl)}{}^{6})$.

B. (PENTAFLUOROPHENYL)
(TETRAHYDROTHIOPHENE)GOLD(I)

$$AuCl(SC_4H_8) + LiC_6F_5 \longrightarrow Au(C_6F_5)(SC_4H_8) + LiCl$$

Procedure

Into a 250-mL round-bottomed two-neck flask provided with a Teflon-coated magnetic stirring bar, is placed a solution of (pentafluorophenyl)lithium (15 mmol) in 60 mL of diethyl ether at -78 °C (prepared as recently described).[7] A suspension of chloro(tetrahydrothiophene)gold(I) (3.2 g, 10 mmol) in 30 mL of diethyl ether is added and the mixture is stirred under dry nitrogen for 10 min. It is slowly (1 h) allowed to warm to -30 °C, permitting excess pressure to be vented, and the stirring is continued for other 15 min. The cooling system is removed and the mixture stirred for an additional 30 min. To hydrolyze the excess of the lithium derivative the stopper is removed and a drop or two of water is added to the solution. The mixture is then filtered through a 5-mm layer of anhydrous magnesium sulfate. The filtrate is concentrated to ~ 8 mL under reduced pressure. Addition of 20 mL of hexane causes a white solid to precipitate, which is filtered off and washed with two 5-mL portions of hexane. Yield: 4.43 g (98%). Exposure to air may produce a mixture of black and white crystals. The black crystals can be removed by redissolving the crystals in diethyl ether, filtering, and removal of the ether by evaporation.

Anal. Calcd. for $C_{10}H_8AuF_5S$: C, 26.55; H, 1.8; Au, 43.55. Found: C, 26.3; H, 1.7; Au, 42.8.

Properties

(Pentafluorophenyl)(tetrahydrothiophene)gold(I) is a white crystalline solid, which is air and moisture stable at room temperature. It decomposes before melting (109 °C). It is very soluble in acetone, benzene, dichloromethane, diethyl ether, nitromethane, and tetrahydrofuran, soluble in ethanol and methanol, and insoluble in hexane.

Its IR spectrum shows strong bands at 1510, 1070 (br), 960 and 802 cm^{-1} due to the pentafluorophenyl group. It is monomeric in chloroform (MW 440, calcd 452) and nonconducting in acetone and nitromethane. Its solutions emits a slight odor of tetrahydrothiophene and slow deposition of metallic gold can be observed. Its tetrahydrothiophene group can readily be displaced by other neutral or anionic ligands[2] (see the syntheses in Sections D, E, and F).

C. TRIS(PENTAFLUOROPHENYL) (TETRAHYDROTHIOPHENE)GOLD(III)

$$Au(C_6F_5)(SC_4H_8) + Tl(C_6F_5)_2Cl \longrightarrow Au(C_6F_5)_3(SC_4H_8) + TlCl$$

■ **Caution.** *Compounds of thallium are toxic. This procedure should be carried out in a well-ventilated hood; protective gloves should be worn to avoid any contact with thallium compounds. Residues containing thallium compounds must be disposed of in accordance to toxic waste disposal procedures.*

Procedure

A 100-mL round-bottomed flask containing a Teflon-coated magnetic stirring bar and equipped with a reflux condenser is charged with 0.90 g (2 mmol) of (pentafluorophenyl)(tetrahydrothiophene)gold(I),   1.15 g   (2 mmol)   of chlorobis(pentafluorophenyl)thallium(III),[6] and 60 mL of toluene. The purity of $Tl(C_6F_5)_2Cl$ is critical to obtain the stated yields.

The mixture is stirred for 45 min at room temperature, and subsequently for 2 h at 80 °C. The precipitated thallium(I) chloride is filtered and the filtrate is evaporated to dryness. The white residue is extracted with 30 mL of diethyl ether, concentrated to ~ 8 mL and addition of 30 mL of hexane causes precipitation of the white title complex. Yield: 1.29 g (82%).

Anal. Calcd. for $C_{22}H_8AuF_{15}S$: C, 33.6; H, 1.05; Au, 25.05. Found: C, 33.65; H, 0.9; Au, 24.8.

Properties

Tris(pentafluorophenyl)(tetrahydrothiophene)gold(III) is a white crystalline solid, which is air and moisture stable at room temperature. It melts at 190 °C. It is soluble in acetone, benzene, chloroform, dichloromethane, diethyl ether and toluene, and is insoluble in aliphatic hydrocarbons. Its IR spectrum shows strong bands at 1510, 1080, 1070, 970, 810, and 790 cm^{-1}, due to the presence of the three C_6F_5 groups. It is monomeric in chloroform (MW 752, calcd. 786) and nonconducting in acetone solution. Its tetrahydrothiophene group can readily be displaced by other ligands, thus allowing the synthesis of other new gold(III) derivatives.[3]

D. (BENZYL)TRIPHENYLPHOSPHONIUM CHLORO(PENTAFLUOROPHENYL)AURATE(I)

$$Au(C_6F_5)(SC_4H_8) + [PhCH_2PPh_3]Cl$$
$$\longrightarrow [PhCH_2PPh_3][Au(C_6F_5)Cl] + SC_4H_8$$

Procedure

A 100-mL round-bottomed flask containing a Teflon-coated magnetic stirring bar is charged with a solution of 0.69 g (1.5 mmol) of (pentafluorophenyl)(tetrahydrothiophene)gold(I) in 30 mL of ethanol; 0.58 g (1.5 mmol) of (benzyl)triphenylphosphonium chloride (from Fluka AG) is added and the mixture is stirred at room temperature for 1 h. Concentration of the solution by removal of ~ 20 mL of solvent under reduced pressure and cooling to 0 °C results in a white precipitate. This is separated by filtration and washed with two 5 mL portions of hexane. Yield: 0.94 g (83%).

Anal. Calcd. for $C_{31}H_{22}AuClF_5P$: C, 49.45; H, 2.95; Au, 26.15. Found: C, 50.1; H, 2.9; Au, 25.2.

Properties

(Benzyl)triphenylphosphonium chloro(pentafluorophenyl)aurate(I) is a white crystalline solid, which is stable at room temperature and melts with decomposition at 123 °C. It is soluble in acetone and dichloromethane, slightly soluble in ethanol and diethyl ether, and insoluble in hexane. Its IR spectrum shows, along with the absorptions due to the pentafluorophenyl group (see Section B), strong bands at 1110, 740, 680, 660, 500, and 485 cm^{-1} arising from the cation. It is conducting in acetone ($110\,\Omega^{-1}\,cm^2\,mol^{-1}$ in $5.10^{-4}\,M$ solution). It reacts[8] with NaH or $Tl(C_5H_5)$ to give the ylide complex $Au(C_6F_5)\{CH(Ph)PPh_3\}$.

E. BIS[1, 2-PHENYLENEBIS(DIMETHYLARSINE)]-GOLD(I) BIS(PENTAFLUOROPHENYL)AURATE(I)

$$2Au(C_6F_5)(SC_4H_8) + 2(pdma) \longrightarrow [Au(pdma)_2][Au(C_6F_5)_2] + 2SC_4H_8$$

Procedure

A 100-mL round-bottomed flask containing a Teflon-coated magnetic stirring bar is charged with a solution of 0.45 g (1 mmol) of (pentafluorophenyl)(tetrahydrothiophene)gold(I) in 50 mL of diethyl ether, to which is added, 0.15 mL (1 mmol) of pdma. After stirring for 5 min at room temperature formation of a white precipitate is observed. Concentration of the solvent to ~ 10 mL causes white crystals to precipitate, which are filtered, washed with two 10 mL portions of hexane, and dried. Yield: 0.52 g (80%).

Anal. Calcd. for $C_{32}H_{32}As_4Au_2F_{10}$: C, 29.55; H, 2.5; Au, 30.3. Found: C, 29.6; H, 2.3; Au, 29.9.

Properties

Bis[1,2-phenylenebis(dimethylarsine)]gold(I) bis(pentafluorophenyl)aurate(I) is a white crystalline solid that melts at 178 °C. It is soluble in acetone and dichloromethane, slightly soluble in diethyl ether, and insoluble in hexane. Its IR spectrum shows, along with those due to the pentafluorophenyl group (see Section B), strong absorptions at 745, 590, 435, 365, and 350 cm^{-1} arising from the bis(arsine). In the cation the bis(arsine) acts as a chelate and therefore the cation contains a tetrahedral four-coordinated gold(I) atom.[9]

F. (PENTAFLUOROPHENYL)-μ-THIOCYANATO(TRIPHENYLPHOSPHINE)DIGOLD(I)

$$(C_6F_5)Au(SC_4H_8) + (NCS)Au(PPh_3)$$

$$\longrightarrow (C_6F_5)Au(NCS)Au(PPh_3) + SC_4H_8$$

Procedure

A 100-mL round-bottomed flask, containing a Teflon-coated magnetic stirring bar is charged with a solution of 0.45 g (1 mmol) of (pentafluoro-phenyl)(tetrahydrothiophene)gold(I)   in   20 mL   of   dichloromethane. (Thiocyanato)(triphenylphosphine)gold(I)[10] (0.52 g, 1 mmol) is added and stirring is continued for 6 h at room temperature. The solution is filtered through a 1-cm layer of kieselguhr to remove the precipitated metallic gold. The filtrate is concentrated to $\sim$ 5 and 30 mL of hexane is added to precipitate a white solid containing $(C_6F_5)Au(NCS)Au(PPh_3)$ and $(NCS)Au(PPh_3)$, which is filtered. The compounds are separated by treatment with 15 mL of diethyl ether, the remaining solid $(NCS)Au(PPh_3)$ is filtered; the filtrate is concentrated to $\sim$ 5 and 30 mL of hexane is added to precipitate the binuclear complex as white crystals. Yield: 0.18 g (20%).

Anal. Calcd. for $C_{25}H_{15}Au_2F_5NPS$: C, 34.05; H, 1.7; N, 1.6; Au, 44.7. Found: C, 33.9; H, 1.75; N, 1.7; Au, 45.6.

Properties

(Pentafluorophenyl)-μ-thiocyanato(triphenylphosphine)digold(I) is a white crystalline solid, which is air and moisture stable at room temperature. It melts under decomposition at 120 °C. It is soluble in acetone, benzene, dichloromethane and diethyl ether, and insoluble in aliphatic hydrocarbons. It is monomeric in benzene (MW 834, calcd. 881) and nonconducting in acetone solution. In its IR spectrum the vibration due to $v_{(C\equiv N)}$ is to be observed as a strong absorption band at 2170 cm^{-1}.

References

1. F. Bonati and G. Minghetti, *Gazz. Chim. Ital.,* **103**, 373 (1973).
2. R. Usón, A. Laguna, and J. Vicente, *J. Organometal. Chem.,* **131**, 471 (1977).
3. R. Usón, A. Laguna, M. Laguna, and E. Fernández, *J. Chem. Soc. Dalton Trans.,* **1982**, 1971.
4. M. R. Awang, G. A. Carriedo, J. A. K. Howard, K. A. Mead, I. Moore, C. M. Nunn, and F. G. A. Stone, *J. Chem. Soc. Chem. Commun.,* **1983**, 964.
5. G. Brauer, *Handbuch der Präparativen Anorganischen Chemie*, Vol. 2, 3rd ed., Verlag, Stuttgart, 1978, p. 1014.

6. E. A. Allen and W. Wilkinson, *Spectrochim. Acta*, **28A**, 2257 (1972).
7. R. Usón and A. Laguna, *Inorg. Synth.*, **21**, 72 (1982).
8. R. Usón, A. Laguna, M. Laguna, and A. Usón, *Inorg. Chim. Acta*, **73**, 63 (1983).
9. R. Usón, A. Laguna, J. Vicente, J. García, P. G. Jones, and G. M. Sheldrick, *J. Chem. Soc. Dalton Trans.*, **1981**, 655.
10. C. Kowala and J. M. Swan, *Aust. J. Chem.*, **19**, 547 (1966). R. Usón, P. Royo, A. Laguna, and J. García, *Rev. Acad. Cienc. Exactas Fis. Quim. Nat. Zaragoza*, **28**, 67 (1973).

Chapter Three

MONONUCLEAR TRANSITION METAL COMPLEXES Part II: COMPLEXES WITH WEAKLY BONDED ANIONS

Preface W. Beck*

In the following procedures, the preparation and reactions are given for complexes that behave as strong electrophilic Lewis acids. Such complexes,

$$L_n M — X$$

$$L = CO, \pi\text{-}C_5H_5$$

$$X = [BF_4]^-, [ClO_4]^-, [PF_6]^-, [PO_2F_2]^-, [AsF_6]^-, [SbF_6]^-,$$
$$[SO_3CF_3]^-$$

contain a "soft" metal in a low oxidation state with strongly bonded π-acceptor ligands and a "hard" labile anionic or neutral ligand[1] which can be easily substituted by other even weakly nucleophilic ligands, usually under very mild conditions. These complexes are precursors of coordinatively and electronically unsaturated compounds and have proved to be excellent starting materials in preparative organometallic chemistry. Complexes with easily dissociating ligands are also possible precursors of catalysts.

Although compounds with weakly coordinated anions have been isolated with some main group elements and with the more electron rich transition

*Institut für Anorganische Chemie Universität München, Meiserstr. 1, 8000 München, Federal Republic of Germany.

metals such as Cu, Zn, or Ni^2 only a few compounds with carbonyl ligands were reported prior to 1978.

A series of organometallic compounds with good leaving groups has been isolated and fully characterized by physical methods (for further examples see the following procedures):

Cr(CO)$_3$(PMe$_3$)(CMe)(FBF$_3$) (Ref. 3), W(CO)$_4$(CNEt$_2$)(FBF$_3$) (Ref. 3), M(CO)$_5$X (Ref. 4) (M = Mn, Re; X = FAsF$_5$, OClO$_3$, O$_2$PF$_2$, FBF$_3$), M(CO)$_3$(π-C$_5$H$_5$)X(Ref. 5) (X = FBF$_3$, FPF$_5$, FAsF$_5$, FSbF$_5$), Cr(NO)$_2$ (π-C$_5$H$_5$)X (Ref. 6) (X = FBF$_3$, FPF$_5$, FAsF$_5$), Fe(CO)$_2$ (π-C$_5$H$_5$)FBF$_3$ (Ref. 7), Ir(Cl (CO) (PPh$_3$)$_2$ (H)FBF$_3$ (Ref. 8).

However, it is not always necessary to isolate these complexes. Often the complexes generated *in situ* or the solvent containing compounds, for example, [(M(CO)$_2$(π-C$_5$H$_5$) (solvent)][BF$_4$] (Ref. 9) (M = Fe, Ru, Os), W(NO)$_2$ (π-C$_5$H$_5$)(BF$_4$) (Ref. 10), and Mo(CO)$_3$ (π-C$_5$H$_5$)(BF$_4$) (Ref. 11) can be used in synthesis.

Various routes have been used by many groups for the preparation of metal complexes with weakly bonded anions. These are

(a) Abstraction of an anionic ligand Y by a cationic Lewis acid A$^+$X$^-$ to give a very stable compound AY and the desired complex. Hereby the Lewis acidity of A$^+$ is transferred to the metal complex

$$L_nM—Y + A^+X^- \rightarrow L_nM—X + AY$$

Y$^-$ = H$^-$, CH$_3^-$, halide, [N$_3$]$^-$
A$^+$ = H$^+$, Ag$^+$, Me$_3$O$^+$, CPh$_3^+$, NO$^+$
X$^-$ = [FBF$_3$]$^-$, [FPF$_5$]$^-$, [FAsF$_5$]$^-$, [FSbF$_5$]$^-$, [OClO$_3$]$^-$, [OSO$_2$CF$_3$]$^-$
AY = H$_2$, CH$_4$, Ag halide, Ph$_3$CH, Ph$_3$CCH$_3$, N$_2$O, respectively

(b) Addition of main group Lewis acids to a coordinated halide

$$L_nM—Y + EX_n \rightarrow L_nM—YEX_n$$

Y = halide
EX$_n$ = for example, BF$_3$, AlCl$_3$, AsF$_5$

(c) Oxidation of dimeric complexes

$$[L_nM]_2 \xrightarrow[\text{solvent}]{-2e} 2[L_nM(\text{solvent})]^+$$

(d) Oxidative addition of AX to low-valent coordinatively unsaturated complexes

$$L_n M + AX \rightarrow L_n M \underset{X}{\overset{A}{\diagdown}}$$

A comprehensive survey on organometallic Lewis acids is given in Ref.[12].

General Remarks

The user should be familiar with the Schlenk technique for handling air and moisture sensitive compounds.[13] Prepurified argon or nitrogen are recommended as inert gas. The argon is dried by use of a column (100-cm length, 5-cm diameter) filled with molecular sieves 4 and 5 Å. Traces of oxygen are removed by another column filled with chromium(II) oxide on silica gel, which was prepared as reported[14] by reduction of CrO_3 on silica gel with carbon monoxide.

- **Caution.** *Chromium(VI) oxide, CrO_3, is carcinogenic. Especially after the drying procedure, inhalation of the fine powder must be avoided by working in a well-ventilated hood. Also the use of the toxic CO gas makes an efficient hood absolutely necessary. The final Cr(II) catalyst is extremely pyrophoric and should* never *be allowed to get into contact with air.*

Solvents are purified as follows: Dichloromethane is passed through a column (200-cm length, 3-cm diameter) with molecular sieve 4 Å, then heated to reflux over P_2O_5 for 1 day and subsequently distilled under argon; it is stored in a (100-cm length, 2-cm diameter) column, filled with molecular sieves 4 Å; pentane is degassed by evacuating and filling with argon for three times; for very sensitive compounds [e.g., $Mo(CO)_3Cp(FBF_3)$] it should be stored over Na–K alloy (prepared by mixing 3 g Na and 9 g K in 250 mL of xylene and refluxing for 1 day) and refluxed and distilled prior to use. The same procedure is used for hexane (Na–K alloy might here be replaced by potassium alone).

- **Caution.** *Potassium and especially Na–K alloy are extremely flammable in moist air and explode with liquid water. Potassium should be cut under paraffin oil and the alloy should be stored under xylene. The alloy can be transferred via a syringe that has been thoroughly dried. tert-Butyl alcohol is recommended for destroying unused portions of both K and Na–K alloy.*

Easy separation of solids from the solution can be achieved by using a centrifuge (Macrofuge C-4, Heraeus-Christ). The centrifuge can be provided with polyethylene blocks each having a hole just fitting the Schlenk tube.

These blocks are usually stored in Dry Ice so that for short periods of time, sufficiently low temperatures during centrifugation can be maintained.

Before the reaction, the Schlenk tube, provided with the magnetic stirring bar, is flamed in a high vacuum for at least 10 min with a Bunsen burner, and cooled in a stream of argon.

■ **Caution.** *Since metal carbonyls and their derivatives are toxic volatile compounds, all operations must be performed in an efficient hood.*

References

1. J. A. Davies and F. R. Hartley, *Chem. Rev.*, **81**, 79 (1981).

2. M. R. Rosenthal, *J. Chem. Educ.*, **50**, 331 (1973); A. P. Gaughan, Jr., Z. Dori, and J. A. Ibers, *Inorg. Chem.*, **13**, 1657 (1974).

3. K. Richter, E. O. Fischer, and C. G. Kreiter, *J. Organomet. Chem.*, **122**, 187 (1976); E. O. Fischer,.D. Wittmann, D. Himmelreich, U. Schubet, and K. Ackermann,*Chem. Ber.*, **115**, 3141 (1982).

4. R. Mews, *Angew. Chem. Int. Ed. Engl.*, **14**, 640 (1975); F. L. Wimmer and M. R. Snow, **31**, 267 (1978); R. Uson, V. Riera, J. Gimeno, M. Laguna, and M. P. Gamasa, *J. Chem. Soc. Dalton Trans.*, **1979**, 966; K. Raab, U. Nagel, and W. Beck, *Z. Naturforsch. Teil B*, **38**, 1466 (1983) and references cited therein.

5. W. Beck and K. Schloter, *Z. Naturforsch. Teil B*, **33**, 1214 (1978); K. Sünkel, U. Nagel, and W. Beck, *J. Organometal. Chem.*, **251**, 227 (1983) and references cited therein.

6. F. J. Regina and A. Wojcicki, *Inorg. Chem.*, **19**, 3803 (1980); G. Hartmann, R. Froböse, R. Mews, and G. M. Sheldrick, *Z. Naturforsch. Teil B*, **37**, 1234 (1982); P. Legzdins, D. T. Martin, Ch. R. Nurse, and B. Wassink, *Organometallics*, **2**, 1238 (1983).

7. B. M. Mattson and W. A. G. Graham, *Inorg. Chem.*, **20**, 3186 (1981).

8. B. Olgemöller, H. Bauer, H. Löbermann, U. Nagel, and W. Beck, *Chem. Ber.*, **115**, 2271 (1982).

9. D. L. Reger, C. J. Coleman, and P. J. Mc Elligott, *J. Organometal. Chem.*, **171**, 73 (1979); E. K. G. Schmidt and C. H. Thiel, *J. Organometal. Chem.*, **209**, 373 (1981); J. K. Hoyano, C. J. May, and W. A. G. Graham, *Inorg. Chem.*, **21**, 3095 (1982); and references cited therein.

10. P. Legzdins and D. T. Martin, *Organometallics*, **2**, 1785 (1983).

11. S. J. La Croce and A. R. Cutler, *J. Am. Chem. Soc.*, **104**, 2312 (1982); T. C. Forschner and A. R. Cutler, *Inorg. Synth.*, **26**, 240 (1989).

12. W. Beck and K. Sünkel, *Chem. Rev.*, **88**, 1405–1421 (1988).

13. D. F. Shriver and M. A. Drezdzon, *The Manipulation of Air Sensitive Compounds*, 2nd ed., McGraw-Hill, New York, 1986.

14. H. L. Krauss and H. Stach, *Z. Anorg. Allgem. Chem.*, **366**, 34 (1969).

18. CARBONYL(η^5-CYCLOPENTADIENYL)-(TETRAFLUOROBORATO)MOLYBDENUM AND TUNGSTEN COMPLEXES

Submitted by WOLFGANG BECK, KLAUS SCHLOTER, KARLHEINZ
SÜNKEL, and GÜNTER URBAN*
Checked by THOMAS FORSCHNER, ALICIA TODARO, and ALAN CUTLER[†]

A. TRICARBONYL(η^5-CYCLOPENTADIENYL)(TETRA-FLUOROBORATO)MOLYBDENUM AND -TUNGSTEN[1a]

An efficient method for the preparation of tetrafluoroborato complexes is hydride abstraction from metal hydrides using triphenylmethylium tetrafluoroborate.[1a] This method has been first reported by Sanders for hydridoruthenium complexes.[2]

$$[Ph_3C][BF_4] + MCp(CO)_3H \longrightarrow MCp(CO)_3(FBF_3) + Ph_3CH$$

$$M = Mo, W$$

In a similar way, the hexafluoroarsenato and hexafluoroantimonato complexes $MCp(CO)_3FEF_5$ (M = Mo, W; E = As, Sb) have been prepared from $MCp(CO)_3H$ and $[Ph_3C][EF_6]$.[1b]

An alternative method for the preparation of $MCp(CO)_3(FBF_3)$(M = Mo, W) is protonation of $MCP(CO)_3CH_3$ by $H[BF_4]\cdot Et_2O$.[1c]

Procedure

Tritylium tetrafluoroborate is commercially available (Fluka AG) and should be freshly recrystallized from dichloromethane or dichloromethane–ethyl acetate prior to use. The hydrido complexes, $MCp(CO)_3H^3$ should be purified by sublimation or by chromatography (neutral alumina, activity 3, pentane eluant) prior to use. All solvents must be rigorously dried and handled under an inert atmosphere, see the preceding general comments.

A quantity of $[Ph_3C][BF_4]$ (0.33 g, 1.0 mmol) is dissolved in 10 mL of CH_2CL_2 in a 50-mL Schlenk tube, under an inert atmosphere. The solution is cooled to $-40\,^\circ$C (using Dry Ice–acetone and a low-temperature thermometer). To this is added $MoCp(CO)_3H$ (Ref. 3) (0.22 g, 0.89 mmol) or

*Institut für Anorganische Chemie der Universität München, Meiserstr. 1, 8000 München 2, Federal Republic of Germany.
[†]Department of Chemistry, Rensselaer Polytechnic Institute, Troy, 12180-3590.

WCp(CO)$_3$H (Ref. 3) (0.30 g, 0.90 mmol). An immediate color change from yellow to purple-red is observed. After stirring for 10 min, 0.2 mL of the solution is syringed into an infrared (IR) solution cell and a spectrum is taken. If a more or less intense band is observed at $\sim 1355\,\mathrm{cm}^{-1}$, indicating the presence of unreacted tritylium salt, small amounts of the corresponding hydrides are then added via a spatula. After stirring for 5 min, the IR spectrum is recorded for another solution aliquot. The addition of hydride is repeated until the IR spectrum of the solution shows no band at $1355\,\mathrm{cm}^{-1}$. As soon as this equivalence point is reached, a sudden color change from dark red to lilac or violet is observed (see the solution in Section A). If this color change does not occur, the presence of moisture can be suspected. In this case the solution may be used for a reaction with stronger ligands than water, otherwise the preparation has to be tried again.

Two procedures are given for the treatment of the solution in Section A. In the first the solution is cooled down to $-60\,°\mathrm{C}$, and 20 mL of hexane is added. Careful evaporation under vacuum to $\sim 20\,\mathrm{mL}$ removes most of the CH$_2$Cl$_2$. The lilac precipitate is isolated by centrifugation ($\sim 2\,\mathrm{min}$ at 1500 rpm) and decanting off the solution. Hexane (20 mL) is added at $-60\,°\mathrm{C}$ and the suspension is stirred for 10 min. Centrifugation, decanting, and washing are repeated three times. Then the product is dried at $-20\,°\mathrm{C}$ for 8 h on a high-vacuum line (10^{-3} torr).

Alternate Procedure for Treatment of the Solution in Section A

The lilac colored reaction mixture is transferred into a second Schlenk flask (100 mL) using a double-ended stainless steel cannula. The second flask contains hexane previously cooled to $-78\,°\mathrm{C}$ (Dry Ice–acetone bath). A lilac colored solid precipitates. The solvent is siphoned off and the solid is washed three times with hexane (20 mL) previously cooled to $-78\,°\mathrm{C}$ and transferred into the flask using the double ended cannula technique. The wash solvent is siphoned off and remaining solid is dried under vacuum (10^{-3} torr, oil pump) at $-40\,°\mathrm{C}$ for 8 h. Yields: for MoCp(CO)$_3$(FBF$_3$): 282–319 mg (85–96%), for WCp(CO)$_3$(FBF$_3$): 357–378 mg (85–90%).

Anal. Calcd. for C$_8$H$_5$BF$_4$MoO$_3$: C, 28.95; H, 1.52. Found: C, 28.27; H, 1.64. Calcd. for C$_8$H$_5$BF$_4$O$_3$W: C, 22.28; H, 1.20. Found: C, 23.63; H, 1.39.

Properties*

---

*See Section B.

B.   DICARBONYL(η^5-CYCLOPENTADIENYL)(TETRAFLUORO-BORATO)(TRIPHENYLPHOSPHINE)MOLYBDENUM AND TUNGSTEN, MCp(CO)$_2$(PPh$_3$)(FBF$_3$)(M = Mo, W)[4]

Substitution of a CO group by a phosphine ligand makes the metal center electron richer and therefore less Lewis acidic. This weakens the coordination of the [BF$_4$]$^-$ ion. In addition, steric interactions with the phosphine ligands, the possibility of cis–trans isomerism in the complexes with "four legged piano stool" geometry,[5] and the introduction of the ^{31}P nucleus as another sensitive NMR probe make this variation of the synthesis described in Section A, an interesting field of further investigation. The preparation described here for the PPh$_3$ compounds, can also be used with other PR$_3$ ligands such as PMe$_3$, PEt$_3$, P(OPh)$_3$, or $\frac{1}{2}$(dppe)[dppe = 1, 2-ethanediyl-bis(diphenylphosphine)].[†]

Dicarbonyl(η^5-cyclopentadienyl)hydrido(triphenylphosphine) molybdenum and -tungsten

$$MCp(CO)_3H + PPh_3 \rightarrow MCp(CO)_2 PPh_3 H + CO$$

$$M = Mo, W$$

Monophosphine substituted carbonylcyclopentadienylhydrido complexes of molybdenum and tungsten have been obtained by protonation of the anions [MCp(CO)$_2$(PR$_3$)]$^-$,[6] or by substitution of CO with phosphines in the hydrides MH (CO)$_3$Cp.[7] The straightforward synthesis of the hydrides MH (CO)$_3$Cp (M = Mo, W)[3] makes the latter procedure preferable, at least for PPh$_3$, P(OPh)$_3$, PMe$_3$, and PEt$_3$, where fast reactions and good yields can always be obtained. For the analogous syntheses of PMe$_3$ or PEt$_3$ substituted hydrides, special precautions for handling these highly toxic, malodorous, and highly inflammable phosphines must be taken.[8]

Procedure

A quantity of freshly sublimed MCp(CO)$_3$H (0.49 g, M = Mo or 0.66 g, M = W, each 2.0 mmol) is dissolved in 15 mL of hexane at room temperature in a 100-mL Schlenk flask. To this is added PPh$_3$ (0.58 g, 2.1 mmol) under vigorous stirring. The Schlenk tube is then connected to a mercury bubbler and a stream of argon (0.5 L min^{-1}) is passed over the solution for 2 h (it is

---

[†]Commonly known as 1, 2-bis(diphenylphosphino)ethane.

not necessary to bubble the argon through the solution). Soon a white precipitate forms, which is isolated by filtration under argon and washed twice with 5 mL of hexane. The product is dried at room temperature for 6 h *in vacuo*. It may be recrystallized from CH_2Cl_2–hexane. Yields: $MoCp(CO)_2(PPh_3)H$, 625 mg, 65%; $WCp(CO)_2(PPh_3)H$, 636 mg (56%).

Properties

The hydrides $MCp(CO)_2(PPh_3)H$ are yellowish-white powders. They are air stable for several minutes exposure as solids, however, for extended storage they should be kept under argon. IR spectra (in CH_2Cl_2): $v_{(CO)} = 1936$, $1856\, cm^{-1}$ (Mo); 1923, $1835\, cm^{-1}$ (W); ^1H NMR [in CD_2Cl_2 (Mo), $CDCl_3$ (W)]: $\delta_{(C_5H_5)} = 5.08$ (Mo); 5.10 ppm (W); $\delta_{(M-H)} = -5.56$ (Mo), -7.06 ppm (W), "doublets" $^2J_{(^{31}P^1H)av} = 47\, Hz$ (Mo), 55 Hz (W); ^{31}P NMR (in CD_2Cl_2): $\delta_{(PPh_3)} = 74.3$ (Mo), 40.9 ppm (W) (relative to H_3PO_4). A fast equilibrium between the cis and trans isomers[5] leads to averaging of the signals and coupling constants at room temperature. Both isomers can be distinguished by low-temperature ^1H NMR [$\delta_{(Mo-H)}$: $-5.33d$ and $-6.14d$; $^2J_{(^{31}P^1H)} = 64$ and 21.4 Hz; $\delta_{(W-H)}$: $-6.90d$, $-7.36d$, $^2J_{(^{31}P^1H)} = 65$ and 22 Hz].

The analogous PMe_3 and PEt_3 containing hydrides tend to form oils and decompose quickly on contact with air; the tungsten compounds are more stable than the molybdenum analogs. Their spectral properties are similar to those of the PPh_3 compounds. The best yields are obtained with the $P(OPh)_3$ ligand, which leads exclusively to the stable cis isomers.[5]

Dicarbonyl(η⁵-cyclopentadienyl)(tetrafluoroborato)(triphenylphosphine)-molybdenum and -tungsten[4]

$$[Ph_3C][BF_4] + MCp(CO)_2(PPh_3)H \longrightarrow MCp(CO)_2(PPh_3)(FBF_3) + Ph_3CH$$

$$M = Mo, W$$

Procedure

Generally, the same guidelines as described in Section A have to be followed. A quantity of $MoCp(CO)_2(PPh_3)H$ (0.45 g, 0.94 mmol) or $WCp(CO)_2(PPh_3)H$ (0.55 g, 0.97 mmol) is added to a solution of $[Ph_3C][BF_4]$ (0.33 g, 1.00 mmol) in 10 mL CH_2Cl_2 at $-40\,°C$ contained in a Schlenk flask (100 mL) equipped with a magnetic stirring bar. The mixture is stirred for 20 min, after which an IR spectrum is recorded of an aliquot (0.2 mL) to inspect the intensity of the band at $1355 \, \text{cm}^{-1}$. Small amounts of the hydride are added until the IR spectrum, recorded at 5-min intervals shows no band at $1355 \, \text{cm}^{-1}$. Usually, a lilac precipitate forms before the equivalence point is reached. The equivalence point is again indicated by a lilac color of the solution. Complete precipitation of the product is obtained by addition of 20 mL of hexane at $-60\,°C$, or by transfer of the complete reaction mixture to another Schlenk flask containing the hexane cooled to $-60\,°C$ (see the procedure in Section A). Isolation of the product is the same as described in Section A. Yields: $MoCp(CO)_2(PPh_3)(FBF_3)\cdot 2CH_2Cl_2$ 632 mg. (86%); $WCp(CO)_2$-$(PPh_3)(FBF_3)$ 556 mg (85%).

Anal. Calcd. for $C_{25}H_{20}BF_4MoO_2P\cdot 2CH_2Cl_2$: C, 44.1; H, 3.29. Found: C, 45.0; H, 3.32.

Properties

All tetrafluoroborato complexes are very sensitive to moisture. Schlenk tubes used for storage therefore have to be heated to $400\,°C$ or more under vacuum for several hours; O-ring stopcocks or similar grease-free stopcocks are superior to the usual ground glass stopcocks. Although the phosphine containing BF_4 complexes are thermally more stable than the unsubstituted compounds, storage at temperatures below $-25\,°C$ under Ar is recommended for all these compounds. They dissolve in CH_2Cl_2 and $CHCl_3$ below $-40\,°C$ without decomposition, while solvents with donor properties like acetone or acetonitrile dissolve these complexes under substitution of tetrafluoroborato ligands by the solvent to give ionic complexes, for example, $[MoCp(CO)_3(\text{acetone})][BF_4]$. They can be characterized by their IR spectra in the region from 1200 to $700 \, \text{cm}^{-1}$ and by their low-temperature ^{19}F and, where appropriate, ^{31}P NMR spectra.[9]

TABLE I.   Spectroscopic Data of $MoCp(CO)_2L(FBF_3)$.

$L = CO$
IR: $v_{(CO)} = 2071, 1988 \, \text{cm}^{-1}$ (in CH_2Cl_2)
$v_{(^{11}BF)} = 1130, 884, 722 \, \text{cm}^{-1}$ (in Nujol)
^{19}F NMR:^{a} $-155d$ ($MoFBF_3$), $-370q$ ($MoFBF_3$), 95 Hz $[^2J_{(F-F)}]$
$L = PPh_3$
IR: $v_{(CO)} = 1991, 1903 \, \text{cm}^{-1}$ (in CH_2Cl_2)
$v_{(^{11}BF)} = 1119, 901, 732 \, \text{cm}^{-1}$ (in Nujol)
^{19}F NMR:^{a} $-155d$ ($MoFBF_3$), $-344q$, $-391q$ ($MoFBF_3$), 90 Hz $[^2J_{(F-F)}]$

$^a\delta$ in ppm, relative to $CFCl_3$, in CD_2Cl_2, $-80\,°C$.

TABLE II.   Spectroscopic Data of WCp(CO)$_2$L(FBF$_3$).

L = CO

IR: $v_{(CO)}$ = 2067, 1975 cm^{-1} (in CH$_2$Cl$_2$)

   $v_{(^{11}BF)}$ = 1149, 874, 704 cm^{-1} (in Nujol)

^{19}F NMR:[a] -153d (WFBF$_3$), -394q (WFBF$_3$), 99 Hz [$^2J_{(F-F)}$]

L = PPh$_3$

IR: $v_{(CO)}$ = 1988, 1963, 1877 cm^{-1} (in Nujol)

   $v_{(^{11}BF)}$ = 1148, 887, 720 cm^{-1} (in Nujol)

^{19}F NMR:[b] -156d (WFBF$_3$), -371q (WFBF$_3$), 98 Hz [$^2J_{(F-F)}$].

[a] δ in ppm, relative to CFCl$_3$, in CD$_2$Cl$_2$, -52 °C.
[b] δ in ppm, relative to CFCl$_3$, in CD$_2$Cl$_2$, -80 °C.

Coordination of the [BF$_4$] ion lowers the T_d symmetry of [BF$_4$]$^-$ and makes the fluorine atoms nonequivalent. Therefore the IR spectra show three instead of one $v_{(^{11}B-F)}$ absorptions[1,4] (Tables I and II); the low-temperature ^{19}F NMR spectra[9] show two distinct fluorine resonances, a high-field quartet (which may be split by coupling to the phosphorus in the PR$_3$ substituted compounds and a doublet at lower field, close to the resonance of free [BF$_4$]$^-$;[13] the ^{31}P NMR spectrum shows at low temperature a pseudodoublet, produced by coupling with the coordinated fluorine (Tables I and II). Compounds MoCp(CO)$_2$(PR$_3$)(FBF$_3$) are obtained as cis and trans isomers. In WCp(CO)$_2$(P(OPh)$_3$)(FBF$_3$) total isomerization from the pure cis hydride to the pure trans-BF$_4$ compound could be followed via NMR.[9]

Reactions of tetrafluoroborato complexes with ethylene, diphenylacetylene and acetone

General Remarks

The tetrafluoroborate ligand of these highly reactive complexes can be easily substituted by a series of N, O, P, and S σ donors[1a,4,10,11] and π donors (see Section C–F).

As described in Sections A and B, a lilac solution of the corresponding tetrafluoroborato complex is prepared at -30 °C in 10 mL of CH$_2$Cl$_2$. Complete reaction of the trityaborptionlium salt is verified by checking for the disappearance of the 1355 cm^{-1} band in the IR spectrum of the solution. This solution is used for the following reactions without isolation of the tetrafluoroborato complex.

C. TRICARBONYL(η^5-CYCLOPENTADIENYL)(η^2-ETHENE)-MOLYBDENUM(1 +) TETRAFLUOROBORATE(1 −)[1a]

$$MoCp(CO)_3(FBF_3) + C_2H_4 \longrightarrow [MoCp(CO)_3(\eta^2\text{-}C_2H_4)][BF_4]$$

The title compound can be obtained in three ways. One method starts from $Mo(C_5H_5)(CO)_3Cl$, which is reacted with C_2H_4 at a pressure of 70 bar in the presence of $AlCl_3$ and consecutive precipitation with ammonium salt.[12] A second method involves β-hydride abstraction from the ethyl group in $MoCp(CO)_3(C_2H_5)$ by $[Ph_3C][BF_4]$.[13] The third method, described here, has the advantage of mild reaction conditions and a good overall yield. Analogous complexes with other olefins have been prepared similarly.[4]

Procedure

Ethylene (1 bar), dried over P_2O_5, is bubbled through a vigorously stirred lilac solution of $MoCp(CO)_3(FBF_3)$ (1 mmol) in CH_2Cl_2 in a Schlenk flask (50 mL) cooled to − 30 °C. With continuous ethylene bubbling, the cooling bath is removed and the flask is permitted to warm up to + 20 °C over a 4-h period. The flow of ethylene is then stopped, and the reaction mixture stirred under argon for another 30 min. The yellow precipitate is isolated by centrifugation or filtration under Ar. After washing four times with 5 mL of CH_2Cl_2, the product is dried 1 h at + 40 °C on a high-vacuum line. It may be recrystallized from acetone–diethyl ether. Yield: 282 mg (79%).

Properties

The compound decomposes on heating at 102 to 108 °C. Infrared spectra (in Nujol): $\nu_{(CO)} = 2104$, 2053, 2001 cm^{-1}; ^{1}H NMR (acetone-d_6) $\delta = 6.35$ ppm (C_5H_5).

D. CARBONYL(η^5-CYCLOPENTADIENYL)BIS(DIPHENYL-ACETYLENE)MOLYBDENUM(1 +) TETRAFLUORO-BORATE(1 −)[1a]

$$MoCp(CO)_3(FBF_3) + 2PhCCPh \longrightarrow [MoCp(CO)(PhCCPh)_2][BF_4] + 2CO$$

Other syntheses of cationic bis(alkyne) complexes of molybdenum and tungsten of the same type include $AgBF_4$ oxidation of the dimer

[MoCp(CO)$_3$]$_2$ in CH$_2$Cl$_2$ in the presence of diphenylacetylene or several other alkynes.[14] Protonation of MoCp(CO)$_3$CH$_3$ with CF$_3$COOH and consecutive addition of 2-butyne in acetonitrile, followed by precipitation with a methanolic solution of [NH$_4$][PF$_6$] gives the corresponding 2-butyne complex.[15] Refluxing a solution of MoCp(CO)$_3$Cl with (HOCH$_2$)CC(CH$_2$OH) leads to an analogous compound.[16] The method described here uses very mild conditions and can be applied also for other alkynes, like 2-butyne or acetylene.[17]

Procedure

Diphenylacetylene (535 mg, 3.0 mmol) is added to a lilac solution of MoCp(CO)$_3$FBF$_3$ (1 mmol in 10 mL CH$_2$Cl$_2$, prepared as described above) at −30 °C under vigorous stirring in a Schlenk flask (50 mL) equipped with a magnetic stirring bar. The flask is connected to a mercury bubbler and flushed by a constant flow of argon or nitrogen gas. After 30 min, the gas flow is stopped and the cooling bath removed. Stirring is continued for 4 days at room temperature, over which time a yellow precipitate is formed. Diethyl ether (20 mL) is added and the yellowish red suspension is filtered under argon. The residue on the frit is extracted three times with 10 mL of CH$_2$Cl$_2$. The combined extracts are evaporated to 5 mL, to which is added diethyl ether (20 mL). The orange-yellow precipitate is isolated by centrifugation or filtration under argon, washed three times with 10-mL aliquots of diethyl ether, and then dried for 1 h *in vacuo* at 40 °C. The product may be recrystallized from CH$_2$Cl$_2$–pentane. Yield: 235 mg (37%).

Anal. Calcd. for C$_{34}$H$_{25}$BF$_4$MoO: C, 64.58; H, 3.99. Found: C, 63.98; H, 4.09.

Properties

The yellow compound is soluble in polar solvents such as CH$_2$Cl$_2$, acetone, or acetonitrile. Although prolonged exposure to air leads to decomposition, the compound can be handled in air for short periods of time. Its IR spectrum in Nujol shows one ν_{12CO} vibration at 2088 cm^{-1} and a weak ν_{13CO} band at 2040 cm^{-1}. Also a weak absorption at 1741 cm^{-1} occurs, which may be due to the $\nu_{(C≡C)}$ band of the coordinated alkyne. The ^{1}H NMR spectrum in CH$_2$Cl$_2$ has a sharp singlet for the C$_5$H$_5$ protons at $\delta = 6.20$ ppm, besides the broad resonance of the phenyl protons of the diphenylacetylene. Interestingly, KBr pellets of the compound several hours after initially formed, show a bathochromic shift of the $\nu_{(CO)}$ band, which is also observed with other cationic alkyne complexes.[17]

E. CARBONYL(η^5-CYCLOPENTADIENYL)(DIPHENYL-ACETYLENE)(TRIPHENYLPHOSPHINE)MOLYBDENUM(1 +) TETRAFLUOROBORATE(1 −)[18]

$$MoCp(CO)_2(PPh_3)(FBF_3) + PhCCPh$$

$$\longrightarrow [MoCp(CO)(PPh_3)(PhCCPh)][BF_4] + CO$$

Green and coworkers[19] prepared the title compound and other related monoalkyne complexes by reaction of the corresponding bis(alkyne) complex with triphenylphosphine (or other phosphines) in good yields. The method described here works for several alkynes, for example, 2-butyne or phenylacetylene, and also for phosphines, for example, PEt_3 or $P(OPh)_3$.

Procedure

Diphenylacetylene (1.78 g, 10.0 mmol) is added to a magnetically stirred lilac suspension of $MoCp(CO)_2PPh_3FBF_3$ (1.0 mmol in 10 mL of CH_2Cl_2, as described previously) in a Schlenk flask (50 mL) cooled to − 30 °C. The flask is connected to a mercury bubbler and purged with argon for 15 min. The gas flow is stopped and the cooling bath is allowed to warm up to room temperature. Stirring is continued for 2 days during which time the flask is purged several times with argon to remove the carbon monoxide evolved in the reaction. Then hexane (20 mL) is added. Stirring is continued for another day at ambient temperature, after which the dark green suspension is filtered under argon. The residue on the filter is washed four times with 15-mL aliquots of hexane and then dried 3 h under vacuum at 25 °C. Yield: 408 mg (57%).

Anal. Calcd. for $C_{38}H_{30}BF_4MoOP$: C, 63.7; H, 4.22. Found: C, 62.8; H, 4.15.

Properties

The title compound is soluble in polar organic solvents, for example, acetone, acetonitrile, or dichloromethane. Although storing under inert gas is recommended, no decomposition can be observed when handled as a solid in air for short periods of time. IR (in CH_2Cl_2): $v_{(^{12}CO)} = 1987\,cm^{-1}$; ^1H NMR (CD_2Cl_2): $\delta_{(C_5H_5)} = 5.77$ ppm, $\delta_{[C-C_6H_5,P(C_6H_5)_3]}$ 8–7 ppm; ^{31}P NMR (in CD_2Cl_2) $\delta_{(PPh_3)} = 54.8$ ppm.

The crystal structure of this compound shows a slightly elongated C≡C bond of the alkyne and the usual deviation from linearity at the two carbon atoms of the triple bond.[18,19]

F. (ACETONE)(TRICARBONYL)(η⁵-CYCLOPENTADIENYL)-MOLYBDENUM(1 +) AND -TUNGSTEN(1 +) TETRAFLUOROBORATE(1 −)[1a,10]

$$MCp(CO)_3(FBF_3) + (CH_3)_2CO \longrightarrow [MCp(CO)_3(OC(CH_3)_2)][BF_4]$$

$$M = Mo, W$$

Procedure

A lilac solution of the tetrafluoroborato complex $MCp(CO)_3(FBF_3)$, M = Mo or W (1 mmol in 10 mL of CH_2Cl_2) is prepared as indicated above in a Schlenk flask (50 mL) and cooled to − 30 °C. To this is added acetone (0.1 mL, 1.38 mmol). An immediate color change to red occurs, and stirring is continued for 3 h. Hexane (15 mL) is then added. The solution is cooled to − 78 °C (Dry Ice) and stored overnight giving a dark red precipitate. This is isolated by centrifugation and washed twice with 19 mL of hexane at 0 °C. Alternatively, the supernatant solution may be removed by a stainless steel cannula fitted with a sintered glass frit. The solids are washed with two aliquotes of cold (0 °C) hexane (19 mL), each removed by use of the stainless steel cannula fitted with the glass frit. The product is then dried for 6 h at 0 °C on a high vacuum line. Yield: $[MoCp(CO)_3(OC(CH_3)_2)][BF_4]$ 350 mg (90%); $[WCp(CO)_3(OC(CH_3)_2)][BF_4]$ 420 mg (88%).

Anal. Calcd. for $C_{11}H_{11}BF_4MoO_4$: C, 33.88; H, 2.84. Found: C, 33.87; H, 2.85. Calcd. for $C_{11}H_{11}BF_4O_4W$: C, 27.65; H, 2.32. Found: C, 26.90; H, 2.45.

Similar acetone complexes can also be prepared from the PPh_3 containing tetrafluoroborate complexes.

Properties

Solutions of the compounds in CH_2Cl_2 or acetone decompose at 20 °C within a short time, especially when traces of water are present. The solid compounds can be stored under argon at − 30 °C for several weeks without decomposition.

IR (CH_2Cl_2): $v_{(CO)} = 2072$, $1987\,cm^{-1}$ (Mo) IR (in Nujol): $v_{(CO)} = 2050$, $1930\,cm^{-1}$ (W); $v_{(M—O=C)} = 1660\,cm^{-1}$ (Mo), $1640\,cm^{-1}$ (W). ¹H NMR (in CD_2Cl_2): $\delta_{(C_5H_5)} = 6.11$ (Mo), 6.19 ppm (W); $\delta_{(CH_3)} = 2.39$ (Mo), 2.43 ppm (W).

References

1. (a) W. Beck and K. Schloter, *Z. Naturforsch. Teil B*, **33**, 1214 (1978); (b) K. Sünkel, U. Nagel, and W. Beck, *J. Organomet. Chem.*, **251**, 227 (1983); (c) M. Appel, K. Schloter, J. Heidrich, and W. Beck, *J. Organomet. Chem.*, **322**, 77 (1987).

2. J. R. Sanders, *J. Chem. Soc. Dalton Trans.*, **1972**, 1333.

3. (a) R. B. King and F. G. A. Stone, *Inorg. Synth.*, **7**, 107 (1963). E. O. Fischer, *Inorg. Synth.*, **7**, 136 (1963); (b) R. B. King, *Organometallic Syntheses*, Vol. 1, Academic Press, New York, 1965, p. 156; (c) W. P. Fehlhammer, W. A. Herrmann, and G. K. Öfele, in *Handbuch der Präparativen Anorganischen Chemie*, Vol. 2, 3rd ed., F. Enke Verlag, Stuttgart, 1981.

4. K. Sünkel, H. Ernst, and W. Beck, *Z. Naturforsch. Teil B*, **36**, 474 (1980).

5. J. W. Faller and A. S. Anderson, *J. Am. Chem. Soc.*, **92**, 5852 (1970).

6. (a) A. R. Manning, *J. Chem. Soc. A*, **1968**, 651; (b) M. J. Mays and S. M. Pearson, *J. Chem. Soc. A*, **1968**, 2291.

7. (a) A. Bainbridge, P. J. Craig, and M. Green, *J. Chem. Soc. A*, **1968**, 2715; (b) P. Kalck, R. Pince, R. Poilblanc, and J. Roussel, *J. Organomet. Chem.*, **24**, 445 (1970).

8. R. T. Markham, E. A. Dietz, Jr., and D. R. Martin, *Inorg. Synth.*, **16**, 153 (1976).

9. K. Sünkel, G. Urban, and W. Beck, *J. Organomet. Chem.*, **252**, 187 (1983); M. Appel and W. Beck, *J. Organomet. Chem.*, **319**, C1 (1987).

10. K. Sünkel, G. Urban, and W. Beck, *J. Organomet. Chem.*, **290**, 231 (1985).

11. K. Schloter and W. Beck, *Z. Naturforsch. Teil B*, **35**, 985 (1980); G. Urban, K. Sünkel, and W. Beck, *J. Organomet. Chem.*, **290**, 329 (1985).

12. E. O. Fischer and K. Fichtel, *Chem. Ber.*, **94**, 1200 (1961).

13. M. Cousins and M. L. H. Green, *J. Chem. Soc.*, **1963**, 889.

14. M. Bottrill and M. Green, *J. Chem. Soc. Dalton Trans.*, **1977**, 2365.

15. P. L. Watson and R. G. Bergman, *J. Am. Chem. Soc.*, **102**, 2698 (1980).

16. J. W. Faller and H. H. Murray, *J. Organomet. Chem.*, **172**, 171 (1979).

17. K. Schloter, K. Sünkel and W. Beck, unpublished results.

18. K. Sünkel, U. Nagel, and W. Beck, *J. Organomet. Chem.*, **222**, 251 (1981).

19. S. R. Allen, P. K. Baker, S. G. Barnes, M. Green, L. Trollope, L. Manojlovic-Muir, and K. W. Muir, *J. Chem. Soc. Dalton Trans.*, **1981**, 873.

19. PENTACARBONYL(TETRAFLUOROBORATO)RHENIUM AND -MANGANESE AND REACTIONS THEREOF

Submitted by WOLFGANG BECK and KLAUS RAAB*
Checked by J. R. SHAPLEY and B. R. WHITTLESEY†

Highly reactive pentacarbonylmanganese and rhenium complexes with weakly coordinated anions include $M(CO)_5(FAsF_5)$ (Ref. 1), $M(CO)_5(OClO_3)$ (Ref. 2), $M(CO)_5(OPOF_2)$ (Ref. 2), $M(CO)_5(OSO_2CF_3)$(Ref. 3), $M(CO)_5OTeF_5$ (Ref. 4) (M = Mn, Re), and $Mn(CO)_5(O_2CCF_3)$ (Ref. 5), which are usually prepared from pentacarbonyl

*Institut für Anorganische Chemie, Universität München, Meiserstr. 1, 8000 München 2, Federal Republic of Germany.
†452 Noyes Laboratory, Box 20, 505S. Mathews, University of Illinois Urbana, IL 61801.

halides and the silver salt of the corresponding anion. In the following procedures, the corresponding tetrafluoroborates and some of their reactions are described. These tetrafluoroborato complexes $M(CO)_5(FBF_3)$ (M = Mn, Re) are accessible from the corresponding methyl complexes and triphenylmethylium tetrafluoroborate[6] or tetrafluoroboric acid, respectively.[7] Interestingly, methyl metal compounds may react with the triphenylmethylium ion by abstraction of the methyl group[6,8] or by abstraction of hydride to give methylene carbene complexes.[9]

A. PENTACARBONYLMETHYLRHENIUM[10]

$$Re_2(CO)_{10} + 2Na(Hg) \longrightarrow 2Na[Re(CO)_5]$$

$$Na[Re(CO)_5] + CH_3I \longrightarrow Re(CH_3)(CO)_5 + NaI$$

Procedure

■ **Caution.** *Pentacarbonylmethylrhenium is a volatile metal carbonyl derivative. Metal carbonyls usually are very toxic and must be handled in a well-ventilated hood.*

The starting material $Re_2(CO)_{10}$ may be purchased either from Strem or from Pressure Chemicals. The reactions are conducted in Schlenk tubes under a dry argon or nitrogen atmosphere. Sodium amalgam is prepared by addition of sodium metal (0.3 g) to 3 mL of mercury under nitrogen.

■ **Caution.** *The dissolution of sodium metal in mercury is an exothermic reaction; therefore sodium must be added in small pieces.*

A quantity of $Re_2(CO)_{10}$(3.00 g, 4.6 mmol) is dissolved in 10 to 12 mL of dry tetrahydrofuran (THF) previously saturated with argon or nitrogen in a Schlenk flask (100 mL). After all the solid has dissolved, the flask is cooled to 0 °C, and the solution is transferred to the sodium amalgam, also in a Schlenk flask (100 mL) equipped with a stirring bar and cooled to 0 °C. The mixture is stirred for 60 min at 0 °C and for another hour at room temperature. A third Schlenk flask (100 mL) is flushed with argon or nitrogen. The red, air sensitive solution obtained in the previous step is transferred away from the excess sodium amalgam into the third Schlenk flask, using Teflon tubing passing from one flask to the next through rubber septa. A light over pressure is applied over the solution to be transferred. The transferred solution is cooled to ~ -25 °C (a Dry Ice–2-propanol bath) (no precipitate of $Na[Re(CO)_5]$ should be formed) and iodine-free iodomethane (0.6 mL, 9.7 mmol) is added dropwise.

■ **Caution.** *Iodomethane is volatile and carcinogenic.*

After stirring for 10 min the solution is warmed up to room temperature.

After stirring for another 75 min the solvent is evaporated at $-20\,°C$ under vacuum. A trap cooled with liquid nitrogen is placed between the Schlenk tube and the pump. To remove the last traces of THF the yellow residue is dried for a short period in an oil pump vacuum. Finally, the yellow residue is sublimed *in vacuo* at 30 to 40 °C for 2 to 3 days. Since $Re(CO)_5(CH_3)$ is volatile, the stopcock between the sublimation apparatus and the pump is opened only briefly several times. Yield: 2.3–2.9 g (60–76%).

Another 1 to 5% of $Re(CO)_5(CH_3)$ can by isolated by adding 300 mL of water to the THF in the trap at room temperature. The formed precipitate is washed with water and dried over P_4O_{10} in a small evacuated desiccator.

Properties

Pentacarbonylmethylrhenium is a colorless, volatile solid. It is air and moisture stable and soluble in most organic solvents. The IR shows CO bands at 2129 (w), 2012 (s), 1975 cm^{-1} (in CH_2Cl_2). The structure of $Re(CO)_5(CH_3)$ has been determined by electron diffraction.[11]

B. PENTACARBONYL(TETRAFLUOROBORATO)RHENIUM(I)[6,7]

Method a

$$Re(CO)_5(CH_3) + H[BF_4]Et_2O \longrightarrow Re(CO)_5(FBF_3) + CH_4 + Et_2O$$

Method b

$$Re(CO)_5(CH_3) + Ph_3C[BF_4] \longrightarrow Re(CO)_5(FBF_3) + H_3CCPh_3$$

Procedure

See *General Remarks* for this chapter for the preparation and handling of tetrafluoroborato complexes.

■ **Caution.** *Pentacarbonylmethylrhenium is a volatile metal carbonyl derivative. Metal carbonyls usually are very toxic and must be handled in a well-ventilated hood. Tetrafluoroboric acid is a very corrosive chemical. Contact with the skin has to be avoided. Gloves should be worn.*

The reagents may be purchased as indicated: Tetrafluoroboric acid from Merck; $[Ph_3C][BF_4]$ from Fluka.

Method a

A quantity of $Re(CO)_5(CH_3)$ (342 mg, 1.0 mmol) is dissolved in 3 mL of dichloromethane in a dried Schlenk flask (50 mL). To the stirred solution tetrafluoroboric acid (54% in diethyl ether, $d = 1.18\,\mathrm{g\,cm^3}$; 138, 5 μL, 1.0 mmol) is added at room temperature using a plastic micropipette. The tetrafluoroboric acid solution in diethyl ether is transferred from the original bottle under an atmosphere of argon. Excess tetrafluoroboric acid should be avoided. After addition of $H[BF_4]\cdot Et_2O$ to the solution a vigorous evolution of methane occurs and a colorless precipitate is formed which—after 20 min—is centrifugated off or collected on a glass frit, washed several times each with 5 mL of dichloromethane, and dried in a high vacuum (2 h). Yield: 380–401 mg (92–97%).

Anal. Calcd. for $C_5BF_4O_5Re$ (MW 413.0): C, 14.54. Found: C, 14.17.

Properties

The colorless compound $Re(CO)_5(FBF_3)$ is very sensitive to moisture. On exposure to moist air pentacarbonyl(trifluorohydroxoborato)rhenium, $Re(CO)_5(HOBF_3)$ is formed. The complex $Re(CO)_5(FBF_3)$ is only sparingly soluble in dichloromethane. The coordinated $[BF_4]^-$ ligand shows the following v_{B-F} stretching bands in the IR spectrum ($v_{10_{B-F}}$): 1203 (m), 1172 (sh), 930 (m), 757 (m); ($v_{11_{B-F}}$): 1162 (s), 1128 (s), 902 (s), 738 cm^{-1} s in Nujol). The three $v_{(CO)}$ bands (in Nujol) at 2165 (w, A_1), 2055 (vs, br, E), 2014 (s, A_1) cm^{-1} are characteristic for the $Re(CO)_5$ group. The complexes $Re(CO)_5(FBF_3)$ and $Re(CO)_5(FAsF_5)$[1] undergo many reactions with anionic and neutral donor ligands, usually under substitution of tetrafluoroborate.[6,12,13]

Ionic complexes are formed under mild conditions with various soft and hard neutral donor molecules:

$$Re(CO)_5(FBF_3) + L \longrightarrow [(OC)_5ReL][BF_4]$$

(L = CO, ethylene, propene, pent-1-ene, butadiene, H_2S, THF, acetone, acetonitrile, nitromethane). The moiety $[Re(CO)_5]^+$ can also be added to a nucleophilic atom of a coordinated ligand, which provides a systematic way to prepare ligand-bridged complexes,[12] for example,

$$(OC)_5Re\!-\!O(H)C\!\!=\!\!O + Re(CO)_5(FBF_3)$$
$$\longrightarrow [(OC)_5ReOC(H)ORe(CO)_5][BF_4]$$
$$[Au(CN)_2]^- + 2Re(CO)_5(FBF_3)$$
$$\longrightarrow [(OC)_5Re\!-\!NCAuCN\!-\!Re(CO)_5]^+[BF_4]^- + [BF_4]^-$$

The complex $Re(CO)_5(FBF_3)$ also reacts with but-2-yne or pent-2-yne via cyclodimerization of the alkyne to give complexes with a coordinated methylenecyclobutene derivative.[7]

The compound $Re(CO)_5(FBF_3)$ is soluble in water to give a solution of $[Re(CO)_5(OH_2)]^+$. A series of water insoluble neutral pentacarbonylrhenium derivatives has been obtained from aqueous solutions of pentacarbonyl-(tetrafluoroborato)rhenium and various salts[12]:

$$[Re(CO)_5(OH_2)]^+ + X^- \xrightarrow[-H_2O]{} Re(CO)_5X$$

$$X^- = Cl^-, Br^-, I^-, NO_2^-, NO_3^-, OOCH^-, NCO^-, SCN^-, SeCN^-, RS^-,$$
$$Au(CN)_2^-, \tfrac{1}{2}Pt(CN)_4^{2-}, \tfrac{1}{2}C_4O_4^{2-}, \tfrac{1}{2}C_4S_4^{2-}$$

C. PENTACARBONYL(η^2-ETHENE)RHENIUM(1 +) TETRAFLUOROBORATE(1 −)[6]

The complexes $[M(CO)_5(\eta^2\text{-}C_2H_4)][AlCl_4]$ (M = Mn, Re) were first prepared by abstraction of the chloride ligand in $M(CO)_5Cl$ using aluminum-trichloride under ethylene pressure.[14] The preparation of $[BF_4]^-$ salts of these cationic pentacarbonylethene complexes of manganese and rhenium proceeds under very mild conditions (1 bar) and gives high yields.[6]

$$Re(CO)_5(FBF_3) + C_2H_4 \longrightarrow [Re(CO)_5(\eta^2\text{-}C_2H_4)][BF_4]$$

Procedure

Moisture has to be carefully excluded. A quantity of $Re(CO)_5(FBF_3)$ (1.21 g, 2.93 mmol) is weighed into a Schlenk tube (100 mL) under a dry argon atmosphere. A magnetic stirring bar is added. The Schlenk tube is connected with a mercury bubbler and evacuated and flushed with ethene several times. [Drying of ethene with molecular sieve 4 Å is not necessary since it is not used in large excess in this procedure]. Dried dichloromethane (10 mL) is added with a pipette under a flush of ethene. The suspension is stirred magnetically for 1 to 2 days. If the mercury bubbler shows reduced pressure the tube is again filled with ethene (1 bar). The complex is centrifugated off or collected on a glass frit, washed with dichloromethane and dried in a high vacuum. Yield: 1,23–1,29 g (95–100%).

The complex can be dissolved in acetone at −20 °C and precipitated by addition of dichloromethane or diethyl ether.

Anal. Calcd. for $C_7H_4BF_4O_5Re$ (MW 441.1): C, 19.06; H, 0.91. Found: C, 19.21; H, 0.78.

The complex $[Re(CO)_5(\eta^2\text{-}C_2H_4)][BF_4]$ may also be obtained directly from the reaction of $[Ph_3C][BF_4]$ with $Re(CO)_5(CH_3)$ in an ethene atmosphere. Reaction time: 3–4 days. Yield: (95–98%).

Properties

The colorless complex is nearly insoluble in dichloromethane and readily soluble in acetone, acetonitrile, and nitromethane. In these solvents ethene is very slowly substituted to give $[Re(CO)_5(\text{solvent})][BF_4]$. IR (in Nujol): 2174(m), 2055(s, br) (ν_{CO}); 1055(vs) (ν_{BF_4}); 1538(vw) $(\nu_{C=C})\,cm^{-1}$. IR (in CH_3NO_2): 2172(m), 2071(s) cm^{-1} (ν_{CO}). ^1H NMR (acetone-d_6, i-TMS): $\delta =$ 5.12 ppm (singlet). The complex $[Re(CO)_5(\eta^2\text{-}C_2H_4)][BF_4]$ has been used for the preparation of the ethene bridged complex $(OC)_5ReCH_2CH_2Re(CO)_5$ via nucleophilic attack of pentacarbonylrhenate(I−) at the coordinated ethene of $[Re(CO)_5(C_2H_4)]^{+}$.[6]

Other alkene complexes $[Re(CO)_5(\text{alkene})][BF_4]$ (alkene = propene, pent-1-ene, buta-1,3-diene) can also be obtained from $Re(CO)_5(FBF_3)$ and alkene.[13]

D. DI-μ_3-(CARBON DIOXIDE)OCTADECACARBONYL-TETRARHENIUM[15]

The classic "Hieber-base reaction"[16] is that of a hydroxide with metal carbonyls, which proceeds by nucleophilic attack of the hydroxide at a carbon atom of a carbonyl ligand to give a carboxy group or consequently carbon dioxide and a metal hydride.[17] Metal carbonyls are catalysts for the water–gas shift reaction.[18] Pentacarbonyl(tetrafluoroborato)rhenium reacts with alkali hydroxide in a similar way; however, due to the coordinatively unsaturated nature of the $[Re(CO)_5]^{+}$ group polynuclear compounds are formed.[15]

$$Re(CO)_5(FBF_3) + OH^- \xrightarrow{\;H_2O\;} [Re(CO)_4(COOH)]_n$$

Tetracarbonyl(carboxy)rhenium

Procedure

The preparation can be carried out in air. A quantity of $Re(CO)_5(FBF_3)$ (0.85 g, 2.06 mmol) is dissolved in water (15 mL) in a Schlenk flask (25 mL). The solution is filtered into a second flask (25 mL) equipped with a magnetic stirring bar. Aqueous NaOH (2.15 mL of a 1 M solution) is added to the filtrate under stirring. After a few minutes the colorless precipitate is collected on a glass frit and washed several times with water. The solid is dried over P_4O_{10} for 2 days and for 20 h in a high vacuum. Yield: 643–693 mg (91–98%).

Anal. Calcd. for C_5HO_6Re: C, 17.50; H, 0.29. Found: C, 17.33; H, 0.36.

Properties

IR (KBr): 2145 (m), 2098 (m), 2073 (m), 2050 (sh), 2030 (vs), 1981 (vs), 1963 (vs), 1900 (vs), 1870 (vs), (v_{CO}); 1458 (s), 1180 (s), 1165 (sh) (v_{CO_2}); 3270 (m, br), 2900–2850 cm^{-1} (v_{OH}, fluorinated Nujol).

Di-μ_3-(carbon dioxide)octadecacarbonyltetrarhenium

A quantity of $[Re(CO)_4(COOH)]_n$ (0.63 g) is dissolved in acetone (40 mL) in a Schlenk flask (100 mL). The pale yellow solution is filtered quickly, if necessary. After a few minutes a colorless precipitate is formed. The suspension is stirred for 30 to 40 min and the solvent is evaporated *in vacuo* to ~ 4 mL. The solid is centrifuged off or collected on a glass frit, washed two times each with 2 mL of acetone, and dried in a high vacuum. Yield: 344 mg (56%).

Anal. Calcd. for $C_{20}O_{22}Re_4$ (MW 1337.0): C, 17.97. Found: C, 18.15.

Properties

The colorless CO_2 bridged complex is stable in air and soluble in THF. It is only slightly soluble in acetone and dichloromethane. IR (KBr): 2147 (m), 2088 (m), 2084 (sh), 2054 (s), 2033 (s), 1987 (s), 1968 (s), 1921 (s), 1898 (w) (v_{CO}); 1379 (s), 1294 (m), 1259 (w) cm^{-1} (v_{CO_2}). The X-ray structure of the complex has been determined.[15]

References

1. R. Mews, *Angew. Chem.*, **87**, 669 (1975); *Angew. Chem. Int. Ed. Engl.*, **14**, 640 (1975); M. Oltmanns and R. Mews, *Z. Naturforsch. Teil B*, **35**, 1324 (1980); G. Hartmann, R. Froböse, R. Mews, and G. M. Sheldrick, *Z. Naturforsch. Teil B*, **37**, 1234 (1982).

2. F. L. Wimmer and M. R. Snow, *Aust. J. Chem.*, **31**, 267 (1978); E. Horn and M. R. Snow, *Aust. J. Chem.*, **33**, 2369 (1980); R. Uson, V. Riera, J. Gimeno, M. Laguna, and M. P. Gamasa, *J. Chem. Soc. Dalton Trans.*, **1979**, 996.

3. W.C. Trogler, *J. Am. Chem. Soc.*, **101**, 6459 (1979); W. C. Trogler, *Inorg. Synth.*, **26**, 113 (1989).

4. K. D. Abney, K. M. Long, O. P. Anderson, and S. H. Strauss, *Inorg. Chem.*, **26**, 2638 (1987).

5. F. A. Cotton, D. J. Darensbourg, and B. W. S. Kolthammer, *Inorg. Chem.*, **20**, 1287 (1981).

6. K. Raab, B. Olgemöller, K. Schloter, and W. Beck, *J. Organomet. Chem.*, **214**, 81 (1981); K. Raab, U. Nagel, and W. Beck, *Z. Naturforsch. Teil B*, **38**, 1466 (1983).

7. K. Raab and W. Beck, *Chem. Ber.*, **117**, 3169 (1984).

8. P. J. Harris, S. A. R. Knox, R. J. McKinney, and F. G. A. Stone, *J. Chem. Soc. Dalton Trans.*, **1978**, 1009.

9. A. T. Patton, C. E. Strouse, C. B. Knobler, and J. A. Gladysz, *J. Am. Chem. Soc.*, **105**, 5804 (1983).

10. W. Hieber and G. Braun, *Z. Naturforsch. Teil B*, **14**, 132 (1959); W. Hieber, G. Braun, and W. Beck, *Chem. Ber.*, **93**, 901 (1960).

11. D. W. H. Rankin and A. Robertson, *J. Organomet. Chem.*, **105**, 331 (1976).

12. K. Raab and W. Beck, *Chem. Ber.*, **118**, 3830 (1985).

13. W. Beck, K. Raab, U. Nagel, and W. Sacher, *Angew. Chem. Int. Ed. Engl.*, **24**, 505 (1985).

14. E. O. Fischer and K. Fichtel, *Chem. Ber.*, **94**, 1200 (1961); E. O. Fischer, K. Fichtel, and K. Öfele, *Chem. Ber.*, **95**, 249 (1962); E. O. Fischer und K. Öfele, *Angew. Chem.*, **73**, 581 (1961); **74**, 76 (1962); A. M. Brodie, G. Hulley, B. F. G. Johnson, and J. Lewis, *J. Organomet. Chem.*, **24**, 201 (1970).

15. W. Beck, K. Raab, U. Nagel, and M. Steimann, *Angew. Chem.*, **94**, 556 (1982); *Angew. Chem. Int. Ed. Engl.*, **21**, 526 (1982).

16. W. Hieber and F. Leutert, *Z. Anorg. Allg. Chem.*, **204**, 145 (1932).

17. Th. Kruck, M. Höfler, and M. Noack, *Chem. Ber.*, **99**, 1153 (1966).

18. D. J. Darensbourg and A. Rokiki, *Organometallics*, **1**, 1685 (1982); R. M. Laine and E. J. Crawford, *J. Mol. Catal.*, **44**, 357 (1988).

20. MANGANESE(I) AND RHENIUM(I) PENTACARBONYL(TRIFLUOROMETHANESULFONATO) COMPLEXES

Submitted by STEVEN P. SCHMIDT,* JAY NITSCHKE,* and WILLIAM C. TROGLER[†]
Checked by SARAH INMAN HUCKETT[‡] and ROBERT J. ANGELICI[‡]

Metal complexes that contain a weakly basic hard donor ligand coordinated to a soft transition metal center exhibit high reactivity. Poor ligands such as

*Department of Chemistry, Northwestern University, Evanston, IL 60201.
[†]Department of Chemistry, University of California, San Diego, La Jolla, CA 92093.
[‡]Department of Chemistry, Iowa State University, Ames, IA 50011.

water, alcohols, and ethers can displace ligands such as $[ClO_4]^-$, $[AsF_6]^-$, and $[BF_4]^-$ from low-valent metal complexes.[1]

- **Caution.** *Perchlorate, $[ClO_4]^-$ complexes are highly explosive and their use is not recommended under any condition.*

The trifluoromethanesulfonato (triflate) ligand, which is known to be an excellent leaving group in organic[2] and coordination[3] chemistry, has not been widely used in synthetic studies of low-valent metal complexes. Stuhl and Muetterties[4] reported the formation (17% yield) of $Mn(O_3SCF_3)[P(OCH(CH_3)_2)_3]_2(CO)_2$ from the reaction between $Mn(\eta^3\text{-}C_3H_5)[P(OCH(CH_3)_2)_3]_2(CO)_2$ and CF_3SO_3H. Humphrey et al.[5] prepared $Fe[\eta^5\text{-}C_5(CH_3)_5](CO)_2(O_3SCF_3)$ in 50% yield from the reaction between $Fe[\eta^5\text{-}C_5(CH_3)_5](CO)_2[CH(C_6H_5)OCH_3]$ and $(CH_3)_3SiO_3SCF_3$. The complex $Mn(CO)_5(O_3SCF_3)$ has been reported[6a] to form in neat HO_3SCF_3 on reaction with $Mn(CO)_5H$ or $[Mn(CO)_5(NC_5H_5)]^+$, and from the reaction[6b] between $Mn(CO)_5Br$ and AgO_3SCF_3. High yield syntheses and physical properties of $M(CO)_5(O_3SCF_3)$ (M = Mn and Re), prepared by metathesis of $M(CO)_5Br$ with AgO_3SCF_3 in CH_2Cl_2 solvent, are described here.

A. PENTACARBONYL(TRIFLUOROMETHANESULFONATO)-MANGANESE(I), $Mn(CO)_5(O_3SCF_3)$

$$Mn(CO)_5Br + AgO_3SCF_3 \xrightarrow{CH_2Cl_2} Mn(CO)_5(O_3SCF_3) + AgBr$$

Procedure

Freshly sublimed bromopentacarbonylmanganese[7] (0.84 g, 3.0 mmol) is placed in a 50-mL Schlenk flask along with a Teflon-coated stirring bar under an atmosphere of nitrogen. Dichloromethane (40 mL freshly distilled from P_4O_{10} under nitrogen) is added through a septum by syringe, and the solution is stirred until all the $Mn(CO)_5Br$ is dissolved. From a Schlenk addition tube 0.98 g (3.8 mmol) of $Ag(O_3SCF_3)$ (Aldrich) is added under a purge of N_2, and stirring is continued for 1 h at room temperature. The $Ag(O_3SCF_3)$ should be weighed under subdued light and the reaction flask wrapped with foil to exclude room light just before addition of silver triflate. Although $Ag(O_3SCF_3)$ is slightly soluble in CH_2Cl_2, the reaction occurs soon after mixing. The fluffy AgBr precipitate is removed by filtration through a fine Schlenk frit (or through a medium frit covered with Kieselguhr) to yield a clear, orange-yellow solution. The solution volume is reduced to ~ 20 mL under vacuum, allowed to warm to room temperature, and 20 mL of hexane (freshly distilled from sodium benzophenone ketyl under N_2) is carefully layered onto the CH_2Cl_2 solution. The reaction flask is wrapped in foil and allowed to stand for 2.5 h after which time large orange-yellow, needle-shaped crystals form. Separation by filtration affords the air- and light-sensitive product $Mn(CO)_5(O_3SCF_3)$. Yield: 0.89 g, 2.6 mmol, (87% yield).

Anal. Calcd. for $Mn(CO)_5(O_3SCF_3)$: C, 20.94; S, 9.32; Mn, 15.98. Found: C, 20.38; S, 9.39; Mn, 15.83.

Properties

Pentacarbonyl(trifluoromethanesulfonato)manganese(I) is an orange-yellow compound that is sensitive to light and air and partially decomposes on sublimation to yield $Mn_2(CO)_{10}$. It dissolves readily in CH_2Cl_2, is insoluble in aromatic and aliphatic hydrocarbons and decomposes, with evolution of CO, in diethyl ether, acetone, and tetrahydrofuran (THF). The infrared spectrum in CH_2Cl_2 exhibits CO stretching absorptions at 2158 (w), 2073 (s), 2042 (w), and 2020 (m) cm^{-1}, consistent with the expected pseudo-C_{4v} symmetry at the metal. The v_{SO} stretching absorptions at 1013 (m), 1179 (m), and 1336 (s) cm^{-1} are as expected[9] for unidentate binding of the triflate ligand, and peaks attributable to v_{C-F} are observed at 1202 (s) and 1234 (m) cm^{-1}. The reaction between $Mn(CO)_5(O_3SCF_3)$ and THF has been shown[10b] to yield $Mn(CO)_3(THF)_2(O_3SCF_3)$ by a first-order pathway. Multiple substitution occurs rapidly after formation of monosubstituted species. Acetonitrile and $P(n-Bu)_3$ substitute triflate by an associative reaction pathway.[10a] Thiophene adds to the complex under reflux[6b] to yield $[Mn(\eta-C_4H_4S)(CO)_3]^+$.

B. PENTACARBONYL(TRIFLUOROMETHANESULFONATO)-RHENIUM(I), $Re(CO)_5(O_3SCF_3)$

$$Re(CO)_5Br + Ag(O_3SCF_3) \xrightarrow[CH_2Cl_2]{} Re(CO)_5(O_3SCF_3) + AgBr$$

Procedure

This complex is prepared from bromopentacarbonyl rhenium[8] (1.22 g, 3.0 mmol) and $Ag(O_3SCF_3)$ (0.98 g, 3.8 mmol) by a procedure analogous to that employed for the manganese compound. The reaction is complete after 2.5 h of stirring at room temperature. After adding 20 mL of hexane to the filtered and concentrated reaction solution at room temperature the CH_2Cl_2 is removed slowly *in vacuo* to precipitate $Re(CO)_5(O_3SCF_3)$ as a white powder. Yield: 1.14 g, 80%, (the checkers obtained 68% yield on the initial filtration and 10% more from the reaction flask).

Anal. Calcd. for $Re(CO)_5(O_3SCF_3)$; C, 15.16; F, 11.99. Found: C, 15.20; F, 12.48.

Properties

Pentacarbonyl(trifluoromethanesulfonato)rhenium(I) is a white solid (stable to room light) with solubility properties similar to those of the manganese derivative. In contrast to the manganese compound, small quantities (0.1 g) of $Re(CO)_5(O_3SCF_3)$ can be sublimed at 55 °C and 0.1 torr. The complex exhibits a parent ion in the mass spectrum at 476 m/e. The IR spectral data in CH_2Cl_2 [ν_{CO} 2166 (w), 2059 (s), 2031 (w), 2004 (m) cm^{-1}; ν_{SO} 1006 (m), 1174 (m), 1343 (m) cm^{-1}; ν_{CF} 1202 (s), 1235 (m)] are similar to those of the manganese analog. Unlike the manganese derivative, $Re(CO)_5(O_3SCF_3)$ is inert toward THF, diethyl ether, acetone, and methanol. Nucleophiles such as CH_3CN or $P(n\text{-Bu})_3$ substitute triflate according to a second-order rate law[10a] and multiple substitution {e.g., formation of $Re(CO)_3[P(n\text{-Bu})_3]_3^+$} occurs at long reaction times. The susceptibility of the triflate complexes to nucleophilic substitution permitted synthesis of $(OC)_5MnRe(^{13}CO)_5$ by facile addition of $KMn(CO)_5$ to $Re(^{13}CO)_5(O_3SCF_3)$ at room temperature.[11]

References

1. (a) F. L. Wimmer and M. R. Snow, *Aust. J. Chem.*, **31**, 267 (1978); (b) R. Usón, V. Rivera, J. Gimeno, M. Laguna, and M. P. Gamasa, *J. Chem. Soc., Dalton Trans.*, **1979**, 996; (c) E. Horn and M. R. Snow, *Aust. J. Chem.*, **33**, 2369 (1980); (d) M. R. Snow and F. L. Wimmer, *Inorg. Chim. Acta*, **44**, L189 (1980); (e) M. Oltmanns and R. Mews, *Z. Naturforsch.*, **35b**, 1324 (1980); (f) I. Mitteil, W. Beck, and K. Schloter, *Z. Naturforsch.*, **33b**, 1214 (1978); (g) B. M. Mattson and W. A. G. Graham, *Inorg. Chem.*, **20**, 3186 (1981); (h) K. Raab, B. Olegmoler, K. Schloter, and W. Beck, *J. Organomet. Chem.*, **214**, 81 (1981); (i) K. Schloter and W. Beck, *Z. Naturforsch.*, **35b**, 985 (1980); (j) E. O. Fischer and F. J. Gammel, *Z. Naturforsch.*, **34b**, 1183 (1979); (k) K. Richter, E. O. Fischer, and C. G. Kreiter, *J. Organomet. Chem.*, **122**, 187 (1976); (l) E. O. Fischer, S. Walz, A. Ruhs, and F. R. Kreissl, *Chem. Ber.*, **111**, 2765 (1978).

2. P. J. Stang and M. R. White, *Aldrichimica Acta*, **16**, 15 (1983).

3. (a) P. A. Lay, R. H. Magnuson, J. Sen, and H. Taube, *J. Am. Chem. Soc.*, **104**, 7658 (1982); (b) N. E. Dixon, W. G. Jackson, M. J. Lancaster, G. A. Lawrance, and A. M. Sargeson, *Inorg. Chem.*, **20**, 470 (1981); (c) D. A. Buckingham, P. J. Cresswell, A. M. Sargeson, and W. G. Jackson, *Inorg. Chem.*, **20**, 1647 (1981); (d) N. E. Dixon, W. G. Jackson, G. A. Lawrance, and A. M. Sargeson, *Inorg. Synth.*, **22**, 103 (1983).

4. L. S. Stuhl and E. L. Muetterties, *Inorg. Chem.*, **17**, 2148 (1978).

5. M. B. Humphrey, W. M. Lamanna, M. Brookhart, and G. R. Husk, *Inorg. Chem.*, **22**, 3355 (1983).

6. (a) W. C. Trogler, *J. Am. Chem. Soc.*, **101**, 6459 (1979); (b) D. A. Lesch, J. W. Richardson, Jr., R. A. Jacobson, and R. J. Angelici, *J. Am. Chem. Soc.*, **106**, 2901 (1984).

7. M. H. Quick and R. J. Angelici, *Inorg. Synth.*, **19**, 160 (1979).

8. S. P. Schmidt, W. C. Trogler, and F. Basolo, *Inorg. Synth.*, **23**, 41 (1985).

9. D. Strope and D. F. Shriver, *Inorg. Chem.*, **13**, 2652 (1974).
10. (a) J. Nitschke, S. P. Schmidt, and W. C. Trogler, *Inorg. Chem.*, **24**, 1972 (1985).
   (b) K. D. Abney, K. M. Long, O. P. Anderson, and S. H. Strauss, *Inorg. Chem.*, **26**, 2638 (1987).
11. S. P. Schmidt, F. Basolo, C. M. Jensen, and W. C. Trogler, *J. Am. Chem. Soc.*, **108**, 1894 (1986).

21. IRIDIUM(III) COMPLEXES WITH THE WEAKLY BONDED ANIONS $[BF_4]^-$ AND $[OSO_2CF_3]^-$

Submitted by WOLFGANG BECK,* HERBERT BAUER,* and BERNHARD OLGEMÖLLER*
Checked by DEVEREAUX A. CLIFFORD[†] and ROBERT H. CRABTREE[†]

trans-Carbonylchlorobis(triphenylphosphine)iridium(I) (Vaska's compound) forms adducts with many substrates such as acids and alkyl halides.[1] The oxidative addition of CH_3SO_3F or $CH_3OSO_2CF_3$ to *trans*-Ir(Cl)(CO)(PPh$_3$)$_2$ gives iridium(III) complexes Ir(Cl)(CO)(PPh$_3$)$_2$(CH$_3$)X X = SO$_3$F, SO$_3$CF$_3$) in which the weakly bonded anion X can be easily substituted by other ligands.[2] Highly reactive iridium(III) complexes are also obtained by oxidative addition of tetrafluoroboric acid and trialkyloxonium tetrafluoroborate to Vaska's compound.[3,4] The oxidative addition of $H_3COSO_2CF_3$ and tetrafluoroboric acid to the nitrogen complex[5] *trans*-Ir(Cl)(N$_2$)(PPh$_3$)$_2$ gives iridium(III) complexes Ir(Cl)(N$_2$)(PPh$_3$)(R)(X) (R = CH$_3$, H; X = SO$_3$CF$_3$, BF$_4$) with two excellent leaving groups (N$_2$ and X).[6,7] They are precursors for 14 electron iridium(III) complexes.

A. CARBONYLCHLOROHYDRIDO(TETRAFLUOROBORATO)-BIS(TRIPHENYLPHOSPHINE)IRIDIUM(III)[3]

$$
\textit{trans}\text{-Ir(Cl)(CO)(PPh}_3)_2 + \text{HBF}_4 \longrightarrow
\begin{array}{c}
\quad\quad H \\
\text{Cl}\diagdown \;|\;\diagup\text{PPh}_3 \\
\quad\quad \text{Ir} \\
\text{Ph}_3\text{P}\diagup\;|\;\diagdown\text{CO} \\
\quad\quad \text{FBF}_3
\end{array}
$$

Procedure

- **Caution.** *Tetrafluoroboric acid is a very corrosive chemical. Contact with skin has to be avoided. Gloves should be worn. The procedure should be carried out in a well-ventilated hood.*

*Institut für Anorganische Chemie, Universität München, Meiserstr. 1, 8000 München 2, Federal Republic of Germany.
[†] Department of Chemistry, Yale University, New Haven CT 06511.

A quantity of *trans*-Ir(Cl)(CO)(PPh$_3$)$_2$ (780 mg, 1.00 mmol)[8] is suspended in 20 mL of dry dichloromethane in a Schlenk tube (100 mL) under dry argon, and 165 mg (0.139 mL, 1.00 mmol) of tetrafluoroboric acid (d = 1, 18 g cm^3, 54% in diethyl ether, Merck), is added by means of a syringe at room temperature. The mixture is stirred magnetically for 15 min. The solid dissolves within a few seconds and the product crystallizes after several min. The precipitation is completed by addition of 60 mL of pentane. The mixture is centrifuged and the solvent is decanted. The solid is washed with five portions of pentane, 20 mL each, and dried *in vacuo*. Yield: 860 mg (99%).

Anal. Calcd. for C$_{37}$H$_{31}$BClF$_4$IrOP$_2$ (MW 868.1): C, 51.19; H, 3.60. Found: C, 50.04; H, 3.53.

Properties

The hydrido(tetrafluoroborato)iridium(III) complex is a white crystalline solid, highly sensitive to air and moisture. It is soluble in dichloromethane. The IR spectrum (Nujol) shows absorptions at 2061 (v_{CO}), 2333 (v_{IrH}), 322 (v_{IrCl}), and at 1137, 910, 730 cm^{-1} (v_{11BF_4}) for the coordinated BF$_4$ ligand. The ^{1}H NMR spectrum (in CH$_2$Cl$_2$) shows a multiplet at δ $-$ 26.5 ppm. The hydride and the BF$_4$ group are in trans position, and the [BF$_4$]$^-$ ligand is coordinated via a fluorine atom as shown by an X-ray structural determination.[3] The BF$_4$ ligand can be easily substituted by neutral σ or π donors to give cationic complexes [Ir(Cl)(CO)(PPh$_3$)$_2$(H)L][BF$_4$] (L = PPh$_3$, CH$_3$CN, H$_2$O, tetrahydrofuran (THF), C$_2$H$_4$).[3] Substitution of the [BF$_4$]$^-$ ligand by various anions gives neutral iridium(III) complexes.[9] The complex is deprotonated by strong bases to give *trans*-Ir(Cl)(CO)(PPh$_3$)$_2$.

B.   CARBONYLCHLORO(METHYL)(TETRAFLUOROBORATO)BIS-(TRIPHENYLPHOSPHINE)IRIDIUM(III)[3]

trans-Ir(Cl)(CO)(PPh$_3$)$_2$ + [(H$_3$C)$_3$O][BF$_4$]

$$\longrightarrow \begin{matrix} & CH_3 \\ Ph_3P & | & FBF_3 \\ & \diagdown Ir \diagup & \\ OC & | & PPh_3 \\ & Cl & \end{matrix} + (H_3C)_2O$$

Procedure

■ **Caution.** *Benzene is a highly toxic solvent. It should be handled in a well-ventilated hood.*

Crystalline and rigorously dry trimethyloxonium tetrafluoroborate (150 mg, 1.02 mmol) is added to a suspension of *trans*-Ir(Cl)(CO)(PPh$_3$)$_2$ (780 mg, 1.00 mmol)[8] in 10 mL of dry benzene in a Schlenk tube (60 mL) and under an argon atmosphere. The mixture is stirred magnetically for $\sim$ 1 week until the yellow solid becomes colorless. After addition of 20 mL of pentane, the mixture is centrifuged and the solvent decanted. The crystalline solid is washed with three portions of dry pentane, 10 mL each, and dried *in vacuo*. Yield: 810–836 mg (92–95%).

Anal. Calcd. for C$_{38}$H$_{33}$BClF$_4$IrOP$_2$(MW 882.1): C, 51.74; H, 3.78. Found: C, 51.28; H, 4.00.

Properties

The methyl(tetrafluoroborato)iridium(III) complex is an air and moisture sensitive white crystalline solid, soluble in dichloromethane. The IR spectrum (in Nujol) shows absorptions at 2070 (v_{CO}), 310 (v_{IrCl}) and 1136, 908 cm^{-1} (v_{11BF_4}).[3]

C. CHLORO(DINITROGEN)HYDRIDO-(TETRAFLUOROBORATO)BIS-(TRIPHENYLPHOSPHINE)IRIDIUM(III)[7]

$$trans\text{-Ir(Cl)(N}_2)(PPh_3)_2 + HBF_4 \longrightarrow$$

Procedure

- **Caution.** *See precautionary note in the procedure in Section A.*

A suspension of *trans*-Ir(Cl)(N$_2$)(PPh$_3$)$_2$ (1.36 g, 1.74 mmol)[5] in 12 mL of dry dichloromethane in a Schlenk tube is cooled to $-25\,°$C, and a solution of tetrafluoroboric acid in diethyl ether (54%, $d = 1.18\,\text{g cm}^{-3}$) (0.25 mL, 1.89 mmol) is added in one portion by means of a plastic micropipette under stirring and under a flush of dry argon. The mixture is stirred for 1 h at $-25\,°$C. A pale yellow precipitate settles and the brown solution is decanted. The solid is washed 3 to 4 times, each with 8 mL of cold ($-25\,°$C) dichloromethane, until the solid becomes colorless. The solid is then washed three times with 10 mL each of cold (-20 to $-25\,°$C) pentane and dried for 8 h at a high vacuum, during which the temperature is raised from -20 to $0\,°$C. Yield: 1.44 g (95%).

Anal. Calcd. for $C_{36}H_{31}BClF_4IrN_2P_2$ (MW 868.1): C, 49.81; H, 3.60; N, 3.23. Found: C, 49.97; H, 4.78; N, 3.12.

Properties

The complex is a colorless air and moisture sensitive solid, which is thermally stable up to $60\,°C$ when all traces of solvent have been removed by careful drying at $0\,°C$ *in vacuo*. Containing traces of dichloromethane, the complex decomposes quickly at room temperature or in moist CH_2Cl_2. The dry complex can be stored in a refrigerator under argon atmosphere for a long period of time without decomposition. In a solution of dichloromethane, nitrogen evolution is observed at temperatures above $0\,°C$. The IR spectrum of the solid (in Nujol) shows absorptions at 2310 (vw, v_{IrH}), 2229 (s, v_{N_2}), 347 (w, v_{IrCl}), and at 1129 (s), 908 (s), 740 (sh) cm^{-1} for the ^{11}B—F stretching vibrations of the coordinated $[BF_4]^-$ ligand. The ^{1}H NMR spectrum (in CD_2Cl_2, 240 K) contains multiplets at $\delta = 7.5$ (phenyl) and -30.3 ppm (IrH). The complex reacts at low temperatures with H_2O, CH_3OH, acetone, THF, CH_3CN, CO, or C_2H_4 by substitution of the $[BF_4]^-$ ligand without loss of dinitrogen to give ionic complexes, for example, $[Ir(Cl)(N_2)(PPh_3)_2(H)L]^+[BF_4]^-$. At higher temperatures or with bidentate ligands [e.g., N,N-diethyldithiocarbamate or valinate, $H_2NCH(CHMe_2)COO^-$] dinitrogen is also displaced.[7]

D. CARBONYLHYDRIDOBIS(TRIFLUOROMETHANESULFONATO) BIS(TRIPHENYLPHOSPHINE)IRIDIUM(III)

$$trans\text{-}Ir(Cl)(CO)(PPh_3)_2 + Ag(OSO_2CF_3)$$

$$\longrightarrow trans\text{-}Ir(OSO_2CF_3)(CO)(PPh_3)_2 + AgCl$$

$$trans\text{-}Ir(OSO_2CF_3)(CO)(PPh_3)_2 + HOSO_2CF_3$$

$$\longrightarrow \begin{array}{c} H \\ OC\diagdown\,|\,\diagup PPh_3 \\ Ir \\ Ph_3P\diagup\,|\,\diagdown OSO_2CF_3 \\ OSO_2CF_3 \end{array}$$

Procedure

■ **Caution.** *Trifluoromethanesulfonic acid is a corrosive chemical. Contact with skin must be avoided; gloves should be worn.*

A magnetic stirring bar, *trans*-Ir(Cl)(CO)(PPh$_3$)$_2$ (520 mg, 0.67 mmol)[8] and dry AgOSO$_2$CF$_3$ (173 mg, 0.67 mmol) are placed in a Schlenk tube (60 mL) under an argon atmosphere. The mixture is dried for 2 h under high vacuum.

After addition of 10 mL of dry dichloromethane, the yellow suspension is stirred magnetically for 3 h. The solution is filtered away from the silver chloride through a Schlenk frit under an argon atmosphere into another dry Schlenk tube. To the yellow solution trifluoromethane sulfonic acid (0.058 mL, 0.66 mmol) (distilled at 43 °C *in vacuo* with an oil pump and stored under argon) is added by means of a micropipette. The mixture is stirred for 1 h. A colorless precipitate forms. Precipitation of the product is completed by addition of 10 mL of pentane. The mixture is centrifuged and the solution decanted. The remaining solid is washed twice with 10 mL of pentane, and dried under high vacuum for 4 h. Yield: 620 mg (89%).

Anal. Calcd. for $C_{39}H_{31}F_6IrO_7P_2S_2$ (MW 1043.9): C, 44.87; H, 2.99; S, 6.14. Found: C, 44.58; H, 3.12; S, 6.74.

Properties

The iridium(III) complex is an air and moisture sensitive white solid, which is slightly soluble in dichloromethane and melts at 245 °C with decomposition. The IR spectrum of the solid (in Nujol) shows absorptions at 2297 (w, ν_{IrH}), 2073 (vs), 2058 (sh) (ν_{CO}) and 1347 (vs), 1319 (vs), 1206 (vs), 1005 (vs), 978 (vs) cm^{-1} for the SO-stretching vibrations of the coordinated sulfonate groups. The ^{1}H NMR spectrum (in CD_2Cl_2) contains signals at $\delta = 7.5$ ppm (multiplet, phenyl) and -20.6 ppm [triplet, IrH, $J_{(^{31}P-H)} = 11.0$ Hz]. In a solution of acetonitrile, the two OSO_2CF_3 ligands are substituted by solvent to give a solution of $[Ir(CO)(H)(PPh_3)_2(CH_3CN)_2]^{2+}(CF_3SO_3^-)_2$. Acetone does not replace the sulfonate ligands.

References

1. L. Vaska and J. W. Diluzio, *J. Am. Chem. Soc.*, **83**, 2784 (1961); L. Vaska, *J. Am. Chem. Soc.*, **88**, 4100, 5325 (1966); H. Singer and G. Wilkinson, *J. Chem. Soc. A*, **1968**, 2516.

2. D. Strope and D. F. Shriver, *Inorg. Chem.*, **13**, 2652 (1974); C. Eaborn, N. Farrell, J. L. Murphy, and A. Pidcock, *J. Chem. Soc. Dalton Trans.*, **1976**, 58.

3. B. Olgemöller, H. Bauer, H. Löbermann, U. Nagel, and W. Beck, *Chem. Ber.*, **115**, 2271 (1982); B. Olgemöller and W. Beck, *Inorg. Chem.*, **22**, 997 (1983).

4. M. Kubota, T. M. McClesky, R. K. Hayashi, and C. G. Webb, *J. Am. Chem. Soc.*, **109**, 7569 (1987).

5. J. P. Collman, N. W. Hoffman, and J. W. Hosking, *Inorg. Synth.*, **12**, 8 (1970); R. J. Fitzgerald and H.-M. Lin, *Inorg. Synth.*, **16**, 42 (1976).

6. L. R. Smith and D. M. Blake, *J. Am. Chem. Soc.*, **99**, 3302 (1977).

7. B. Olgemöller, H. Bauer, and W. Beck, *J. Organomet. Chem.*, **213**, C 57 (1981); H. Bauer and W. Beck, *J. Organomet. Chem.*, **308**, 73 (1986).

8. K. Vrieze, J. P. Collman, G. T. Sears, Jr., and M. Kubota, *Inorg. Synth.*, **11**, 101 (1968).

9. B. Olgemöller and W. Beck, *Chem. Ber.*, **114**, 2360 (1981).

22. DIHYDRIDOBIS(SOLVENT)BIS(TRIPHENYL-PHOSPHINE)IRIDIUM(III) TETRAFLUOROBORATES

Submitted by ROBERT H. CRABTREE,* MICHELLE F. MELLEA,* and JEAN M. MIHELCIC*
Checked by AYUSMAN SEN† and VENKATASURYANARYANA CHEBOLU†

Complexes between organometallic or metal hydride complexes and hard bases have attracted attention recently.[1] In many cases the reaction solvent plays the role of a Lewis base; such complexes can be catalytically active by solvent ligand dissociation.[2,3] We describe here the complexes $[IrH_2S_2(PPh_3)_2][BF_4]$, where $S = Me_2CO$,[4,5] H_2O,[5] or $S_2 = o$-diiodobenzene.[6] The oxygen-donor complexes are very labile[7] and are even capable of dehydrogenating alkanes.[8] The acetone complex seems to have been reported first by Araneo et al.[9] in 1965, but they formulated it as $[IrH_2L_2]^+$. The aqua[5] and halocarbon[6] complexes were reported by us much more recently, the latter being the first crystallographically characterized halocarbon complex.

Several of the syntheses described in this chapter start from the well-known $[Ir(cod)(PPh_3)_2][BF_4]$, $(cod = 1,5$-cyclooctadiene), which can easily be obtained,[4,10] from the commercially available $[Ir(cod)Cl]_2$ (Strem.).

Our own synthesis is a modification of the ones that have been described but in which we use $Ag[BF_4]$ to abstract chloride ion, this prevents the formation of $Ir(cod)Cl(PPh_3)$ as a side product.

A. (η^4-1,5-CYCLOOCTADIENE)BIS(TRIPHENYLPHOSPHINE)-IRIDIUM(I) TETRAFLUOROBORATE(1 −)

$$[Ir(C_8H_{12})Cl]_2 + 2P(C_6H_5)_3 + Ag[BF_4]$$
$$\longrightarrow [Ir(C_8H_{12})(P(C_6H_5)_3)_2][BF_4] + AgCl$$

Precautions against the admission of air must be taken in this preparation; an N_2 or Ar atmosphere is satisfactory. A 100-mL Schlenk tube is equipped with a rubber septum and a magnetic stirrer. To an evacuated flask filled with N_2 were added 700 mg of dichlorobis(η^4-1,5-cyclooctadiene)diiridium(I) (1.04 mmol), 546 mg of triphenylphosphine (2.09 mmol), and 224 mg (1.15 mmol) of silver tetrafluoroborate. Degassed (but not necessarily dry) methanol (50 mL) is added and the mixture stirred at room temperature for

---

*Department of Chemistry, Yale University, 225 Prospect Street, New Haven, CT 06520.
†Department of Chemistry, Pennsylvania State University, University Park, PA 16820.

24 h. This step is best done in the dark by covering the flask with aluminum foil. The solvent is then completely removed *in vacuo* and 20 mL of degassed dichloromethane added to the solid residue. The resulting mixture is filtered in an inert atmosphere, preferably through Celite®, and the filtrate collected. The dichloromethane is reduced to ~ 2 mL *in vacuo* and 20 mL of degassed diethyl ether slowly added through a syringe to precipitate the red product, which is filtered and washed with three portions of 5 mL of diethyl ether. The crude material is then recrystallized from the minimum volume of dichloromethane by the addition of diethyl ether to give a fine red powder, and filtered and washed as above. Yield: 1.8 g (90%).

Anal. Calcd. for $C_{44}H_{42}P_2BF_4Ir \cdot CH_2Cl_2$: C, 56.22; H, 4.61. Found: C, 56.10; H, 4.71.

Properties

The complex is air stable in the solid state but slightly air sensitive in solution. It is soluble in CH_2Cl_2, $CHCl_3$ and Me_2CO, but insoluble in H_2O, C_6H_6, and alkanes. Other characteristics are given in the literature.[4,10]

B.   BIS(ACETONE)DIHYDRIDOBIS(TRIPHENYL-PHOSPHINE)IRIDIUM(III) TETRAFLUOROBORATE(1 −)

$$[Ir(C_8H_{12})\{P(C_6H_5)_3\}_2][BF_4] + 2(CH_3)_2CO + 3H_2$$
$$\longrightarrow [IrH_2\{(CH_3)_2CO\}_2\{P(C_6H_5)_3\}_2][BF_4] + C_8H_{16}$$

Precautions against the admission of air were taken. A 100-mL Schlenck tube is equipped with a rubber septum and a magnetic stirrer, distilled (but not especially dried) acetone (10 mL) is then added, followed by 500 mg (0.543 mmol) of (η^4-1, 5-cyclooctadiene)bis(triphenylphosphine)iridium(I) fluoroborate(1 −).[10] The flask is evacuated briefly (10 s) and N_2 introduced. The mixture is cooled to 0 °C by immersion in an ice bath, and magnetically stirred. A long steel needle is passed through the septum, and by this means a gentle stream of hydrogen (~ 10 mL min^{-1}) is introduced into the mixture, a second needle, (not dipping in the solution) serving as an exhaust for the spent gases. The red solution becomes colorless or yellow. After 30 min, this treatment is stopped, and the gas entry needle is removed. Without removing the flask from the ice bath, the hydrogen from the flask, or the exhaust needle from the septum, 25 mL of diethyl ether is slowly added to the stirred solution with a syringe. Over a few minutes the crude product precipitates. The mixture is left for 3 h to complete the precipitation. This light tan material is recrystallized by dissolving in degassed dichloromethane (10 mL) and adding diethyl ether (~ 25 mL) as an upper layer. If the flask is placed in a cold room for several days, crystals appear. Alternatively, one can stir the mixture, in which case a microcrystalline colorless powder precipitates. The product is removed by filtration, washed with diethyl ether (3 × 5 mL), and dried *in vacuo*. Yield: 510 mg (93%).*

Anal. Calcd. for $C_{42}H_{44}O_2P_2F_4BIr \cdot CH_2Cl_2$: C, 51.30; H, 4.60. Found: C, 51.25; H, 4.60%.

Properties

The complex is air stable and soluble in CH_2Cl_2, Me_2CO, and $CHCl_3$. It is insoluble in H_2O, C_6H_6, and alkanes. It is best identified by its strong (C=O) stretching absorbtions at 1713 and 1666 cm^{-1} and the (Ir—H) stretching mode at 2257 cm^{-1} (Nujol). The ^1H NMR (CD_2Cl_2, 25 °C) shows an IrH resonance at $-27.8\,\delta$ [triplet, $^2J_{(P,H)} = 17$ Hz] and a Me_2CO resonance at 1.8 δ. The complex has also been crystallographically characterized.[11] In contrast to many other acetone complexes,[2] it is not moisture sensitive; water seems to bind more weakly than the ketone. It is catalytically active for alkene hydrogenation[12] and isomerization[13] and for cyclohexene aromatization.[14]

C. DIAQUADIHYDRIDOBIS(TRIPHENYLPHOSPHINE)-IRIDIUM(III) TETRAFLUOROBORATE(1 −)

$$[Ir(C_8H_{12})\{P(C_6H_5)_3\}_2][BF_4] + 2H_2O + 3H_2$$
$$\longrightarrow [IrH_2(H_2O)_2\{P(C_6H_5)_3\}_2][BF_4] + C_8H_{16}$$

To a 100-mL Schlenck Tube equipped as in the preparation in Section B, are added distilled water (60 mL) and 500 mg of (η^4-1, 5-cyclooctadiene)bis-(triphenylphosphine)iridium(I) tetrafluoroborate (0.543 mmol). The flask is evacuated briefly and filled with N_2. Hydrogen is gently bubbled through the solution as in the previous preparation for up to 3 h or until the original red suspension has turned to a yellow suspension. The time taken for this heterogeneous reaction is very sensitive to the exact conditions. The yellow solid is removed by filtration and recrystallized from dichloromethane (10 mL). Droplets of water are often present at this stage. These are removed by careful decantation of the organic phase with a hypodermic syringe. To the resulting organic solution is added 20 mL of olefin-free pentanes, bp 30–40 °C

---

*The checkers found 47% yield.

(Aldrich). The white complex precipitates, is filtered, and washed with pentanes ($3 \times 5\,mL$). Yield: $325\,mg$ (65%).*

Anal. Calcd. for $C_{36}H_{36}P_2F_4BO_2Ir\cdot CH_2Cl_2$: C, 47.96, H, 4.13. Found: C, 47.70; H, 4.13.

Properties

The aqua complex closely resembles the acetone complex described above. The chief distinguishing features are a triplet resonance at $-29.8\,\delta\,[^2J_{(P,H)} = 16\,Hz]$, an H_2O resonance at $2.4\,\delta$, and a $v_{(Ir-H)}$ vibration in the IR at $2280\,cm^{-1}$ (w). The H_2O vibrations appear at 3650 (m) and $1610\,cm^{-1}$ (m).

D. (1, 2-DIIODOBENZENE)DIHYDRIDOBIS(TRIPHENYL-PHOSPHINE)IRIDIUM(III) TETRAFLUOROBORATE(1 −)

$$[IrH_2\{(CH_3)_2CO\}_2\{P(C_6H_5)_3\}_2][BF_4] + C_6H_4I_2$$
$$\longrightarrow [IrH_2(C_6H_4I_2)\{P(C_6H_5)_3\}_2][BF_4] + 2(CH_3)_2CO$$

Bis(acetone)dihydridobis(triphenylphosphine)iridium(III)   tetrafluoroborate(1 −) ($40.5\,mg$, $0.044\,mmol$), as prepared in Section C, is dissolved in $20\,mL$ of degassed dichloromethane in a 250-mL Schlenck tube. To this is added 1, 2-diiodobenzene (degassed, $500\,mg$, $1.5\,mmol$) and the mixture stirred for 10 min at room temperature. Diethyl ether ($50\,mL$) is added to precipitate the crude product, which is then washed with diethyl ether ($3 \times 5\,mL$) and dried *in vacuo*. This material is recrystallized by dissolving in degassed CH_2Cl_2 ($5\,mL$) and adding degassed Et_2O ($15\,mL$) slowly with stirring. Yield: $41.7\,mg$ (79%).*

Anal. Calcd. for $C_{42}H_{36}I_2F_4BIr\cdot\frac{1}{4}CH_2Cl_2$: C, 43.87; H, 3.18. Found: C, 43.77; H, 3.23%.

Properties

The complex can be identified from the ^1H NMR resonace at $-16.5\,\delta$ [triplet $^2J_{(P,H)} = 13\,Hz$] and by the IR absorption at $2217\,cm^{-1}$ (w). It has been characterized crystallographically[6] and has normal Ir—I covalent bonds [2.726(2) and 2.745(1)Å]. More recently the analogous bis(iodomethane) complex[6] has also been crystallographically characterized.[15]

---

*The checkers obtained 71%.
*The checkers found 54% yield.

Acknowledgements

We thank the Petroleum Research Fund (M. F. M.) and National Science Foundation (J. M. M.) for funding, and Professor J. W. Faller and Dr. Brigitte Segmuller for performing the crystal structures of the complexes in Sections B and D.

References

1. W. Beck, Z. *Naturforsch Teil B*, **33**, 1214 (1978).
2. J. A. Davies and F. R. Hartley, *Chem. Rev.*, **81**, 79 (1981).
3. A. Sen and T.-W. Lai, *J. Am. Chem. Soc.*, **103**, 4627 (1981).
4. J. R. Shapley, R. R. Schrock, and J. A. Osborn, *J. Am. Chem. Soc.*, **91**, 2816 (1969).
5. R. H. Crabtree, P. C. Demou, D. Eden, J. M. Mihelcic, . P. Parnell, J. M. Quirk, and G. E. Morris, *J. Am. Chem. Soc.*, **104**, 6994 (1982).
6. R. H. Crabtree, J. W. Faller, M. F. Mellea, and J. M. Quirk, *Organometallics*, 1, 1361 (1982).
7. O. W. Howarth, C. H. McAteer, P. Moore, and G. E. Morris, *J. Chem. Soc. Dalton Trans.*, 1981, 1481.
8. R. H. Crabtree, J. M. Mihelcic, and J. M. Quirk, *J. Am. Chem. Soc.*, **101**, 7738 (1979).
9. A. Araneo, S. Martinengo, and P. Pasquale, *Rend. Ist. Lomb. Sci. Lett. Parte Sen. Atti. Uffi.*, **99**, 797 (1965).
10. M. Green, T. A. Kuc, and S. H. Taylor, *J. Chem. Soc. A*, 1971 2334.
11. R. H. Crabtree, G. G. Hlatky, C. P. Parnell, B. E. Segmuller, and R. J. Uriarte, *Inorg. Chem.*, 23, 354, (1984).
12. R. H. Crabtree, H. Felkin, and G. E. Morris, *J. Organometal. Chem.*, **141**, 205 (1977).
13. D. Baudry, M. Ephritikine, and H. Felkin, *Nouv. J. Chim.*, **2**, 355 (1978).
14. R. H. Crabtee and C. P. Parnell, *Organometallics*, **4**, 519, (1985).
15. M. J. Burk, R. H. Crabtree, and B. Segmuller, manuscript in preparation (1985).

23. *cis*-CHLOROBIS(TRIETHYLPHOSPHINE)(TRIFLUORO-METHANESULFONATO)PLATINUM(II)

Submitted by WOLFGANG BECK, BERNHARD OLGEMÖLLER, and LUITGARD OLGEMÖLLER*
Checked by S. CHALOUPKA and L. M. VENANZI†

Palladium(II) and platinum(II) ions are considered as soft Lewis acids, and hard oxygen and nitrogen donors are only weakly bonded to these metals.

*Institut für Anorganische Chemie, Meiserstr. 1, 8000 München 2, Federal Republic of Germany.
†Laboratorium für Anorganische Chemie, ETH, Universitätsstrasse 6, 8092 Zürich, Switzerland.

Such complexes are highly reactive and have been implicated in catalytic cycles.[1] A trifluoromethanesulfonato complex of platinum(II) is described in the following sections.[2] The trifluoromethanesulfonato (triflate) anion has been widely used as a leaving group in organic synthesis.[3]

$$cis\text{-Pt(PEt}_3)_2\text{Cl}_2 + \text{HOSO}_2\text{CF}_3 \longrightarrow \begin{array}{c} \text{Et}_3\text{P} \quad\quad \text{Cl} \\ \diagdown \quad \diagup \\ \text{Pt} \\ \diagup \quad \diagdown \\ \text{Et}_3\text{P} \quad\quad \text{OSO}_2\text{CF}_3 \end{array} + \text{HCl}$$

Procedure

■ **Caution.** *Trifluoromethanesulfonic acid is a corrosive chemical. Contact with the skin should be avoided and gloves should be worn.*

A quantity of cis-Pt(PEt$_3$)$_2$Cl$_2$ (500 mg, 1.0 mmol)[4] is suspended in 20 mL of dry pentane in a Schlenk tube fitted with a mercury bubbler under a dry inert atmosphere (N$_2$ or Ar) and cooled to $-78\,°C$. To this a quantity of freshly distilled trifluoromethanesulfonic acid (150 mg, 0.088 mL, 1.0 mmol) is added by means of a micropipette. The mixture is stirred magnetically at $-20\,°C$ for 8 h. The pentane phase is decanted after centrifugation and the solid is washed three times, each with 20 mL of pentane at ambient temperature and dried *in vacuo*. Yield: 588–616 mg (96–100%).

Anal. Calcd. for C$_{13}$H$_{30}$ClF$_3$O$_3$PtS (MW 615.9): C, 25.35; H, 4.91. Found: C, 25.60; H, 5.17.

Properties

cis-Chlorobis(triethylphosphine)(trifluoromethanesulfonato)platinum(II) is a white, hygroscopic solid. The infrared (IR) spectrum of the solid in Nujol shows $v_{(SO)}$ absorptions for the coordinated sulfonate ligand at 1312 and 1226 cm^{-1}. ^{31}P NMR (CD$_2$Cl$_2$): $\delta = 18.5$ ppm, $^1J_{(^{195}\text{Pt}-^{31}\text{P})} = 3507$ Hz.

The complex reacts with various σ and π donors, L (e.g., L = phosphine, acetonitrile, and ethene) to give the ionic complexes [Pt(PEt$_3$)$_2$(Cl)L]$^+$O$_3$SCF$_3^-$. With anionic chelate lignads (e.g., L = α-aminoacidate) complexes of the type [Pt(PEt$_3$)$_2$(chelate)]$^+$O$_3$SCF$_3^-$ are formed. The complex dimerizes in THF or acetone to give [(Et$_3$P)$_2$Pt(μ-Cl)$_2$Pt(PEt$_3$)$_2$]$^{2+}$[O$_3$SCF$_3^-$]$_2$.[2,5] Chloro-bridged complexes of this type have been obtained previously by other routes.[6]

References

1. Review: J. A. Davies and F. R. Hartley, *Chem. Rev.*, **81**, 79 (1981).
2. B. Olgemöller. L. Olgemöller, and W. Beck, *Chem. Ber.*, **114**, 2971 (1981).
3. P. J. Stang, M. Hannack, and L. R. Subramanian, *Synthesis*, **1982**, 85.
4. G. W. Parhall, *Inorg. Synth.*, **13**, 27 (1972).
5. L. Olgemöller and W. Beck, unpublished results.
6. W. P. Fehlhammer, Enke, W. A. Herrmann, and G. K. Öfele, *Handbuch der Präparativen Anorganischen Chemie*, Vol. 3, Enke Verlag Stuttgart 1981, p. 2013 and references therein.

24. ACETONITRILE COMPLEXES OF SELECTED TRANSITION METAL CATIONS

Submitted by RICHARD R. THOMAS* and AYUSMAN SEN*
Checked by WOLFGANG BECK† and REICH LEIDL†

A necessary requirement of homogeneous catalysts is that they have the ability to create vacant coordination sites by the dissociation of weakly held ligands, thereby allowing the metal to interact with the substrates. Transition metal cations incorporating weakly coordinating solvent molecules and having noncoordinating counteranions should meet this criterion. Indeed, the three cationic acetonitrile solvated transition metal complexes, the syntheses of which are described in this chapter, have been shown to catalyze a variety of organic transformations.[1] In addition, because of the high lability of the coordinated acetonitriles, these complexes serve as convenient synthetic precursors to other transition metal compounds. The synthesis of $[Pd(CH_3CN)_4][BF_4]_2$ has been previously reported.[2]

Since these cationic, weakly solvated transition metal complexes are reactive towards atmospheric moisture, all manipulations should be performed using standard inert atmosphere techniques.

A. TETRAKIS(ACETONITRILE)PALLADIUM(II) BIS[TETRAFLUOROBORATE(1 −)]

$$Pd + 2(NO)[BF_4] + 4CH_3CN \longrightarrow [Pd(CH_3CN)_4][BF_4]_2 + 2NO$$

■ **Caution.** *Since the preparation described here results in the evolution of gas from the reaction solution, the apparatus must contain a volume sufficient to*

*Department of Chemistry, Pennsylvania State University, University Park, PA 16802.
†Institut für Anorganische Chemie, Universität München 8000 München 2, Federal Republic of Germany.

accommodate this gas so that the internal pressure does not exceed atmospheric. In the present case, the gas was released through a mercury-filled bubbler into a well-ventilated hood.

Procedure

A Schlenk-type apparatus, illustrated in Fig. 1 is used for the synthesis. This apparatus is similar to that employed in other syntheses[3] but allows for multiple filtrations of the reaction solution and manipulation of the complex in a completely sealed system. In a nitrogen glove box, a 100-mL round-bottomed flask (B) is charged with 1.0 g (9.4 mmol) of palladium sponge (99.95%, Johnson Matthey) and 2.2 g (18.8 mmol) of nitrosyl tetrafluoroborate (Aldrich, sublimed 220 °C, 10^{-3} torr). This flask is connected to a medium porosity filter frit (C) that is attached to another 100-mL round-bottomed flask, via a ground glass joint. The entire apparatus is attached to a vacuum manifold through a high-vacuum Teflon valve (E) with a ground glass joint. The apparatus is evacuated to 10^{-3} torr, and 50 mL of acetonitrile is added by vacuum distillation from a phosphorus(V) oxide slurry. Upon warming to 25 °C, visible evolution of NO gas is noted from the metal surface.

The reaction is allowed to stir until no metal remains (12 h). The light yellow solution is filtered by closing the ground glass valve (D), inverting the apparatus, and cooling the round-bottomed flask (A) to liquid nitrogen temperature. After filtration the ground glass joint (D) is opened to allow for pressure equalization in the apparatus. The solution is then concentrated to 10 mL in vacuum to yield a light yellow precipitate. A 40-mL quantity of diethyl ether is added by vacuum distillation, from sodium benzophenone ketyl, to effect further precipitation. The suspension is filtered in a manner previously described and the solid is washed by further vacuum distillation of two 5-mL portions of diethyl ether. The precipitate is dried under vacuum (10^{-3} torr, 25 °C) overnight to yield 3.0 to 4.0 g (75–96%) of pale yellow tetrakis(acetonitrile)palladium(II) bis[tetrafluoroborate(1 −)].

Anal. Calcd. for $C_8H_{12}N_4F_8B_2Pd$: C, 21.7; H, 2.7; N, 12.6. Found: C, 21.8; H, 2.9; N, 12.3 (authors); C, 21.6; H, 4.0; N, 12.5 (checkers).

Properties

The compound $[Pd(CH_3CN)_4][BF_4]_2$ is a moisture-sensitive pale yellow crystalline material. It is very soluble in acetonitrile and nitromethane but virtually insoluble in less polar organic solvents such as diethyl ether, chloroform, and benzene. The ^1H NMR spectrum in CD_3NO_2 exhibits a

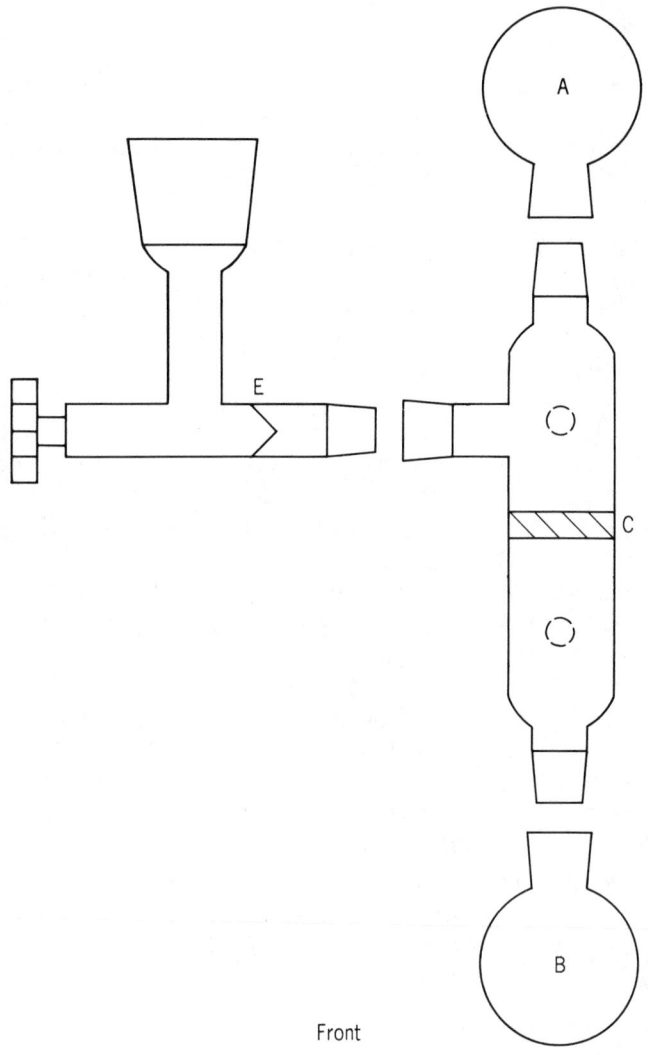

Front

Fig. 1. Modified Schlenk apparatus, front and side views.

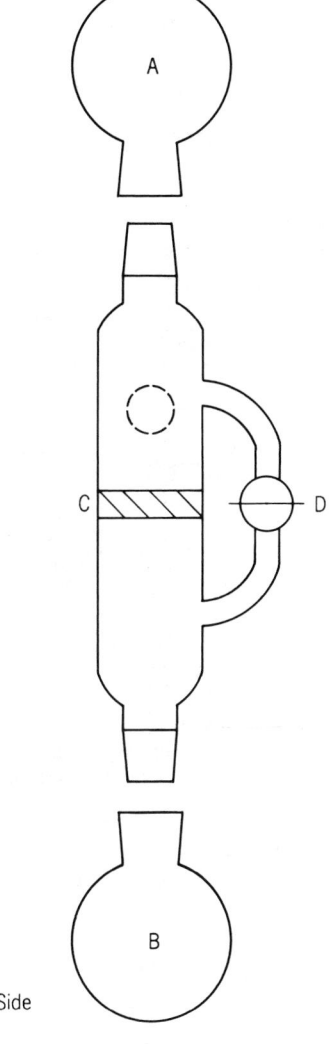

A

C ⟍⟍⟍ D

B

Side

Fig. 1. (*Continued*)

singlet at 2.65 ppm. The IR spectrum of its Nujol mull shows a coordinated —
$C \equiv N$ band at 2335 cm^{-1}.

B. *cis*-TETRAKIS(ACETONITRILE)DINITROSYLMOLYBDE-
NUM(II) BIS[TETRAFLUOROBORATE(1 −)]

$$Mo(CO)_6 + 2(NO)[BF_4] + 4CH_3CN$$
$$\longrightarrow [Mo(NO)_2(CH_3CN)_4][BF_4]_2 + 6CO$$

- **Caution.** *Since the preparation described here results in the evolution of gas from the reaction solution, the apparatus must contain a volume sufficient to accommodate this gas so that the internal pressure does not exceed atmospheric. In the present case, the gas was released through a mercury-filled bubbler into a well-ventilated hood.*

Procedure

The apparatus and techniques used are similar to those for the synthesis of tetrakis(acetonitrile)palladium(II) bis[tetrafluoroborate(1 −)]. A 100-mL round-bottomed flask (B) is charged with 2.0 g (7.6 mmol) of hexacarbonylmolybdenum (Alfa) and 1.77 g (15.2 mmol) of nitrosyl tetrafluoroborate (Aldrich, sublimed 220 °C, 10^{-3} torr). Dry acetonitrile (50 mL) is added and the solution is allowed to stir at 25 °C until no hexacarbonylmolybdenum remains (6 h).

The resulting dark emerald green solution is filtered and the filtrate is concentrated to near dryness in vacuum. Diethyl ether (40 mL) is added and the solution is vigorously stirred to cause precipitation of the dark green molybdenum complex. The solution is filtered and the solid washed with two 5-mL portions of diethyl ether. The precipitate is dried in vacuum (10^{-3} torr, 25 °C) to afford 3.36 g (90%) of dark emerald green *cis*-tetrakis(acetonitrile)-dinitrosylmolybdenum(II) bis[tetrafluoroborate(1 −)].

Anal. Calcd. for $C_8H_{12}N_6O_2F_8B_2Mo$: C, 19.5; H, 2.5; N, 17.0. Found: C, 19.0; H, 2.4; N, 16.9.

Properties

The compound $[Mo(NO)_2(CH_3CN)_4][BF_4]_2$ is a moisture-sensitive dark crystalline emerald green substance. It is very soluble in acetonitrile and nitromethane but virtually insoluble in less polar organic solvents such as diethyl ether, chloroform, and benzene. The ^{1}H NMR spectrum in CD_3NO_2 exhibits two singlets of equal intensity at 2.65 and 2.55 ppm, respectively, which is consistent with a cis arrangement of the two NO groups on the metal atom. The IR spectrum of its Nujol mull shows absorptions at 2330 and 2300 cm^{-1} (coordinated —C≡N), 1862, 1755, and 1724 cm^{-1} (coordinated NO) and may show weak absorptions at 2040 and 1957 cm^{-1} due to coordinated CO (impurity).

C. *cis*-TETRAKIS(ACETONITRILE)DINITROSYLTUNGSTEN(II) BIS[TETRAFLUOROBORATE(1 —)]

$$W(CO)_6 + 2(NO)[BF_4] + 4CH_3CN$$

$$\longrightarrow [W(NO)_2(CH_3CN)_4][BF_4]_2 + 6CO$$

■ **Caution.** *Since the preparation described here results in the evolution of gas from the reaction solution, the apparatus must contain a volume sufficient to accommodate this gas so that the internal pressure does not exceed atmospheric. In the present case, the gas was released through a mercury-filled bubbler into a well-ventilated hood.*

Procedure

In a procedure identical to that described in Section B, the reaction is performed using 2.5 g (7.1 mmol) of hexacarbonyltungsten (Alfa) and 1.66 g (14.2 mmol) of nitrosyl tetrafluoroborate (Aldrich, sublimed 220 °C, 10^{-3} torr) An emerald green complex is isolated to yield 3.95 g (96%) of *cis*-tetrakis-(acetonitrile)dinitrosyltungsten(II) bis[tetrafluoroborate(1 —)].

Anal. Calcd. for $C_8H_{12}N_6O_2F_8B_2W$: C, 16.4; H, 2.1. Found: C, 16.5; H, 2.3.

Properties

The compound $[W(NO)_2(CH_3CN)_4][BF_4]_2$ is a moisture sensitive dark emerald green crystalline substance. It is very soluble in acetonitrile and nitromethane but virtually insoluble in less polar organic solvents such as diethyl ether, chloroform, and benzene. The ^{1}H NMR spectrum in CD_3NO_2 exhibits two singlets of equal intensity at 2.65 and 2.55 ppm, respectively, which is consistent with a cis arrangement of the two NO groups on the metal atom. The IR spectrum of its Nujol mull shows absorptions at 2332 and 2300 cm^{-1} (coordinated —C≡N), 1862, 1820, 1772, and 1728 cm^{-1} (coordinated NO) and may show weak absorptions at 2030 and 1940 cm^{-1} due to coordinated CO (impurity).

References

1. (a) A. Sen and T.-W. Lai, *Inorg. Chem.*, **23**, 3257 (1984); (b) A. Sen and R. R. Thomas, *Organometallics*, **1**, 1251 (1982); (c) A. Sen and T.-W. Lai, *Organometallics*, **1**, 415 (1982); (d) A. Sen and T.-W. Lai, *J. Am. Chem. Soc.*, **103**, 4627 (1981); (e) W. A. Nugent and F. W. Hobbs, *J. Org. Chem.*, **48**, 5364 (1983).

2. R. F. Schramm and B. B. Wayland, *Chem. Commun.*, 898 (1968). $[Pd(CH_3CN)_4]^{2+}$ may also be generated by the reaction of $Pd(CH_3CN)_2Cl_2$ with two equivalents of $AgClO_4$, see: F. R. Hartley, S. G. Murray, W. Levason, H. E. Soutter, and C. A. McAuliffe, *Inorg. Chim. Acta*, **35**, 265 (1979).

3. For example, M. Ghedini and G. Dolcetti, *Inorg. Synth.*, **21**, 104 (1982). We thank Professor J. E. Bercaw of California Institute of Technology for providing us with the design of the present apparatus. For an early version of this apparatus, see: R. W. Parry, D. R. Schultz, and P. R. Girardot, *J. Am. Chem. Soc.*, **80**, 1 (1958).

25. THE FORMATION OF THE HYDRIDO(METHANOL)BIS(TRIETHYL-PHOSPHINE)PLATINUM(II) CATION AND ITS REACTIONS WITH UNSATURATED HYDROCARBONS

Submitted by CARLA SORATO and LUIGI M. VENANZI*
Checked by HENRY E. BRYNDZA and SONBINH NGUYEN†

Solvato complexes of platinum(II) of the type $trans$-$[PtY(solvent)L_2]^+$ (Y = hydride, alkyl, or aryl; solvent = alcohol or ketone; L = tertiary phosphine or arsine) have been known since 1961.[1] They are obtained by halogen abstraction from the corresponding halo complexes $trans$-$[PtXYL_2]$ in the presence of the desired solvent.[2] The methanol complex is also rapidly and quantitatively formed when $trans$-$[PtH(NO_3)(PEt_3)_2]$ is dissolved in this solvent.[2]

These solvato complexes are reactive species that have been characterized only in solution. The coordinated solvent is labile and readily replaced by stronger donors.

Because of the donor atom lability, the solvato complexes behave as coordinatively unsaturated species to the extent that they facilitate reactions such as alkene and alkyne insertion reactions and cleavage of B—C bonds. The cation $trans$-$[PtH(CH_3OH)(PEt_3)_2]^+$ is obtained from $trans$-$[PtHCl(PEt_3)_2]$ by the general method just outlined.[2] Its normal alkene and alkyne insertion reactions with methyl acrylate,[3] diphenylacetylene,[4] and

*Laboratorium für Anorganische Chemie ETH Zürich, Universitätstrasse 6, CH-8092 Zürich, Switzerland.
†Central Research & Development Dept., E.I. du Pont de Nemours & Co., Inc., Wilmington, DE 19898.

norbornadiene[3] have been previously described. It has also been reported that *cyclo*-octa-1, 5-diene reacts with a methanolic solution of *trans*-[PtH(NO$_3$)(PEt$_3$)$_2$] giving the expected η^1, η^2-cyclooctenyl complex.[3] It has subsequently been shown that when this reaction is carried out at room temperature the corresponding (η^3) π-allylic complex is obtained.[5] However, more recent investigations have shown that the η^1, η^2-species is indeed formed if the reaction is carried out at low temperature ($-60\,°C$) and that this species subsequently rearranges to the π-allylic product.[6]

The reaction of *trans*-[PtH(CH$_3$OH)(PEt$_3$)$_2$]$^+$ with Na[BPh$_4$], which yields *trans*-[(PEt$_3$)$_2$(Ph)Pt(μ-H)PtH(PEt$_3$)$_2$][BPh$_4$] (see Ref. 7) shows several uncommon features: (a) [BPh$_4$]$^-$ acts as a phenylating agent[8-10] and (b) a μ-hydrido complex is formed.

General Comments on the Procedures

All these reactions are best carried out (a) under an atmosphere of nitrogen and (b) with mechanical stirring. General safety precautions should be maintained, namely, the use of eye protection in the laboratory, proper gloves to avoid skin contact with chemicals, and provision of a hood or dry box with adequate ventilation for all chemical manipulations.

The stoichiometries of reactions 2 to 5 are based on the amounts of reagents used in reaction 1 and all yields are calculated on the basis of the platinum complex used in reaction 1. The complex *trans*-[PtHCl(PEt$_3$)$_2$] is prepared according to the procedure given in Ref. 11.

The NMR spectra cited in Section A were recorded using Bruker HX-90 and WM-250 spectrometers. A positive sign of the chemial shift denotes a resonance downfield of the reference.

A. *trans*-HYDRIDO(METHANOL)BIS(TRIETHYL-PHOSPHINE)PLATINUM(II) BIS(TRIFLUORO-METHANESULFONATE)(1 −)

1.   *trans*-[PtHCl(PEt$_3$)$_2$] + CH$_3$OH + Ag(CF$_3$SO$_3$)

⟶ *trans*-[PtH(CH$_3$OH)(PEt$_3$)$_2$](CF$_3$SO$_3$) + AgCl

Solution A

Procedure

To a Schlenk flask (5 mL) is added a quantity of silver trifluoromethanesulfonate (200 mg, 8.8 mmol); care should be taken during the weighing of this salt as it is hygroscopic. Methanol (1.5 mL, absolute) is added to the salt to give a clear solution. This solution is slowly transferred by cannula or syringe to a well-stirred suspension of *trans*-chlorohydridobis(triethylphosphine)platinum(II) (364 mg, 0.8 mmol) in methanol (6 mL) contained in a second Schlenk flask (50 mL). A nitrogen flow through the side arm provides an inert atmosphere. Silver chloride precipitates and is filtered away. The solution contains *trans*-hydrido(methanol)bis(triethylphosphine)platinum(II)-trifluoromethanesulfonate (Solution A). Yield: 69% (based on the amount of platinum complex used and calculated by integration of the ^{1}H hydride signal against an external standard).

A solid sample of the product can be obtained by evaporating the solvent under reduced pressure, keeping the temperature below $-20\,°C$.

Properties

The methanol solution of *trans*-hydrido(methanol)bis(triethyl-phosphine)platinum(II) slowly decomposes with formation of metallic platinum, if not kept under inert atmosphere.

The ^{1}H NMR spectrum, in dichloromethane-d_2 solution, shows a main signal at δ 26.27 ppm for the hydride, flanked by its [195]Pt satellites ($^1J_{\text{Pt,H}}$ = 1587 Hz), a methanol signal, $\delta_{(\text{CH}_3)}$ 3.44 ppm, and two multiplets $\delta_{(\text{CH}_2)}$ 1.96 ppm and $\delta_{(\text{CH}_3)}$ 1.44 ppm, respectively, for the ethyl group of triethyl-phosphine. The methanol signal is only slightly shifted with respect to that of free methanol as coordinated and free methanol are in rapid exchange.

The $^{31}P\{^1H\}$ NMR spectrum of the methanol solution, recorded using a capillary filled with acetone-d_6 as lock, shows a signal at δ 27.8 ppm flanked by its [195]Pt satellites ($^1J_{\text{Pt,P}}$ = 2707 Hz).

B. μ-HYDRIDO-HYDRIDOPHENYLTETRAKIS-(TRIETHYLPHOSPHINE)DIPLATINUM(II) TETRAPHENYLBORATE(1 −)

2.   *trans*-[PtH(CH$_3$OH)(PEt$_3$)$_2$](CF$_3$SO$_3$) + 2Na[BPh$_4$]

$\longrightarrow$ [(PEt$_3$)$_2$(Ph)Pt(μ-H)PtH(PEt$_3$)$_2$][BPh$_4$]*

Procedure

A Schlenk flask (20 mL) is charged with sodium tetraphenylborate(1 −) (548 mg, 1.6 mmol) and methanol (10 mL) and stirred until all the solid has dissolved.

This solution is transferred to a solution of the *trans*-hydrido(methanol)bis(triethylphosphino)platinum(II) cation, Solution A, prepared as described in Section A. Stirring is continued for 16 to 18 h while the

*No exact stoicheiometry can be given as the boron-containing product(s) have not been identified.

solution is kept under a flow of nitrogen. The white product, which gradually precipitates, is filtered off from the yellow mother liquor, dried under vacuum, and recrystallized by dissolving it in 1 to 2 drops of chloroform, adding 1 to 2 mL of methanol, and placing the solution in the refrigerator. The white crystals, which slowly appear, are filtered off and dried under vacuum. Yield: 0.66 g (66%), dec: 136 °C.

Properties

The product is stable both as a solid and in solution, if kept in an inert atmosphere at or below 0 °C; its crystal structure has been reported.[7]

In the infrared (IR) spectrum the $\nu_{(Pt-H)}$ of the terminal platinum hydrogen stretching vibration appears as a sharp band of medium intensity at 2160 cm^{-1}. The $\nu_{(Pt-D)}$ band of the corresponding deuteride occurs at 1580 cm^{-1}.

The ^{1}H NMR spectrum was recorded in a chloroform-d_1 solution. The hydride region of the spectrum shows two sets of multiplets, each with platinum satellites, corresponding to the bridging and to the terminal hydrides with the following parameters.

$\delta_{(H_A)}$-9.45 ppm $^1J_{Pt_A H_A}$; $^1J_{Pt_B H_A}$; $\sim$ 500 Hz

$\delta_{(H_B)}$-11.54 ppm $^1J_{Pt_B,H_B}$;   1135 Hz; $^3J_{Pt_A,H_B}$;   105 Hz

The ethyl groups of nonequivalent triethylphosphines appear as two groups of signals with $\delta_{(CH_2)}$: 2.04 ppm and 1.65 ppm; $\delta_{(CH_3)}$: 1.17 ppm and 1.04 ppm, respectively.

In the ^{31}P {^{1}H} NMR spectrum in a chloroform-d_1 solution two main signals appear. Each signal is flanked by two pairs of ^{195}Pt satellites. One pair is due to the coupling of ^{31}P with ^{195}Pt over one bond, the other pair to the coupling with ^{195}Pt over three bonds. The relevant parameters are

$\delta_{(P_A)}$ 9.3 ppm $^1J_{Pt_A,P_A}$;   2644 Hz $^3J_{Pt_B,P_A}$;   13.9 Hz

$\delta_{(P_B)}$ 19.3 ppm $^1J_{Pt_B,P_B}$;   2598 Hz $^3J_{Pt_A,P_B}$;   14.2 Hz

C. *cis*-(3-METHOXY-3-OXO-κO-PROPYL-κC^1)-BIS(TRIPHENYLPHOSPHINE)PLATINUM(II) TETRAPHENYLBORATE(1 −)

3. $trans$-[PtH(CH$_3$OH)(PEt$_3$)$_2$](CF$_3$SO$_3$)

$+ $ CH$_2$:CHCOOCH$_3$ + Na[BPh$_4$]

$\longrightarrow$ cis-[Pt(CH$_2$CH$_2$COOCH$_3$)(PEt$_3$)$_2$][BPh$_4$]

$+ $ Na(CF$_3$SO$_3$) + CH$_3$OH

Procedure

A quantity of sodium tetraphenylborate(1 −) (548 mg, 1.6 mmol) and methanol (10 mL) are placed in a Schlenk flask (20 mL) and stirred until all the solid has dissolved. Solution A, prepared as described in Section A, is treated first with 1 mL of methyl acrylate and then with the tetraphenylborate(1 −) solution. A nitrogen flow is passed through a side arm to provide an inert atmosphere.

The oil that separates from the mixture subsequently crystallizes. The product is collected by filtration, washed with methanol, and recrystallized from a dichloromethane: methanol mixture (1:1). Yield: 0.36 g (54%), Mp: 112 °C (dec occurs above 150 °C).

Properties

The product is air stable both in solid state and in solution. The IR spectrum shows a broad band of medium intensity at 1710 cm^{-1} corresponding to the ester group and a sharp strong band at 1605 cm^{-1} due to a carbonyl group coordinated to platinum in a chelate system.

The ^{1}H NMR spectrum, recorded in a chloroform-d_1 solution, shows the two nonequivalent methylene groups of the propionate as multiplets at δ 1.38 and δ 2.97 ppm, respectively, and a singlet at δ 3.7 ppm corresponding to a methyl group bound to oxygen. The ethyl groups of the two nonequivalent triethylphosphines appear as complex multiplets centered at $\delta_{(CH_2)}$ 1.75 ppm and $\delta_{(CH_3)}$ 1.10 ppm.

The $^{31}P\{^1H\}$ NMR spectrum for a chloroform-d_1 solution, shows two pairs of ^{31}P signals flanked by their respective ^{195}Pt satellites. This pattern is indicative of a compound with two nonequivalent triethylphosphines in cis position to each other. The relevant parameters are

$\delta_{(P_A)}$ 22.5 ppm $^1J_{P_A,Pt}$;   1860 Hz

$\delta_{(P_B)}$ 7.0 ppm $^1J_{P_B,Pt}$;   4414 Hz $^2J_{P_A,P_B}$;   13.5 Hz

D. (η^3-CYCLOOCTENYL)BIS(TRIETHYLPHOSPHINE)-PLATINUM(II) TETRAPHENYLBORATE(1 –)

4.   *trans*-[PtH(CH$_3$OH)(PEt$_3$)$_2$](CF$_3$SO$_3$) + 1, 5-cod

    + Na[BPh$_4$] $\longrightarrow$ *cis*-[Pt(η^3-C$_8$H$_{13}$)(PEt$_3$)$_2$][BPh$_4$]

    + Na(CF$_3$SO$_3$) + CH$_3$OH

    cod = 1, 5-cyclooctadiene

Procedure

To Solution A, prepared as described in Section A, is added a quantity of cyclooocta-1, 5-diene   (0.21 mL,   1.7 mmol).   A   solution   of   sodium tetraphenylborate(1 –)(274 mg, 0.8 mmol) in methanol (5 mL) is prepared in a separate Schlenk flask (50 mL). Under a flow of N$_2$, the sodium tetraphenylborate(1 –) solution is added to the stirred mixture of Solution A containing the cyclooocta-1, 5-diene. A white solid precipitates within a few minutes and is collected by filtration.

The dried product is dissolved in 1 to 2 drops of dichloromethane and then 1 to 2 mL of methanol are added. The product separates out as white crystals. Yield: 0.52 g (79%).

Properties

The product is air stable in the solid state. In solution gradual decomposition sets in after 2 to 3 days.

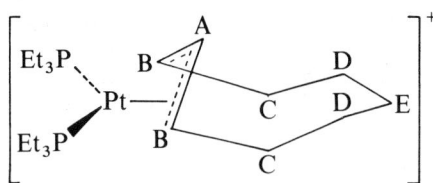

The ^1H NMR spectrum, recorded in a chloroform-d_1 solution, shows three sets of signals: (a) the aliphatic protons of the ethyl groups and those of the carbocycle, (b) the protons of the phenyl groups, and (c) the allylic protons of the carbocycle. Some of their parameters are given below:

$\delta_{(H_A)}$ 4.8 ppm $^1J_{H_A,H_B}$;   8.1 Hz $^2J_{Pt,H_A}$;   52.3 Hz

$\delta_{(H_B)}$ 4.4 ppm

The $^{31}P\{^1H\}$ NMR spectrum, also recorded in a chloroform-d_1 solution, shows one signal at $\delta\,8.7$ ppm flanked by its platinum satellites ($^1J_{Pt,P} = 3664$ Hz), indicating that the two triethylphosphines are magnetically equivalent.

The ^{13}C NMR spectrum shows seven inequivalent carbon atoms. The relevant parameters of the carbocycle are δ, ppm (J Hz): 110.5 ($J_{Pt,C} = 6$); 76.8 (overlaps with the $CDCl_3$ triplet); 32.2 ($J_{Pt,C} = 23$); 29.1 ($J_{Pt,C} = 7$); 29.5 ($J_{Pt,C} = 10$). The carbon atoms of the ethyl groups show the following parameters: $\delta_{(CH_2)}$ 19.2, ($J_{P,C} = 34$); $\delta_{(CH_3)}$ 8.4, ($J_{P,C} = 23$).

E.   *trans*-CHLORO (*cis*-1,2-DIPHENYLETHENYL)BIS(TRIETHYL-PHOSPHINE)PLATINUM(II)

5.   *trans*-[PtH(CH$_3$OH)(PEt$_3$)$_2$](CF$_3$SO$_3$) + PhC⋮CPh

$\longrightarrow$*trans*-[Pt(C(Ph):CHPh)(CH$_3$OH)(PEt$_3$)$_2$](CF$_3$SO$_3$);

trans-[Pt(C(Ph):CHPh)(CH$_3$OH)(PEt$_3$)$_2$](CF$_3$SO$_3$)

+ LiCl $\longrightarrow$ *trans*-[PtCl(C(Ph):CHPh)(PEt$_3$)$_2$]

+ Li(CF$_3$SO$_3$) + CH$_3$OH

Procedure

A solution of diphenylacetylene (130 mg, 1.6 mmol) in 5 mL of methanol is added to Solution A, prepared as described in Section A. Stirring is continued for 5 min. The solution changes color from yellow to orange.

A solution of lithium chloride (68 mg, 1.6 mmol) in 2 mL of methanol is prepared and added to the orange solution described earlier. A white product gradually precipitates and is filtered away after ~ 30 min.

The product is recrystallized by dissolving it in 1 to 2 mL of methanol under warming. Upon cooling, a solid reprecipitates after ~ 30 min. The solid is filtered and dried under vacuum. Yield: 0.32 g, (59%), dec: 157 °C.

Properties

The product is air stable both in solid state and in solution.

Its ^1H NMR spectrum, recorded in a chloroform-d_1 solution, shows three sets of signals: (a) the aliphatic protons of the ethyl group of triethylphosphine, (b) the protons of the phenyl groups, and (c) the olefinic proton flanked by its platinum satellites. The relevant parameters are δ, ppm (J Hz): $\delta_{(CH_3)}$, 1.10; $\delta_{(CH_2)}$, 1.75 and 1.98 (two signals are observed because of *virtual coupling* effects);[12] $\delta_{(H_\beta)}$ 6.75 ($^3J_{Pt,H_\beta} = 97.4$); $\delta_{(phenyl\ protons)}$ 7.0, 7.21, and 7.45.

The $^{31}P\{^1H\}$ NMR spectrum, also recorded in a chloroform-d_1 solution, shows one main signal centered at 12.2 ppm flanked by its ^{195}Pt satellites ($^1J_{Pt,P} = 2856$ Hz).

The $^{31}C\{^1H\}$ NMR spectrum shows 12 inequivalent carbon atoms. The relevant parameters are δ, ppm (J Hz): (a) diphenyl, vinyl group $\delta_{(C_\alpha)}$ 120.6 ($^2J_{P,C_\alpha} = 18$; $\delta_{(C_\beta)}$ 132.2 ($^3J_{P,C_\beta} = 9.5$); $\delta_{(C_1)}$ 146.9; $\delta_{(C_2)}$ 131.1 ($^3J_{Pt,C_2} = 47$); $\delta_{(C_3)}$ 128.1; $\delta_{(C_4)}$ 125.8 ($^5J_{Pt,C_4} = 9$); $\delta_{(C_5)}$ 141.7; $\delta_{(C_6)}$ 128.4 ($^4J_{Pt,C_6} = 9$); $\delta_{(C_7)}$ 127.3; $\delta_{(C_8)}$ 124.8; (b) triethylphosphine: $\delta_{(CH_3)}$ 8.1; $\delta_{(CH_2)}$ 13.9.

References

1. F. Basolo, J. Chatt, H. B. Gray, R. G. Pearson, and B. L. Shaw, *J. Chem. Soc.*, **1961**, 2207.

2. H. C. Clark and H. Kurosawa, *J. Chem. Soc. Dalton Trans.*, **1971**, 957.

3. A. J. Deeming, B. F. G. Johnson, and J. Lewis, *J. Chem. Soc. Dalton Trans.*, **1973**, 1848.

4. N. M. Boag, M. Green, D. M. Grove, J. A. K. Howard, J. L. Spencer, and G. A. Stone, *J. Chem. Soc. Dalton Trans.*, **1980**, 2170.

5. L. M. Venanzi, *Coord. Chem. Rev.*, **20**, 99 (1980).

6. C. Sorato and L. M. Venazi, unpublished observations.

7. G. Bracher, D. M. Grove, P. S. Pregosin, and L. M. Venanzi, *Angew. Chem., Int. Ed. Engl.*, **17**, 778 (1978).

8. H. C. Clark and K. R. Dixon, *J. Am. Chem. Soc.*, **91**, 596 (1969).

9. H. C. Clark and J. D. Ruddick, *Inorg. Chem.*, **9**, 1226 (1970).

10. H. C. Clark and L. E. Manzer, *Inorg. Chem.*, **10**, 2669 (1971).

11. J. Chatt and B. L. Shaw, *J. Chem. Soc. Dalton Trans.*, **1962**, 5075.

12. J. G. Verkade, *Coord. Chem. Rev.*, **9**, 1 (1972).

Chapter Four

MONONUCLEAR TRANSITION METAL COMPLEXES Part III: METALLACYCLIC COMPLEXES

*Preface   Ekkehard Lindner**

Metallacyclic complexes are of increasing interest because in many transition metal catalyzed organic syntheses carried out in homogeneous phase, they occur as highly reactive intermediates.[1] In the case of typical examples like valence isomerization, metathesis reactions of alkenes and alkynes, oligomerization or cyclooligomerization and Ziegler polymerization of olefines, metallacycloalkanes can be assigned special importance. Catalytic active transition metal fragments react with two, three, or four alkene functions to give five-, seven-, or even nine-membered heterocycles. The stereochemistry of the metallacyclopentane/bis(alkene) complex interaction was subject of a theoretical analysis.[2] The catalytic efficiency of metallacycloalkanes depends on the easiness of the M—C bond cleavage, which is the result of reductive elimination of the organic substrate or of β-hydrogen transfer. Also α- or β-C—C bond rupture was detected. Heterocycles with an aliphatic carbon frame and a metal adjacent donor atom are suitable model compounds to study special catalytic steps and structural properties.[3]

   In connection with the activation of C—H bonds the cyclometalation has become a very general reaction, intensively investigated by Kaesz and coworkers[4] and reviewed by Bruce.[5] In transition metal complexes an organic ligand reacts with the transition metal resulting in the formation of a metal—

*Institut für Anorganische Chemie, Universität Tübingen, D-7400 Tübingen 1, Federal Republic of Germany.

carbon σ bond. Metalations of phenyl substituted ligands are described most frequently in this context and designated as ortho metalation.

Not less attractive is the cyclooligomerization of alkynes leading to benzene derivatives, *N*-heterocycles or cyclooctatetraene.[1a,b,6] The cyclotrimerization of alkynes proceeds via metallacyclopropenes and metallacyclopentadienes as intermediates. The latter are distinguished by their remarkable thermal stability. Subsequently, in dependence on the nature of the alkyne, either metallacycloheptatrienes or metallabicycloheptadienes are formed. Cotrimerization of alkynes with nitriles affords pyridine derivatives. The cyclocotrimerization of (η^2-phosphinothioyl)metal complexes with alkynes in an analogous reaction leads to bicycloheptadienes containing phosphorus and sulfur. The similarity of this heteroanalogous cyclotrimerization can be traced back to the alkynelike behavior of the P=S group.[7]

Another interesting feature is the synthesis of benzo- and naphthoquinones succeeding via the reductive elimination from metallacycloheptadienediones. The latter are formed by an alkyne insertion into metallacyclopentenediones.[8]

This volume of *Inorganic Syntheses* includes a representative collection of metallacyclic complexes the significance of which is emphasized in this preface.

References

1. (a) G. Wilke, *Pure Appl. Chem.*, **50**, 677 (1978); (b) K. Itoh, in *Fundamental Research in Homogeneous Catalysis*, Vol. 3, M. Tsutsui (ed.), Plenum, New York, 1979, p. 865 ff; (c) R. J. Puddephatt, *Coord. Chem. Rev.*, **33**, 149 (1980); (d) R. J. Puddephatt, *Comments Inorg. Chem.*, **2**, 69 (1982); (e) E. Lindner, *Adv. Heterocycl. Chem.*, **39**, 237 (1986).

2. (a) A. Stockis and R. Hoffmann, *J. Am. Chem. Soc.*, **102**, 2952 (1980); (b) E. Lindner, E. Schauss, W. Hiller, and R. Fawzi, *Angew. Chem. Int. Ed. Engl.*, **23**, 711 (1984).

3. (a) E. Lindner, G. Funk, and S. Hoehne, *Chem. Ber.*, **114**, 2465 (1981); (b) E. Lindner, G. Funk, and S. Hoehne, *Chem. Ber.*, **114**, 3855 (1981).

4. (a) R. J. McKinney, G. Firestone, and H. D. Kaesz, *Inorg. Chem.*, **14**, 2057 (1975); (b) R. J. McKinney and H. D. Kaesz, *J. Am. Chem. Soc.*, **97**, 3066 (1975); (c) S. S. Crawford, G. Firestone, and H. D. Kaesz, *J. Organomet. Chem.*, **91**, C57 (1975).

5. M. I. Bruce, *Angew. Chem. Int. Ed. Engl.*, **16**, 73 (1977).

6. (a) H. Bönnemann, *Angew. Chem. Int. Ed. Engl.*, **17**, 505 (1978); **24**, 248 (1985); (b) Y. Wakatsuki and H. Yamazaki, *J. Chem. Soc. Dalton Trans.*, **1978**, 1278; (c) R. G. Bergman, *Pure Appl. Chem.*, **53**, 161 (1981); (d) Y. Wakatsuki, O. Nomura, K. Kitaura, K. Morokuma, and H. Yamazaki, *J. Am. Chem. Soc.*, **105**, 1907 (1983).

7. (a) E. Lindner, A. Rau, and S. Hoehne, *Angew. Chem. Int. Ed. Engl.* **20**, 787 (1981); **20**, 788 (1981); (b) E. Lindner, C.-P. Krieg, W. Hiller, and R. Fawzi, *Angew. Chem. Int. Ed. Engl.*, **23**, 523 (1984).

8. (a) L. S. Liebeskind, S. L. Baysdon, and M. S. South, *J. Am. Chem. Soc.*, **102**, 7397 (1980); (b) L. S. Liebeskind, S. L. Baysdon, M. S. South, S. Iyer, and J. P. Leeds, *Tetrahedron*, **41**, 5839 (1985); (c) L. S. Liebeskind, S. L. Baysdon, V. Goedken, and R. Chidambaram, *Organometallics*, **5**, 1086 (1986).

26.   MAIN GROUP METAL (Li OR Mg) *o*-PHENYLENEDIMETHYL [*o*-C$_6$H$_4$($\bar{\text{C}}$H$_2$)$_2$ OR *o*-C$_6$H$_4${$\bar{\text{C}}$H(SiMe$_3$)}$_2$] TRANSFER REAGENTS

Sumbitted by M. F. LAPPERT,* T. R. MARTIN,* and C. L. RASTON†
Checked by R. A. ANDERSEN‡

$$C_6H_4(CH_2Cl)_2 + 2Mg \xrightarrow{\text{thf}} o\text{-}C_6H_4\{CH_2MgCl(thf)_m\}_2$$

1

$$o\text{-}C_6H_4\{CH_2MgCl(thf)_m\}_2$$

1

$$\xrightarrow{-40\,^\circ C} \tfrac{1}{3}[\{Mg(CH_2C_6H_4CH_2\text{-}o\text{-}\mu)(thf)_2\}_3] + \tfrac{2}{3}MgCl_2$$

2

$$\tfrac{1}{3}[\{Mg(CH_2C_6H_4CH_2\text{-}o\text{-}\mu)(thf)_2\}_3]$$

2

$$\xrightarrow{-\text{thf}} \tfrac{1}{n}[\{Mg(CH_2C_6H_4CH_2)(thf)\}_n]$$

3

$$o\text{-}C_6H_4\{CH_2MgCl(thf)_m\}_2 + 2SiMe_3Cl$$

1

$$\xrightarrow{\text{thf}} o\text{-}C_6H_4(CH_2SiMe_3)_2 + 2MgCl_2(thf)_n$$

$$o\text{-}C_6H_4(CH_2SiMe_3)_2 + 2CH_3(CH_2)_3Li(tmeda)$$

$$\xrightarrow{\text{hexane}} [\{Li(tmeda)\}_2\{o\text{-}C_6H_4(CHSiMe_3)_2\}] + 2C_4H_{10}$$

4

The *o*-xylylene moiety has recently gained some prominence in metal complexes and reagents to transfer this moiety provide the potential for obtaining as yet unknown *o*-xylylene metal complexes. This moiety may behave in either (a) a chelating or (b) a bridging manner. With regard to (a), the ligand may be described as (i) a dicarbanion, functioning in an η^2-bidentate mode, for example, in the metallaindanes [M(CH$_2$C$_6$H$_4$CH$_2$-*o*)(η-C$_5$H$_5$)$_2$]

*School of Chemistry and Molecular Sciences, University of Sussex, Brighton, NB1 9QJ, United Kingdom.
†Division of Science and Technology, Griffith University, Nathan, Brisbane, Queensland, Australia 4111.

(M = Ti, Zr, Hf, or Nb), **5**;[1] (ii) the neutral dimethylenecyclohexadiene operating as an η^4-ligand, for example, in [Fe(CH$_2$C$_6$H$_4$CH$_2$-o-η^4)(CO)$_3$] (Ref. 2) or [Ru(CH$_2$C$_6$H$_4$CH$_2$-o-η^4)(PMe$_2$Ph)$_3$];[3] or (iii) an intermediate between (i) and (ii), for example, in [W(CH$_2$C$_6$H$_4$CH$_2$-o)$_3$], **6**,[4] or in one [the other behaving as (i)] of the chelating ligands of [{W(CH$_2$C$_6$H$_4$CH$_2$-o)$_2$O}$_2$Mg(thf)$_4$], **7**. As for (b), examples are the 15-membered macrocycle [{Mg(CH$_2$C$_6$H$_4$CH$_2$-o-μ)(thf)$_2$}$_3$], **2**,[5] [{Co(μ-CH$_2$C$_6$H$_4$CH$_2$-o)(η-C$_5$H$_5$)(μ-CO)}$_2$],[6a] and [Pt$_2$(CH$_2$C$_6$H$_4$CH$_2$-o-μ)Me$_4$(Ph$_2$PCH$_2$PPh$_2$-μ)];[6b] in all of these the *o*-xylylene group may be regarded as a bridging dicarbanion, whereas in [{Fe(CO)$_3$}$_2${C$_6$H$_4$(CH$_2$)$_2$-o-η^6-μ}],[7] it functions as a bis(η^3-allyl) formally involving the bis(allylic) dicarbanion (CH$_2$$=\overline{\text{C}=\text{CHCH}}=$ $\overline{\text{CHCH}}=$C=CH$_2$).[2-8] For compounds **5** to **7**, as well as [$\overline{\text{Pt(CH}_2\text{C}_6\text{H}_4\text{CH}_2}$-o)(cod)] where cod is 1,5-cyclooctadiene,[1] the di-Grignard **1** derived from *o*-C$_6$H$_4$(CH$_2$Cl)$_2$, or a chloride-free analog (especially **3**), are the starting material of choice.

The organodilithium reagent **4** is a convenient transfer reagent for the α-, α'-bis(trimethylsilyl)-*o*-xylylene ligand, and has been used for this purpose for obtaining *meso*-[$\overline{\text{M}\{\text{CH(SiMe}_3)\text{C}_6\text{H}_4\text{CHSiMe}_3}$-o}($\eta$-C$_5H_5$)$_2$] (M = Ti, Zr, or Hf),[9] *meso,meso*-Sn[$\overline{\text{CH(SiMe}_3)\text{C}_6\text{H}_4\text{CHSiMe}_3}$-o]$_2$,[10] and *meso*-Sn[$\overline{\text{CH(SiMe}_3)\text{C}_6\text{H}_4\text{CHSiMe}_3}$-o]. A feature of this ligand is the chirality at the α and α' positions, when, in principle, the derived metal chelate complex is found as either the meso or rac diastereoisomer; the former appears to be thermodynamically preferred.

The di-Grignard reagent **1** is accessible in high yield (90–96%) only under rather stringent conditions:[5] involving (a) the use of *o*-bis(chloromethyl)benzene rather than the dibromide analog, (b) care as to the source of the magnesium powder, (c) the use of tetrahydrofuran (thf) as the solvent, (d) high dilution, typically 0.075 *M*, and (e) the use of freshly purified *o*-bis(chloromethyl)benzene. As for (d), use of concentrations greater than ~ 0.1 *M* results in a much diminished yield of **1**. When a dilute thf solution of **1** is concentrated beyond ~ 0.1 *M* and set aside for several days at ambient temperature colorless needles of a magnesium chloride-free macrometallacycle (**2**) are obtained, which readily loses solvent to afford **3**.

Addition of chlorotrimethylsilane to a solution of the di-Grignard reagent **1** is a convenient route to *o*-bis[(trimethylsilyl)methyl]benzene. The yield is much higher than that obtained[11] by the *in situ* trapping Grignard reaction of *o*-bis(bromomethyl)benzene, magnesium, and chlorotrimethylsilane. The organodilithium reagent **4** is prepared in high yield by direct lithiation of *o*-C$_6$H$_4$(CH$_2$SiMe$_3$)$_2$ by LiBu-*n*-tmeda (tmeda = *N, N, N', N'*-tetramethyl-1,2-ethanediamine);[12] use of the diamine tmeda is crucial for the selective, high-yield reaction.

A. THE DI-GRIGNARD REAGENT OF *o*-BIS(CHLOROMETHYL)BENZENE, 1

Procedure

■ **Caution.**   *o-Bis(chloromethyl)benzene is a lachrymatory substance and should be handled with care in a well-ventilated hood.*

All operations including those in Sections B and C should be carried out in an atmosphere of purified nitrogen and solvents should be free from oxygen or moisture. Magnesium powder (1.20 g, 49 mmol) (May and Baker or Alpha, 50 mesh) is placed in a dry 500-mL three-necked flask equipped with gas inlet, a 250-mL dropping funnel, and a filtration tube attached to a 250-mL Schlenk flask (Fig. 1). The magnesium is suspended in 5 mL of thf to which 0.1 mL of 1, 2-dibromoethane is added.

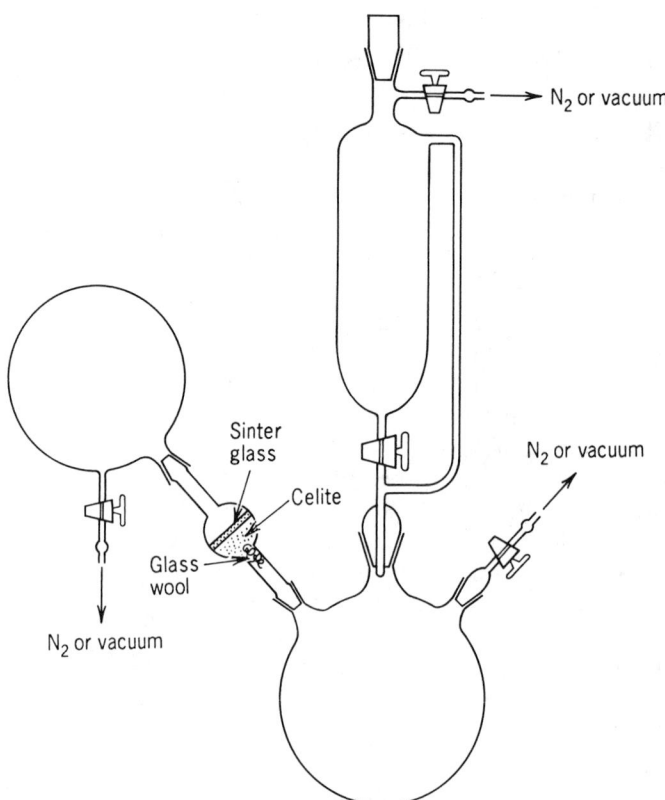

Fig. 1.   Reactor for preparation of the di-Grignard reagent solution.

■ **Caution.** *1,2-Dibromoethane is a carcinogen. Use a well-ventilated hood.*

Warming of the mixture is necessary until evolution of bubbles of ethene are evident, after which stirring is continued for ~ 5 min. The thf is then removed using a syringe and replaced by 12.5 mL of fresh thf. To this mixture is added dropwise, with stirring, a solution of *o*-bis(chloromethyl)benzene (2.14 g, 12.2 mmol) in 150 mL of thf over a period of 3.5 h, during which time the turbid solution may be green or pale yellow. Stirring is continued for ~ 15 h at ~ 20 °C, after which the suspension is filtered, and the concentration of the yellow di-Grignard filtrate **1** determined by quenching 1-mL aliquots with 0.1 *M* HCl and back titrating with 0.1 *M* NaOH. A typical concentration is 0.073 *M*, 96% yield.

The di-Grignard solution (**1**) is extremely air sensitive but it can be kept under nitrogen for prolonged periods. If stored at − 40 °C overnight, colorless needles of **2** are obtained. Removal of the solvent at − 40 °C, using a syringe, followed by washing with diethyl ether and drying *in vacuo* at − 10 to − 20 °C for up to a week, affords a material of stoichiometry {MgCH₂C₆H₄CH₂-*o*}-(thf), **2**. Yield of this material is typically 40 to 55%, mp 165 °C with decomposition.

Properties

Complex **2** decomposes rapidly in air, is only slightly soluble in thf, and is insoluble in many other inert organic solvents. The proton NMR spectrum (thf-d_8)[5] consists of a singlet at $\delta = 1.29$ ppm (rel. to SiMe₄)(4H, C₂H₂) and multiplets at $\delta = 6.48$ and $\delta = 6.08$ ppm (center of each AA′BB′ multiplet, 4H, C*H*). The X-ray structure of the magnesium–chloride-free macrocycle **2** has been determined.[5]

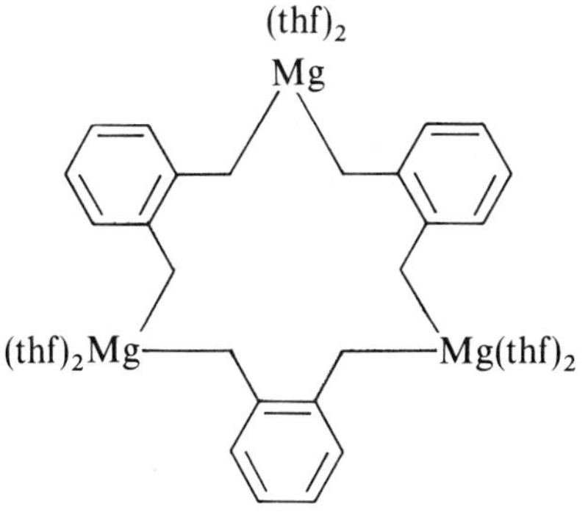

2

B.   *o*-BIS[(TRIMETHYLSILYL)METHYL]BENZENE

Procedure

■ **Caution.** *Chlorotrimethylsilane is volatile and toxic, and all operations involving its use should be conducted in an efficient hood.*

A thf solution of the di-Grignard reagent 1, prepared according to the scale and the procedure described in Section A (up to the filtration step), is cooled in an ice bath and chlorotrimethylsilane (4.4 mL, 35 mmol) is introduced, by means of a syringe, into the dropping funnel with the stopcock closed. (For this synthesis the filter tube and Schlenk flask, depicted in Fig. 1, are not required.) The chlorotrimethylsilane is slowly added under rapid stirring over a 2-h period and the mixture is stirred for a further 2 h at ambient temperature. Excess chlorotrimethylsilane and solvent are then removed, using a rotary evaporator. The product is extracted from the solid residue using three 20-mL portions of hexane, and to the combined extracts is added 5 mL of 1.0 M HCl. The mixture is then shaken and the hexane layer separated and dried with anhydrous sodium carbonate. The solution is then filtered and concentrated using a rotary evaporator. The resulting pale yellow liquid boils at 60 to 62 °C/0.1 torr. It is collected as 2.17 g of the colorless liquid, *o*-$C_6H_4(CH_2SiMe_3)_2$,   71%   based   on   the   initial   amount   of   *o*-bis(chloromethyl)benzene. (Checker reports yield of 1.5 g and boiling range 78–82 °C/0.1 torr.)

Properties

o-Bis[(trimethylsilyl)methyl]benzene is an air-stable, odorless liquid of low volatility. The proton NMR spectrum $(CDCl_3)^{12}$ consists of singlets at $\delta = 0.00$ ppm [18 H, $Si(CH_3)_3$] and $\delta = 2.03$ ppm (4 H, CH_2), and a multiplet at $\delta = 6.9$ (4 H, CH).

C.   μ-[(α, α′, 1,2-*h*: α, α′, 1,2-*h*)-*o*-BIS[(TRIMETHYLSILYL)METHYL]-BENZENE]BIS(*N, N, N′, N′*-TETRAMETHYL-1, 2-ETHANEDIAMINE)-DILITHIUM*

Procedure

■ **Caution.** *Butyllithium can be spontaneously combustible upon contact with air.*

Under an inert atmosphere butyllithium in hexane (purchased in bulk, 120 L, from Metallgesellschaft) (or may be purchased in smaller quantities from

---

**o-Bis[(tetramethylethylenediamine)-trimethylsilylmethyllithio]phenylene, 4.

Aldrich)(25.5 mL, 1.6 M, 44 mmol) is added to a stirred ice-cooled solution of tmeda (6.6 mL, 44 mmol) and o-bis[(trimethylsilyl)methyl]benzene (5.0 g, 20 mmol) in a 50-mL Schlenk flask. The color of the solution changes from pale yellow to orange and, after allowing the solution to warm to ambient temperature, stirring is discontinued. The solution is then set aside for 12 h, during which time massive yellow crystals are deposited. Using a syringe, the supernatant liquid is removed, and the residual product washed three times with 5-mL portions of hexane. It is then dried *in vacuo* at room temperature for 15 min. Yield: 7.41 g (75%).

Properties

The product is a yellow, crystalline solid (mp 164–165 °C), which is rapidly decomposed in air, being pyrophoric when finally divided, but is stable for prolonged periods under nitrogen. It is soluble in benzene, toluene, thf, or diethyl ether. The proton NMR spectrum $(C_6D_6)^{12}$ consists of singlets at $\delta = 0.45$ ppm [18 H, Si(CH_3)₃], $\delta = 0.96$ ppm (2H, CHSi), $\delta = 1.85$ ppm (8 H, NCH_2) and $\delta = 1.95$ ppm (24 H, NCH_3), and AA'BB' multiplets centered at $\delta = 6.53$ and $\delta = 7.11$ ppm (4 H, CH). The X-ray structure of the crystalline reagent **4** has been determined.[12]

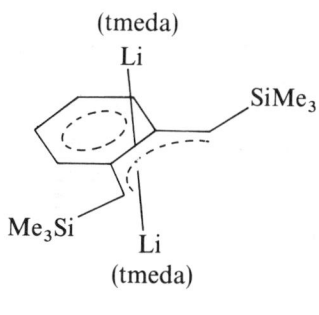

4

References

1. G. S. Bristow, M. F. Lappert, T. R. Martin, J. L. Atwood, and W. E. Hunter, *J. Chem. Soc. Dalton Trans.,* **1984**, 399.

2. W. R. Roth and J. D. Meier, *Tetrahedron Lett.,* **1967**, 2053.

3. S. D. Chappell, D. J. Cole-Hamilton, A. M. R. Galas, and M. B. Hursthouse, *J. Chem. Soc. Dalton Trans.,* **1982**, 1867.

4. M. F. Lappert, C. L. Raston, G. L. Rowbottom, B. W. Skelton, and A. H. White, *J. Chem. Soc. Dalton Trans.,* **1984**, 883.

5. M. F. Lappert, T. R. Martin, C. L. Raston, B. W. Skelton, and A. H. White, *J. Chem. Soc. Dalton Trans.,* **1982**, 1959.

6. (a) W. H. Hersh and R. G. Bergman, *J. Am. Chem. Soc.,* **103**, 6992 (1981); (b) A. T. Hutton, B. Shabanzadeh, and B. L. Shaw, *J. Chem. Soc. Chem. Commun.,* **1982**, 1343.

7. R. Victor and R. Ben-Shoshan, *J. Organomet. Chem.,* **80**, C1 (1974).

8. An interesting variant on (ii), found for the o-$C_6Me_4(CH_2)_2$ group, is provided by [Ru$\{\eta^4$-$C_6Me_4(CH_2)_2$-$o\}(\eta$-$C_6Me_6)$] in which the exocyclic methylenes are not involed in bonding to the metal (J. W. Hull and W. L. Gladfelter, *Organometallics,* **1**, 1716 (1982).

9. M. F. Lappert and C. L. Raston, *J. Chem. Soc. Chem. Commun.,* **1980**, 1284; M. F. Lappert, C. L. Raston, B. W. Skelton, and A. H. White, *J. Chem. Soc. Dalton Trans.,* **1984**, 893.

10. M. F. Lappert, W.-P. Leung, C. L. Raston, A. J. Thorne, B. W. Skelton, and A. H. White, *J. Organomet. Chem.,* **233**, C28 (1982).

11. H. Bock, and A. Alt, *J. Am. Chem. Soc.,* **92**, 1569 (1970); H. Bock and W. Kaim, *Chem. Ber.,* **111**, 3552 (1978).

12. M. F. Lappert, C. L. Raston, B. W. Skelton, and A. H. White, *J. Chem. Soc. Chem. Commun.,* **1982**, 14.

27. CYCLOMETALLATED ORGANOLITHIUM COMPOUNDS

Submitted by JOHANN T. B. H. JASTRZEBSKI* and GERARD VAN KOTEN*
Checked by M. F. LAPPERT,[†] P. C. BLAKE,[†] and D. R. HANKEY[†]

Cyclometallated organolithium compounds with one σ C—Li bond and at least one heteroatom–Li bond are valuable starting materials in organometallic[1] and organic synthesis.[2] These organolithium species are made either by a lithium–halogen exchange reaction[3,4] or via direct metallation when activating ligands such as tetramethylethylenediamine (tmeda) are used.[5] Further reactions are mostly carried out with the solutions or suspensions prepared *in situ.* This is, however, a great disadvantage in organometallic synthesis since other species present (R—X, LiX, tmeda, etc.) can either react further with the desired organometallic product (thereby lowering the yield) or hamper product isolation.

We describe here a series of organolithium compounds that can be prepared easily as pure crystalline solids. The synthesis involves a heteroatom assisted lithiation reaction[6] of the parent hydrocarbon using butyllithium and can, moreover, be scaled up without difficulty.

Under an inert atmosphere these extremely air-sensitive compounds are, in contrast to their solutions, almost indefinitely stable. The recording of

*Laboratory of Organic Chemistry, University of Utrecht, Dept. Metal Mediated Synthesis Padualaan 8, 3584 CH, Utrecht, The Netherlands.
[†]School of Chemistry and Molecular Science, University of Sussex, Brighton, BN1 9QJ, United Kingdom.

^{1}H NMR spectra in suitable solvents (see procedures) under an N_2 atmosphere provides the best method of checking their purity.

The organolithium compounds described here have proved to be valuable in the synthesis of a wide range of cyclometallated compounds, many of which are of current interest.[7] The precise site of lithiation in these reagents has been determined from investigations of the air-stable products that they form with Me_3SiCl and Me_3SnCl.[8]

General proceudre

■ **Caution.** *Organolithium compounds are extremely air and moisture sensitive and therefore all manipulations described here should be carried out under a dry nitrogen atmosphere using Schlenk-tube techniques.[9] Solvents are dried preferably with sodium metal and distilled under nitrogen. Pure isolated organolithium compounds are pyrophoric in air.*

The following detailed procedure is appropriate for the preparation and isolation of all the organolithium compounds on a 25-mmol scale. Precise information concerning reaction times, temperatures, quantity of reagents, and so on is given later for each specific synthesis.

A 100-mL Schlenk tube, fitted with a rubber serum cap and containing a magnetic stirring bar, is connected to a standard nitrogen gas–vacuum pump system and filled with nitrogen. Using syringe techniques the amine (which has been distilled and stored under N_2) is added to the Schlenk tube followed by solvent(s) and a solution of butyllithium in hexane. An $\sim 2.5\,M$ solution of butyllithium in hexane is obtainable from Aldrich. Before each synthesis the molarity of the solution is determined by methods described in Ref. 10. It is advised that the mixture be stirred until the reaction is complete, in order to prevent formation of the desired lithium product as large crystals which have reduced reactivity. In most preparations the product will then precipitate. However, in the case that it does not, the solvent and volatiles should be removed *in vacuo*. Pentane is added to form a suspension. The serum cap is removed and replaced by a joined glass frit filter (15–40 μ, about 3 cm in diameter) fitted with a vacuum tap on each side of the frit and connected to a second Schlenk tube. During this exchange, two strong strems of nitrogen are led through the system in order to ensure exclusion of air. Inversion of the whole system allows filtration of the mixture and, while keeping the solid under nitrogen, the upper (reaction) Schlenk tube is removed, the solid washed with pentane (3 × 10 mL), and a storage Schlenk tube connected to the filter. The lower Schlenk tube containing the mother liquor and washings is removed and the frit is closed by a rubber stopper. Then the solid product left on the filter, is deried *in vacuo*. Finally, the system is inverted again so that the dry material can be easily transferred to the storage Schlenk tube.

■ **Caution.** *Residual organolithium either in solution or on the filter should be carefully decomposed under nitrogen with a toluene–ethanol 10:1 mixture.*

A. [2-[(DIMETHYLAMINO)METHYL]PHENYL]LITHIUM

$$4 \left\langle \begin{array}{c} CH_2-NMe_2 \\ \end{array} \right\rangle + 4n\text{-BuLi} \longrightarrow \left[\begin{array}{c} CH_2-NMe_2 \\ \hline \quad -Li \end{array} \right]_4 + 4n\text{-BuH}$$

Procedure

This is a modification of the methods by Hauser and coworkers[3] and Viswanathan and Wilkie.[11] Following the general procedures just described, N,N-dimethylbenzylamine (3.37 g, 25 mmol) is allowed to react in diethyl ether (50 mL) with ~2.5 M of butyllithium in hexane (25 mmol, ~10 mL) for 36 h at room temperature after which the product is filtered off. Yield: 3 g (87%). ^{1}H NMR in THF-d_8 (δ ppm): NMe$_2$ 2.10 (6H), NCH$_2$ 3.40 (2H), H$_6$(aryl) 7.90 (1H) and other aryl-H 6.80 m (3H).

Properties

[2-[(Dimethylamino)methyl]phenyl]lithium is a white crystalline solid and pyrophoric in air. An X-ray structure determination reveals a tetranuclear structure in which each aryl moiety is four-center two-electron bound via C-1 to one Li$_3$ face of a Li$_4$ tetrahedron. The remaining free coordination site on each Li atom is occupied by the heteroatom containing substituent CH$_2$(Me$_2$)N.[12] It is insoluble in hydrocarbons and diethyl ether, but very soluble in THF in which the tetramer breaks down into a dimeric species containing two three-center two-electron bound aryl groups. This was established by ^{13}C NMR in THF-d_8 showing the resonance of C-1 as a seven line pattern arising from coupling to two equivalent ^{7}Li ($I = \frac{3}{2}$) atoms.[12,14]

B. [2-[(DIMETHYLAMINO)METHYL]-5-METHYLPHENYL]-LITHIUM

$$4 \left\langle \begin{array}{c} CH_2-NMe_2 \\ \end{array} \right\rangle_{Me} + 4n\text{-BuLi} \longrightarrow \left[\begin{array}{c} CH_2-NMe_2 \\ \hline \quad -Li \\ Me \end{array} \right]_4 + 4n\text{-BuH}$$

Procedure

Following the general procedures just described, N,N-4-trimethylbenzylamine (3.73 g, 25 mmol) in diethyl ether (50 mL) is allowed to react with ~ 2.5 M of n-butyllithium in hexane (25 mmol, ~ 10 mL) for 36 h at room temperature. After evaporating the solution *in vacuo* and adding pentane (20 mL) the solid is filtered off. Yield: 3 g (77%). ^{1}H NMR in C_6D_6 (δ ppm): NMe$_2$ 1.30 (3H) and 1.90 (3H); NCH$_2$ AB pattern 2.95 (1H) and 4.50 (1H), J_{AB} 12 Hz; Me(aryl) 2.30 (3H); H$_6$(aryl) 8.15 (1H), other aryl-H 7.00 (2H).

Properties

[2-[(Dimethylamino)methyl]-5-methylphenyl]lithium is a white crystalline solid and is pyrophoric in air. Furthermore, it is soluble in hydrocarbons and ethers. Molecular weight determinations in benzene have established a tetrameric structure.[12] In THF, like [2-[(dimethylamino)-methyl]phenyl]lithium, this tetramer breaks down to a dimeric species.

C.   [2-[(DIMETHYLAMINO)PHENYL]METHYL]LITHIUM

Procedure

This is a modification and extension of the preparation given by Manzer.[13] Using the general procedures just described, a quantity of N,N-2-trimethylaniline (3.37 g, 25 mol) is placed in a Schlenk tube, and to this is added hexane (50 mL), diethyl ether (2 mL), and ~ 2.5 M of butyllithium in hexane (24 mmol, ~ 10 mL). The serum cap is replaced by a reflux condenser, which is connected to a nitrogen gas line and an oil bubbler that is suitably designed to prevent build up of internal pressure. The mixture is refluxed for 10 h. After cooling to room temperature, the precipitate is filtered off under inert atmosphere. Yield: 2.8 g (79%). ^{1}H NMR in THF-d_8 (δ ppm): NMe$_2$ 2.50 (6H); CH$_2$Li 1.48 (2H); aryl-H 5.65 (1H), 6.30 (2H) and 7.00 (1H).

Properties

[2-[(Dimethylamino)phenyl]methyl]lithium is a pale yellow crystalline solid pyrophoric in air that is insoluble in hydrocarbons but readily soluble in THF.

To date there is little information concerning the nuclearity of this Li compound either in solution or in the solid state.

D. (DIETHYL ETHER)[8-(DIMETHYLAMINO)-1-NAPHTHYL]LITHIUM

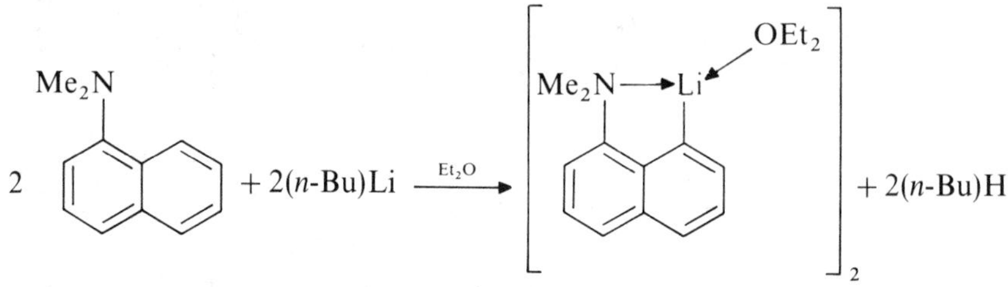

Procedure

Using the general procedure just described, N,N-dimethyl-1-naphthaleneamine (4.28 g, 25 mmol) in diethyl ether (40 mL) is reacted with a $\sim 2.5\,M$ hexane solution of butyllithium (25 mmol, $\sim 10\,\text{mL}$) at room temperature for 36 h after which the product is filtered off. Yield: 4.3 g (70%). ^{1}H NMR in C_6D_6 (δ ppm): Me(Et$_2$O) 1.05 (6H); CH$_2$(Et$_2$O) 3.20 (4H); NMe$_2$ 1.95 (6H); Aryl-H 7.05 (2H), 7.30 (1H), 7.65 (2H) and 8.15 (1H).

Properties

(Diethyl ether)[8-(dimethylamino)-1-naphthyl]lithium is a very oxygen- and moisture-sensitive yellow crystalline solid. An X-ray structure determination[14] reveals a dimeric complex in which each of the 8-(dimethylamino)-1-naphthyl groups is bound to two Li atoms via C-1 and via the lone pair of the NMe$_2$ nitrogen atom, to one of these Li atoms. The fourth coordination site of each Li atom is occupied by a coordinated diethyl ether molecule. ^{1}H, ^{13}C, and ^{7}Li NMR studies[12,14] indicate that in solution an equilibrium exists between the dimeric diethyl ether complex and an ether-free organolithium species whose structure and aggregation is still under investigation. This white, ether-free species can be isolated as a microcrystalline solid in almost quantitative yield by simply removing the solvent *in vacuo* from a benzene solution of the etherate.

References

1. Compare I. Omae. *Chem. Rev.*, **79**, 287 (1979) and *J. Organometal. Chem. Library*, **18**, 1 (1986); G. R. Newkome, W. E. Puckett, V. K. Gupta, and G. E. Kiefer, *Chem. Rev.*, **86**, 451 (1986).

2. D. W. Slocum and D. I. Sugerman, *Adv. Chem. Ser.*, 130 (1974); E. M. Kaiser and D. W. Slocum, *Organic Reactive Intermediates*, S. P. McManus (ed.), Academic Press, New York, 1973, Chapter 5.

3. Compare F. N. Jones, M. F. Zinn, and C. R. Hauser, *J. Org. Chem.*, **28**, 663 (1963).

4. M. I. Bruce, *Angew. Chem.*, **89**, 75 (1977).

5. Compare R. E. Ludt, G. P. Crowther, and C. R. Hauser, *J. Org. Chem.*, **35**, 1288 (1970).

6. H. W. Gschwend and H. R. Rodriquez, *Org. React.*, **26**, 1 (1979).

7. Compare L. E. Manzer, *J. Am. Chem. Soc.*, **100**, 8068 (1978); F. A. Cotton and G. N. Mott, *Inorg. Chem.*, **20**, 3896 (1981); G. van Koten and J. G. Noltes, *Comprehensive Organometallic Chemistry*, Vol. 2, Pergamon, New York, 1982, Chapter 14; C. Arlen, M. Pfeffer, O. Bars, and D. Grandjean, *J. Chem. Soc. Dalton Trans.*, 1535 (1983).

8. G. van Koten, J. T. B. H. Jastrzebski, J. G. Noltes, W. M. G. F. Pontenagel, J. Kroon, and A. L. Spek, *J. Am. Chem. Soc.*, **100**, 5021 (1978); J. T. B. H. Jastrzebski, G. van Koten, C. T. Knaap, A. M. M. Schreurs, J. Kroon, and A. L. Spek, *Organometallics*, **5**, 1551 (1983).

9. D. F. Shriver, *The Manipulation of Air-Sensitive Compounds*, McGraw-Hill, New York, 1969; R. B. King, *Organometallic Synthesis*, Vol. 1., J. J. Eisch and R. B. King (eds.), Academic Press I, New York, 1965.

10. H. Gilman and F. K. Cartledge, *J. Organometal. Chem.*, **2**, 447 (1964).

11. C. T. Viswanathan and C. A. Wilkie, *J. Organometal. Chem.*, **54**, 1 (1973).

12. J. T. B. H. Jastrzebski, G. van Koten, M. Konijn, and C. H. Stam, *J. Am. Chem. Soc.*, **104**, 5490 (1982); *J. Organometal. Chem.*, **353**, 133, 145 (1988).

13. L. E. Manzer, *J. Organometal. Chem.*, **135**, C6 (1977).

14. J. T. B. H. Jastrzebski, G. van Koten, C. Arlen, and M. Pfeffer, *J. Organometal. Chem.*, **246**, C75 (1983); A. A. H. van der Zeÿden, and G. van Koten, *Recl. Trav. Chim. Pays-Bas*, **107**, 431 (1988).

28. METALLATION OF AROMATIC KETONES WITH PENTACARBONYLMETHYLMANGANESE(I)

Submitted by R. J. MCKINNEY* and S. S. CRAWFORD*
Checked by K. OBERDORF† and J. L. SPENCER†

The internal metallation of aromatic groups on ligands of transition metal complexes has been the subject of widespread studies primarily because of interest in C—H activation.[1,2] Most studies have involved either nitrogen or phosphorus donor ligands, whereas very few have explored metallation with oxygen donor ligands. Perhaps this disparity may be traced to the relatively weak bonding interaction of oxygen-donor ligands in general towards late transition metal complexes as compared to nitrogen and phosphorus-donor

*Department of Chemistry, University of California, Los Angeles, CA 90024. Present Address for R. J. McKinney: Central Research & Development Department, E. I. du Pont de Nemours & Company, Experimental Station, Wilmington, DE 19880-0328.
†Department of Inorganic Chemistry, University of Bristol, Bristol BS8 1TS, United Kingdom.

ligands. However, the directed metalation of molecules with oxygen function-ality by thallium salts has illustrated the practical utility that metallation may bring to organic synthesis.[3] Here we describe the metallation of aromatic ketones by methylmanganesepentacarbonyl. Section A is the reaction with acetophenone.[4] Section B is a rather unique metallation of an aromatic ketone, which results from cyclometallation of triphenylphosphine.[5] The product of Section B is of further interest because it contains an extended planar π system that includes the two manganese atoms, which are not bonded directly to each other.[6]

Preparation of the starting materials $Mn(CO)_5(CH_3)$ and the phosphine-substituted derivative $[cis\text{-}Mn(CO)_4(PPh_3)(CH_3)]$ follow those from the literature.[7,8] Solvents are dried over 4A molecular sieves and purged with nitrogen before use (except for chromatography). Silica gel for chroma-tography was obtained from Baker.[‡] All reactions should be carried out under an inert atmosphere but after cooling may be worked up in air (however, solutions of the products slowly decompose in air).

▪ **Caution.** *Because of the high toxicity of carbon monoxide, these reactions should be carried out in an efficient fume hood.*

A. PENTACARBONYLMETHYLMANGANESE(I) AND (2-ACETYLPHENYL-C, O)TETRACARBONYLMANGANESE

1. $Mn_2(CO)_{10} \xrightarrow[(2)CH_3I]{(1)Na/Hg} Mn(CO)_5CH_3$

Procedure

1. A 500-mL three-neck round-bottomed flask equipped with an overhead mechanical stirrer and nitrogen bubbler is flushed with nitrogen and charged

‡The checkers, apparently using a higher activity silica gel (BDH 60–120 mesh), found product retention and degradation on columns to be a problem. We have also used Florisil (Matheson, Coleman and Bell) for an even lower activity solid phase. Improved yields were also observed when chromatography was carried out under N_2 with dried and distilled solvents.

with mercury (50 mL). An amalgam is prepared by adding sodium (6 g, 0.26 mol) in very small pieces while stirring.

■ **Caution.** *Amalgam formation is very exothermic. Do not continue adding sodium until each piece has reacted.*

After the amalgam has cooled, anhydrous, deoxygenated diethyl ether (200 mL) and $Mn_2(CO)_{10}$ (20 g, 51 mmol) are added and the mixture stirred vigorously for 2 h. The gray-green solution is separated from the excess amalgam with the aid of a large syringe (100 mL) and transferred to a separate 500-mL three-neck round-bottomed flask equipped with a nitrogen bubbler, magnetic stirring bar, and a pressure-equalizing dropping funnel. The dropping funnel is charged with iodomethane, CH_3I (10 mL, 0.16 mol) in deoxygenated diethyl ether (40 mL). The mixture is cooled in an ice bath and the CH_3I added slowly with stirring. The solution is allowed warm to ambient temperature for a half hour and then the solvent and excess CH_3I are evaporated under reduced pressure (water aspiration). The white crystalline product, air stable as a solid, is purified by sublimation at 40 °C/0.1 torr onto a probe cooled to 0 or − 78 °C in a closed system (because of the high volatility of the product at room temperature). Yield: ∼ 15 g [70% based on $Mn_2(CO)_{10}$].

2. A 200-mL round-bottomed flask equipped with a reflux condenser, nitrogen bubble, and magnetic stirring bar is flushed with nitrogen and then charged with pentacarbonylmethylmanganese (1.0 g, 4.8 mol), acetophenone (∼ 4 mL, 34 mmol), and octane (100 mL). The solution is heated under reflux for 0.5 h, cooled to room temperature, and applied directly to a silica gel chromatography column (5 × 50 cm).[‡] Elution with hexane gives first a yellow band of $Mn_2(CO)_{10}$ (∼ 10% yield) followed by another yellow band. Solvent is removed from the second band under reduced pressure giving the yellow crystalline product (0.7 g). The complex may be recrystallized from hexane–dichloromethane mixtures by slowly evaporating the mixture under a nitrogen flow.

Anal. Calcd. for $C_{12}H_7O_5Mn$: C, 50.4; H, 2.45. Found: C, 50.5; H, 2.69.

Properties

The yellow crystalline product is air stable as a solid but will slowly decompose in solution under air. It is soluble in all organic solvents. It is characterized by carbonyl IR absorptions at 2082 (m), 1997 (vs), and 1947 (s) cm^{-1}; and proton NMR spectra (60 MHz, $CDCl_3$) δ 2.6 (s), 7.17 (td), 7.41 (td), 7.83 (dd), 8.09 (dd).

[‡]See footnote on p. 156.

B. OCTACARBONYL-1κ⁴*C*, 2κ⁴*C*-μ-[CARBONYL-1κ*C*:2κ*O*-[(3-DIPHENYLPHOSPHINO-1κ*P*)-*o*-PHENYLENE:2κ*C*²]]-DIMANGANESE, 5

1. $Mn(CO)_5CH_3 + PPh_3 \longrightarrow$ [*cis*-$Mn(CO)_5(CH_3)(PPh_3)$]

2. [*cis*-$Mn(CO)_5(CH_3)(PPh_3)$] $\longrightarrow$

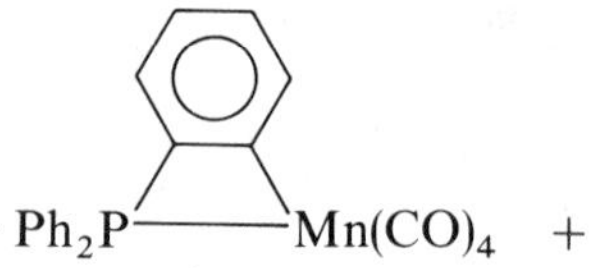
Ph2P
Mn(CO)4

1

\+

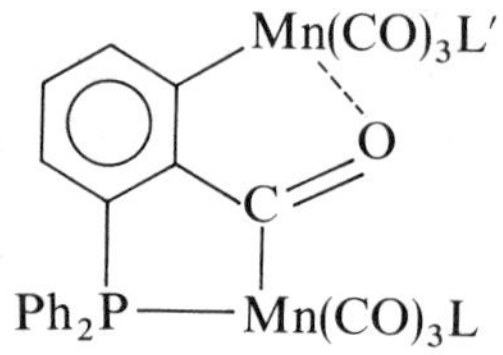
Mn(CO)3L′
O
C
Ph2P
Mn(CO)3L

2, L = PPh_3, L′ = CO
3, L = CO, L′ = PPh_3

3. **1** + CO $\longrightarrow$

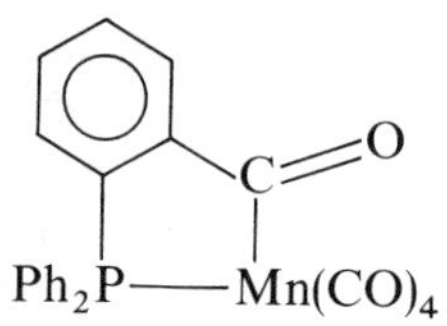
O
C
Ph2P
Mn(CO)4

4

4. **4** + $Mn(CO)_5CH_3 \longrightarrow$

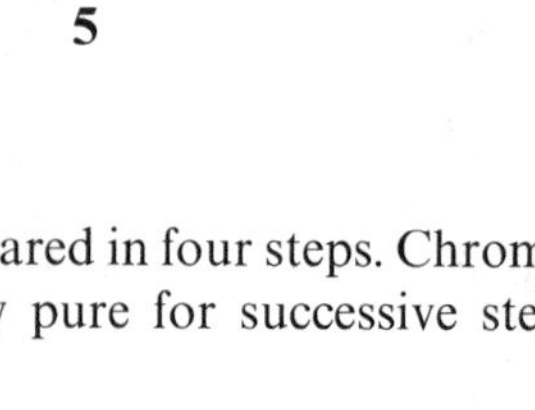
Mn(1)(CO)4
O
C
Ph2P
Mn(2)(CO)4

\+ CH_4 + CO

5

Procedure

The desired metallation product **5** is prepared in four steps. Chromatographically separated products are sufficiently pure for successive steps without recrystallization.

1. A 200-mL round-bottomed flask equipped with a reflux condenser, nitrogen bubbler, and magnetic stirring bar is flushed with nitrogen and charged with $Mn(CO)_5CH_3$ (2.1 g, 10 mmol), triphenylphosphine (2.6 g,

10 mmol), and dry, deoxygenated tetrahydrofuran (100 mL). The mixture is heated under reflux for 20 h. The solvent is removed under reduced pressure, and the amber colored product recrystallized from diethyl ether–hexane under nitrogen. Yield: 3.5 g (80%).

2. A 200-mL round-bottomed flask equipped with a reflux condenser, nitrogen bubbler, and magnetic stirring bar is flushed with nitrogen and charged with *cis*-tetracarbonyl(methyl)(triphenylphosphine)manganese (1.0 g; 2.2 mmol) and deoxygenated toluene (100 mL). The solution is heated under reflux for 3 h. The amber solution is cooled and the solvent removed under reduced pressure. The brownish residue is applied to a column (5 × 50 cm) in a 10% benzene–hexane solution.[‡] The products are eluted with gradually increasing percentage of benzene–hexane mixtures.

■ **Caution.** *Owing to the toxicity of benzene, the chromatographic separation should be carried out in an efficient fume hood.*

The first pale yellow band elutes as a virtually colorless solution, which upon removal of solvent and recrystallization from diethyl ether–hexane, yields snow white crystals (0.5 g, 1.2 mmol) of the metallation product, **1**.

The second bright yellow band separates from the third yellow band with 15% benzene–hexane. Removing the solvent and recrystallizing the residue from dichloromethane–hexane yields the secondary metallation products **2** (0.08 g, 0.10 mmol) and **3** (0.12 g, 0.14 mmol), respectively. These are the phosphine-substituted derivatives of the secondary metallation product described below. The yield of these two products may be maximized by heating the starting complex in the absence of solvent to 145 °C (whereupon it melts) for 15 to 20 min.

3. A clean stainless steel Hoke cylinder (∼ 100 mL) is charged with $\overline{Ph_2P—C_6H_4—M}n(CO)_4$ (**1**) (0.5 g, 1.2 mmol) and benzene (30 mL), sealed and degassed by the free–thaw method. The cylinder is pressurized with carbon monoxide (50 psig), closed and the temperature is lowered to − 198 °C with liquid nitrogen. The cylinder is removed from the liquid nitrogen and pressurized again to 50 psig of CO while still cold. This procedure brings the pressure in the cylinder at room temperature to ∼ 100 psig without the use of any special pressure equipment. The cylinder is heated in an oven at 90 °C for 5 h. Upon cooling, the cylinder is vented, emptied into a round-bottomed flask, and the solvent removed under reduced pressure. The residue is chromatographed on a column using 30% dichloromethane–hexane eluent.[‡] Two pale yellow bands [of $Mn_2(CO)_{10}$ and unreacted **1**] preceded the bright yellow product band. Recrystallization from diethyl ether–hexane yields yellow crystals of the acylated product (**4**) (0.45 g, 1.0 mmol).

4. A 100-mL round-bottomed flask equipped with a reflux condenser, nitrogen bubbler, and magnetic stirring bar is flushed with nitrogen and then

[‡]See footnote on p. 156.

charged with **4** (0.30 g, 0.66 mmol), pentacarbonylmethylmanganese (0.50 g, 2.4 mmol), and deoxygenated methylcyclohexane (50 mL). The solution is heated under reflux for 2.5 h, then cooled and applied directly to a silica gel column (2.5 × 50 cm).[‡]

■ **Caution.** *Owing to the toxicity of benzene, the chromatographic separation should be carried out in an efficient fume hood.*

Products are eluted with benzene–hexane mixtures. The compounds $Mn_2(CO)_{10}$ and $Mn(CO)_5CH_3$ precede the bright yellow product band. Removal of solvent and recrystallization from dichloromethane–hexane gives yellow crystals of **5** (0.32 g, 0.51 mmol).

Anal. Calcd. for **1**, $C_{22}H_{14}MnO_4P$; and the dimers **2** and **3**, $C_{44}H_{28}Mn_2O_8P_2$: C, 61.68; H, 3.29. Found for **1**: C, 61.65; H, 3.46; **2**: C, 62.23; H, 3.44; **3**: C, 62.08; H, 3.43. *Anal.* Calcd. for **4**, $C_{23}H_{14}MnO_5P$: C, 60.52; H, 3.07. Found for **4**: C, 60.26; H, 3.13. *Anal.* Calcd. for **5**, $C_{27}H_{13}Mn_2O_9P$: C, 52.12; H, 2.11. Found for **5**: C, 52.13; H, 2.28.

Properties

Complexes **1** to **5** are air stable as solids but decompose slowly in solution. They are moderately soluble in most organic solvents. They are characterized by very distinctive IR carbonyl stretching patterns (cyclohexane solutions) v_{CO} **1**, 2067 (m), 1987 (sh), 1983 (s), 1951 (s); **2**, 2074 (m), 2024 (w), 1992 (s), 1984 (sh), 1946 (s), 1935 (s), 1926 (sh), ($v_{C=O-M}$) 1460; **3** 2073 (m), 2011 (vs), 2006 (sh), 1990 (m), 1968 (s), 1936 (s), 1904 (m), 1899 (m), (v_{CO-M}) 1470; **4**, 2068 (m), 2005 (m), 1976 (s), 1963 (s) (CCl_4) 1618; **5**, 2085 (m), 2071 (m), 2014 (s), 1998 (vs), 1992 (sh), 1977 (s), 1942 (s), ($v_{C=O-M}$) 1473 cm^{-1}.

References

1. The Specialists' Periodical Reports: *Organometallic Chemistry*, Vol. 2 (1972)–Vol. 7 (1977) review the literature on the subject of cyclometalation and contain over 350 references. Other references include A. J. Carty, *Organomet. Chem. Rev. Sect. A*, **7**, 191, (1972); M. I. Bruce and B. L. Goodall, *Chemistry of Azo, Hydrazo, and Azoxy Compounds*, S. Patai (ed.) Wiley, New York, 1974, Chapter 9.
2. G. W. Parshall, *Acct. Chem. Res.*, **8**, 113 (1975); **3**, 139 (1970).
3. A. McKillop and E. C. Taylor, *Chem. Br.*, **9**, 4 (1973).
4. R. J. McKinney, G. Firestein, and H. D. Kaesz, *Inorg. Chem.* **14**, 2057 (1975); C. B. Knobler, S. S. Crawford, and H. D. Kaesz, *Inorg. Chem.*, **14**, 2062 (1975).
5. R. J. McKinney and H. D. Kaesz, *J. Am. Chem. Soc.*, **97**, 3066 (1975).
6. B. T. Huie, C. B. Knobler, G. Firestein, R. J. McKinney, and H. D. Kaesz, *J. Am. Chem. Soc.*, **99**, 7852, 7862 (1977).

[‡]See footnote on p. 156.

7. R. D. Closson, J. Kozikowski, and T. H. Coffield, *J. Org. Chem.* **22**, 598 (1957); R. B. King, *Organometallic Synthesis*, J. J. Eisch and R. B. King eds., Academic Press, New York, 1965, p. 147.

8. C. S. Kraihanzel and P. Maples, *J. Am. Chem. Soc.*, **87**, 5267 (1965).

29. CYCLO-COTRIMERIZATION OF PHOSPHINOTHIOYLS WITH ALKYNES. PHOSPHINOTHIOYL-CONTAINING MANGANESE ANALOGS OF CYCLOPENTADIENES AND BICYCLOHEPTADIENES

Submitted by E. LINDNER,* A. RAU,* and V. KÄSS*
Checked by MARTIN A. BENNETT† and HORST NEUMANN†

In the course of the cyclotrimerization of alkynes and the cyclo-cotrimerization of alkynes with olefines, metallacyclopentadienes or metal-lacyclopentenes occur as intermediates.[1,2] As investigations on model compounds have shown, these intermediates further react with an alkyne to give metallabicycloheptadienes or, in dependence on the electronic nature of the alkyne, metallacycloheptatrienes, and metallacycloheptadienes, respectively.[3-9] These are transformed into the corresponding benzene or cyclohexadiene derivatives either spontaneously (catalytically) or after addition of certain ligands or oxidizing agents, accompanied by the elimination of the metal residue. The transfer of this principle to the preparation of heterocycles was hitherto successful only in the case of the cobalt catalyzed cyclo-cotrimerization of acetylenes with nitriles to pyridine compounds.[10] Of special interest is the cyclo-cotrimerization of the $R_2P{=}S$ function with alkynes, which proceeds via different intermediates and can be used for the synthesis of S, O, and/or P containing heterocycles. Because of the similar atomic radii of phosphorus and sulfur and their comparable electronegativity values, the $P{=}S$ system behaves like an alkyne.[11]

The reactions being described start with an (η^2-phosphinothioyl)manganese complex of the type $(OC)_4Mn(\eta^2{-}S{=}PR_2)$. If R is not a bulky substituent, these complexes are subject to an immediate dimerization to give $[(OC)_4Mn(\mu{-}S{=}PR_2)]_2$. Between the monomeric and dimeric species exists a dissociative equilibrium.[12]

*Institut für Anorganische Chemie, Universität Tübingen, D-7400 Tübingen 1, Federal Republic of Germany.
†Research School of Chemistry, Australian National University, Box 4, P.O. Canberra, A.C.T. 2600, Australia.

General Procedures

All reactions and manipulations should be performed under an atmosphere of dry nitrogen either in a dry box or using Schlenk tube techniques. Filtered and degassed solvents should be used for the chromatographic separations.

A. DIMETHYLPHOSPHINE SULFIDE

$$(CH_3)_2 \underset{\substack{\| \\ S}}{P} - \underset{\substack{\| \\ S}}{P}(CH_3)_2 + NaOH \longrightarrow (CH_3)_2HPS + (CH_3)_2PSONa$$

Procedure

A 100-mL reaction flask, which is equipped with a magnetic stirrer, a reflux condenser, and a mercury excess pressure outlet, is flushed thoroughly with nitrogen and then charged with 4.0 g (0.1 mol) of NaOH in 50 mL of degassed water. To this solution is added in one portion [SP(CH$_3$)$_2$]$_2$ (18.6 g, 0.1 mol)[13] at 20 °C. The mixture is vigorously stirred and quickly heated to 90 °C. After stirring for 1 h at this temperature the solution is cooled to room temperature and extracted three times with 25 mL of carefully degassed trichloromethane. The combined trichloromethane solutions are dried over degassed Na$_2$SO$_4$ and subsequently the solvent is removed under vacuum. The product is purified by distillation (bp 50 °C/0.5 torr). Yield: 4.5–6.5 g (47–69%).[14]

Properties

Colorless, air-sensitive liquid. Soluble in all organic solvents. IR spectrum (CCl$_4$, cm^{-1}): 2348 (st) [$\nu_{(P-H)}$]; (Film): 595 (st) [$\nu_{(P=S)}$]. ^{1}H NMR [CDCl$_3$ rel. tetramethylsilane (TMS)]: δ 1.82 (dd, $^2J_{PH} = 14.8$, $^3J_{HH} = 4.7$ Hz, CH$_3$); 6.94 (dsept, $^1J_{PH} = 447$, $^3J_{HH} = 4.7$ Hz, H). ^{31}P {^{1}H} NMR (CHCl$_3$, rel. 1% H$_3$PO$_4$): δ 4.1 (s).

B. OCTACARBONYLBIS-μ-(DIMETHYLPHOSPHINOTHIOYL-*P:S*)DIMANGANESE

$$2BrMn(CO)_5 + 2(CH_3)_2HPS \xrightarrow[-2[C_2H_5(i\text{-}C_3H_7)_2NH]Br, \ -2CO]{2C_2H_5(i\text{-}C_3H_7)_2N}$$

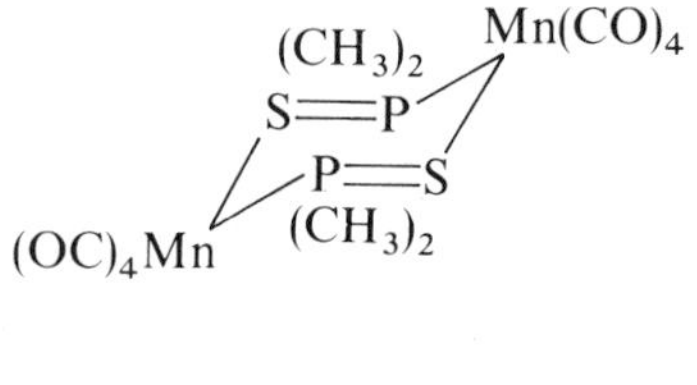

Procedure

A 20-mL Schlenk tube, which is equipped with a magnetic stirrer, a reflux condenser, a pressure-equalizing addition funnel, and a mercury excess pressure outlet, is flushed thoroughly with nitrogen and then charged with 1.096 g (4.0 mmol) of $BrMn(CO)_5$ in 50 mL of diisopropyl ether. To the stirred solution is added dropwise a solution of $(CH_3)_2HPS$ (376 mg, 4.0 mmol) and of $C_2H_5(i\text{-}C_3H_7)_2N$ (517 mg, 4.0 mmol, 0.697 mL) in 50 mL of diisopropyl ether at 60 °C. The reaction mixture is then stirred for 1 h at 60 °C. The solvent is removed under vacuum at room temperature and the residue is suspended in 10 mL of ice-cold methanol. The yellow compound is filtered (P4), washed with 10 mL of ice-cold methanol and dried under vacuum. From the methanol mother liquor additional product crystallizes at − 40 °C. The crude product is recrystallized from methanol at − 40 °C. Yield: 411 mg (39%).

Anal. Calcd. for $C_{12}H_{12}Mn_2O_8P_2S_2$: C, 27.71; H, 2.33; Mn, 21.12; S, 12.33; MW 520.18. Found: C, 27.81; H, 2.56; Mn, 21.77; S, 12.01; MS (EI, 70 eV, 200 °C): m/e 520 [M^+].

Properties

The purified product consists of orange-red, roughly octahedral crystals, thermally stable to 110 °C. It is moderately soluble in nonpolar solvents and is soluble in polar solvents. The solubility in methanol depends remarkably on the temperature of the solvent. In the carbonyl region of the IR spectrum (hexane) eight $C\equiv O$ bands are observed: 2088 (w), 2075 (s), 2018 (m), 2011 (s-vs), 2007 (vs), 1995 (m-s), 1968 (m-s), 1961 (s) cm^{-1}. ^1H NMR (CDCl$_3$, rel. TMS): δ 2.0 (d, $^2J_{PH}$ = 7.5 Hz, CH$_3$). $^{31}P\{^1H\}$ NMR (toluene, rel. 1% H$_3$PO$_4$, − 40 °C): δ 39.8 (s). According to an X-ray structural analysis, the compound crystallizes from hexane in the orthorhombic space group *Pbca* with $a = 1123.0(1)$, $b = 1556.5(3)$, $c = 1125.9(2)$ pm, $Z = 4$, $d_{calcd} = 1.75\,g\,cm^{-3}$, $d_{obs} = 1.75\,g\,cm^{-3}$ (Mn—S = 242.4, Mn—P = 232.4, P$=$S = 206.6 pm).[15]

C. TETRACARBONYL[2-(DIMETHYLPHOSPHINOTHIOYL)-1, 2-BIS(METHOXYCARBONYL)ETHENYL-C′, S]MANGANESE

$$[(OC)_4Mn(\mu\text{-}S\!=\!P(CH_3)_2)]_2 + 2C_2R_2 \xrightarrow{\text{THF}} 2(OC)_4Mn\diagup\begin{smallmatrix}S\!\!\nwarrow\!\!P(CH_3)_2\\ \\ \diagdown R\\ R\end{smallmatrix}$$

$$R = CO_2CH_3$$

■ **Caution.** *This procedure requires the use of highly toxic carbon monoxide and should be carried out in a well-ventilated hood.*

Procedure

A 200-mL Schlenk tube, which is equipped with a magnetic stirrer, a reflux condenser, and a mercury excess pressure outlet, is flushed thoroughly with nitrogen and then charged with a solution of $[(OC)_4Mn(\mu-S=P(CH_3)_2)]_2$ (300 mg, 0.58 mmol) and $C_2(CO_2CH_3)_2$ (330 mg, 2.32 mmol) in 100 mL of tetrahydrofuran (THF). The reaction mixture is heated with an oil bath to 66 °C over a period of 20 min in an atmosphere of carbon monoxide. The reaction is followed by means of thin layer chromatography, and is interrupted after ~ 15 min by evaporation of the solvent under vacuum when the amount of $(OC)_4\overline{Mn-S=P(CH_3)_2C(CO_2CH_3)=\overset{\rceil}{C}(CO_2CH_3)}$ $[R_f = 0.805$ (dichloromethane/ethyl acetate = 2:1), 0.723 (dichloromethane/ethyl acetate = 10:1)] has achieved a maximum. The purification is carried out by chromatography with CH_2Cl_2 on a short silica gel column ($L = 5$ cm, $\phi = 2$ cm). Solvent is then removed under vacuum. Further purification is achieved by means of *Middle Pressure Liquid Chromatography* on a silica gel column with CH_2Cl_2 (third fraction) [*MPLC*: Lobar "Fertigsäule," seize B (310-25) LiChroprep Si 60 (40–60 μm) (Merck); Duramat dosing pump of CFG; UV detector type 6, recorder UA 5, and multiplexer 1133 of ISCO]. Final recrystallization is from petroleum ether (60–90 °C)/CHCl₃. Yield: 100–240 mg (21–52%).

Anal. Calcd. for $C_{12}H_{12}MnO_8PS$: C, 35.84; H, 3.01; Mn, 13.66; S, 7.97; MW, 402.2. Found: C, 36.13; H, 3.16; Mn 13.61; S, 8.25; MS (*EI*, 70 eV): m/e 402 [M⁺].

Properties

The product consists of yellow crystals (mp 164–168 °C with dec). It is soluble in polar organic solvents and somewhat unstable under atmospheric oxygen. The compound will react with activated alkynes. In the IR spectrum (CCl_4) four C≡O absorptions at 2087 (m), 2012 (vs), 2008 (s-vs), 1963 (s-vs), and two closely separated bands at 1525 and 1520 cm⁻¹ (KBr) (C=C) and one absorption at 555 (w) cm⁻¹ (polyethylene)(P=S) are observed. ³¹P {¹H} NMR (CH_2Cl_2, rel. 1% H_3PO_4): δ 76.1 (s). ¹H NMR ($CDCl_3$, rel. TMS): δ 2.03 [d, $^2J_{PH} = 13.7$ Hz, $P(CH_3)_2$]; 3.76 (s), 3.86 (s) (OCH₃). The unit cell constants as determined from single-crystal X-ray diffraction are as follows: $a = 2434.9(6)$, $b = 1438.2(4)$, $c = 1032.1(9)$ pm, $\beta = 108.99(4)°$, $Z = 8$, $d_{calcd} = 1.56$ g cm⁻³, $V = 3.418 \times 10^9$ pm³. From CH_2Cl_2/hexane the compound crystallizes in the monoclinic space group $C2/c$ (P=S = 199.6, C=C = 135.6, Mn—S = 240.6, Mn—C = 204.9 pm).[16]

D. TRICARBONYL[η⁴-3,6-DIHYDRO-3,4,5,6-TETRAKIS(METHOXYCARBONYL)-2,2-DIMETHYL-2H-1,2-THIAPHOSPHORIN-2-IUM-3,6-DIYL]MANGANESE

$$R = CO_2CH_3$$

Procedure

A 100-mL Schlenk tube, which is equipped with a magnetic stirrer, a reflux condenser, and a mercury excess pressure outlet, is flushed thoroughly with nitrogen. Subsequently a quantity of $(OC)_4Mn—S=P(CH_3)_2—C(CO_2CH_3)=C(CO_2CH_3)$ (140 mg, 0.35 mmol) and dimethyl 2-butynedioate (77 mg, 0.54 mmol) are placed in this Schlenk tube, to which is added 50 mL of THF. The mixture is brought to reflux (66 °C) over a period of 15 min and then magnetically stirred for 2 h at this temperature. After cooling to room temperature the solvent is removed under vacuum. Subsequently the compound is dissolved in 30 mL of dichloromethane/ethyl acetate = 1:1 and purified by chromatography on a short silica gel column ($L = 5$ cm, $\phi = 2$ cm) with dichloromethane/ethyl acetate = 1:1. The work up by MPLC is carried out as in the case of the procedure used in Section B with trichloromethane on a silica gel column. To avoid MPLC, alternatively a 30-cm column with silica gel 60 (dichloromethane/ethyl acetate = 10:1) can be used for purification of the product. The yellow oil that remains after evaporating the solvent under vacuum crystallizes within 12 h. Yield: 160 mg (89%).

The same compound is obtained directly by bringing a solution of $[(OC)_4Mn(\mu\text{-}S=P(CH_3)_2)]_2$ (190 mg, 0.37 mmol) and dimethyl 2-butynedioate (408 mg, 2.87 mmol) in 100 mL of THF to reflux (66 °C) over a period of 20 min with magnetic stirring. The heating is continued for 30 min at this temperature. After cooling to room temperature the solvent is removed under vacuum. Further working up is as described above. Yield: 220–356 mg (58–94%).

Anal. Calcd. for $C_{17}H_{18}MnO_{11}PS$: C, 39.55; H, 3.51; Mn, 10.64; S, 6.21; MW, 516.3. Found: C, 39.26; H, 3.36; Mn, 10.47; S, 6.51; MS (*EI*, 70 eV): *m/e* 516 [M⁺].

Properties

The metallacyclic complex forms yellow crystals that decompose at 130 °C. It is soluble in polar organic solvents, is remarkably thermally stable and also stable towards atmospheric oxygen. The action of Ce(IV) complexes on the metallacyclic compound results in the cleavage of the $Mn(CO)_3$ and $P(CH_3)_2$ residues under formation of the substituted thiophene, $\overline{CRCRCRCRS}$ (R = CO_2CH_3) (see the next procedure). The IR spectrum of the manganese complex (CCl_4) is characterized by three C≡O bands at 2028 (vs), 1957 (s), and 1944 (s-vs) as well as by a P=S band at 530 (w) (polyethylene) (or 525 cm^{-1} in KBr). ^{1}H NMR ($CDCl_3$, rel. TMS): δ 2.03 [d, $^2J_{PH}$ = 12.7 Hz, $P(CH_3)_2$], 2.27 (d, $^2J_{PH}$ = 13.4 Hz, $P(CH_3)_2$]; δ 3.75 (s), 3.82 (s), 3.85 (s) (OCH_3). ^{31}P {^{1}H} NMR (CH_2Cl_2, rel. 1% H_3PO_4): δ 74.4(s). From cyclohexane/dichloromethane the compound crystallizes in the triclinic space group $P\bar{1}$ with a = 902.3(2), b = 1692.1(3), c = 867.0(2) pm, α = 102.01(2)°, β = 94.87(2)°, γ = 90.68(2)°, Z = 2, d_{calcd} = 1.44 g cm^{-3}, d_{obs} = 1.43 g cm^{-3} (P=S = 203.8 pm).[17]

E. TETRAMETHYL THIOPHENETETRACARBOXYLATE

$$R = CO_2CH_3$$

Procedure

The reaction is carried out in a 50-mL Schlenk tube, which is equipped with a magnetic stirrer. To a solution of the metallacyclic complex $(OC)_3Mn[\overline{CR—CR=CR—CR—P(CH_3)_2S}]$ (R = CO_2CH_3) (258 mg, 0.5 mmol) in 10 mL of toluene/methanol = 10:1 $(NH_4)_2[Ce(NO_3)_6]$ (1370 mg, 2.5 mmol) is added in portions until the violent CO elimination has finished. Subsequently the reaction mixture is magnetically stirred for 5 min and the solvent is removed under vacuum. The residue is extracted with toluene and the solution is separated from the insoluble material by filtration (P3). This procedure is followed by evaporation of the solvent under vacuum. The residue is then dissolved in 1 mL of dichloromethane/ethyl acetate = 5:1 and the product is purified by chromatography with a silica gel column ($L = 5$ cm, $\phi = 2$ cm). After evaporation of the solvent under vacuum the thiophene is recrystallized from methanol. Yield: 150 mg (95%).

Anal. Calcd. for $C_{12}H_{12}O_8S$: C, 45.57; H, 3.82; S, 10.14; MW, 316.3. Found: C, 45.37; H, 3.75; S, 10.28; MS (*EI*, 70 eV): *m/e* 316 [M^+].

Properties

The melting point of the compound is 125 °C. Its ^1H NMR spectrum in C_6D_6 shows two signals at δ 3.24 and 3.59 for the OCH_3 groups (rel. TMS).[17]

F. TRICARBONYL[η^4-2, 5-DIHYDRO-2, 3, 4, 5-TETRAKIS(METHOXYCARBONYL-1, 1-DIMETHYL-1*H*-PHOSPHOL-1-IUM-2, 5-DIYL]MANGANESE

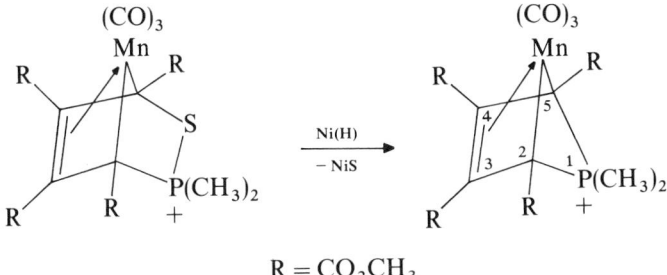

$$R = CO_2CH_3$$

Procedure

A 200-mL Schlenk tube, which is equipped with a magnetic stirrer and a mercury excess pressure outlet, is flushed thoroughly with nitrogen and then charged with a suspension of neutral Raney nickel (1.00 g, 17.03 mmol) in 100 mL of methanol. To this suspension is added the metallacyclic complex $(OC)_3 Mn[\overline{CR - CR = CR - CR - P(CH_3)_2 = S}]$ ($R = CO_2CH_3$) (200 mg, 0.39 mmol). The reaction mixture is magnetically stirred for 2 h. Subsequently the Raney nickel is centrifuged off and the solvent is removed under vacuum. The residue is dissolved in dichloromethane/ethyl acetate = 2:1. Chromatographic purification is carried out on a short silica gel column ($L = 5$ cm, $\phi = 2$ cm). After removing the solvent under vacuum the yellow compound crystallizes within 12 h. Yield: 180 mg (96%).

Anal. Calcd. for $C_{17}H_{18}MnO_{11}P$: C, 42.17; H, 3.75; Mn, 11.34; MW, 484.2. Found: C, 42.55; H, 3.65; Mn, 11.22; MS (*EI*, 70 eV): m/e 484 [M^+].

Properties

The yellow phosphole complex melts in the range from 188 to 192 °C (with sublimation) and without noticeable decomposition. The compound is soluble in polar organic solvents and remarkably stable both thermally and towards atmospheric oxygen. The IR spectrum (CCl_4) contains three $C≡O$ absorptions at 2028 (vs), 1958 (s), and 1946 (s-vs) cm^{-1}. ^1H NMR ($CDCl_3$, rel. TMS): δ 1.51 [d, $^2J_{PH} = 11.70$ Hz, $P(CH_3)_2$], 2.36 [d, $^2J_{PH} = 14.1$ Hz, $P(CH_3)_2$]; δ 3.68 (s), 3.89 (s) (OCH_3). $^{31}P\{^1H\}$ NMR (CH_2Cl_2, rel. 1% H_3PO_4): δ 51.6 (s). The ring contracted heterocycle crystallizes from methanol in the monoclinic space group $P2_1/m$ with $a = 842.6(2)$, $b = 1622.1(4)$, $c = 864.8(2)$ pm, $\beta = 116.62(2)°$, $Z = 2$, $d_{calcd} = 1.52$ g cm^{-3}, $d_{obs} = 1.52$ g cm^{-3}.[17]

Analogous Complexes

Similar complexes and compounds mentioned in Sections A to F can also be prepared with other PR_2 ($R = C_2H_5$, *cyclo*-C_6H_{11}, C_6H_5 etc.) and 3d transition metal fragments [$L_nM = C_3F_7Fe(CO)_3$, $(OC)_2PR_3Co, (\eta^5$-$C_5H_5)Ni$] and other activated alkynes C_2R_2, for example, $R = CO_2C_2H_5$, $CO_2C_6H_{11}$, CF_3.[11,12,15−21]

References

1. H. Suzuki, K. Itoh, Y. Ishii, K. Simon, and J. A. Ibers, *J. Am. Chem. Soc.*, **98**, 8494 (1976).
2. L. D. Brown, K. Itoh, H. Suzuki, K. Hirai, and J. A. Ibers, *J. Am. Chem. Soc.*, **100**, 8232 (1978).
3. R. Burt, M. Cooke, and M. Green, *J. Chem. Soc. A*, **1970**, 2981.
4. Y. Wakatsuki, K. Aoki, and H. Yamazaki, *J. Am. Chem. Soc.*, **96**, 5284 (1974).
5. J. J. Eisch and J. E. Galle, *J. Organomet. Chem.*, **96**, C23 (1975); H. Hoberg and W. Richter, *J. Organomet. Chem.*, **195**, 355 (1980).
6. Y. Wakatsuki and H. Yamazaki, *J. Organomet. Chem.*, **139**, 169 (1977).
7. D. R. McAlister, J. E. Bercaw, and R. G. Bergman, *J. Am. Chem. Soc.*, **99**, 1666 (1977).
8. Y. Wakatsuki, K. Aoki, and H. Yamazaki, *J. Am. Chem. Soc.*, **101**, 1123 (1979).
9. P. Caddy, M. Green, E. O'Brien, L. E. Smart, and P. Woodward, *J. Chem. Soc. Dalton Trans.*, **1980**, 962.
10. H. Bönnemann, *Angew. Chem. Int. Ed. Engl.*, **17**, 505 (1978); **24**, 248 (1985).
11. E. Lindner, C.-P. Krieg, W. Hiller, and R. Fawzi, *Angew. Chem. Int. Ed. Engl.*, **23**, 523 (1984).
12. E. Lindner, C.-P. Krieg, W. Hiller, and R. Fawzi, *Chem. Ber.*, **118**, 1398 (1985).
13. J. E. Bercaw and G. W. Parshall, *Inorg. Synth.*, **23**, 199 (1985).
14. K. Sasse, in *Methoden der Organischen Chemie*, J. Houben, Th. Weyl, and E. Müller (eds.), 4th ed., Vol. XII/1, Thieme, Stuttgart, 1963, p. 212.

15. E. Lindner, C.-P. Krieg, S. Hoehne, and A. Rau, *Z. Naturforsch.*, **B36**, 1487 (1981).
16. E. Lindner, A. Rau, and S. Hoehne, *Chem. Ber.*, **114**, 3281 (1981).
17. E. Lindner, A. Rau, and S. Hoehne, *J. Organomet. Chem.*, **218**, 41 (1981).
18. E. Lindner, F. Bouachir, and W. Hiller, *J. Organomet. Chem.*, **210**, C37 (1981).
19. E. Lindner, F. Bouachir, and S. Hoehne, *Chem. Ber.*, **116**, 46 (1983).
20. E. Lindner, K. E. Frick, R. Fawzi, W. Hiller, and M. Stängle, *Chem. Ber.*, **121**, 1075 (1988).
21. E. Lindner, K. E. Frick, M. Stängle, R. Fawzi, and W. Hiller, *Chem. Ber.*, **122**, 53 (1989).

30. TETRACARBONYL{[2-(DIPHENYLPHOSPHINO)PHENYL]HYDROXYMETHYL-*C, P*}MANGANESE

$(CO)_5MnH$ + [2-(diphenylphosphino)benzaldehyde] ⟶ $(CO)_4Mn$—$C(OH)H$[Ph$_2$P-phenyl] + CO

Submitted by G. D. VAUGHN* and J. A. GLADYSZ*
Checked by K. YOUNGDAHL,[†] Y. K. PARK,[†] and M. Y. DARENSBOURG[†]

Homogeneous α-hydroxyalkyl complexes, L$_n$MCH(R)OH, have been proposed as catalytic intermediates in the conversion of CO–H$_2$ gas mixtures to oxygen-containing organic molecules.[1] However, isolable examples of metal α-hydroxyalkyl complexes are scarce.[2] We have discovered that incorporation of the α-hydroxyalkyl moiety into a metallacycle provides a marked stability enhancement.[3]

Described below is one of the three types of syntheses of metallacyclic α-hydroxyalkyl complexes that we have developed.[3] It is simple to execute and employs two starting materials that have been reported in previous volumes of *Inorganic Syntheses*.[4,5]

Procedure

■ **Caution.** *Starting materials and products should all be considered toxic. Since CO is evolved during the reaction an efficient hood must be used.*

*Department of Chemistry, University of Utah, Salt Lake City, UT 84112.
†Department of Chemistry, Texas A & M University, College Station, TX 77843.

All operations are performed under nitrogen except for the column preparation. Acetone and toluene are Fisher certified grade. Ethyl acetate and hexane are Fisher HPLC (high performance liquid chromatography) grade. Diethyl ether is anhydrous reagent grade (Aldrich). Solvents are purged with nitrogen or freeze–pump–thaw degassed before use. The $(CO)_5MnH$ should be colorless or very light yellow for optimum results.[4]

An NMR tube (5-mm diameter) is charged with 0.207 g (0.713 mmol) of 2-(diphenylphosphino)benzaldehyde.[5] Then a solution of 0.152 g (0.776 mmol) of $(CO)_5MnH$ in 1.5 mL of toluene is added via syringe. These operations are conveniently performed in a glove box. The tube is fitted with a septum and is vigorously shaken. A small needle connected to a hood vented oil bubbler is inserted into the septum to release the CO that evolves vigorously. After the CO evolution subsides somewhat, the disappearance of the aldehyde ^1H NMR doublet ($J_{HP} = 5$ Hz, $\sim \delta$ 10.5) can be monitored. After 10 h, the aldehyde is consumed.

A silica gel column (37 × 3 cm) is packed in 85:15 (v/v) hexane–ethyl acetate. The column is then made air tight and is purged with air-free 85:15 hexane–ethyl acetate. The contents of the tube (and a subsequent acetone rinse of the tube) are directly applied to the top of the column. The reaction mixture is then eluted from the column with 85:15 hexane–ethyl acetate. A yellow $Mn_2(CO)_{10}$ fraction elutes first, followed by a light yellow product fraction. Solvent is removed from the latter under oil pump vacuum to give 0.259 g (0.565 mmol, 79%) of $(CO_4\overline{MnP(C_6H_5)_2(2\text{-}C_6H_4}\overline{C}HOH)$ as a light yellow oil. Yields range typically from 75 to 84%. The oil is taken up in toluene and is passed through a Pasteur pipet containing silica gel. The toluene is removed under oil pump vacuum. The residue is taken up in 4 mL of diethyl ether. Then 5 mL of hexane is added, and the mixture is concentrated under oil pump vacuum until a cloud point is reached. The flask is then capped and allowed to stand for 1 h, and is then placed in a refrigerator (2 °C) for 12 h. Lemon yellow crystals form and are collected and washed with hexane to give 0.211 g (0.460 mmol, 65%) of $(CO)_4\overline{MnP(C_6H_5)_2(2\text{-}C_6H_4}\overline{C}HOH)$.

Properties

The product is air sensitive but it is stable at room temperature under an inert atmosphere. However, it is best stored cold. The crystals decompose under N_2 at 121.5 to 123.5 °C.

Anal. Calcd. for $C_{23}H_{16}MnO_5P$: C, 60.28; H, 3.52. Found: C, 60.35; H, 3.70. The ^1H NMR spectrum in CD_2CN exhibits a complex phenyl resonance at δ 7.82–7.73 (m, 1H) and 7.64–7.12 (m, 13H), a methine resonance at δ 6.28 (br s, 1H), and a hydroxyl resonance at δ 3.73 (br s, 1H). The ^{13}C {^1H,

^{31}P} NMR spectrum (ppm, CD_3CN, $-20\,^{\circ}C$) is: 219.0 (br), 218.6 (br), 217.3 (br), 165.3, 135.8, 134.2, 133.9, 132.4, 132.1, 131.9, 131.6, 131.4, 130.0, 126.9, 126.8, 78.6. The ^{31}P{^{1}H} NMR spectrum (ppm downfield from external H_3PO_4, CD_3CN) is 77.4. Principal IR absorptions (cm^{-1}, CH_3CN) are ν_{OH} 3496 (w, br); ν_{CO} 2059 (m), 1967 (vs), 1949 (s). The significant fragments in the mass spectrum are (m/e, 15 eV): 458 (M$^+$, 0.4%), 430 (M$^+$— CO, 9.4%), 402 (M$^+$—2CO, 1.0%), 374 (M$^+$—3CO, 22.2%), 346 (M$^+$—4CO, 100%), and 316 (M$^+$—4CO—CH$_2$O, 52.3%). These properties are similar to those of the analogous rhenium complex.[3b]

References

1. (a) B. D. Dombek, *Adv. Cat.*, **32**, 325 (1983); (b) D. R. Fahey, *J. Am. Chem. Soc.*, **103**, 136 (1981); (c) J. W. Rathke and H. M. Feder, in *Catalysis of Organic Reactions*, W. R. Moser (ed.), Marcel Dekker, 1981, p. 219.

2. See, *inter alia*: (a) T. Blackmore, M. I. Bruce, P. J. Davidson, M. Z. Iqbal, and F. G. A. Stone, *J. Chem. Soc. A*, **1970**, 3153; (b) C. P. Casey, M. A. Andrews, D. R. McAlister, and J. E. Rinz, *J. Am. Chem. Soc.*, **102**, 1927 (1980); (c) J. R. Sweet and W. A. G. Graham, *J. Am. Chem. Soc.*, **104**, 2811 (1982); (d) G. R. Clark, C. E. L. Headford, K. Marsden, and W. R. Roper, *J. Organomet. Chem.*, **231**, 335 (1982); (e) B. B. Wayland, B. A. Woods, and V. M. Minda, *J. Chem. Soc., Chem. Commun.*, **1982**, 634; (f) D. L. Thorn and T. H. Tulip, *Organometallics*, **1**, 1580 (1982); (g) C. Lapinte and D. Astruc, *J. Chem. Soc., Chem. Commun.*, **1983**, 430; (h) Y. C. Lin, D. Milstein, and S. S. Wreford, *Organometallics*, **2**, 1461 (1983); (i) G. O. Nelson, *Organometallics*, **2**, 1474 (1983).

3. (a) J. C. Selover, G. D. Vaughn, C. E. Strouse, and J. A. Gladysz, *J. Am. Chem. Soc.*, **106**, 1455 (1986); (b) G. D. Vaughn, C. E. Strouse, and J. A. Gladysz, *J. Am. Chem. Soc.*, **106**, 1462 (1986); (c) G. D. Vaughn and J. A. Gladysz, *J. Am. Chem. Soc.*, **106**, 1473 (1986).

4. R. B. King and F. G. A. Stone, *Inorg. Synth.*, **7**, 196 (1963).

5. J. E. Hoots, T. B. Rauchfuss, and D. A. Wrobleski, *Inorg. Synth.*, **21**, 175 (1981).

31. CYCLOMETALLATION REACTIONS

Submitted by MICHAEL I. BRUCE,* MICHAEL J. LIDDELL,*
and GEOFF N. PAIN*
Checked by MARTIN A. BENNETT† and HORST NEUMANN†

The following syntheses detail the preparation of complexes containing typical examples of cyclometallated ligands. One of the most extensively studied

*Department of Physical and Inorganic Chemistry, University of Adelaide, Adelaide, South Australia 5000, Australia.
†Research School of Chemistry, Australian National University, G.P.O. Box 4, Canberra, A.C.T. 2600, Australia.

systems is derived from azobenzene (diphenyldiazene)[1] the first examples of which, containing nickel[2] and palladium or platinum,[3] were described in the early 1960s. The present examples encompass metallation by palladium or manganese complexes (with elimination of HCl or alkane, respectively), the latter to give mono- and binuclear complexes, and transfer of the metallated ligand from palladium to cobalt by reaction with an anionic metal carbonyl.[4] Several polycyclic nitrogen heterocycles have been metallated, one of the most useful being benzo[*h*]quinoline.[5] The synthesis of the ruthenium derivative of benzo[*h*]quinoline using $Ru_3(CO)_{12}$ provides a route to a complex containing two cyclometallated ligands. In this example, loss of hydrogen from the ring carbon atom occurs by combination with a metal carbonyl fragment, which appears as the polynuclear complex $Ru_4(\mu\text{-H})_4(CO)_{12}$. Similar eliminations of metal carbonyl hydrides are assumed to occur in other reactions of metal carbonyls, for example, that of $Mn_2(CO)_{10}$ and azobenzene, from which $\overline{Mn(C_6H_4N} = N\,Ph)(CO)_4$ is obtained in yields never exceeding 50%, the rest of the manganese probably forming $MnH(CO)_5$. Triphenyl phosphite is one of the most widely metallated *P*-donor ligands, and its metallation by base-induced elimination of HCl is also described in this chapter.

General Procedures

The syntheses can be carried out in a 100-mL Schlenk tube, containing a stirring bar and equipped with a reflux condenser and nitrogen inlet and outlet (or bypass). The reactions are normally run with the exclusion of oxygen, although with the exception of the cobalt complex, work-up procedures (solvent removal on a rotary evaporator, chromatography, crystallization) do not require an inert atmosphere. Petroleum ether cited in the various procedures below is in the boiling range from 40 to 60 °C.

A.   BENZYLPENTACARBONYLMANGANESE[6]

$$Na[Mn(CO)_5] + PhCH_2Cl \longrightarrow Mn(CH_2Ph)(CO)_5 + NaCl$$

Procedure

A filtered solution of $Na[Mn(CO)_5]$, prepared from 2.42 g (6.21 mmol) of $Mn_2(CO)_{10}$ (Strem) in freshly distilled dry tetrahydrofuran (THF) (50 mL),[7] is treated with benzyl chloride (1.70 g, 13.49 mmol). The mixture is stirred for 2 h at room temperature and then refluxed for 10 min. After cooling, the solution is filtered through a pad of filter aid (2 cm), which is then washed with three 10-mL portions of dry diethyl ether. The solvent is then removed from the combined filtrate and washings under reduced pressure (rotary evaporator) until ∼ 5 mL remain. Silica gel (3 g, 200–325 mesh, Ajax) is then added and the mixture evaporated to dryness. The coated silica gel is loaded on to a second pad of silica gel (5×3 cm) and the product is washed through with five 20-mL portions of petroleum ether. The solvent is removed at reduced pressure (rotary evaporator, followed by brief exposure to 0.1-mm oil pump vacuum) to leave an off-white crystalline solid, which can be used directly in the next preparation. Yield: 2.83 g, 9.89 mmol (80%). If necessary the product may be further purified by dissolving in a minimum amount of petroleum ether and chromatographing on a column of silica gel (20×3 cm, 100–200 mesh, Merck) using petroleum ether as eluant. The largest colorless band (which can be detected either by *brief* viewing of the column in UV light, or by evaporation of aliquots of the column eluate) is collected and the solvent removed at reduced pressure to give a white solid.

Anal. Calcd. for $C_{12}H_7MnO_5$: C, 50.4; H, 2.4; Found: C, 49.7; H, 2.3.

Properties

Benzylpentacarbonylmanganese is a white crystalline solid (mp 37–38 °C), which is soluble in most organic solvents to give colorless solutions, which are readily oxidized by air, first turning yellow [formation of $Mn_2(CO)_{10}$] and then depositing dark brown solids.[6] The infrared (IR) spectrum of a freshly prepared solution in cyclohexane contains principal $\nu_{(CO)}$ bands at 2106 (w), 2041 (vw), 2015 (vs), 2007 (vs) and 1990 (m) cm^{-1}. The ^{1}H NMR spectrum in CCl_4 contains resonances at δ 2.36 (2H, CH$_2$) and 7.17 (5H, Ph). The mass spectrum contains a parent ion at m/z 195, five ions at 28 amu intervals formed by successive loss of the five CO groups, and a strong ion at m/z 91 ($[C_7H_7]^+$). The complex is chemically quite reactive, undergoing the usual migratory insertion reactions with SO_2, for example,[8] and reacting readily with active organic hydrogen atoms eliminating toluene and forming σ-bonded organomanganese complexes. In this regard, it has been found to have superior reactivity, compared with either $Mn_2(CO)_{10}$ or $MnMe(CO)_5$, in cyclometallation reactions of, for example, $PMe_2(CH_2Ph)$ or $AsMe_2(CH_2Ph)$.[9]

B. TETRACARBONYL[2-(PHENYLAZO)PHENYL-C^1, N^2]-MANGANESE AND μ-(AZODI-2, 1-PHENYLENE-C^1, N^2:$C^{1'}$,N^1)-OCTACARBONYLDIMANGANESE

1. $Mn(CH_2Ph)(CO)_5 + PhN{=}NPh$

   $\longrightarrow \overline{Mn(C_6H_4N}{=}NPh)(CO)_4 + PhMe + CO$

2. $2Mn(CH_2Ph)(CO)_5 + PhN{=}NPh$

   $\longrightarrow (OC)_4\overline{Mn(C^1\text{-}C_6H_4N} \quad \overline{NC_6H_4\text{-}C^{1'}})Mn(CO)_4 + PhMe + CO$

Procedure

1.   *Tetracarbonyl[2-(phenylazo)phenyl-C^1, N^2]manganese.*   After   being
flushed with nitrogen, a Schlenk tube is loaded with $Mn(CH_2Ph)(CO)_5$
(800 mg, 2.79 mmol), azobenzene (510 mg, 2.80 mmol), and octane (50 mL), and
the mixture is heated at reflux point (oil bath, 170–180 °C) for 2 h with
magnetic stirring. As the black dimanganese complex (see below) is formed
initially the reaction should be monitored by thin layer chromatography
(TLC) [silica gel (Merck) using petroleum ether–dichloromethane 2:1]
to check that reaction is complete. This is indicated by the presence of only one
orange band ($R_f \sim 0.5$), and the reaction should be stopped at this stage to
prevent more complicated purification procedures having to be used.
Excessive reaction times result in decreased yields. After cooling, the orange
solution is filtered through a 2-cm pad of filter aid and solvent is removed from
the filtrate under reduced pressure (rotary evaporator). The residue is
extracted with two 10-mL portions of petroleum ether at 50 °C, and the
combined extracts are chromatographed directly on a column of Florisil (10
× 3.5 cm), initially packed in petroleum ether. Elution with petroleum ether
affords a red fraction, which after removal of solvent (rotary evaporator) gives
the pure complex as red crystals. Yield: 700 mg, 2.01 mmol (72%).

Anal. Calcd. for $C_{16}H_9MnN_2O_4$: C, 55.2; H, 2.6; N, 8.0. Found: C, 55.1; H, 2.5;
N, 7.9.

2.   *μ-(Azodi-2,1-phenylene-C^1, N^2:$C^{1'}$, N^1)-octacarbonyldimanganese.*   A
mixture of $Mn(CH_2Ph)(CO)_5$ (800 mg, 2.79 mmol), azobenzene (255 mg,
1.40 mmol) and octane (50 mL) is placed in a Schlenk tube flushed with
nitrogen. The mixture is then heated at reflux point for 15 min (oil bath
temperature 170–180 °C) with magnetic stirring. After cooling, the black
solution is filtered through a 2-cm pad of filter aid, which is then washed with
two 10-mL portions of petroleum ether. Solvent is removed from the
combined filtrate and washings to leave a black crystalline solid. This is
extracted with the minimum amount of dichloromethane (~ 1 mL) and
petroleum ether (~ 3 mL) is added. The extracts are placed on a column of
Florisil (30 × 3.5 cm) packed in petroleum ether. Elution with petroleum ether
gives a black band, which after removal of solvent leaves a black crystalline
solid. Yield: 590 mg, 1.15 mmol (83%).

Anal. Calcd. for $C_{20}H_8Mn_2N_2O_8$: C, 46.7; H, 1.6; N, 5.5. Found: C, 46.8; H,
1.5; N, 5.6.

Properties

The compound $\overline{Mn(C_6H_4N}=NPh)(CO)_4$ forms red crystals (mp 112–
113 °C), which are stable in air, although solutions slowly decompose in air.[10]

It is very soluble in all organic solvents, but insoluble in water. The IR spectrum of a cyclohexane solution shows $v_{(CO)}$ at 2076 (m), 2004 (s), 1996 (vs), and 1961 (vs) cm^{-1}, and the ^{1}H NMR spectrum (CDCl$_3$; TMS internal standard) contains several multiplets in the aromatic region δ 7.2–8.1. The mass spectrum contains a parent ion at m/z 348, together with *five* lower-mass multiplets at 28 amu intervals, resulting from stepwise loss of N$_2$ and four CO groups.

The binuclear complex, (OC)$_4$$\overline{Mn C_6 H_4 N}$=$\underline{NC_6 H_4 Mn}(CO)_4$, forms dark green crystals (mp 136–144 °C), which in thin section are orange-green dichroic.[11] The complex is stable in air as a solid, and also for some hours in solution. It is soluble in most organic solvents, but not in water. In the $v_{(CO)}$ region the IR spectrum of a cyclohexane solution has bands at 2076 (m), 2015 (vs), 2000 (sh), 1994 (vs), and 1960 (vs) cm^{-1}, and the ^{1}H NMR spectrum (CDCl$_3$ solution; TMS internal standard) contains four multiplets in the region δ 7.2–8.6. The mass spectrum of the compound (by the checkers) shows parent ion peak at m/z 514 with eight lower mass multiplets at 28 amu intervals.

C. BIS-[μ-CHLORO-[2-(PHENYLAZO)PHENYL-C^1,N^2] PALLADIUM][12]

$$2Li_2[PdCl_4] + 2PhN{=}NPh$$

$$\longrightarrow [\overline{Pd(\mu\text{-}Cl)(C_6H_4N{=}NPh)}]_2 + 2HCl + 4LiCl$$

Procedure

A 0.1 M solution of Li$_2$[PdCl$_4$] is prepared by stirring LiCl(789 mg, 17.7 mmol) with PdCl$_2$ (1769 mg, 10 mmol; Johnson Mathey) in 100 mL of methanol for 5 h. The solution is then filtered through Celite$^®$ and kept in a sealed flask. Azobenzene (564 mg, 3.1 mmol) is dissolved in 20 mL of methanol and the solution is filtered into a 100-mL Schlenk flask. To this solution is added 31 mL of the 0.1 M Li$_2$[PdCl$_4$] solution, and the mixture is then refluxed for 2 h. After cooling, the precipitated red-orange solid is removed by filtration, washed with three 10-mL portions of methanol, and dried in vacuum (0.1 mm). The red microcrystalline solid may be used directly in the next procedure. Yield: 990 mg, 1.53 mmol (98%). The complex may be recrystallized with difficulty from hot benzene.

Anal. Calcd. for C$_{24}$H$_{18}$Cl$_2$N$_4$Pd$_2$: C, 44.61; H, 2.81; N, 8.67. Found: C, 44.97; H, 2.91; N, 8.42.

Properties

The compound $[\overline{Pd(\mu\text{-}Cl)(C_6H_4N\!:\!\overset{\_}{N}Ph)}]_2$ forms air-stable red-orange crystals [mp 272–275 °C (dec)] with only low solubility in organic solvents. The far IR spectrum[13] contains bands at 548 (s), 523 (s), 431 (m), 419 (m), 368 (m), 320 (m), 233 (m), 200 (m), and 176 (m) cm^{-1}, together with characteristic $\nu_{(PdCl)}$ bands at 337 (s) and 262 (s) cm^{-1}. Although relatively insoluble, this complex reacts readily with tertiary phosphines, to give $\overline{PdCl(C_6H_4N\!=\!}$ $\overset{\_}{N}Ph)(PR_3)_2$, with thallium acetylacetonate to form $\overline{Pd(acac)(C_6H_4N\!=\!\overset{\_}{N}Ph)}$, and with thallium cyclopentadienide to give $\overline{Pd(C_6H_4N\!=\!\overset{\_}{N}Ph)}(\eta\text{-}C_5H_5)$.[10,14]

D. TRICARBONYL[2-(PHENYLAZO)PHENYL-C^1,N^2]COBALT[10]

$$[\overline{Pd(C_6H_4N\!=\!\overset{\_}{N}Ph)Cl}]_2 + 2Na[Co(CO)_4] \longrightarrow$$

$$2\overline{Co(C_6H_4N\!=\!\overset{\_}{N}Ph)}(CO)_3 + 2NaCl + 2Pd + 2CO$$

Procedure

A filtered solution of sodium tetracarbonylcobaltate(1 −) is prepared by adding $Co_2(CO)_8$ (430 mg, 1.26 mmol; Strem) under nitrogen to 25 mL of dry, oxygen-free THF in a 100-mL two-necked Schlenk flask containing 5 mL of 1% sodium amalgam. The mixture is stirred for 30 min under a nitrogen atmosphere to give a colorless solution. The spent amalgam is removed through a pressure-equalized outlet by inverting the Schlenk flask. The solution of $Na[Co(CO)_4]$ is then carefully decanted via the pressure-equalized outlet through a 1-cm pad of filter aid into a second 100-mL Schlenk flask containing solid $[\overline{Pd(\mu\text{-}Cl)(C_6H_4N\!=\!\overset{\_}{N}Ph)}]_2$ (800 mg, 1.24 mmol) and a magnetic stirrer bar. The mixture immediately turns red and after stirring for 1 h, solvent is removed from the red solution (0.1-mm oil pump vacuum using an appropriately sized trap cooled to − 78 °C). The solid residue is extracted at room temperature with two 15-mL portions and five 5-mL portions of petroleum ether (boiling range 30–40 °C). The combined extracts are filtered through a 2-cm pad of filter aid, which is then washed with three 5-mL portions of petroleum ether. Evaporation (0.1-mm oil pump vacuum) to ∼ 5 mL, cooling to − 78 °C until an oil separates, and warming to − 50 °C results in crystallization of the product. The supernatant liquid is removed using a syringe to leave a red crystalline solid, which is dried in vacuum (0.1 mm). Further successive crystallization from pentane and methanol gives an analytically pure sample. Yield: 550 mg, 1.69 mmol (68%).

Anal. Calcd. for $C_{15}H_9CoN_2O_3$: C, 55.4; H, 3.1; N, 8.6. Found: C, 55.2; H, 3.3; N, 8.7.

Properties

This complex forms red crystals (mp 64.0–64.5 °C), which slowly decompose in air; solutions decompose more rapidly. It is very soluble in all organic solvents and sublimes readily at reduced pressures (50 °C/0.1 mm). The IR spectrum of a cyclohexane solution contains $v_{(CO)}$ bands at 2075 (s), 2025 (s), and 2005 (s) cm^{-1}.

The complex reacts readily with ligands such as tertiary phosphines, phosphites, or arsines to give substitution of a CO ligand on the cobalt atom. With CO, however, the complex reacts to give 2-phenylindazolone and 3-phenyl-2, 4(1*H*, 3*H*)-quinazolinedione,[15] whereas the reaction with CO and hexafluoro-2-butyne affords an anilinoquinoline, probably via an intermediate complex in which the alkyne and CO have inserted into the Co—C bond.[16]

E. *cis*-BIS(BENZO[*h*]QUINOLIN-10-YL-C^{10},N^1)-DICARBONYLRUTHENIUM(II)

$$Ru_3(CO)_{12} + C_{13}H_9N \longrightarrow Ru(CO)_2(C_{13}H_8N)_2$$
$$+ Ru_4H_4(CO)_{12} + \cdots$$

Procedure

A mixture of $Ru_3(CO)_{12}$ (350 mg, 0.55 mmol),* benzo[*h*]quinoline (617 mg, 3.44 mmol), 1, 2-dimethoxyethane (2 mL), and octane (15 mL) is heated at reflux for 4 h with magnetic stirring, after which time a pale yellow precipitate has deposited from the brown solution. After cooling, solvent is removed (rotary evaporator) to leave a mixture of the crude complex and yellow $Ru_4H_4(CO)_{12}$. The latter is removed by stirring the solid residue in a mixture of methanol (3 mL) and diethylamine (1 mL) for 30 min under nitrogen. The remaining solid is removed by filtration, using a sintered glass funnel, and washed with methanol (3 × 1 mL) and petroleum ether (3 × 3 mL). After drying *in vacuo*, the pure complex is obtained as very pale yellow microcrystals. Yield: 602 mg (71%).

---

*Dodecacarbonyltriruthenium can be obtained commercially (Strem Chemicals Inc., Newburyport, MA 01950) or made by direct carbonylation[17,18] of $RuCl_3 \cdot nH_2O$ (Strem, or Johnson Mathey Chemicals Ltd, Royston SG8 5HE, United Kingdom).

Anal. Calcd. for $C_{28}H_{16}N_2O_2Ru$: C, 65.5; H, 3.1; N, 5.45. Found: C, 65.3; H, 2.9; N, 5.5.

Properties

The complex $Ru(CO)_2(C_{13}H_8N)_2$ forms pale yellow air-stable crystals (mp 348–352 °C (sealed tube under argon) to a black liquid). It decomposes slowly in solution with the formation of a dark green material, which is presently uncharacterized. It is only slightly soluble in petroleum ether and cold methanol, but more soluble in aromatic hydrocarbons, diethyl ether, chloroform, dichloromethane, and acetone. The IR spectrum of a dichloromethane solution contains two strong absorptions at 2020 and 1954 cm^{-1}, characteristic of a *cis*-$M(CO)_2$ moiety. The ^1H NMR spectrum (CDCl$_3$ solution; TMS internal standard) is complex, and contains several multiplets between δ 6.9–8.5. In the solid state, the ruthenium has approximate octahedral coordination, the two metallated nitrogen heterocyclic ligands being arranged so that the two mutually cis CO groups lie trans to the two nitrogen atoms.

F. η^5-CYCLOPENTADIENYL)[2-[(DIPHENOXY-PHOSPHINO)OXY]PHENYL-*C*, *P*](TRIPHENYL PHOSPHITE-*P*)RUTHENIUM(II)

1. $RuCl(PPh_3)_2(\eta\text{-}C_5H_5) + 2P(OPh)_3$

$\longrightarrow RuCl[P(OPh)_3]_2(\eta^5\text{-}C_5H_5) + 2PPh_3$

2. $RuCl[P(OPh)_3]_2(\eta^5\text{-}C_5H_5) + NH(C_6H_{11})_2$

$\longrightarrow Ru[\overline{(C_6H_4O)P}(OPh)_2][P(OPh)_3](\eta^5\text{-}C_5H_5)$

$+ [NH_2(C_6H_{11})_2]Cl$

Procedure

1.   *Chloro(η^5-cyclopentadienyl) (triphenylphosphite-P)ruthenium(II).*   A Schlenk tube is flushed with dry nitrogen and loaded with $RuCl(PPh_3)_2(\eta^5\text{-}C_5H_5)$ (1.152 g, 1.59 mmol),[19] triphenyl phosphite (1.026 g, 3.31 mmol), and decalin (10 mL). The mixture is heated at reflux for 30 min with magnetic stirring. After cooling, the mixture is added to a column of Florisil (2.5 × 8 cm), initially made up in petroleum ether. The column is eluted with petroleum ether (~ 50 mL), which removes decalin and triphenylphosphine. Subsequent elution with dichloromethane separates a yellow fraction, which is evaporated (rotary evaporator) to give yellow $RuCl[P(OPh)_3]_2(\eta^5\text{-}C_5H_5)$ (1.225 g, 94%),

which is sufficiently pure to be used directly in the next stage of the preparation.

2. *(η^5-Cyclopentadienyl)[2-[(diphenoxyphosphino)oxy]phenyl-C, P](triphenyl phosphite-P)ruthenium(II)*. A Schlenk tube (100 mL) is loaded with RuCl[P(OPh)$_3$]$_2$(η^5-C$_5$H$_5$) (942 mg, 1.14 mmol), dicyclohexylamine (276 mg, 1.52 mmol), and decalin (6 mL). After flushing with nitrogen, the mixture is heated at reflux for 4 h with magnetic stirring. After cooling, the mixture is filtered through cotton wool to remove precipitated dicyclohexylammonium chloride. The solid is washed with petroleum ether (5 mL), and the combined filtrate and washings are added to a Florisil column (10 × 2.5 cm), initially packed in petroleum ether. Washing with petroleum ether (~ 25 mL), followed by elution with a 4:1 petroleum ether–dichloromethane mixture gives a pale yellow fraction. Removal of solvent (rotary evaporator) gives a pale yellow glassy solid, which is redissolved in the minimum quantity of warm methanol. Cooling to − 30 °C overnight gives the pure complex as a pale yellow crystalline solid. Yield: 495–790 mg (55–88%).

Anal. Calcd. for C$_{41}$H$_{34}$O$_6$P$_2$Ru: C, 62.7; H, 4.4. Found: C, 62.5; H, 4.3.

Properties

The complex forms pale yellow crystals (mp 116–118 °C), which are stable in air for prolonged periods. It is very soluble in common organic solvents, but insoluble in water. The ^{1}H NMR spectrum of a CS$_2$ solution (TMS internal standard) contains resonances at δ 4.08 [t, $J_{(HP)}$ 1.0 Hz], assigned to the C$_5$H$_5$ protons, and 7.13 (m), for the aromatic protons.

References

1. M. I. Bruce and B. L. Goodall, in *The Chemistry of the Hydrazo, Azo and Azoxy Groups*, S. Patai (ed.), Wiley, London, 1975, p. 259.
2. J. P. Kleiman and M. Dubeck, *J. Am. Chem. Soc.*, **85**, 1544 (1963).
3. A. C. Cope and R. W. Siekman, *J. Am. Chem. Soc.*, **87**, 3272 (1965).
4. R. F. Heck, *J. Am. Chem. Soc.*, **90**, 313 (1968).
5. M. I. Bruce, B. L. Goodall, and F. G. A. Stone, *J. Organomet. Chem.*, **60**, 343 (1973); B. N. Cockburn, D. V. Howe, T. Keating, B. F. G. Johnson, and J. Lewis, *J. Chem. Soc. Dalton Trans.*, **1973**, 404; M. Nonoyama, *Bull. Chem. Soc. Jpn.*, **47**, 767 (1974).
6. R. D. Closson, T. H. Coffield, and J. Kozikowski, *J. Org. Chem.*, **22**, 598 (1957); F. Calderazzo, K. Noack, and U. Scherer, *J. Organomet. Chem.*, **8**, 517 (1967).
7. A. T. T. Hsieh and M. J. Mays, *Inorg. Synth.*, **16**, 61 (1976); D. Drew, M. Y. Darensbourg and D. J. Darensbourg, *J. Organomet. Chem.*, **85**, 73 (1975).
8. S. E. Jacobson and A. Wojcicki, *J. Organomet. Chem.*, **72**, 113 (1974).

9. R. L. Bennett, M. I. Bruce, and F. G. A. Stone, *J. Organomet. Chem.*, **94**, 65 (1975).

10. R. F. Heck, *J. Am. Chem. Soc.*, **90**, 313 (1968); M. I. Bruce, M. Z. Iqbal, and F. G. A. Stone, *J. Chem. Soc. A*, **1970**, 3204.

11. R. L. Bennett, M. I. Bruce, B. L. Goodall, and F. G. A. Stone, *Aust. J. Chem.*, **27**, 2131 (1974).

12. A. C. Cope and R. W. Siekman, *J. Am. Chem. Soc.*, **87**, 3272 (1965).

13. B. Crociani, T. Boschi, R. Pietropaolo, and U. Belluco, *J. Chem. Soc. A*, **1970**, 531.

14. D. L. Weaver, *Inorg. Chem.*, **9**, 2250 (1970); T. Joh, N. Hagihara, and S. Murahashi, *J. Chem. Soc. Jpn.*, **88**, 786 (1967).

15. S. Murahashi and S. Horiie, *Bull. Chem. Soc. Jpn.*, **33**, 78, 88 (1960).

16. M. I. Bruce, B. L. Goodall, and F. G. A. Stone, *J. Chem. Soc. Dalton Trans.*, **1975**, 1651.

17. M. I. Bruce, C. M. Jensen, and M. L. Jones, *Inorg. Synth.*, **26**, 000 (1989).

18. J. M. Patrick, A. H. White, M. I. Bruce, M. J. Beatson, D. St C. Black, G. B. Deacon, and N. C. Thomas, *J. Chem. Soc. Dalton Trans.*, **1983**, 2121.

19. M. I. Bruce, C. Hameister, A. G. Swincer, and R. C. Wallis, *Inorg. Synth.*, **21**, 79 (1982).

32.   (η^6-HEXAMETHYLBENZENE)RUTHENIUM COMPLEXES CONTAINING HYDRIDE AND ORTHO-METALLATED TRIPHENYLPHOSPHINE

Submitted by M. A. BENNETT* and J. L. LATTEN*
Checked by W. D. JONES†

The hexamethylbenzene complex $RuHCl(\eta^6\text{-}C_6Me_6)(PPh_3)$ has been shown to catalyze hydrogen transfer to olefins both from hydrogen gas and from secondary alcohols.[1] It also catalyzes the hydrogenation of benzene to cyclohexane.[1] It is best prepared as described here by reducing $[RuCl_2(\eta^6\text{-}C_6Me_6)]_2$ (Ref. 2) in the presence of an approximately threefold excess of triphenylphosphine with 2-propanol and anhydrous sodium carbonate. Similar treatment of the monomeric complex $RuCl_2(\eta^6\text{-}C_6Me_6)(PPh_3)$ in the absence of free triphenylphosphine gives poorer yields, and the use of hydridic reducing agents, such as $Li[AlH_4]$ or $Na[BH_4]$, tends to give the dihydrido complex $RuH_2(\eta^6\text{-}C_6Me_6)(PPh_3)$.[3] The complex $RuHCl(\eta^6\text{-}C_6Me_6)(PPh_3)$ reacts readily with methyllithium at $-78\,°C$ to form an *ortho*-metallated triphenylphosphine complex $\overline{RuH(o\text{-}C_6H_4PPh_2)}(\eta^6\text{-}C_6Me_6)$, which may arise by internal oxidative addition to the metal of a C—H bond in a 16-electron ruthenium(0) species, $Ru(\eta^6\text{-}C_6Me_6)(PPh_3)$. This in turn could be formed either by reductive elimination of methane from an undetected

---
*Research School of Chemistry, Australian National University, Canberra, A.C.T., Australia 2601.
†Department of Chemistry, University of Rochester, River Station, Rochester, NY 14627.

hydrido(methyl) complex $RuH(CH_3)(\eta^6$-$C_6Me_6)(PPh_3)$ or by base-promoted elimination of HCl from $RuHCl(\eta^6$-$C_6Me_6)(PPh_3)$.

A. CHLORO(η^6-HEXAMETHYLBENZENE)(HYDRIDO)-(TRIPHENYLPHOSPHINE)RUTHENIUM (II)

$$\{RuCl_2[\eta^6\text{-}C_6(CH_3)_6]\}_2 + 2P(C_6H_5)_3 + 2(CH_3)_2CHOH + Na_2CO_3$$
$$\longrightarrow 2RuHCl[\eta^6\text{-}C_6(CH_3)_6][P(C_6H_5)_3]$$
$$+ (CH_3)_2CO + 2NaCl + H_2O + CO_2$$

Procedure

■ **Caution.** *Owing to the toxicity of aromatic hydrocarbons, this procedure should be carried out in a well-ventilated hood.*

All manipulations must be carried out under an inert atmosphere with use of standard Schlenk techniques and solvents that have been freshly distilled under nitrogen or argon. 2-Propanol is dried by heating overnight under reflux over calcium hydride and similarly distilled under nitrogen or argon. A 100-mL two-necked round-bottomed flask fitted with a gas inlet and a magnetic stirring bar is charged with $[RuCl_2[\eta^6\text{-}C_6(CH_3)_6]]_2$ (0.45 g, 0.67 mmol),[2] triphenylphosphine (0.9 g, 3.4 mmol) and powdered, anhydrous sodium carbonate (0.45 g, 4.2 mmol). The flask is sealed with a septum and is evacuated and refilled with nitrogen or argon several times. Anhydrous 2-propanol (30 mL) is added by means of a syringe, the septum is replaced with a reflux condenser and gas outlet, and the suspension is heated under reflux for 15 h. The yellow solution containing suspended solid is allowed to cool and is evaporated to dryness under reduced pressure. The residue is extracted with toluene until the extract is colorless ($\sim$ 50 mL is required), and the combined extracts are evaporated to approximately half-volume under reduced pressure. Hexane is added until the solution just becomes turbid, and the solution is refrigerated overnight. The fine yellow crystals of the product that separate are removed by filtration and dried *in vacuo*. Care must be exercised to avoid loss of material when handling the product due to its fine, powdery nature.

A second crop of product may be obtained by evaporating the filtrate under reduced pressure and adding hexane. The yields in different preparations vary from 0.50 to 0.64 g (65–84%). The analytical sample is recrystallized from tetrahydrofuran (THF)–hexane.

Anal. Calcd. for $C_{30}H_{34}ClPRu$: C, 64.10; H, 6.10; Cl, 6.31; P, 5.51. Found: C, 63.25; H, 6.08; Cl, 6.77; P, 5.60.

Properties

The complex $RuHCl[\eta^6\text{-}C_6(CH_3)_6][P(C_6H_5)_3]$ is air sensitive in the solid state and forms very air-sensitive solutions in benzene, dichloromethane, and THF. It is insoluble in hexane. The infrared (IR) spectrum in a KBr disk shows a band at $1925\,cm^{-1}$ due to $\nu_{(Ru-H)}$. The ^1H NMR spectrum in C_6D_6 at 200 HMz contains a singlet at δ 1.70 due to the $C_6(CH_3)_6$ protons and a doublet at δ -8.34 $(^2J_{PH} = 56\,Hz)$ due to the hydride proton, in addition to complex multiplets in the region δ 6.9–7.1 and 7.7–8.0 due to the aromatic protons of triphenylphosphine. The $^{31}P\,\{^1H\}$ NMR spectrum in C_6D_6 at 80.98 MHz consists of a singlet at δ 57.1 relative to external 85% H_3PO_4.

B. [2-(DIPHENYLPHOSPHINO)PHENYL-C^1, P]- (η^6-HEXAMETHYLBENZENE)(HYDRIDO)RUTHENIUM(II)

$$RuHCl[\eta^6\text{-}C_6(CH_3)_6][P(C_6H_5)_3] + CH_3Li \longrightarrow LiCl + CH_4 +$$

$(C_6H_5)_2$

Procedure

■ **Caution.** *Extreme care must be exercised when handling methyllithium, which is pyrophoric and reacts violently with both air and water.*

All manipulations must be carried out in an inert atmosphere using standard Schlenk techniques and solvents that have been freshly distilled under nitrogen or argon. Methyllithium ($\sim 5\%$ solution in diethyl ether obtained from Ega Chemie or Aldrich) is standardized before use.[4]

A 50-mL two-necked round-bottomed flask fitted with a gas inlet and magnetic stirring bar is charged with $RuHCl(\eta^6\text{-}C_6Me_6)(PPh_3)$ (0.53 g, 0.94 mmol), sealed with a septum, and evacuated and refilled several times with nitrogen or argon. Toluene (10 mL) is then added by means of a syringe, and the suspension is cooled in a Dry Ice–ethanol bath. Methyllithium (2.5 mmol, ~ 1.5–2.0 mL of ~ 1.5 M solution in diethyl ether) is added dropwise, and the suspension is stirred for 5 min. The suspension is allowed to warm to room temperature and is stirred for 2 h. The yellow-brown solution is again cooled in a Dry Ice–ethanol bath and treated dropwise with methanol (2 mL) to decompose the excess of methyllithium (gas evolution). The thick yellow slurry is allowed to warm to room temperature, and the resulting pale yellow solution is evaporated to dryness under reduced pressure. The residue is extracted with toluene (~ 15 mL), and the solution is filtered. The extract is evaporated to $\sim$ half-volume, hexane is added until the solution just becomes turbid, and the solution is refrigerated overnight. The fine yellow crystals are removed by filtration and dried *in vacuo*. A second crop may be obtained by evaporation of the filtrate under reduced pressure and addition of hexane. The yield is 0.29 g (58%). The analytical sample was recrystallized from THF–hexane.

Anal. Calcd. for $C_{30}H_{33}PRu$: C, 68.55; H, 6.33; P, 5.89. Found: C, 68.74; H, 6.51; P, 6.17.

Properties

The yellow, air-sensitive crystals of $\overline{RuH(o\text{-}C_6H_4P}Ph_2)(\eta^6\text{-}C_6Me_6)$ are soluble in benzene, THF, and dichloromethane and insoluble in hexane. The IR spectrum in a KBr disk shows a band at 1940 cm^{-1} due to $\nu_{(Ru-H)}$ and bands at 1550, 1410, and 720 cm^{-1}, which are characteristic of *ortho*-metallated triphenylphosphine.[5,6] The ^{1}H NMR spectrum in C_6D_6 at 200 MHz contains a singlet at δ 1.95 due to the $C_6(CH_3)_6$ protons, a doublet at $\delta -7.58$ ($^2J_{PH} = 41$ Hz) due to the hydride proton, and complex multiplets in the region δ 6.7–8.1 due to the aromatic protons of *ortho*-metallated triphenylphosphine. The ^{31}P$\{^1$H$\}$ NMR spectrum in C_6D_6 at 80.98 MHz consists of a singlet at $\delta -8.4$ ppm, the upfield shift relative to that of $RuHCl(\eta^6\text{-}C_6Me_6)(PPh_3)$ being characteristic of the cyclometallated four-membered ring.[7]

References

1. M. A. Bennett, T.-N. Huang, and T. W. Turney, *J. Chem. Soc. Chem. Commun.*, **1978**, 582.

2. M. A. Bennett, T.-N. Huang, T. W. Matheson, and A. K. Smith, *Inorg. Synth.*, **21**, 74 (1982).

3. H. Werner and H. Kletzin, *J. Organometal. Chem.*, **228**, 289 (1982).

4. H. Gilman and F. K. Cartledge, *J. Organometal. Chem.*, **2**, 447 (1964).

5. M. A. Bennett and D. L. Milner, *J. Am. Chem. Soc.*, **91**, 6983 (1969).

6. D. J. Cole-Hamilton and G. Wilkinson, *J. Chem. Soc. Dalton Trans.*, **1977**, 797.

7. P. E. Garrou, *Chem. Rev.*, **81**, 229 (1981).

33. AN OSMIUM CONTAINING BENZENE ANALOG, Os(CSCHCHCHCH)(CO)(PPh₃)₂, CARBONYL(5-THIOXO-1,3-PENTADIENE-1,5-DIYL-C^1, C^5, S)-BIS(TRIPHENYLPHOSPHINE)OSMIUM, AND ITS PRECURSORS

Submitted by GREGORY P. ELLIOTT,* NICOLA M. MCAULEY,* AND
WARREN R. ROPER*
Checked by PATRICIA A. SHAPLEY[†]

The preparation of metallacycles from the reaction of low-valent metal centers with acetylenes commonly gives metallacyclopentadiene complexes. However, the presence of a thiocarbonyl ligand attached to the metal can result in a six-membered ring in which the carbon from the thiocarbonyl has inserted into the ring. The complex Os(CSCHCHCHCH)(CO)(PPh₃)₂ is the first reported[1] metallabenzene analog. Its crystal structure shows a planar six-membered ring with no significant alternation of bond lengths, thus supporting the idea of electron delocalization in the ring. The synthesis described here requires five steps beginning with $(NH_4)_2[OsCl_6]$. The first step requires heating for 5 days, the remaining steps can be completed in 2 days. All reactions should be conducted under nitrogen. The products are air-stable unless otherwise stated.

A. DICHLOROTRIS(TRIPHENYLPHOSPHINE)OSMIUM(II)

This compound was originally prepared by Hoffman and Caulton.[2]

$$(NH_4)_2[OsCl_6] + 5PPh_3 \xrightarrow[\;H_2O\;]{(CH_3)_3COH} OsCl_2(PPh_3)_3$$

$$+ 2NH_4Cl + 2PPh_3Cl_2$$

Procedure

Ammonium hexachloroosmate(IV) $(1.0\,g)$[3] and triphenylphosphine $(4.2\,g)$ are combined into a solution of 1,1-dimethylethanol $(150\,mL)$ and water $(60\,mL)$ in a 500-mL flask equipped with a nitrogen gas inlet and a magnetic stirring bar. The mixture is frozen to $-78\,°C$ and evacuated in three successive freeze–thaw cycles, the gas being replaced by rigorously purified nitrogen. The

*Department of Chemistry, The University of Auckland. Auckland, New Zealand.
[†]Department of Chemistry, University of Illinois, Urbana, IL 61801.

mixture is then refluxed for 3 h, the suspension turning deep green. Failure to exclude all traces of oxygen rigorously could require 3 to 5 days to achieve the deep green color. In that case the reaction may proceed via an orange suspension, in which case the longest reaction period is required. The mixture is allowed to cool to room temperature, and green crystals are collected on a sintered glass filter, washed with ethanol (4 × 25 mL) and hexane (4 × 25 mL), and allowed to dry at room temperature. Yield: 2.28 g (95.5%).

Anal. Calcd. for $C_{54}H_{45}Cl_2OsP_3$: C, 61.89; H, 4.30; Cl, 6.59. Found: C, 61.15; H, 4.54; Cl. 6.78%.

Properties

The green crystals (mp 144–146 °C) are soluble in dichloromethane, acetone, and benzene and insoluble in methanol, ethanol, and hexane. The solid is stable in air but solutions are unstable. Recrystallization is unnecessary for the reaction in Section B.

B. DICHLORO(THIOCARBONYL)TRIS(TRIPHENYL-PHOSPHINE)OSMIUM(II)

$$OsCl_2(PPh_3)_3 + PPh_3 + CS_2 \longrightarrow OsCl_2(CS)(PPh_3)_3 + PPh_3S$$

■ **Caution.** *Carbon disulfide is extremely flammable and toxic and should be handled in an efficient fume hood.*

Dichlorotris(triphenylphosphine)osmium(II) (2.0 g) and triphenylphosphine (1.0 g) are combined under inert atmosphere or in a dry box in a 100-mL flask equipped with a nitrogen gas inlet and a magnetic stirring bar. To the solids are added 50 mL of toluene [previously degassed by the freeze–thaw method and repressurized by nitrogen (< 6 ppm oxygen)] and 5 mL of carbon disulfide.

The mixture is heated under nitrogen to reflux temperature with stirring. The color immediately changes from green to red-brown with a tan solid beginning to precipitate. The mixture should be cooled at this point. The solvent volume is reduced to 20 mL on a rotary evaporator under reduced pressure and absolute ethanol (5 mL) is added. Further reduction of the solvent volume until crystallization is complete and filtration gives white or tan needles, which are washed with ethanol (4 × 25 mL) and hexane (4 × 25 mL) and dried at room temperature. Yield: 2.02 g (98%).

Anal. Calcd. for $C_{55}H_{45}Cl_2OsP_3S$: C, 60.49; H, 4.15; P, 8.51. Found: C, 60.17; H, 4.43; P, 8.55%.

Properties

The white crystals (mp 201–205 °C) are soluble in dichloromethane, acetone, benzene, and chloroform and insoluble in ethanol, methanol, and hexane. Both solid and solution are air stable. Recrystallization is from dichloromethane brought about by addition of ethanol. The IR spectrum (Nujol mull between KBr plates) shows a strong band at 1290 cm^{-1} due to thiocarbonyl.[4]

C. DIHYDRIDO(THIOCARBONYL)TRIS(TRIPHENYL-PHOSPHINE)OSMIUM(II)[5]

$$OsCl_2(CS)(PPh_3)_3 + 2NaOH + CH_3OCH_2CH_2OH$$

$$\longrightarrow OsH_2(CS)(PPh_3)_3 + 2NaCl + 2H_2O + CH_3OCH_2C(O)H$$

Procedure

Dichloro(thiocarbonyl)tris(triphenylphosphine)osmium(II) (2.0 g), triphenylphosphine (0.1 g) and sodium hydroxide (0.5 g) are combined with 2-methoxyethanol (20 mL) in a 100-mL flask, under a nitrogen atmosphere. The mixture is heated to reflux with stirring and maintained at reflux for 20 min. The resulting suspension is cooled on ice. Filtration gives white crystals, which are redissolved in dichloromethane and filtered through a Celite® pad. Ethanol is added, and the dichloromethane is removed under reduced pressure. Filtration gives white crystals, which are washed with absolute ethanol (4 × 25 mL) and hexane (4 × 25 mL) and dried at room temperature. Yield: 1.22–1.72 g (65–91%).

Anal. Calcd. for $C_{55}H_{47}OsP_3S$: C, 64.56; H, 4.67; P, 9.08. Found: C, 64.59; H, 5.00; P, 8.95%.

Properties

The white crystals (mp 174–176 °C) are soluble in dichloromethane, acetone, benzene, and chloroform and insoluble in ethanol, methanol, and hexane. Both solid and solution are air stable, and recrystallization is from dichloromethane–ethanol. The IR spectrum (Nujol mull between KBr plates) shows strong bands at 2070 and 1895 cm^{-1} due to Os—H stretching vibrations, a band at 1233 cm^{-1} due to C—S stretching, and weak bands at 800 and 775 cm^{-1} due to Os—H deformation. ^{1}H NMR (CDCl$_3$, 37 °C) shows multiplets centered at δ − 10.43 and − 7.32.

D. CARBONYL(THIOCARBONYL)TRIS(TRIPHENYL-PHOSPHINE)OSMIUM(0)[6]

$$OsH_2(CS)(PPh_3)_3 + HClO_4 + H_2O$$

$$+ \xrightarrow{\text{MeOH}} [OsH(H_2O)(CS)(PPh_3)]ClO_4 + H_2 \uparrow$$

$$\downarrow_{CO}$$

$$[OsH(CO)(CS)(PPh_3)_3]ClO_4$$

$$\downarrow_{NaOH}$$

$$Os(CO)(CS)(PPh_3)_3$$

▪ **Caution.** *Carbon monoxide is highly toxic and therefore should be vented into an efficient fume hood.*

▪ **Caution.** *Perchlorates are potentially explosive and should be treated with care.*

Procedure

Dihydrido(thiocarbonyl)tris(triphenylphosphine)osmium (1.2 g) is dissolved in dichloromethane (30 mL) in a 100-mL flask equipped with magnetic stirring bar. To this is added a solution of perchloric acid (70%, w/w, 0.6 mL) in methanol (30 mL). Hydrogen gas is evolved. After gas evolution has ceased, carbon monoxide is bubbled through the solution for 5 min and the flask stoppered under carbon monoxide and left for a further 0.5 h. The dichloromethane is removed, and the resulting methanolic solution is diluted to a total volume of 50 mL with addition of further methanol. The solution is degassed by passing a stream of nitrogen through it for 10 min. Crushed sodium hydroxide (0.6 g) is added to the hot solution, and heating under reflux is continued for 20 min. The solution is then cooled to room temperature. Filtration gives tan or orange crystals, which are washed with methanol (4 × 25 mL) and hexane (4 × 25 mL) and dried under vacuum at room temperature. Yield: 1.00–1.20 g (81–97.5%).

Anal. Calcd. for $C_{56}H_{45}OOsP_3S$: C, 64.11; H, 4.32. Found: C, 63.75; H, 4.86%.

Properties

The tan crystals (mp 159–161 °C) are soluble in acetone and benzene and insoluble in ethanol, methanol, and hexane. The solid can be handled briefly in air (up to ~ 2 h) but preferably should be stored under nitrogen. The IR spectrum (Nujol mull between KBr plates) shows strong bands at 1890 and 1230 cm^{-1} due to $v_{(CO)}$ and $v_{(CS)}$.

E. CARBONYL(5-THIOXO-1,3-PENTADIENE-1,5-DIYL-C^1, C^5, S)-BIS(TRIPHENYLPHOSPHINE)OSMIUM, Os(CSCHCHCHCH)(CO)(PPh$_3$)$_2$

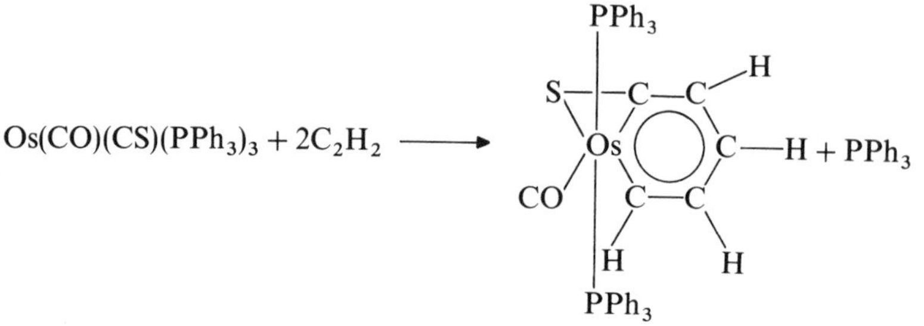

$$Os(CO)(CS)(PPh_3)_3 + 2C_2H_2 \longrightarrow$$

$+ \text{PPh}_3$

Procedure

- **Caution.** *Acetylene gas (*C_2H_2*) is potentially explosive under pressure.* Carbonyl(thiocarbonyl)tris(triphenylphosphine)osmium (1.0 g) is dissolved in dry toluene (50 mL) in a 100-mL flask and the solution heated to 70 °C with acetylene gas bubbling through the solution for 20 min. The solution initially changes color from tan to green black, and then to dark red brown. The solution is allowed to cool to room temperature, and hexane (50 mL) is slowly added with stirring. A brown, floccular solid (0.3 g) is removed by filtration and discarded. Reduction of the solvent volume under reduced pressure gives a brown oil, which is redissolved in a minimum volume of dichloromethane and placed on a 2.5-cm silica gel column (~ 7-cm length) with dichloromethane as eluant. A dark red-brown band is collected from the column. Addition of ethanol and removal of dichloromethane gives dark brown crystals, which are collected by filtration, washed with hexane (4 × 25 mL), and dried at room temperature. Yield: 0.25–0.38 g (31–35%).

Anal. Calcd. for C$_{42}$H$_{34}$OOsP$_2$S: C, 60.13; H, 4.08. Found: C, 59.58; H, 4.70%.

Properties

The dark red-brown crystals (mp 200 °C) are soluble in dichloromethane, benzene, and acetone and insoluble in ethanol and hexane. Both solid and solution are air-stable, and recrystallization is from dichloromethane–

ethanol. The IR spectrum (Nujol mull between KBr plates) shows strong bands at 1890 and 1387 cm^{-1} and other bands at 1530, 1312, 1245, 1195, 1184, and 610 cm^{-1}. The ^{1}H NMR (CDCl$_3$, 37 °C) shows multiplets at 7.28 (unresolved) and 13.95 δ (dq, $^3J_{HH} = 9.4$, $^4J_{HH} \approx {}^3J_{PH} \approx 1.6$ Hz). The weak metal–sulfur bond is broken, and the six-membered ring is retained in reactions of Os($\overline{\text{CSCHCHCHC}}$H)(CO)(PPh$_3$)$_2$ with methyl iodide and carbon monoxide:

Os($\overline{\text{CSCHCHCHCH}}$)(CO)(PPh$_3$)$_2$

$\xrightarrow{\text{MeI}}$ Os($\overline{\text{C[SMe]CHCHCHC}}$H)(CO)(I)(PPh$_3$)$_2$

Os($\overline{\text{CSCHCHCHCH}}$)(CO)(PPh$_3$)$_2$

$\xrightarrow{\text{CO}}$ Os($\overline{\text{C[S]CHCHCHCH}}$)(CO)$_2$(PPh$_3$)$_2$

References

1. G. P. Elliott, W. R. Roper, and J. M. Waters, *J. Chem. Soc. Chem. Commun.*, **1982**, 811.
2. P. R. Hoffman and K. G. Caulton, *J. Am. Chem. Soc.*, **97**, 4221 (1975).
3. F. P. Dwyer and J. W. Hogarth, *Inorg. Synth.*, **5**, 206 (1957).
4. T. J. Collins and W. R. Roper, *J. Organometal. Chem.*, **139**, C57 (1977).
5. T. J. Collins and W. R. Roper, *J. Organometal. Chem.*, **159**, 73 (1978).
6. T. J. Collins, K. R. Grundy, and W. R. Roper, *J. Organometal. Chem.*, **231**, 161 (1982).

34. (1,3-BUTADIENE-1,4-DIYL)(η^5-CYCLOPENTADIENYL)-(TRIPHENYLPHOSPHINE)COBALT* WITH VARIOUS SUBSTITUENTS

Submitted by Y. WAKATSUKI[†] and H. YAMAZAKI[†]
Checked by E. LINDNER[‡] and A. BOSAMLE[‡]

The η^5-cyclopentadienylcobalt unit, [Co(η^5-C$_5$H$_5$)], is believed to be a key active species in catalytic cyclooligomerization of acetylenes or cooligomerization of acetylenes with other unsaturated reactants. It can be generated from several precursors, [Co(η^5-C$_5$H$_5$)(CO)$_2$],[1] [Co(η^5-C$_5$H$_5$)(η^4-diene)],[2]

*(1,3-Butadiene-1,4-diyl)cobalt≡Cobaltacyclopentadiene.
[†]The Institute of Physical and Chemical Research, Saitama 351-01, Japan.
[‡]Institut für Anorganische Chemie, Universität Tübingen, 1400 Tübingen 1, Federal Republic of Germany.

or $[Co(\eta^5\text{-}C_5H_5)(PPh_3)_2]$.[3] Among others, the route from the bis(triphenylphosphine) complex differs markedly from that of the other precursors. In this route, intermediate metallacycles are stabilized by the coordination of PPh_3 and can be isolated when substituted acetylenes are used, usually as very stable crystalline complexes. Unsubstituted acetylene also gives the metallacycle, but the yield is low.[4] The metallacycles thus prepared serve as reagents for the preparation of highly substituted-benzene or substituted-heterocyclic compounds, difficult to obtain by any other preparative methods.[5]

When disubstituted acetylenes are employed, the η^2-acetylene complex, $[Co(\eta^5\text{-}C_5H_5)(PPh_3)(RC{\equiv}CR)]$, can be isolated. In such cases, one can construct a cobaltacyclopentadiene with two different acetylene units by addition of other acetylenes, or one can synthesize cobaltacyclopentene complexes[6] by addition of olefins.

The starting bis(triphenylphosphine) complex is prepared from readily available $[CoCl(PPh_3)_3]$, which was first reported by Aresta, Rossi, and Sacco.[7] Of the three methods described in the original literature, $[CoCl(PPh_3)_3]$ is most conveniently prepared by $Na[BH_4]$ reduction of $CoCl_2 \cdot 6H_2O$ in the presence of PPh_3, as given in the procedure in Section A.

A.   CHLOROTRIS(TRIPHENYLPHOSPHINE)COBALT

$$CoCl_2 \cdot 6H_2O + Na[BH_4] + 3PPh_3 \longrightarrow [CoCl(PPh_3)_3]$$
$$+ NaCl + BH_3 + 6H_2O$$

Procedure

A 9.6-g sample of cobalt(II) chloride hexahydrate and a 32.0-g portion of triphenylphosphine are placed in a 1-L two-necked flask equipped with a magnetic stirring bar. One neck of the flask is connected to a nitrogen inlet and all air is flushed from the system by a nitrogen stream. A 600-mL quantity of ethanol is added and the heterogeneous solution is degassed by bubbling nitrogen for several minutes. The mixture is stirred vigorously for 30 min at 60 to 70 °C to ensure the formation of a blue-colored fine powder of $CoCl_2(PPh_3)_2$. The mixture is cooled to ~ 30 °C and under vigorous stirring 1.28 g of sodium tetrahydroborate(1$-$) is added in 10 portions during ~ 10 min. The color of the mixture turns from blue to dark green and finally to dark brown. After the reduction reaction is complete the brown precipitate is separated by filtration, which can be performed in air. The powder of $[CoCl(PPh_3)_3]$ is washed with ethanol until the filtrate has no blue color and then with water. It is washed again with ethanol and finally with hexane to accelerate the drying. After evaporation of hexane *in vacuo*, it can be used immediately for the following procedure. Although $[CoCl(PPh_3)_3]$ has been described to be fairly stable to the air in the solid state,[7] it is decomposed by storage in air over several days. When stored under an inert atmosphere in a refrigerator, it is indefinitely stable. Yield: 24.0 g (67%).

Anal. Calcd. for $C_{54}H_{45}ClCoP_3$: Cl, 4.0; P, 10.5; Co, 6.7. Found: 4.0; P, 10.4; Co, 6.6.

B. (η⁵-CYCLOPENTADIENYL)BIS(TRIPHENYL-PHOSPHINE)COBALT

$$CoCl(PPh_3)_3 + NaC_5H_5 \longrightarrow Co(\eta^5\text{-}C_5H_5)(PPh_3)_2$$
$$+ PPh_3 + NaCl$$

Procedure

All manipulations must be carried out in an inert atmosphere using degassed solvents.[8] The reaction is carried out in a 300-mL single-necked round-bottomed flask equipped with a nitrogen or preferably an argon bypass and a magnetic stirring bar. Chlorotris(triphenylphosphine)cobalt (24 g, 27.2 mmol) is placed in the flask, and the apparatus is purged with nitrogen by pumping and refilling several times. Toluene (160 mL) is introduced with a syringe. To the stirred suspension a tetrahydrofuran (THF) solution of NaC_5H_5 (Ref. 9) in slight excess (14 mL, 2 mmol mL^{-1}) is added dropwise with a syringe. The brown suspension turns to a dark-red homogeneous solution immediately after the addition. The stirring is continued 30 min more and then excess NaC_5H_5 is hydrolyzed by adding a 10% aqueous solution of NH_4Cl (~ 20 mL). The stirring is stopped and the lower aqueous layer is removed with a syringe. (Checkers recommend addition of 30 mL of degassed water and the use of a separatory funnel to remove the aqueous phase.) The organic layer is transferred to another 300-mL flask containing anhydrous Na_2SO_4 and the mixture is allowed to stand 1 h to remove water. After filtration, solvent is removed under reduced pressure and with swirling to a final volume of ~ 50 mL. Hexane (40 mL) is added slowly, whereupon the crystals begin to form. The flask is stoppered and warmed to 60 °C for 2 h on a water bath in order to grow large crystals. More hexane (30 mL) is added, and the mixture is allowed to stand overnight. The supernatant solution is removed and the remaining large black crystals are washed twice with 10-mL portions of hexane and dried *in vacuo*. The crystals thus formed contain a $\frac{1}{2}M$ amount of hexane as solvent of crystallization and may contain some PPh_3.

An effort to get more crystals from the mother liquid is often unsuccessful. Yield: 9.9–13 g (53–69%). For analytical purpose the product is recrystallized from toluene–hexane.

Anal. Calcd. for $C_{44}H_{42}CoP_2$: C, 76.4; H, 6.1; Co, 8.5. Found: C, 75.0; H, 5.7; Co, 8.8.

Properties

The complex $[Co(\eta^5\text{-}C_5H_5)(PPh_3)_2\cdot\frac{1}{2}(C_6H_{14})]$ forms black crystals (mp 135–140 °C, dec, sealed tube, argon), which are air sensitive and should be stored under an inert atmosphere in a refrigerator. Large crystals can be handled in air, but fine crystals (powder) deteriorate. The compound is soluble in aromatic solvents, THF, chloroform, and dichloromethane and insoluble in aliphatic hydrocarbons. The solution is very air sensitive. The ^1H NMR spectrum (in C_6D_6) has a singlet at δ 4.39 ppm for the C_5H_5 protons. The spectrum also shows the incorporation of $\frac{1}{2}M$ hexane in the crystal. When the product is recrystallized from benzene or toluene just by concentration, the crystals will contain 1 M benzene or toluene in place of the hexane.[10]

C. (η^5-CYCLOPENTADIENYL)[1,1'-(η^2-1,2-ETHYNEDIYL)DI-BENZENE](TRIPHENYLPHOSPHINE)COBALT* AND (η^5-CYCLOPENTADIENYL)(METHYL 3-PHENYL-η^2-2-PROPYNOATE)(TRIPHENYLPHOSPHINE)COBALT†

$$Co(\eta^5\text{-}C_5H_5)(PPh_3)_2 + PhC\equiv CR$$

$$\longrightarrow Co(\eta^5\text{-}C_5H_5)(PPh_3)(PhC\equiv CR) + PPh_3$$

$$R = Ph, CO_2Me$$

Procedure

All solvents must be deoxygenated before use by bubbling nitrogen through them. All reactions must be carried out in an atmosphere of nitrogen. An inert atmosphere is desirable for the column chromatography of the product but not strictly necessary. Alumina of activity grade II to III (or Sumitomo KCG-30) should be used for chromatography. The product is decomposed if the alumina is too active. In that case, the alumina must be deactivated with water

---

*[1,1'-(η^2-1,2-ethynediyl)dibenzene] = η^2-diphenylacetylene.
†(Methyl 3-phenyl-η^2-2-propynoate) = η^2-methyl phenylpropiolate.

prior to use. Activity "Super I" alumina may be treated by 3 to 6% water by weight in order to lower its activity.

Into a 200-mL single-necked flask fitted with a nitrogen bypass and equipped with a magnetic stirring bar is placed $Co(\eta^5\text{-}C_5H_5)(PPh_3)_2\cdot\frac{1}{2}(C_6H_{14})$ (5.4 g, 7.8 mmol). After the flask is thoroughly purged with nitrogen, toluene (35 mL) is introduced by syringe with stirring. 1,1'-(1,2-Ethynediyl)dibenzene (diphenylacetylene, 1.34 g, 7.5 mmol) is then added to the solution and the mixture is allowed to stand for 2 h, during which time the color changes to dark green. Hexane (75 mL) is slowly added. After the solution has stood for several hours, fine dark green-brown crystals of the diphenylacetylene complex precipitate. This is separated by decantation, washed with hexane, and dried *in vacuo*. The crude $Co(\eta^5\text{-}C_5H_5)(PPh_3)(PhC\equiv CPh)$ thus prepared, 3.6 g (82% yield), is sufficiently pure for use in the next steps. If pure complex is desired the crude product is recrystallized from toluene–hexane.

Anal. Calcd. for $C_{37}H_{30}CoP$: C, 78.72; H, 5.36. Found: C, 78.59; H, 5.37.

The methyl 3-phenyl-2-propynoate (methyl phenylpropiolate) complex is prepared in a similar apparatus by the dropwise addition of a toluene solution (10 mL) containing the acetylene (0.9 g, 5.63 mmol) to a stirred solution of $Co(\eta^5\text{-}C_5H_5)(PPh_3)_2\cdot\frac{1}{2}(C_6H_{14})$ (4.05 g, 5.85 mmol) in toluene (50 mL). After 1 h at room temperature the mixture is concentrated under reduced pressure to ~ 30 mL and is absorbed on a chromatography column of alumina activity grade II to III (4 × 20 cm). [Checkers recommend complete solvent removal prior to chromatography, with the residues taken up in the minimum amount of CH_2Cl_2. The CH_2Cl_2 solution is divided into two equal portions, each of which is subjected to column chromatography on alumina of activity grade II to III (column 4 × 15 cm).] Triphenylphosphine and unreacted free acetylene are washed out by elution with toluene–hexane (1:1). A brown band is eluted with toluene or toluene–dichloromethane. Increasing amount of dichloromethane in the eluent is necessary with the more active alumina. The chocolate colored eluate is concentrated under reduced pressure to ~ 10 mL and hexane (30 mL) is added slowly. The fine crystals that form are separated by decantation, washed with hexane, and dried *in vacuo*. Yield: 1.82–2.25 g (57–70%).

Anal. Calcd. for $C_{33}H_{28}CoO_2P$: C, 72.53; H, 5.16. Found: C, 72.64, H, 5.20.

Properties

Analogous complexes of 3-phenyl-2-propynenitrile (cyanophenylacetylene) and dimethyl 2-butynedioate (dimethyl acetylenedicarboxylate) are prepared

TABLE I. Yields and Some Physical Properties for $[Co(\eta^5\text{-}C_5H_5)(PPh_3)(R^1 C\equiv CR^2)]$ Complexes.

R^1	R^2	Color	mp(°C) (dec)	Yield (%)	IR Frequencies $\nu_{C\equiv C}$ KBr(cm^{-1})	NMR Values (C$_6$D$_6$) δ(ppm)	
						C_5H_5	OCH$_3$
Ph	Ph	Dark green	140	82	1820	5.17	
Ph	CO$_2$Me	Chocolate brown	131–133	70	1820	4.74	3.57
Ph	CN	Green brown	156–157	48	1785	4.52	
CO$_2$Me	CO$_2$Me	Orange brown	147–148	15	1822	4.70	3.40

by a similar procedure to that used to obtain the methyl 3-phenyl-2-propynoate (methyl phenylpropiolate) complex. Low yield of the dimethyl butynedioate (dimethyl acetylenedicarboxylate) complex is due to a by-product formed by hydrogen migration.[11] Some properties of the acetylene complexes are summarized in Table I. Their crystals are moderately stable in air but should be stored under an inert atmosphere in a refrigerator. They are soluble in aromatic and polar organic solvents but only sparingly soluble in aliphatic hydrocarbons.

D. (η⁵-CYCLOPENTADIENYL)(2, 3-DIMETHYL-1, 4-DIPHENYL-1, 3-BUTADIENE-1, 4-DIYL)(TRIPHENYL-PHOSPHINE)COBALT*

$$Co(\eta^5\text{-}C_5H_5)(PPh_3)_2 + 2PhC\equiv CMe \longrightarrow$$

$$(\eta^5\text{-}C_5H_5)(PPh_3)Co \quad + PPh_3$$

Procedure

The reaction is carried out in a nitrogen atmosphere in a 200-mL single-necked flask fitted with a nitrogen bypass and equipped with a magnetic stirring bar. The solvent used is deoxygenated by bubbling nitrogen through it.

A toluene solution (5 mL) of 1-propynylbenzene (1-phenylpropyne, 600 mg, 5.17 mmol) is added dropwise to a stirred solution of $Co(\eta^5\text{-}C_5H_5)(PPh_3)_2 \cdot \frac{1}{2}(C_6H_{14})$ (1.82 g, 2.63 mmol) in toluene (30 mL). Stirring is continued for 10 to 20 h. The dark brown solution, which now can be exposed to air for several minutes, is concentrated under reduced pressure to ~ 5 mL. It is chromatographed on alumina (activity grade II–III, 3.5 × 23 cm). An inert atmosphere can be used but not strictly necessary. The column is first eluted with hexane and then with toluene–hexane (1:3, 50 mL), whereupon a broad brown band develops. This brown band is eluted with toluene. The solvent is evaporated from the eluate under reduced pressure. The residue is then dissolved in a minimum amount of dichloromethane and hexane (~ 10 mL) is added dropwise to give dark brown crystals, which are separated by decantation and washed with hexane [mp 179–181 °C (dec)]. Repeated

---

*(2,3-Dimethyl-1,4-diphenyl-1,3-butadiene-1,4-diyl)cobalt≡3,4-dimethyl-2,5-diphenylcobaltacyclopentadiene.

TABLE II. Some Properties of Cobaltacyclopentadiene Complexes, [Co—CR^1=CR^2—CR^3=CR^4(η^5-C$_5$H$_5$)(PPh$_3$)].

R^1	R^2	R^3	R^4	mp(°C) (dec)	Yield (%)	NMR Values (CDCl$_3$) δ (ppm)		
						C$_5$H$_5$	C—Me^a	O—Me
Me	Me	Me	Me^b	166–168	38	4.81^c	2.02, 1.24^c	
Ph	Ph	Ph	Ph	193–194	84	4.82		
CO$_2$Me	CO$_2$Me	CO$_2$Me	CO$_2$Me	216–217	13	5.06		3.56, 3.45
CH$_2$OMe	CH$_2$OMe	CH$_2$OMe	CH$_2$OMe	105–106	34	4.97		3.21, 3.16
Ph	Me	Me	Ph	174–176	44	4.64	1.61 (1 Hz)	
Ph	CO$_2$Me	Ph	CO$_2$Me	215–217	38	4.92		3.06, 3.03
Ph	CO$_2$Me	CO$_2$Me	Ph	218–219	21	4.75		3.32
CO$_2$Me	Me	CO$_2$Me	Me	158–160	48	4.94	1.65 (2 Hz) 2.47	3.57, 3.40
CO$_2$Me	Me	Me	CO$_2$Me	192–194	9	4.98	1.50 (2 Hz)	3.46

^{a} J_{PH} values in parentheses.
^{b} Reacted at 70°C for 2 h.
^{c} Measured in CD$_2$Cl$_2$.

crystallization of the mother liquid gives a total yield of 661 to 843 mg (35–44%). The NMR spectrum and elemental analysis show the crystals thus formed contain a 0.8 M amount of CH_2Cl_2.

Anal. Calcd. for $C_{41.8}H_{37.6}Cl_{1.6}CoP$: C, 73.12; H, 5.52; Cl, 8.26. Found: C, 73.12; H, 5.57; Cl, 8.58. When recrystallized from benzene–hexane, the crystals contain a 1 M amount of benzene (mp 174–176 °C). *Anal.* Calcd. for $C_{47}H_{41}CoP$: C, 81.14; H, 5.94. Found: C, 81.02; H, 6.08.

The other complexes listed in Table II are prepared by similar procedures. The isomers formed from methyl 3-phenyl-2-propyonate (methyl phenyl-propiolate) and methyl 2-butynoate (methyl methylpropiolate) are separated by column chromatography with toluene–THF (20:1) as eluent.

Properties

Some properties of cobaltacyclopentadiene complexes, prepared in this way, are listed in Table II. The crystals are dark brown to orange brown in color and air stable. They can be stored in air. They are soluble in aromatic and polar organic solvents such as chloroform and THF, but not in aliphatic hydrocarbons. Their solutions are moderately stable to air. In general, these cobaltacyclopentadiene complexes are more stable when they contain more electronwithdrawing substituents.

E. [1,4-BIS(METHOXYCARBONYL)-2-METHYL-3-PHENYL-1,3-BUTADIENE-1,4-DIYL]-(η^5-CYCLOPENTADIENYL)-(TRIPHENYLPHOSPHINE)COBALT* AND [1,3-BIS(METHOXYCARBONYL)-2-METHYL-4-PHENYL-1,3-BUTADIENE-1,4-DIYL]-(η^5-CYCLOPENTADIENYL)-(TRIPHENYLPHOSPHINE)COBALT[†]

$$Co(\eta^5\text{-}C_5H_5)(PPh_3)(PhC{\equiv}CCO_2Me) + MeC{\equiv}CCO_2Me \longrightarrow$$

*[1,4-Bis(methoxycarbonyl)-2-methyl-3-phenyl-1,3-butadiene-1,4-diyl]cobalt≡2,5-dicarbo-methoxy-3-phenyl-4-methylcobaltacyclopentadiene.
[†][1,3-Bis(methoxycarbonyl)-2-methyl-4-phenyl-1,3-butadiene-1,4-diyl]cobalt≡2-phenyl-3,5-dicarbomethoxy-4-methylcobaltacyclopentadiene.

TABLE III. Some Properties of Cobaltacyclopentadiene Complexes, $[\overline{Co{-}CR^1{=}CR^2{-}CR^3{=}CR^4}(\eta^5\text{-}C_5H_5)(PPh_3)]$, from Two Different Acetylenes.

R¹	R²	R³	R⁴	mp(°C) (dec)	Yield (%)	NMR Values (CDCl₃) δ(ppm)		
						C_5H_5	C—Me[a]	O—Me
Ph	Ph	H	Ph	180–182	85	4.82		
Ph	Ph	Me	Ph	169–171	67	4.72	1.59(1 Hz)	
Ph	Ph	Me	Me[b]	193–195	38	4.70[c]	1.92, 1.64[c]	
Ph	Ph	H	CO₂Me	149–151	64	4.89		3.43
Ph	Ph	CO₂Me	CO₂Me	119–121	48	4.95		3.46, 3.20
Ph	Ph	CH₂OMe	CH₂OMe	174–176	40	4.85		3.18, 3.10
Ph	Ph	Me	CO₂Me	180–182	68	4.89	1.82(2 Hz)	3.17
Ph	Ph	CO₂Me	Ph	210	82	4.90		3.22
Ph	Ph	Ph	CO₂Me	217–218	13	5.07		2.96
Ph	CO₂Me	Me	Ph	194–196	48	4.66	1.53(1.4 Hz)	3.24
CO₂Me	Ph	Me	Ph	209–210	24	4.84	1.22	2.88
Ph	CO₂Me	Me	CO₂Me	183–185	35	4.89	1.86(2 Hz)	3.30, 3.16
CO₂Me	Ph	Me	CO₂Me	196–197	21	5.03	1.63(2 Hz)	3.34, 3.16
CO₂Me	Ph	Ph	Fc[d]	174–176	25	5.20		2.86
Ph	CO₂Me	Ph	Fc	168–169	24	4.92		2.81

[a] J_{PH} values in parentheses.
[b] Prepared at 50 °C, reaction time = 6 h.
[c] Measured in CD₂Cl₂.
[d] Fc = ferrocenyl group.

Procedure

The reaction is carried out under a nitrogen atmosphere in a 200-mL single-necked flask fitted with a nitrogen bypass and equipped with a magnetic stirring bar. The toluene used is deoxygenated by bubbling nitrogen through it.

A toluene solution (5 mL) of $MeC \equiv CCO_2Me$ (108 mg, 1.1 mmol) is added dropwise to a stirred mixture of $Co(\eta^5\text{-}C_5H_5)(PPh_3)(PhC \equiv CCO_2Me)$ (546 mg, 1 mmol), PPh_3 (131 mg, 0.5 mmol), and toluene (30 mL). The solution is allowed to stand 10 to 20 h at room temperature. The reaction mixture, which now is moderately stable to air, is concentrated under reduced pressure to ~4 mL and chromatographed on alumina (activity grade II–III, 3.5 × 18 cm). An inert atmosphere is not necessary. The column is first eluted with benzene and then with dichloromethane to give a purple band of by-product, a diene complex.[3] On further elution with CH_2Cl_2–THF (20:1), two brown bands separate. After the first red-brown band is eluted, the second yellow-brown broad band is eluted with CH_2Cl_2–THF (10:1). Solvent is removed from these two fractions under reduced pressure. Each residue is dissolved in dichloromethane (~0.3 mL), to which is slowly added hexane (~2 mL) and the mixture is allowed to stand several hours. The dark red crystals that are formed are separated by decantation, washed with hexane, and dried *in vacuo*. More crystals are obtained from the mother liquid. The first fraction gives 135 mg (21% yield) of the 1,4-bis(methoxycarbonyl) isomer (mp 196–197 °C), and the second fraction gives 222 mg (34.5% yield) of the 1,3-bis(methoxy-carbonyl) isomer (mp 183–185 °C).

Anal. Calcd. for $C_{38}H_{34}CoO_4P$: C, 70.81; H, 5.32. Found: C, 70.59; H, 5.60 (the former isomer) and C, 70.67; H, 5.24 (the latter isomer).

Analogs listed in Table III have been prepared by this method.

Properties

For general properties of cobaltacyclopentadiene, see Section D. Yield, melting point, and NMR data of the cobaltacyclopentadienes, prepared from two different acetylenes, are listed in Table III. The regioselectivity of substituents has been explained by their bulkiness.[12]

References

1. (a) K. P. C. Vollhardt, *Acc. Chem. Res.*, **10**, 1 (1977); (b) L. M. Bushnell, E. R. Evitt, and R. G. Bergmann, *J. Organometal. Chem.*, **157**, 445 (1978).

2. (a) Y. Wakatsuki and H. Yamazaki, *Synthesis*, **1976**, 26; (b) H. Bönnemann, *Angew. Chem.*, **90**, 517 (1978).

3. H. Yamazaki and Y. Wakatsuki, *J. Organometal. Chem.*, **139**, 157 (1977).

4. H. Yamazaki and Y. Wakatsuki, *J. Organometal. Chem.*, **272**, 251 (1984).

5. (a) Y. Wakatsuki and H. Yamazaki, *J. Chem. Soc. Chem. Commun.*, **1973**, 280; (b) Y. Wakatsuki, T. Kuramitsu, and H. Yamazaki, *Tetrahedron Lett.*, **1974**, 4549.

6. Y. Wakatsuki, K. Aoki, and H. Yamazaki, *J. Am. Chem. Soc.*, **101**, 1123 (1979).

7. M. Aresta M. Rossi, and A. Sacco, *Inorg. Acta*, **3**, 227 (1969).

8. D. F. Shriver, *The Manipulation of Air-Sensitive Compounds*, McGraw-Hill, New York, 1969.

9. (a) J. M. Birmingham, *Adv. Organometal. Chem.*, **2**, 365 (1964); (b) R. B. King and F. G. A. Stone, *Inorg. Synth.*, **7**, 101, 108, 113 (1963).

10. (a) H. Yamazaki and N. Hagihara, *Bull. Chem. Soc. Jpn*, **44**, 5781 (1971); (b) R. V. Rinze, J. Lorberth, H. Nöth, and B. Stutte, *J. Organometal. Chem.*, **19**, 399 (1969).

11. H. Yamazaki, Y. Wakatsuki, and K. Aoki, *Chem. Lett.*, **1979**, 1041.

12. Y. Wakatsuki, O. Nomura, K. Kitaura, K. Morokuma, and H. Yamazaki, *J. Am. Chem. Soc.*, **105**, 1907 (1983).

35.   AN IRIDIUM(III) COMPLEX CONTAINING CYCLOMETALLATED TRIPHENYLPHOSPHINE FORMED BY ISOMERIZATION OF AN IRIDIUM(I) TRIPHENYL-PHOSPHINE COMPLEX

Submitted by M. A. BENNETT* and J. L. LATTEN*
Checked by L. J. AYERS† and D. L. THORN†

Planar d^8 complexes such as $IrCl(CO)(PPh_3)_2$ (Vaska's compound) and $RhCl(PPh_3)_3$ (Wilkinson's catalyst) have played a key role in the study of oxidative addition reactions of simple molecules to metal centers and in the development of homogeneous catalysis.[1] The iridium(I) complex $IrCl(PPh_3)_3$ (Ref. 2) resembles $IrCl(CO)(PPh_3)_2$ and $RhCl(PPh_3)_3$ in its ability to add molecules such as H_2, HCl, acyl chlorides, 1-alkynes, and silanes,[2,3] but differs from them in undergoing a spontaneous internal oxidative addition of one of the ortho-C—H bonds of the triphenylphosphine ligands. The product is an octahedral hydrido–iridium(III) complex $Ir(o\text{-}C_6H_4\overline{PPh_2})HCl(PPh_3)_2$ (**1**), which is isomeric with $IrCl(PPh_3)_3$ and contains one *ortho*-metallated triphenylphosphine.

*Research School of Chemistry, Australian National University, Canberra, A.C.T., Australia 2601.
†Central Research & Development Dept., E.I. du Pont de Nemours & Co., Inc., Experimental Station, Bldg. 328, Wilmington, DE 19898.

Although the dinitrogen complex $IrCl(N_2)(PPh_3)_2$ reacts with triphenyl-phosphine to give $IrCl(PPh_3)_3$,[4] the latter is most conveniently made by displacement of cyclooctene from the readily available dimer $[IrCl(\eta^2\text{-}C_8H_{14})_2]_2$.

A.   CHLOROTRIS(TRIPHENYLPHOSPHINE)IRIDIUM(I)

$$[IrCl(\eta^2\text{-}C_8H_{14})_2]_2 + 6P(C_6H_5)_3 \longrightarrow 2IrCl[P(C_6H_5)_3]_3 + 4C_8H_{14}$$

Procedure

All manipulations are carried out under nitrogen using freshly degassed solvents. The triphenylphosphine should be recrystallized from hexane before use.

A 100-mL two-necked round-bottomed flask fitted with a magnetic stirring bar and gas inlet is charged with freshly prepared $[IrCl(\eta^2\text{-}C_8H_{14})_2]_2$ (0.60 g, 0.67 mmol)[5] and triphenylphosphine (2.1 g, 8.0 mmol). The flask is sealed with a septum and is evacuated and refilled with nitrogen several times. Petroleum ether (60–80 °C) (50 mL) is added and the orange suspension is stirred for 8 h. The resulting orange precipitate is removed by filtration through a Schlenk frit, washed with diethyl ether (2 × 5 mL), and dried *in vacuo* for 4 h. The yield of $IrCl(PPh_3)_3$ is 1.2 to 1.3 g (89–95%).

Anal. Calcd. for $C_{54}H_{45}ClIrP_3$: C, 63.93; H, 4.47; Cl, 3.49; P, 9.16. Found: C, 63.77; H, 4.52; Cl, 3.53; P, 9.00.

Properties

The complex $IrCl(PPh_3)_3$ is a fine, orange, microcrystalline solid, which can be handled in air very briefly. It is soluble in benzene, dichloromethane, chloroform, and acetone to give very air-sensitive solutions in which rapid ortho-metallation occurs. It is sparingly soluble in diethyl ether and ethanol. The far IR spectrum (Nujol mull) shows a band at 282 cm^{-1} due to $v_{(Ir-Cl)}$. The $^{31}P\{^1H\}$ NMR spectrum in CD_2Cl_2 at 80.98 MHz, measured at -30 °C, is an A_2B (almost A_2X) pattern with $\delta_A = 24.3$, $\delta_B = 15.9$ ppm relative to external 85% H_3PO_4 and $^2J_{AB} = 23$ Hz. On standing, these peaks are replaced by those due to the *ortho*-metallated isomer **1**. Solutions that have been exposed to air often show an AB_2 (almost AX_2) pattern at $\delta_B = -7.6$. $\delta_A = -18.9$ ($^2J_{AB} = 14$ Hz), which may be due to a dioxygen adduct. The ^1H NMR spectrum of $IrCl(PPh_3)_3$ in C_6D_6 shows only a series of multiplets at δ 6.7–7.2 and 7.3–7.9 due to the aromatic protons of triphenylphosphine.

B.   (*OC*-6-53)-CHLORO[(2-DIPHENYLPHOSPHINO)PHENYL-C^1,*P*]HYDRIDOBIS(TRIPHENYLPHOSPHINE)IRIDIUM(III)[6]

$$\text{IrCl}[\text{P}(C_6H_5)_3]_3 \xrightarrow{\ C_6H_{12}\ }$$

OC-6-53

Procedure

All manipulations are carried out in an inert atmosphere using solvents that have been freshly distilled under nitrogen or argon. A 50-mL two-necked, round-bottomed flask equipped with a gas inlet and a magnetic stirring bar is charged with $\text{IrCl}(\text{PPh}_3)_3$ (0.5 g, 0.5 mmol) and is sealed with a septum. It is evacuated and refilled with nitrogen or argon several times, and cyclohexane (10 mL) is added by means of a syringe. Under positive flow of inert gas, the septum is replaced with a reflux condenser and gas outlet. The solution is heated under reflux for 2 h, the color of the suspension changing from orange to cream after ∼ 1 h. The suspension is allowed to cool, and the cream solid is removed by filtration through a Schlenk frit, the filtrate being discarded. Yield: 0.42 g (70%).

The product may be recrystallized from a toluene–hexane solvent mixture described as follows. However, NMR spectra of crude and recrystallized material are identical, so the recrystallization may not be needed for most purposes. The solid is dissolved in several increments of toluene (1 × 20 mL; 2 × 3 mL) by warming to 40° C and the solution is evaporated under reduced pressure to approximately half-volume. Hexane (∼ 3 mL) is added until the solution becomes cloudy and the solution is then refrigerated overnight. The fine, pale cream crystals of the product are removed by filtration and dried *in vacuo*. A second crop can be collected by further evaporation of the filtrate and addition of hexane. Yield: 0.28–0.32 g (47–62%).

Anal. Calcd. for $C_{54}H_{45}ClIrP_3$: C, 63.93; H, 4.47; Cl, 3.49; P, 9.16. Found: C, 63.28; H, 4.41; Cl, 3.89; P, 8.57.

Properties

The complex $\overline{\text{Ir}(C_6H_4\text{PPh}_2)}\text{HCl}(\text{PPh}_3)_2$, **1**, is somewhat air sensitive as a solid. It dissolves readily in benzene, toluene, dichloromethane, and acetone.

The IR spectrum shows a strong band at $2235\,cm^{-1}$ (KBr disk) or $2242\,cm^{-1}$ (Nujol mull) typical of $\nu_{(Ir-H)}$ for hydride trans to chloride in octahedral iridium(III) complexes,[7] and there are also bands at 1565, 1435, and $723\,cm^{-1}$ characteristic of *ortho*-metallated triphenylphosphine.[2,8] The far IR spectrum (Nujol mull) has a band at $254\,cm^{-1}$ typical of $\nu_{(Ir-Cl)}$ for hydride trans to chloride in octahedral iridium(III) complexes.[7] The ^1H NMR spectrum in C_6D_6 at 200 MHz has a double doublet of doublets due to the hydride proton centered at $\delta\,-18.36(^2J_{PH} = 19.4,\,14.5,\,10.3\,Hz)$ in addition to the aromatic proton multiplets in the region from δ 6.6 to 8.0. The $^{31}P\,\{^1H\}$ NMR spectrum in C_6D_6 at 80.98 MHz consists of an AMX pattern arising from the three nonequivalent ^{31}P nuclei: $\delta_A = 3.1$, $\delta_M = -2.5$, $\delta_X = -69.6$ (all in ppm relative to 85% H_3PO_4), $^2J_{AM} = 10.7\,Hz$, $^2J_{MX} = 17.3\,Hz$, $^2J_{AX} = 376\,Hz$. The marked upfield shift of P_X is characteristic of phosphorus in a cyclometallated four-membered ring.[9] The structure for complex 1 deduced from spectroscopic data has been confirmed by a single crystal X-ray study of the analogous bromo complex.[10]

References

1. J. P. Collman and L. S. Hegedus, *Principles and Applications of Organotransition Metal Chemistry*, University Science Books, Mill Valley, CA 1980.

2. M. A. Bennett and D. L. Milner, *J. Am. Chem. Soc.*, **91**, 6983 (1969).

3. M. A. Bennett, R. Charles, and T. R. B. Mitchell, *J. Am. Chem. Soc.*, **100**, 2737 (1978); M. A. Bennett, R. Charles, and P. J. Fraser, *Aust. J. Chem.*, **30**, 1201, 1213 (1977).

4. J. P. Collman, M. Kubota, F. D. Vastine, J. Y. Sun, and J. W. Kang, *J. Am. Chem. Soc.*, **90**, 5430 (1968).

5. J. P. Herde, J. C. Lambert, and C. V. Senoff, *Inorg. Synth.*, **15**, 18 (1974); the submitters successfully used $(NH_4)_2IrCl_6$ in place of hydrated $IrCl_3$ in this preparation, even though the salt is insoluble in water. This is more convenient since iridium is recovered from residues in the form of $(NH_4)_2IrCl_6$.

6. T. E. Sloan, *Topics in Stereochemistry*, Vol. 12, G. Geoffroy, N. L. Allinger, and E. L. Eliel (eds.), Wiley, New York, 1981, p. 1.

7. J. P. Jesson, in *Transition Metal Hydrides*, E. L. Muetterties, (ed.) Marcel Dekker, New York, 1971, Chapter 4.

8. D. J. Cole-Hamilton and G. Wilkinson, *J. Chem. Soc. Dalton Trans.*, 797 (1977).

9. P. E. Garrou, *Chem. Rev.*, **81**, 229 (1981).

10. K. von Deuten and L. Dahlenburg, *Cryst. Struct. Commun.*, **9**, 421 (1980).

36. NICKEL-CONTAINING CYCLIC AMIDE COMPLEXES

Submitted by T. YAMAMOTO*
Checked by H. M. BÜCH and P. Binger†

Reactions of α, β- and β, γ-unsaturated amides, 2-propenamide (acrylamide), 2-methyl-2-propenamide (methacrylamide), *trans*-2-butenamide (crotonamide), N, 2-dimethylpropenamide (N-methylmethacrylamide), 2-methyl-N-phenyl-2-propenamide (N-phenylmethacrylamide), and 3-butenamide with bis(1, 5-cyclooctadiene)nickel, Ni(cod)$_2$, in the presence of a basic and bulky tertiary phosphine, for example, tricyclohexylphosphine or bis(1, 1-dimethylethyl)ethylphosphine (di-*tert*-butylethylphosphine), afford nickel-containing five- and six-membered cyclic amide complexes, respectively. The complexes have Ni—NRCO⎤ rings, and they form corresponding imides when treated with carbon monoxide. The basic and bulky tertiary phosphine ligand of the complex can be replaced by various other tertiary phosphine ligands such as PEt$_3$ and 1, 2-ethanediylbis(diphenylphosphine) [1, 2-bis(diphenylphosphino)ethane] to give a variety of Ni-containing cyclic amide complexes. The procedure described herein is based on two original papers.[1,2]

All reactions and manipulations should be carried out under an atmosphere of dry nitrogen or argon using Schlenk tube techniques.[3a] Solvents are dried over Na wire, distilled under an atmosphere of nitrogen, and stored under an atmosphere of nitrogen.

■ **Caution.** *Allylcyanide (3-butenenitrile) is toxic. The procedure for preparing 3-butenamide below should be carried out in a well-ventilated hood.*

Starting Materials

Bis(1, 5-cyclooctadiene)nickel, Ni(cod)$_2$, is prepared according to the literature,[4] as is tricyclohexylphosphine, tcyp.[5] The latter must be recrystallized from rigorously dried diethyl ether. Methacrylamide (2-methyl-2-propenamide) was purchased from Tokyo Kasei (99% purity, authors) or from Ega Chemie (98–99% purity, checkers). Allylcyanide (3-butenenitrile) was purchased from Tokoyo Kasei (99% purity, authors) or from Merck-Schuchardt (98% purity, checkers).

3-Butenamide is prepared according to the literature.[6] A mixture of 3-butenenitrile (10 mL), H$_2$O (2.2 mL), and conc H$_2$SO$_4$ (6.9 mL) is prepared at

*Research Laboratory of Resources Utilization, Tokyo Institute of Technology, 4259 Nagatsuta, Midori-ku, Yokohama 227, Japan.
†Max-Planck Insitute für Kohlenforschung, Kaiser-Wilhelm-Platz 1, D-4330 Mulheim a.d. Ruhr, Federal Republic of Germany.

0 °C, warmed to 70 °C and stirred for 90 min. The mixture is then cooled to 0 °C and neutralized. Extraction by ethyl acetate gives 3-butenamide (2 g). Checkers ran the reaction with 40.9 g (50 mL, 0.61 mmol) allylcyanide (3-butenenitrile), 11 mL of H_2O and 34.5 mL conc H_2SO_4. Extraction by ethyl acetate gave 14.2 g of 3-butenamide. Additional extraction of the remaining water layer by $CHCl_3$ gave another 12.6 g. Combined yields: 51.6%, mp 74 °C (after recrystallization from EtOH–Et_2O). IR(KBr): 3360, 3190 (NH_2), 1660 (C=O), 1635 (C=C), 990, 910 cm^{-1} (=C—H).

A. [2-METHYLPROPANAMIDATO(2 –)-C^3, *N*](TRICYCLOHEXYL-PHOSPHINE)NICKEL(II), $\overline{NiCH_2CH(CH_3)CON}H$(tcyp)

$$Ni(cod)_2 + CH_2{=}C(CH_3)CONH_2 + tcyp$$

$$\longrightarrow \overline{NiCH_2CH(CH_3)CON}H(tcyp) + 2cod$$

$$tcyp = tricyclohexylphosphine$$

Procedure

A tetrahydrofuran (THF) (20 mL) solution of a mixture of 1.7 g (6.2 mmol) of bis(1, 5-cyclooctadiene)nickel, $Ni(cod)_2$, 1.7 g (6.2 mmol) of tricyclohexylphosphine, and 1.1 g (13 mmol) of 2-methyl-2-propenamide (methacrylamide) is stirred for 6 h at room temperature in a nitrogen-filled 50-mL Schlenk tube equipped with a magnetic stirring bar. A greenish-yellow precipitate results. The reaction mixture is cooled to – 30 °C, and after 2 h the solvent containing

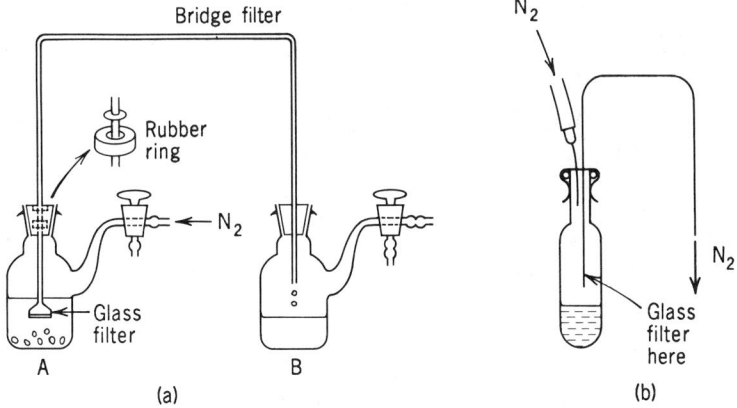

Fig. 1. Filtration by bridge filters. (*a*) from Ref. 7; (*b*) from Ref. 3.

the cod and part of the starting materials is removed using a bridge filter[3b] or a D-3 frit equipped with two inlets (one above and one below the frit) (see Fig. 1). To the precipitate remaining in the Schlenk tube is added 50 mL of diethyl ether, and the suspension is stirred for 1 h at room temperature. The supernatant diethyl ether solution is removed using the bridge filter technique. The remaining solid is then dissolved in 50 mL of toluene at 60 °C. A small amount of an insoluble material is separated by filtration at 60 °C and the resulting solution is cooled to room temperature to give yellow crystals of the product. The crystals are separated by filtration and dried under vacuum. Yield: 1.8 g (68%).

Anal. Calcd. for $C_{22}H_{40}NNiOP$: C, 62.28; H, 9.50; N, 3.30. Found: C, 62.27; H, 9.62; N, 3.42.

Properties

The product is a greenish-yellow crystalline solid that is air sensitive (especially in solution) and decomposes at 239 °C without melting. The tcyp ligand is easily replaced by triethylphosphine (PEt$_3$) and 1, 2-ethanediyl-bis(diphenylphosphine) (dppe) to give, $\overline{NiCH_2CH(CH_3)CON}H(PEt_3)$ and $\overline{NiCH_2CH(CH_3)CON}H(dppe)$, respectively. X-ray crystallographic analysis of the PEt$_3$ complex shows a tetranuclear structure formed through inter-molecular coordination of the amide group to Ni.[8] The low solubility of $\overline{NiCH_2CH(CH_3)CON}H(tcyp)$ in organic solvents may be attributed to its existence in a similar polynuclear complex whose infrared (IR) spectrum (KBr disk) contains a strong, sharp peak at 1572 cm^{-1}; the $^{31}P\{^1H\}$ NMR spectrum (in toluene-d_8) shows a multiplet at $\sim$ 32 ppm from external H_3PO_4. The product reacts with carbon monoxide to give succinimide (quantitatively) and Ni(CO)$_3$(tcyp).

B. [BUTANAMIDATO(2−)-C^4, *N*](TRICYCLOHEXYL-PHOSPHINE)NICKEL(II), $\overline{NiCH_2CH_2CH_2CON}H(tcyp)$

$$Ni(cod)_2 + CH_2{=}CHCH_2CONH_2 + tcyp$$

$$\longrightarrow \overline{NiCH_2CH_2CH_2CON}H(tcyp) + 2cod$$

Procedure

The previous recommendation concerning the use of an inert atmosphere should be observed. A THF (10 mL) solution containing 300 mg (1.1 mmol) of

$Ni(cod)_2$, 310 mg (1.1 mmol) of tcyp, and 94 mg (1.1 mmol) of 3-butenamide is stirred under an inert atmosphere at room temperature in a 30-mL Schlenk tube equipped with a magnetic stirring bar. A yellow precipitate is obtained. After 24 h, the solution is removed by filtration using the bridge filter technique.[3b] The remaining precipitate is washed with diethyl ether (3 times, 10 mL each) and dried under vacuum to yield 130 mg (28%) of the product. A higher overall yield of 72% can be obtained at higher scale. By applying the original procedure on 4.3 mmol of starting materials in 40-mL solvent, 62% yield is obtained; an additional 10% yield is obtained by removal of solvent from the filtrate under vacuum, followed by addition of diethyl ether and filtration.

Anal. Calcd. for $C_{22}H_{40}NNiOP$: C, 62.28; H, 9.50; N, 3.30. Found: C, 61.87; H, 9.68; N, 3.10.

The product is a yellow solid that is air sensitive. It decomposes at 210 °C without melting. It is soluble in hot toluene but the solubility is low. It is insoluble in THF, diethyl ether, acetone, CH_2Cl_2, pyridine, and methanol. The IR spectrum (KBr disk) contains a strong, sharp peak at 1560 cm^{-1}. The product reacts with carbon monoxide affording glutarimide in a 66% yield and $Ni(CO)_3(tcyp)$.

References

1. T. Yamamoto, K. Igarashi, S. Komiya, and A. Yamamoto, *J. Am. Chem. Soc.*, **102**, 7448 (1980).

2. K. Sano, T. Yamamoto, and A. Yamamoto, *Chem. Lett.*, **1982**, 695.

3. (a) D. F. Shriver, *The Manipulation of Air-Sensitive Compounds*, McGraw-Hill, 1969; (b) A glass transfer tube equipped with a glass filter at the end of the tube is used. The solution is transferred by means of a pressure differential (cf. Fig. 7.25, p. 157 of Ref. 3a).

4. B. Bogdanović, M. Kröner, and G. Wilke, *Liebigs Ann. Chem.*, **699**, (1966), 1.

5. K. Issleib and A. Brack, *Z. Anorg. Chem.*, **277**, (1954), 258.

6. R. G. Jones, *J. Am. Chem. Soc.*, **73**, 5610 (1951).

7. T. Yamamoto, T. Saito, and A. Yamamoto, *Kabunshi Jikkengaku*, Vol. 4, S. Kambara, Ed., Kyoritsu, Tokyo, 1983, p. 226.

8. Y. Kushi et al., unpublished result.

37. SIX-MEMBERED CYCLOPALLADATED COMPLEXES OF 2-BENZYLPYRIDINE

Submitted by K. HIRAKI* and Y. FUCHITA*
Checked by M. PFEFFER†

Cyclopalladated complexes provide a potential use for regiospecifically controlled organic syntheses.[1] Various substituents can be introduced onto the palladated carbon position by reactions of cyclopalladated complexes with various reagents.[2-4] Cyclopalladation of aryl-substituted nitrogen or phosphorus bases has been performed mainly by use of $[PdCl_4]^{2-}$ or $PdCl_2(NCPh)_2$ in the presence or absence of an acetate salt, but this method appears to have limit in applicability.[5,6] Here, we describe a preparation of a rare six-membered cyclopalladated complex of 2-benzylpyridine by use of palladium(II) acetate,[7] which seems to be a better starting material for syntheses of cyclopalladated species.[6,7] Only five reports of six-membered cyclopalladated complexes have appeared.[7,8]

A. BIS(μ-ACETATO)BIS[2-(2-PYRIDINYLMETHYL)PHENYL-C^1, N]DIPALLADIUM(II)‡

$$2Pd(CH_3CO_2)_2 + 2C_5H_4N\!-\!CH_2C_6H_5 \longrightarrow$$

$$[\{Pd(C_6H_4\!-\!CH_2C_5H_4N)(\mu\!-\!CH_3CO_2)\}_2]$$

*Department of Industrial Chemistry, Faculty of Engineering, Nagasaki University, Bunkyo-machi, Nagasaki, Japan 852.
†Laboratoire de Chimie de Coordination, Universite Louis Pasteur, ERA 670 CNRS, 4 Rue Blaise Pascal, F-67070 Strasbourg Cédex, France.
‡2-Pyridinylmethyl = 2-picolyl.

■ **Caution.** *The procedure should be carried out in a well-ventilated hood owing to the toxic nature of 2-benzylpyridine.*

Procedure

Palladium(II) acetate was purchased from Nippon Engelhard, and used without further purification. 2-Benzylpyridine (0.46 g, 2.72 mmol) is added to a solution of palladium(II) acetate (0.60 g, 2.67 mmol) in acetic acid (50 mL). This solution is stirred at 25 °C for 24 h affording a pale yellow precipitate. This precipitate is collected on a medium-porosity fritted glass filter and washed with water, methanol, and diethyl ether. Yield: 0.80 g (89%), mp 268 °C (dec under air in a capillary tube).

Anal. Calcd. for $C_{28}H_{26}N_2O_4Pd_2$: C, 50.39; H, 3.93; N, 4.20. Found: C, 49.79; H, 3.95; N, 3.96. This complex can be purified by dissolving in dichloromethane (70 mL) and slowly adding twice its volume of hexane.

Properties

Bis(μ-acetato)bis[2-(2-pyridinylmethyl)phenyl-C^1, N]dipalladium(II) is stable in air and water. It is moderately soluble in dichloromethane and chloroform and sparingly soluble in hexane, diethyl ether, and acetone. The infrared (IR) spectrum shows characteristic bands due to bridging acetato ligand at 1585 and 1420 cm^{-1}. The ^{1}H NMR spectra measured in dichloromethane-d_2 show temperature dependency, indicating a inversion motion of the six-membered boat form ring {i.e., the [2-(2-pyridinylmethyl)phenyl-C^1, N]palladium moiety} above -35 °C.[7] The low-temperature limiting spectrum at -50 °C exhibits two methylene signals at δ 3.15 [q, $\Delta\delta = 0.15$, $^2J_{(HH)} = 14$ Hz] and 4.17 [q, $\Delta\delta = 0.66$, $^2J_{(HH)} = 14$ Hz], and two acetato-methyl signals at δ 1.94 (s) and 1.97 (s).[7]

B. DI-μ-CHLORO-BIS[2-(2-PYRIDINYLMETHYL)-PHENYL-C^1, N]DIPALLADIUM(II)

$$[\{\overset{\frown}{Pd}(C_6H_4-CH_2-C_5H_4\overset{\frown}{N})(\mu\text{-}CH_3CO_2)\}_2] + 2LiCl$$

$$\longrightarrow [\{\overset{\frown}{Pd}(C_6H_4-CH_2-C_5H_4\overset{\frown}{N})(\mu\text{-}Cl)\}_2] + 2CH_3CO_2Li$$

Procedure

In a 50-mL round-bottomed flask containing a magnetic bar are placed bis(μ-acetato)bis[2-(2-pyridinylmethyl)phenyl-C^1, N]dipalladium(II)   (see

Section A) (0.30 g, 0.45 mmol) and 20 mL of acetone. To this suspension, an aqueous solution (10 mL) of lithium chloride (0.076 g, 1.8 mmol) is added. The reaction mixture is stirred for 2 days at room temperature. The resulting pale yellow solid is filtered through a medium-porosity fritted glass and washed with aqueous methanol (50%, 20 mL). Yield: 0.25 g (90%).

Anal. Calcd. for $C_{24}H_{20}Cl_2N_2Pd_2$: C, 46.48; H, 3.25; N, 4.52. Found: C, 46.46; H, 3.39; N, 4.39.

Properties

This complex is stable and almost insoluble in common organic solvents. It has a mp 248 °C (dec under air in a capillary tube). The complex reacts with styrene in refluxing *m*-xylene in the presence of triethylamine to form *trans*-2-[[2-(2-phenylethenyl)phenyl]methyl]pyridine in 44% yield.[7]

C. (3,5-DIMETHYLPYRIDINE)CHLORO[2-(2-PYRIDINYLMETHYL)PHENYL-*C*^{1},*N*]PALLADIUM(II)*

$$[\{\overline{Pd(C_6H_4—CH_2—C_5H_4N)}(\mu\text{-Cl})\}_2] + 2C_5H_3Me_2N$$

$$\longrightarrow 2[\overline{Pd(C_6H_4—CH_2—C_5H_4N)}Cl(C_5H_3Me_2N)]$$

■ **Caution.** *The procedure should be carried out in a well-ventilated hood owing to the toxic nature of 3,5-dimethylpyridine.*

Procedure

3,5-Dimethylpyridine (0.057 g, 0.53 mmol) is added to a dichloromethane (15 mL) suspension of di-μ-chloro-bis[2-(2-pyridinylmethyl)phenyl-*C*^{1}, *N*-]di-palladium(II) (see Section B) (0.15 g, 0.24 mmol) in a 50-mL round-bottomed flask. A clear solution, formed immediately, is stirred for 5 h at room temperature. After the solution is concentrated to $\sim$ 5 mL under reduced pressure, addition of diethyl ether (10 mL) produces white crystals. Yield: 0.15 g (72%), mp 257 °C (dec under air in a capillary tube).

Anal. Calcd. for $C_{19}H_{19}N_2ClPd \cdot \frac{1}{5}CH_2Cl_2$: C, 53.11; H, 4.50; N, 6.45. Found: C, 53.57; H, 4.68; N, 6.43. The residual dichloromethane in this complex is

---

*Common name is *trans*-2-(2-picolyl)stilbene.
*3,5-Dimethylpyridine is 3,5-lutidine.

retained after drying *in vacuo* for 10 h at 40 °C. Dichloromethane is detected in the ^{1}H NMR spectrum.

Properties

This complex is soluble in dichloromethane, chloroform, benzene, and acetone, and sparingly soluble in hexane and diethyl ether. The ^{1}H NMR spectra of this complex indicate that the six-membered palladated 2-(2-pyridinylmethyl)phenyl-C^1, N moiety shows a fluxional motion above -30 °C. In the low-temperature limiting spectrum measured in chloroform-d_1 at -35 °C, the methylene protons are observed at δ 4.45 as an AB quartet [$\Delta\delta$ = 0.93, $^2J_{(HH)} = 14$ Hz], which changes to a sharp singlet at δ 4.51 in the high-temperature limiting spectrum at 55 °C. The cyclopalladated structure of the 2-(2-pyridinylmethyl)phenyl-C^1, N moiety is confirmed by the fact that the aromatic protons appear as two sets of ABCD patterns at -35 °C.[7]

References

1. M. I. Bruce, *Angew. Chem. Int. Ed. Engl.*, **16**, 73 (1977).
2. J. M. Thompson and R. F. Heck, *J. Org. Chem.*, **40**, 2667 (1975).
3. R. A. Holton, *Tetrahedron Lett.*, **1977**, 355.
4. R. A. Holton and K. J. Natalie, Jr., *Tetrahedron Lett.*, **1981**, 267.
5. T. Izumi, H. Watabe, and A. Kasahara, *Bull. Chem. Soc. Jpn.*, **54**, 1711 (1981).
6. Y. Fuchita, K. Hiraki, T. Yamaguchi, and T. Maruta, *J. Chem. Soc. Dalton Trans.*, **1981**, 2405.
7. K. Hiraki, Y. Fuchita, and K. Takechi, *Inorg. Chem.*, **20**, 4316 (1981).
8. R. A. Holton and R. V. Nelson, *J. Organomet. Chem.*, **201**, C35 (1980).

38. CYCLOPALLADATED COMPOUNDS

Submitted by MICHEL PFEFFER*
Checked by ANIL B. GOEL[†]

It is well known that nitrogen-containing ligands are readily metallated by palladium(II). The first reaction of this type was described in 1965 by Cope et al.[1] for azobenzene. Since then numerous examples of similar reactions have been described in the literature.[2] These cyclometallated compounds have been shown to be reactive intermediates in organic synthesis, especially in the

*Laboratoire de Chimie de Coordination, Université Louis Pasteur, UA 416 CNRS, 4 rue Blaise Pascal, F-67070 Strasbourg Cédex, France.
†Ashland Chemical Co., P.O. Box 2219, Columbus, OH 43216.

preparation of heterocyclic products.[3] Moreover, these compounds have been successfully used as starting materials for the synthesis of complexes of transition metals with unusual properties like those that are able to bind reversibly a molecule of carbon dioxide,[4] or to give high yields of heteropoly-metallic compounds via reactions with carbonylmetallate anions.[5] The preparations of compounds with N, N-dimethylbenzylamine (N, N-dimethyl-benzenemethaneamine) and 8-methylquinoline described here follow the methods of Cope and Friedrich[6] and that of Deeming and Rothwell,[7] respectively.

A.   DI-μ-CHLORO-BIS[(2-DIMETHYLAMINO)METHYL]PHENYL-C,N)DIPALLADIUM(II)

$$Li_2PdCl_4 + C_6H_5CH_2N(CH_3)_2 + NEt_3 \longrightarrow$$

$$\tfrac{1}{2}[\overline{PdC_6H_4CH_2N}(CH_3)_2(\mu\text{-Cl})]_2 + NEt_3HCl + LiCl$$

Lithium tetrachloropalladate(II), can be synthesized by treating $PdCl_2$ with two equivalents of LiCl in boiling water. After dissolution of the $PdCl_2$, the water is removed *in vacuo* affording $Li_2[PdCl_4]$ as a brown solid. The $Li_2[PdCl_4]$ (13.12 g, 50 mmol) is dissolved in methanol (300 mL) at room temperature and N, N-dimethylbenzenemethaneamine (7.42 g, 55 mmol) is added to the well-stirred solution. After ~ 5 min, a cream precipitate begins to form. Then NEt_3 (0.05 g, 50 mmol) dissolved in methanol (50 mL) is added dropwise, the addition taking 1 h. The mixture is stirred for 5 h, after which time a bright yellow precipitate is obtained, along with a pale yellow supernatant solution. The precipitate is removed by filtration, washed with methanol (3 × 50 mL) and diethyl ether (2 × 50 mL), and dried *in vacuo*. A 12.95 g yield 94% of the desired product, is thus obtained, which is sufficiently pure for most purposes. The following method of purification can be applied to most of the known chloro-bridged cyclopalladated dimers.

To a suspension of the cyclopalladated compound (12.95 g, 47 mmol) in acetone (300 mL) is added an excess of LiCl (6.77 g, 150 mmol). This mixture is heated with vigorous stirring until the precipitate has dissolved. It is likely that with an excess of LiCl the salt $Li[\overline{PdC_6H_4CH_2N}(CH_3)_2Cl_2]$ is formed, which is stable only in acetone in the absence of water. Addition of water to this solution affords immediately the chloro-bridged dimer. The yellow solution is then quickly filtered over a short column of Celite® (3 cm) to retain the finely divided metallic palladium and the column washed with 50 mL of acetone. The solution thus obtained is poured into a beaker containing 400 mL of water affording a yellow precipitate. The more quickly this operation is done the better the yield of the reaction; the overall time of this purification should not exceed 0.5 h. The yellow precipitate is filtered, washed with water (100 mL), methanol (50 mL), and diethyl ether (100 mL) and dried *in vacuo*. Yield: 12.55 g (97%). This synthesis can be performed on smaller scales affording similar yields of the product.

Anal. Calcd. for $C_9H_{11}ClNPd$: C, 39.14; H, 4.35; N, 5.07. Found: C, 39.49; H, 4.33; N, 5.03.

Properties

The air stable, lemon yellow crystalline compound is slightly soluble in chloroform, acetone, or tetrahydrofuran (THF). Its IR spectrum (KBr pellet) shows typical absorptions for an ortho disubstituted phenyl ring at 745 and 736 cm^{-1}. Its ^{1}H NMR spectrum [CDCl$_3$, 200 MHz] shows three singlets: $\delta = 3.92$ (CH$_2$), 2.855 and 2.838 ppm, (N(CH$_3$)$_2$).

It is a useful starting material for organic synthesis. It reacts readily with alkenes (such as methyl vinyl ketone, styrene, methyl acrylate, etc.), carbon monoxyde, isocyanides, and acyl chlorides to afford good yields of organic products by selective formation of carbon–carbon bonds.[8]

B. DI-μ-CHLORO-BIS(8-QUINOLYL METHYL-*C,N*)-DIPALLADIUM(II)

$$\tfrac{2}{3}[Pd(O_2CCH_3)_2]_3 + 2NC_9H_6CH_3 \longrightarrow$$

$$[\overline{PdCH_2C_9H_6N}(\mu\text{-}O_2CCH_3)]_2 + 2CH_3CO_2H$$

$$[\overline{PdCH_2C_9H_6N}(\mu\text{-}O_2CCH_3)]_2 + 2LiCl \longrightarrow$$

$$[\overline{PdCH_2C_9H_6N}(\mu\text{-}Cl)]_2 + 2CH_3CO_2Li$$

8-Methylquinoline (2.86 g, 20 mmol) is added to a stirred suspension of palladium acetate (4.48 g, 20 mmol) in dichloromethane (150 mL) in a 250-mL beaker. The mixture is stirred for 4 h at room temperature. The brown solution is then filtered and the solution is dried *in vacuo*. The oily residue thus obtained is washed with diethyl ether (10 mL) and then with pentane (4 × 50 mL) to give 5.9 g (96%) of the acetato-bridged dimer as an orange powder. It can be crystallized from a chloroform–diethyl ether solution as deep yellow crystals.

To a solution of this product (4.1 g, 13.3 mmol) in 500 mL of acetone is added an excess of lithium chloride (2.8 g, 66 mmol). The mixture is gently heated with stirring (it is generally not necessary to boil the acetone) until the yellow precipitate of the chloro-bridged dimer, which is formed at the beginning of the reaction, is redissolved. The mixture is filtered to remove the lithium acetate and 500 mL of water is quickly added to the orange solution to give a pale yellow precipitate. This is removed by filtration and washed like the preceding compound. To ensure the removal of unreacted acetato-bridged dimer, the precipitate can be washed with dichloromethane (20 mL) before drying it *in vacuo*. Yield: 2.9 g (76%).

Anal. Calcd. for $C_{10}H_8ClNPd$: C, 42.27; H, 2.82; N, 4.93. Found: C, 41.93; H, 2.81; N, 4.76.

Properties

This compound is almost insoluble in all the common organic solvents and thus cannot be characterized in solution. The strongest absorptions in the IR spectrum (KBr pellet) are at 1502, 813, and 775 cm^{-1}. Like the preceding compound, it readily reacts with N- or P-containing ligands to produce more soluble monomeric species by bridge-cleavage reactions.[2]

Although its Pd—C bond is less reactive towards insertion reactions than that of the previous compound it has been found that two diphenylacetylene molecules can be inserted into the bond affording a new organometallic species containing a nine-membered cyclopalladated ring.[3c] It has, moreover, proved to be an interesting starting material for obtaining organometallic compounds with unusual properties such as CO_2 activation,[4] or for the synthesis of heteronuclear clusters in which a carbonylmetalate anion can be bridging two or three Pd(II) centers.[9]

References

1. A. C. Cope and R. W. Siekman, *J. Am. Chem. Soc.*, **87**, 3272 (1965).
2. M. I. Bruce, *Angew. Chem. Int. Ed. Engl.*, **16**, 73 (1977); J. Dehand and M. Pfeffer, *Coord. Rev.*, **18**, 327 (1976); I. Omae, *Chem. Rev.*, **79**, 287 (1979).
3. (a) C. H. Chao, D. W. Hart, R. Bau, and R. F. Heck, *J. Organomet. Chem.*, **179**, 301 (1979). (b) I. R. Girling and D. A. Widdowson, *Tetrahedron Lett.*, **1982**, 4281. (c) A. Bahsoun, J. Dehand, M. Pfeffer, M. Zinsius, S. E. Bauaoud, and G. Le Borgne, *J. Chem. Soc. Dalton Trans.*, **1979**, 547.
4. P. Braunstein, D. Matt, Y. Dusausoy, J. Fischer, A. Mitschler, and L. Ricard, *J. Am. Chem. Soc.*, **103**, 5115 (1981).
5. M. Pfeffer, J. Fischer, A. Mitschler, and L. Ricard, *J. Am. Chem. Soc.*, **102**, 6338 (1980).
6. A. C. Cope and E. C. Friedrich, *J. Am. Chem. Soc.*, **90**, 909 (1968).
7. A. J. Deeming and I. P. Rothwell, *J. Organomet. Chem.*, **205**, 117 (1981).
8. A. D. Ryabov, *Synthesis*, **1985**, 233.
9. P. Braunstein, J. Fischer, D. Matt, and M. Pfeffer, *J. Am. Chem. Soc.*, **106**, 410 (1984).

Chapter Five

POLYNUCLEAR TRANSITION METAL COMPLEXES

*Preface J. R. Shapley**

Most of the syntheses presented in this chapter resulted from a solicitation for useful procedures involving "cluster" compounds. As such, they represent a cross section of recent activity in a field that has been characterized by metal–metal bonded polynuclear compounds with predominantly carbonyl ligands. Current trends in synthesis with this type of compound are to higher nuclearity, to mixtures of transition metals or transition metals with post-transition metals, and to the inclusion of novel main group ligands (carbide, nitride, sulfide, etc.). These trends are well illustrated in this collection. In many cases the definition of a cluster has been taken to include binuclear compounds with metal–metal bonds as well, and several examples of such compounds are also included here. Using this expanded definition as a guide, syntheses of clusters appearing in Volume 1–25 of *Inorganic Syntheses* are summarized in Table I. Two additional syntheses of binuclear compounds, Nos. 41 and 44, are included in this chapter for convenience, even though the compounds involved do not contain metal–metal bonds.

Interest in the chemistry of transition metal clusters is expanding rapidly in various directions, as indicated by three recent books[1-3] and a selection of recent review articles.[4-6] The concept of a cluster is being broadened to signify a species with properties between a mononuclear complex and a bulk solid material, irrespective of direct metal–metal bonding. Thus, a cluster is a

*Professor J. R. Shapley, School of Chemical Sciences, University of Illinois, Urbana, IL 61801.

215

TABLE I.   Bi- and Polynuclear Metal–Metal Bonded Transition Metal Complexes.[a]

Complex	References
V	
V_2 in polysiloxane	**22**:116
Cr, Mo, W	
$Cr_2(O_2CMe)_4$	**8**:125
$[Cr(CO)_3(\eta^5\text{-}C_5H_5)]_2$	**7**:104
$[Cr(NO)_2(\eta^5\text{-}C_5H_5)]_2$	**19**:211
$[Cr(CO)_3(\eta^5\text{-}C_5H_5)]_2Hg$	**7**:104
$[PPN]_2[Cr_2(CO)_{10}]$	**15**:88
$K[(Cr(CO)_5)_2(\mu\text{-}H)]$	**23**:28
$Mo_2(NMe_2)_6$	**21**:54
$Mo_2Cl_2(NMe_2)_4$	**21**:56
$Mo_2(O_2CC_6H_4R)_4 R = H,$ Cl, Me, OMe	**13**:87
$[NH_4]_5[Mo_2Cl_9]\cdot H_2O$	**19**:129
$Cs_3[Mo_2X_8H], X = Cl, Br$	**19**:129
$Mo_2Cl_4(MeSCH_2CH_2SMe)_2$	**19**:131
$Mo_2Br_4(C_4H_5N)_4$	**19**:131
$Mo_2Br_4[P(n\text{-}Bu)_3]_4$	**19**:131
$Mo_2(O_2CCH_2CH_2CH_2CH_3)_4$	**19**:133
$Mo_2Br_2(O_2CPh)_2[P(n\text{-}Bu)_3]_2$	**19**:133
$[PPN]_2[Mo_2(CO)_{10}]$	**15**:88
$[Mo(CO)_3(\eta^5\text{-}C_5H_5)]_2$	**7**:107
$[Mo_2(H_2O)_8][O_3SCF_3]_4$	**23**:131
$[Mo(\eta^5\text{-}C_5H_5)(CO)_2]_2(\mu\text{-}H)(\mu\text{-}t\text{-}Bu_2P)$	**25**:168
$[Mo(\eta^5\text{-}C_5H_5)(CO)_2]_2(\mu\text{-}H)(\mu\text{-}AsMe_2)$	**25**:169
$[Mo_3O_4(H_2O)_9]A_4 A^- =$ $ClO_4^-, O_3SC_6H_4CH_3^-$	**23**:136
$[Mo_3(Et_2PS_2)_3(S_2)_3S][Et_2PS_2]$	**23**:120
$Mo_3(Et_2PS_2)_4S_4$	**23**:121
$Na_2[Mo_6Cl_8(OR)_6], R = Me, Et$	**13**:99
$[PPN]_2[W_2(CO)_{10}]$	**15**:88
$K[(W(CO)_5)_2(\mu\text{-}H)]$	**23**:29
Mn, Re	
$Mn_3(CO)_{12}H_3$	**12**:43
$[(n\text{-}Bu)_4N]_2[Re_2Cl_8]$	**13**:82, **23**:116
$Re_2(O_2CMe)_4X_2, X = Cl, Br$	**13**:85
$Re_2(O_2CPh)_4X_2, X = Cl, Br$	**13**:87
Re_3Cl_9	**1**:182, **12**:193
Re_3Br_9	**10**:58

TABLE I.   (*Continued*)

Complex	References
$Re_3(CO)_{12}H_3$	**17**:66
$Re_4(CO)_{12}H_4$	**18**:60

Fe, Ru, Os

$Fe_2(CO)_9$	**8**:178
$Na_2[Fe_2(CO)_8]$	**24**:157
$[Fe(CO)_2(\eta^5\text{-}C_5H_5)]_2$	**7**:110
$Fe_3(CO)_{12}$	**7**:193, **8**:181
$Na_2[Fe_3(CO)_{11}]$	**24**:157
$[PPN]_2[Fe_3(CO)_{11}]$	**20**:222, **24**:159
$[PPN]_2[Fe_4(CO)_{13}]$	**21**:66
$[n\text{-}Bu_4N]_2[Fe_4S_4(SPh)_4]$	**21**:35
$[Me_4N]_2[Fe_4S_4(S\text{-}t\text{-}Bu)_4]$	**21**:36
$[n\text{-}Bu_4N]_2[Fe_4Se_4(SPh)_4]$	**21**:36
$[n\text{-}Bu_4N]_2[Fe_4Se_4(S\text{-}t\text{-}Bu)_4]$	**21**:37
$Fe_4(\eta^5\text{-}C_5H_5)_4S_6$	**21**:42
$[Fe_4(\eta^5\text{-}C_5H_5)_4S_5][PF_6]_2$	**21**:44
$Fe_4(\eta^5\text{-}C_5H_5)_4S_5$	**21**:45
$[Ru(CO)_2(\eta^5\text{-}C_5H_5)]_2$	**25**:180
$[Ru(\eta^5\text{-}C_5H_5)]_2(CO)(\mu\text{-}CO)\{\mu\text{-}\eta^1{:}\eta^3\text{-}C(O)C_2Ph_2\}$	**25**:181
$[Ru(CO)(\eta^5\text{-}C_5H_5)]_2(\mu\text{-}CO)(\mu\text{-}CH_2)$	**25**:182
$[Ru(CO)(\eta^5\text{-}C_5H_5)]_2(\mu\text{-}CO)(\mu\text{-}C=CH_2)$	**25**:183
$\{[Ru(CO)(\eta^5\text{-}C_5H_5)]_2(\mu\text{-}CO)(\mu\text{-}CCH_3)\}[BF_4]$	**25**:184
$[Ru(CO)(\eta^5\text{-}C_5H_5)]_2(\mu\text{-}CO)(\mu\text{-}CHCH_3)$	**25**:185
$Ru_3(CO)_{12}$	**13**:92, **16**:45, 47
$[Et_4N][Ru_3(CO)_{11}H]$	**24**:168
$Ru_3(CO)_{10}(NO)_2$	**16**:39
$[PPN][Ru_3(CO)_{10}NO]$	**22**:163
$Os_2(CO)_6(\mu\text{-}I)_2$	**25**:188
$Os_2(CO)_8I_2$	**25**:190
$Os_3(CO)_{12}$	**13**:92
$Os_3(CO)_{10}(NO)_2$	**16**:39
$[PPN][Os_3(CO)_{10}(\mu\text{-}H)(\mu\text{-}CO)]$	**25**:193

Co, Rh, Ir

$Co_2(CO)_8$	**2**:238, **5**:190
$[Co(CO)_2(\mu\text{-}t\text{-}Bu_2P)]_2$	**25**:177
$Co_3(CO)_9(\mu_3\text{-}CR)R = H,$	
Ph, Co_2Et, C(O)NHMe, Cl, $CO_2\text{-}t\text{-}Bu$	**20**:224
$Rh_2(O_2CMe)_4$	**13**:90
$[Rh(CO)(t\text{-}Bu_2PH)]_2(\mu\text{-}t\text{-}Bu_2P)(\mu\text{-}H)$	**25**:170

TABLE I.   (*Continued*)

Complex	References
[Rh(PMe$_3$)(μ-t-BuPH)]$_2$	**25**:174
Rh$_4$(CO)$_{12}$	**17**:115, **20**:209
Rh$_6$(CO)$_{16}$	**16**:49
K$_2$[Rh$_6$(C)(CO)$_{15}$]	**20**:212
Na$_2$[Rh$_{12}$(CO)$_{30}$]	**20**:215
Ir$_4$(CO)$_{12}$	**13**:95
Ni, Pd, Pt	
[Ni(PEt$_3$)$_2$(μ-t-BuPH)]$_2$	**25**:176
Pd$_2$(Ph$_2$PCH$_2$PPH$_2$)$_2$Cl$_2$	**21**:47
Pt microcrystals	**24**:238
Mixed Metal	
FeRu$_3$(CO)$_{13}$H$_2$	**21**:58
[PPN][FeRu$_3$(CO)$_{13}$H]	**21**:60
FeOs$_3$(CO)$_{13}$H$_2$	**21**:63
RuOs$_3$(CO)$_{13}$H$_2$	**21**:64
[PPN][Ru$_3$Co(CO)$_{13}$]	**21**:61
RuCo$_3$(CO)$_{12}$H	**25**:164
Os$_3$Co(η^5-C$_5$H$_5$)(CO)$_9$(μ-CO)(μ-H)$_2$	**25**:196
Os$_3$Co(η^3-C$_5$H$_5$)(CO)$_9$(μ-H)$_3$	**25**:197
Os$_3$Co(η^5-C$_5$H$_5$)(CO)$_9$(μ-H)$_4$	**25**:197

^{a}These complexes appear in *Inorganic Synthesis*, Vols. 1–25.

molecule, with the implied possibility of structural definition at the atomic level, but it also shows evidence for collective properties, either chemical or physical, derived from interactions between more than one metal center. From this perspective there is no doubt that the field of cluster chemistry will continue to grow, and we can expect that additional syntheses of cluster compounds will appear in future volumes of this series.

References

1. M. Moskovits (ed.), *Metal Clusters*, Wiley, New York, 1986.
2. B. C. Gates, L. Guczi, and H. Knözinger (eds.), *Metal Clusters in Catalysis* (Vol. 29 in Studies in Surface Science and Catalysis) Elsevier, Amsterdam, 1986.
3. L. Que (ed.), *Metal Clusters in Proteins*, (ACS Symposium Series No. 372), American Chemical Society, Washington, D.C., 1988.

4. E. Sappa, A. Tiripicchio, A. J. Carty, and G. E. Toogood, Butterfly cluster complexes of the group VIII transition metals, *Prog. Inorg. Chem.*, **35**, 437 (1987).

5. K. C. C. Kharas and L. F. Dahl, Ligand-stabilized metal clusters: Structure, bonding, fluxionality, and the metallic state, *Adv. Chem. Phys.*, **70**, 1 (1988).

6. P. Lemoine, Progress in cluster electrochemistry, *Coord. Chem. Rev.*, **83**, 169 (1988).

39. DITUNGSTEN TETRACARBOXYLATES

Submitted by D. J. SANTURE* and A. P. SATTELBERGER*†
Checked by F. A. COTTON‡ and W. WANG‡

The molybdenum(II) carboxylates, $Mo_2(O_2CR)_4$, are key derivatives containing molybdenum–molybdenum quadruple bonds.[1] First described by Wilkinson and coworkers in the early 1960s,[2] they remain the key starting materials in $(Mo_2)^{4+}$ chemistry and they have been the objects of numerous physical and theoretical studies.[1] The most commonly used method of preparation is the reaction between molybdenum hexacarbonyl and a carboxylic acid in an inert solvent such as 1,2-dichlorobenzene (DCB) [eq. (1)].[3] Excellent yields of $Mo_2(O_2CR)_4$ can be obtained by

$$2Mo(CO)_6 + 4HO_2CR \xrightarrow[\Delta]{DCB} Mo_2(O_2CR)_4 + 2H_2 + 12CO \qquad (1)$$

this procedure. A logical extension of the molybdenum(II) carboxylate synthesis is the substitution of $W(CO)_6$ for $Mo(CO)_6$ to prepare tungsten(II) carboxylates. Such reactions have been performed many times, under varied conditions, in numerous laboratories. They provide trinuclear tungsten(IV) complexes and there is no indication that $W_2(O_2CR)_4$ dimers are ever formed.[4]

In 1981, some 16 years after the crystal structure of $Mo_2(O_2CCH_3)_4$ was published,[5] the synthesis and structure of the first tungsten(II) carboxylate was reported. Abandoning procedures that involved potential oxidizing agents, it was discovered[6] that the sodium amalgam reduction of tungsten tetrachloride in tetrahydrofuran (THF) in the presence of sodium trifluoroacetate provided $W_2(O_2CCF_3)_4$ in good yield. The "reducing salt method" is also applicable to the syntheses of the 2,2-dimethylpropanoate (pivalate)[7,8] and benzoate[9,10] derivatives, but not to tungsten(II) acetate or formate. The acetate can be prepared by metathesis of the trifluoroacetate dimer with tetrabutylammonium acetate[8] or by treatment of $W_2Et_2(NME_2)_4$ with acetic anhydride.

* Department of Chemistry, The University of Michigan, Ann Arbor, MI 48109.
† Present address: Los Alamos National Laboratory, Los Alamos, NM 87545.
‡ Department of Chemistry, Texas A & M University, College Station, TX 77843.

The formate complex $W_2(O_2CH)_4$ has recently been isolated by photolysis of $W_2(CH_2Ph)_2(O_2CH)_4$ in benzene–cyclohexadiene.[12] In the following section, we describe the syntheses of $W_2(O_2CCF_3)_4$, $W_2(O_2C\text{-}t\text{-}Bu)_4$, and $W_2(O_2CCH_3)_4$, which were developed in our laboratory.

General Procedures and Techniques

Due to oxygen and moisture sensitivity of the starting materials and products, all operations are carried out under rigorously dry and oxygen-free atmospheres (nitrogen or argon) using Schlenk or dry box techniques. We prefer dry box techniques[13,14] for both the transfer of solids and the workup of reaction mixtures, and employ a Vacuum Atmospheres dry box equipped with a high-capacity purification system (MO-40H) and a Dri-Cold freezer maintained at $-40\,^\circ\text{C}$. The checkers used Schlenk and cannula techniques[13,14] for all manipulations except the transfer of solids (dry box) and obtained comparable yields. Toluene and THF are dried and freed from dissolved molecular oxygen by distillation from a solution of the solvent, benzophenone and sodium or potassium. Hexane is distilled from a solution of the solvent and butyllithium. Dichloromethane and chlorobenzene are distilled from CaH_2 and P_4O_{10}, respectively. Vacuum sublimations of $W_2(O_2CCF_3)_4$ are most conveniently carried out using a water-cooled, conical McCarter sublimator (Lab Glass 10433-104 or equivalent).

Starting Materials

Although commercially available (Aldrich), polymeric tungsten tetrachloride is routinely prepared by McCarley's method,[7,15] which is given below. Tungsten hexachloride and tungsten hexacarbonyl (Strem) are used without further purification. Sodium–mercury amalgam (0.5 wt%) is prepared in a 500-mL two-necked round-bottomed flask equipped with a N_2 gas inlet and a stopper, by adding 6.83 g (297 mmol) of sodium metal in small pieces (~ 0.1–$0.2\,\text{cm}^3$) to 100 mL of mercury.

- **Caution.** *The reaction of sodium with mercury is highly exothermic. This preparation should be performed in a well-ventilated fume hood.*

The addition of each piece of sodium is carried out under N_2. The stopper is replaced after each addition and the flask is swirled until the sodium dissolves. After all of the sodium is added, the amalgam is allowed to cool to room temperature. The density of the amalgam is $\sim 13.53\,\text{g mL}^{-1}$ at 25°C. Sodium trifluoroacetate and sodium 2,2-dimethylpropanoate ($NaO_2C\text{-}t\text{-}Bu$) are available commercially (Aldrich). These hygroscopic salts are dried by placing 25 g of the appropriate salt in 150 mL of reagent grade benzene and distilling off the benzene–water azeotrope using a Dean-Stark trap (Lab Glass 9130-102 or equivalent). The suspension is then filtered in a nitrogen atmosphere, the salt is washed with diethyl ether, and dried under high vacuum (10^{-5} torr) for 8 h at 25 °C. Commercially available tetrabutylammonium acetate (Tridom/Fluka) is azeotropically dried with benzene, filtered, washed with diethyl ether, and dried under high vacuum (10^{-5} torr, 25 °C, 8 h).

■ **Caution.** *Benzene is a suspected carcinogen. It should only be used in a well-ventilated fume hood and gloves should be worn at all times.*

Procedures

A. TUNGSTEN TETRACHLORIDE [WCl_4]$_x$

$$2WCl_6 + W(CO)_6 \xrightarrow[\Delta]{C_6H_5Cl} 3WCl_4 + 6CO$$

In a dry box, an oven-dried 500-mL three-necked round-bottomed flask equipped with an N_2 gas inlet, a precision mechanical stirrer, and a stopper is charged with 39.66 g (100 mmol) of WCl_6 and 18 g (51 mmol) of $W(CO)_6$. The sealed flask is removed from the dry box, connected to a source of prepurified N_2, and bubbler for pressure release. The stopper is replaced with a rubber septum and ~ 300 mL of purified chlorobenzene is added via cannula techniques. Carbon monoxide evolution starts immediately.

■ **Caution.** *Because of the known toxicity of carbon monoxide, this reaction should be conducted in a well-ventilated fume hood.*

After the solvent is added, the septum is replaced with a reflux condenser, topped with an N_2–CO gas outlet, which is connected to a mineral oil bubbler. The reaction mixture is stirred at reflux for 12 h, and a slow N_2 purge is maintained throughout this period. After cooling to room temperature, the gray-green powder is isolated by filtration, washed with 2 × 50 mL of fresh chlorobenzene, followed by 4 × 50 mL of hexane, and dried *in vacuo* (10^{-5} torr, 25 °C, 8 h). It is important to free the product of all traces of chlorobenzene in order to obtain clean reductions in the following syntheses. Yield: 47 g (or 96% based on WCl_6).

Anal. Calcd. for WCl_4: Cl, 43.5. Found: Cl, 43.2.

Properties

Tungsten tetrachloride is an air and moisture-sensitive gray-green solid, which may be stored indefinitely under a dry nitrogen atmosphere. It is insoluble in common organic solvents. In addition to its use as a precursor for compounds containing tungsten–tungsten quadruple bonds,[6-10] $[WCl_4]_x$ is an excellent starting material for the preparation of mononuclear tungsten(IV) compounds of the type $WCl_4(PR_3)_{2,3}$, $WCl_4(SR_2)_2$, and $WCl_4(NCR)_2$.[15,16]

B. TETRAKIS(TRIFLUOROACETATO)DITUNGSTEN(II). $W_2(O_2CCF_3)_4$ (W$\overset{4}{-}$W)

$$2WCl_4 + 4Na–Hg + 4NaO_2CCF_3 \xrightarrow{THF} W_2(O_2CCF_3)_4 + 8NaCl$$

In a dry box, an oven-dried 500 mL three-necked round-bottomed Morton flask (Lab Glass 7561-100 or equivalent) equipped with an N_2 gas inlet, a precision mechanical stirrer, and a stopper is charged with 16 g (49 mmol) of powdered tungsten tetrachloride, 15 g (110 mmol) of powdered NaO_2CCF_3, ~400 mL of cold (0 °C) THF, and 35 mL of 0.5 wt.% sodium amalgam (103 mmol Na). The sealed flask is removed from the dry box, connected to a source of prepurified N_2 and bubbler for pressure release, and placed in an ice–water bath. The mixture is stirred vigorously for 20 min at 0 °C and a further 2 h at room temperature. During this time the color changes from grey-green to blue-green to yellow-brown. The stirring assembly is then replaced with a stopper and the flask is removed to a dry box where the contents are decanted from the mercury and filtered through a 1 to 2-in. layer of Celite® on a 150–mL medium porosity sintered glass frit. The reaction flask is rinsed with fresh THF (2 × 50 mL), and the combined filtrate and washings are reduced to dryness under vacuum. If the residue is tacky at this point, 50 mL of hexane are added, and the solvent is stripped again leaving a powdery yellow-brown solid. The latter is extracted with hot hexane–toluene (4:1, 4 × 100 mL), and the extracts are filtered though a medium porosity sintered glass frit. The tan solid on the frit is discarded and the solvent is removed from the deep yellow filtrate *in vacuo*. The crude yellow product is then transferred to a large water-cooled sublimator. The latter is sealed, removed from the dry box, and connected to a source of high vacuum (at least 10^{-5} torr). The importance of a good vacuum cannot be overemphasized. The yield, if any, falls dramatically if the vacuum is lower than that specified. The product is heated for 1 h at 60 °C to remove residual hexane and THF and then sublimed at 130 °C for 8 h. After removing the water, the sublimator is returned to the dry box where the yellow crystalline product is scraped off the probe and weighed. Yield: ~12 g (or 60% based on WCl_4).

Anal. Calcd. for $W_2C_8F_{12}O_8$: C, 11.72; F, 27.81. Found: C, 11.56; F, 27.53.

Properties

Tetrakis(trifluoroacetato)ditungsten(II) is a bright yellow, diamagnetic solid that is highly air sensitive. It is very soluble in diethyl ether, THF, toluene, and dichloromethane and moderately soluble in hexane. The 84.26 MHz ^{19}F NMR spectrum (in C_6D_6) consists of a single sharp resonance at $\delta - 70.1$ versus $CFCl_3$ (δ 0.0). Other spectroscopic data and physicochemical properties are reported in the literature.[17]

C. TETRAKIS(2, 2-DIMETHYLPROPANOATO)DITUNGSTEN(II). $W_2(O_2C\text{-}t\text{-}Bu)_4(W \overset{4}{-} W)$

$$2WCl_4 + 4Na\text{--}Hg + 4NaO_2C\text{-}t\text{-}Bu \xrightarrow{\text{THF}} W_2(O_2C\text{-}t\text{-}Bu)_4 + 8NaCl$$

The initial part of the synthesis is identical to that described for $W_2(O_2CCF_3)_4$, with similar precautions for the exclusion of air and moisture. Fifteen grams (121 mmol) of powdered $NaO_2C\text{-}t\text{-}Bu$ are used and the product is isolated by solution techniques. The powdery yellow-brown residue left after THF removal is extracted with hot (70 °C) toluene (5 × 50 mL) and the extracts are filtered through a 150-mL medium porosity sintered glass frit. The tan solid on the frit is discarded, and the yellow-brown filtrate is reduced to ~ 70 mL with mild heating under vacuum. The suspension is cooled to − 40 °C for 8 h, and the crude yellow product is collected, by filtration, on a 150-mL medium porosity sintered glass frit. The filtrate is discarded and the yellow solid is washed (2 × 30 mL) with hexane. The product is recrystallized by dissolving it in hot (70 °C) toluene (10 mL g^{-1} of solid) and slowly cooling to room temperature and subsequently to − 40 °C. The bright yellow needles are removed by filtration, washed once with 20 mL of cold (− 40 °C) toluene, twice with 30 mL of hexane, and dried *in vacuo*. Yield: ~ 13 g (or 68% based on WCl_4).

Anal. Calcd. for $W_2C_{20}H_{36}O_8$: C, 31.11; H, 4.70; W, 47.62. Found: C, 31.22; H, 4.73; W, 47.54.

Properties

Tetrakis(2, 2-dimethylpropanoato)ditungsten(II) is a very air-sensitive, bright yellow, diamagnetic solid. It is very soluble in diethyl ether, THF, and dichloromethane, moderately soluble in toluene and slightly soluble in hexane. The 360 MHz ^1H NMR spectrum (in C_6D_6) shows a single sharp methyl resonance at δ 1.40. Other spectroscopic data and physicochemical properties are reported in the literature.[8]

D.   TETRAKIS(ACETATO)DITUNGSTEN(II). $W_2(O_2CCH_3)_4$- $(W \overset{4}{-} W)$

$$W_2(O_2CCF_3)_4 + 4[(C_4H_9)_4N](O_2CCH_3)$$

$$\xrightarrow{\text{PhCH}_3} W_2(O_2CCH_3)_4 + 4[(C_4H_9)_4N]O_2CCF_3$$

Inside a dry box, freshly sublimed $W_2(O_2CCF_3)_4$ (1.0 g, 1.22 mmol) is dissolved in 5 mL of 60 °C toluene with stirring. A suspension of 1.5 g (5 mmol) of powdered tetrabutylammonium acetate in 10 mL of 25 °C toluene is added, in one portion, to the tungsten solution. Precipitation of a bright yellow solid is observed within seconds after mixing. The mixture is stirred vigorously for a total of 30 s and then filtered rapidly through a 30-mL medium porosity sintered glass frit. The yellow solid on the frit is washed successively with dichloromethane (3 × 10 mL) and hexane (2 × 10 mL) and then dried *in vacuo*. Yield: 0.65–0.70 g (80–90%).

Anal. Calcd. for $W_2C_8H_{12}O_8$: C, 15.91, H, 2.00. Found: C, 16.12; H, 2.08.

Properties

Tetrakis(acetato)ditungsten(II) is a very air-sensitive, bright yellow, diamagnetic solid. It is moderately soluble in THF and acetonitrile but solutions in the latter solvent decompose after several hours. The 360 MHz ¹H NMR spectrum (in THF-d_8) consists of a single methyl resonance at δ 2.91. The mass spectrum (solid probe, electron impact, 40–70-eV ionizing voltage) shows the parent ion multiplet at m/e 604. Other spectroscopic data and physicochemical properties are described in the literature.[8]

References

1. F. A. Cotton and R. A. Walton, *Multiple Bonds Between Metal Atoms*, Wiley, New York, 1982.
2. T. A. Stephenson, E. Bannister, and G. Wilkinson, *J. Chem. Soc.*, **1964**, 2538.
3. R. E. McCarley, J. L. Templeton, T. J. Colburn, V. Katovic, and R. J. Hoxmeier, in *Inorganic Compounds with Unusual Properties*, R. B. King (ed.), Advances in Chemistry Series **150**, Washington, DC, 1976, pp. 318–334.
4. A. Bino, F. A. Cotton, Z. Dori, S. Koch, H. Küppers, M. Millar, and J. C. Sekutowski, *Inorg. Chem.*, **17**, 3245 (1978).

5. D. Lawton and R. Mason, *J. Am. Chem. Soc.*, **87**, 921 (1965).

6. A. P. Sattelberger, K. W. McLaughlin, and J. C. Huffman, *J. Am. Chem. Soc.*, **103**, 2880 (1981).

7. R. R. Schrock, L. G. Sturgeoff, and P. R. Sharp, *Inorg. Chem.*, **22**, 2801 (1983).

8. D. J. Santure, J. C. Huffman, and A. P. Sattelberger, *Inorg. Chem.*, **24**, 371 (1985).

9. F. A. Cotton and W. Wang, *Inorg. Chem.*, **21**, 3860 (1982).

10. F. A. Cotton and W. Wang, *Inorg. Chem.*, **23**, 1604 (1984).

11. M. H. Chisholm, H. T. Chiu, and J. C. Huffman, *Polyhedron*, **3**, 759 (1984).

12. M. H. Chisholm, D. L. Clark, J. C. Huffman, W. G. Van Der Sluys, E. M. Kober, D. L. Lichtenberger, and B. E. Bursten, *J. Am. Chem. Soc.*, **109**, 6796 (1987).

13. D. F. Shriver and M. A. Drezdzon, *The Manipulation of Air-Sensitive Compounds*, 2nd ed., Wiley, New York, 1986.

14. A. L. Wayda and M. Y. Darensbourg, *Experimental Organometallic Chemistry*, ACS Symposium Series **357**, American Chemical Society, Washington, DC, 1987.

15. M. A. Schaefer-King and R. E. McCarley, *Inorg. Chem.*, **12**, 1972 (1973).

16. P. R. Sharp, *Organometallics*, **3**, 1217 (1984).

17. D. J. Santure, K. W. McLaughlin, J. C. Huffman, and A. P. Sattelberger, *Inorg. Chem.*, **22**, 1877 (1983).

40. DIPHENYLPHOSPHINO-BRIDGED DIMANGANESE COMPLEXES

Submitted by J. A. IGGO* and M. J. MAYS[†]
Checked by A. WOJCICKI[‡] and SHIN-GUANG SHYU[‡]

The complex $Mn_2(\mu\text{-}H)(\mu\text{-}PPh_2)(CO)_8$ was first prepared in 2% yield from the reaction of PPh_2Cl with $[Mn(CO)_5]^-$.[1] It has been subsequently synthesized by Hayter[2] in 12% yield from the reaction of $Mn_2(CO)_{10}$ with P_2Ph_4. The complex can be readily deprotonated to give the dimanganese anion, $[Mn_2(\mu\text{-}PPh_2)(CO)_8]^-$, and this anion reacts with complexes of other metals to give a variety of heterometallic clusters.[3,4] The complex $Mn_2(\mu\text{-}H)(\mu\text{-}PPh_2)(CO)_8$ is prepared in 80% yield by the reaction of $Mn_2(CO)_{10}$ with PPh_2H in undried decahydronaphthalane (decalin). This synthesis is described together with the preparation of the anion and the heterometallic clusters $Mn_2(\mu\text{-}AuPPh_3)(\mu\text{-}PPh_2)(CO)_8$ and $Mn_2(\mu\text{-}HgCl)(\mu\text{-}PPh_2)(CO)_8$, which may be derived from the anion. Analogous complexes of molybdenum and iron can be obtained, respectively, from $[(\eta^5\text{-}C_5H_5)Mo(CO)_3]_2$ and $[(\eta^5\text{-}C_5H_5)Fe(CO)_2)]$ using

*Department of Inorganic Chemistry, University of Liverpool P.O. Box 147, Liverpool, L69 3BX, United Kingdom.
[†]University Chemical Laboratory, Lensfield Road, Cambridge, CB2 1EW United Kingdom.
[‡]Department of Chemistry, Ohio State University, Columbus, OH 43210.

similar procedures,[5] as can other hetero metallic complexes containing copper, silver, rhodium, and iridium bonded to the dimanganese moiety.[3]

A. (μ-DIPHENYLPHOSPHINO)-μ-HYDRIDO-BIS (TETRACARBONYLMANGANESE) (Mn—Mn)

$$Mn_2(CO)_{10} + PPh_2H \xrightarrow[\text{decalin}]{150\,^\circ C} Mn_2(\mu\text{-}H)(\mu\text{-}PPh_2)(CO)_8 + 2CO$$

Procedure

■ **Caution.** *Due to the highly toxic nature of released CO and diphenyl phosphine, the reaction should be performed in a well-ventilated hood.*

For the following reaction, $Mn_2(CO)_{10}$ is obtained from Strem and PPh_2H from Aldrich.

In a 500-mL three-necked flask fitted with a reflux condenser, thermometer, nitrogen inlet, and magnetic stirrer is placed 400-mL of undried decalin (used as received from the supplier, Aldrich). Nitrogen is bubbled through the solvent for 15 min after which time the nitrogen inlet is transferred to a bypass on top of the reflux condenser (Fig. 1) and the third neck of the flask is stoppered. Decacarbonyldimanganese (3.12 g, 8 mmol) is added and the

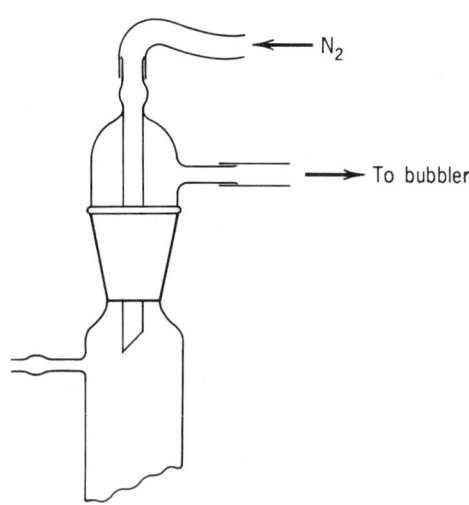

Fig. 1.   Nitrogen inlet and by-pass.

temperature is raised from 145 to 155 °C after which PPh_2H (1.44 mL, 8.0 mmol) is quickly syringed into the stirred solution via the third neck of the flask. A dense cloud of white fumes immediately forms above the surface of the solvent. These disperse slowly (from 2–5 min) during which time the color of the solution changes from golden-orange to orange-red. The temperature is maintained at 145 to 155 °C, and stirring is continued for 1 h. The orange-red to red-brown solution is then allowed to cool to ~ 60 °C, at which temperature the solvent is removed under vacuum to leave a yellow solid. A yellow coloration is observed in the distillate due to unreacted $Mn_2(CO)_{10}$. The yellow solid is redissolved in ~ 35 mL of CH_2Cl_2 and the solution is filtered through a medium porosity sintered glass frit to remove a small quantity of the insoluble complex, $Mn_2(\mu-PPh_2)_2(CO)_8$, which is formed as a by-product in the reaction. Silica (20 g, 70–230 mesh, Merck) is added to the filtrate and, after evaporation of the solvent on a rotary evaporator, the solids are placed at the top of a silica (70–230 mesh) chromatography column 5-cm diameter by 30-cm length. Elution of the column with hexane–dichloromethane (10:1) may give a weak yellow band due to a little unreacted $Mn_2(CO)_{10}$ followed by an intense yellow band due to the major product, $Mn_2(\mu-H)(\mu-PPh_2)(CO)_8$. On evaporation of the solvent (rotary evaporator), the desired complex is obtained as analytically pure, yellow microcrystals (3.32 g, 80%).

The reaction may be scaled up or down; preparations starting from 0.1 to 5.0 g of $Mn_2(CO)_{10}$ can be carried out with substantially no change in yield or purity of the final product. If the concentrations of reactants are increased, however, an impurity of $Mn_2(\mu-H)(\mu-PPh_2)(CO)_7(PPh_2H)$ (Ref. 6) is formed, which is difficult to separate from the major product by chromatography. Use of dried decalin results in a lower yield (60–70%) of the complex.

Anal. Calcd. for $C_{20}H_{11}Mn_2O_8P$: C, 46.2; H, 2.1; P, 6.0. Found: C, 46.3; H, 2.1; P, 5.9.

Properties

The compound $Mn_2(\mu-H)(\mu-PPh_2)(CO)_8$ is a yellow, air stable, but slightly light sensitive solid. It is soluble in most common organic solvents to give solutions that decompose slowly on exposure to air. The carbonyl infrared (IR) spectrum in cyclohexane solution shows absorbances at 2093 (m), 2064 (s), 2011 (s), 2000 (m), and 1966 (s) cm^{-1}; checkers report 2098 (m), 2063 (m), 2005 (s), and 1968 (s) cm^{-1}. The ^1H NMR spectrum in CD_2Cl_2 solution shows a resonance at $\delta - 16.6$ [d, $^2J_{(PH)} = 35.5$ Hz] assigned to the metal-bonded hydrogen; checkers report $\delta - 16.6$ ppm, $J_{(PH)} = 32$ Hz.

The complex $Mn_2(\mu-H)(\mu-PPh_2)(CO)_8$ undergoes carbonyl substitution reactions with phosphite, phosphine, or isocyanide ligands, either thermally or photochemically, to give mono-, di-, or tetra-substituted species.[6] It also reacts with small organic molecules such as alkynes and butadiene.[7,8]

B.   μ-NITRIDO-BIS(TRIPHENYLPHOSPHORUS)(1 +) (μ-DIPHENYLPHOSPHINO)BIS(TETRACARBONYL-MANGANATE) (Mn—Mn) (1 −)

$$Mn_2(\mu\text{-}H)(\mu\text{-}PPh_2)(CO)_8 + Na[BH_4]$$

$$\longrightarrow Na[Mn_2(\mu\text{-}PPh_2)(CO)_8] + \tfrac{1}{2}B_2H_6 + H_2$$

$$Na[Mn_2(\mu\text{-}PPh_2)(CO)_8] + [PPN]Cl$$

$$\longrightarrow [PPN][Mn_2(\mu\text{-}PPh_2)(CO)_8] + NaCl$$

■ **Caution.** *Due to the toxic nature of CO, which may be released in reactions of metal carbonyls, and due to the toxic nature of B_2H_6 and flammable nature of B_2H_6 and H_2, this reaction must be carried out in a well-ventilated hood.*

Deprotonation of $[Mn_2(\mu\text{-}H)(\mu\text{-}PPh_2)(CO)_8]$ may be achieved using KOH–EtOH, Na[BH$_4$], NaH, or LiBu.[3] The reaction with Na[BH$_4$] in MeCN is described here since it affords the best yield of the desired anionic complex.

Procedure

In a 250-mL two-necked flask fitted with a nitrogen inlet, reflux condenser, heating mantle, and magnetic stirrer is placed 125 mL of MeCN. Nitrogen is bubbled through the solution for ~ 15 min after which time the nitrogen inlet is replaced by a stopper and the N_2 flow transferred to a bypass on top of the condenser, see Fig. 1. The stopper is temporarily removed to allow the addition of $[Mn_2(\mu\text{-}H)(\mu\text{-}PPh_2)(CO)_8]$ (0.52 g, 1 mmol) and NaBH$_4$ (0.076 g, 2 mmol) and the solution is then brought to reflux and stirred for 30 min. A quantity of [PPN]Cl (0.57 g, 1 mmol) is then added, followed by 10 mL of distilled water. The product is extracted into CH_2Cl_2 and the organic layer is dried with Na$_2$SO$_4$. The solution is then taken to dryness under reduced pressure and washed several times with hexane to remove unreacted $[Mn_2(\mu\text{-}H)(\mu\text{-}PPh_2)(CO)_8]$ until the washings are colorless. Recrystallization of the residue by diffusion of hexane into a dichloromethane solution affords orange blocks of $[PPN][Mn_2(\mu\text{-}PPh_2)(CO)_8]$ (0.725 g, 69%).

Anal. Calcd. for $C_{56}H_{40}Mn_2NO_8P_3$: C, 63.5; H, 3.8; P, 8.7. Found: C, 63.4; H, 3.7; P, 8.7.

Properties

The compound $[PPN][Mn_2(\mu\text{-}PPh_2)(CO)_8]$ is an orange, air-stable solid. It is soluble in tetrahydrofuran, acetone, and chlorinated solvents, moderately soluble in alcohols and toluene and insoluble in water and hydrocarbon solvents. Solutions of $[PPN][Mn_2(\mu\text{-}PPh_2)(CO)_8]$ are surprisingly stable as compared to most other transition metal carbonyl anions, decomposing only slowly (days) on exposure to air and/or moisture. The carbonyl IR spectrum of the anion in dichloromethane shows absorptions at 2037 (m), 1947 (s), 1941 (vs), 1914 (w), 1888 (m), and 1872 (m).[3]

C. HETEROMETALLIC TRINUCLEAR CLUSTERS

1. **Octacarbonyl-1$\kappa^4 C$, 2$\kappa^4 C$-(μ-diphenylphosphino-1:2κP)-(triphenyl-phosphine-3κP)-*triangulo*-dimanganesegold**

$$[Au(Ph_3)][NO_3] + [PPN][Mn_2(\mu\text{-}PPh_2)(CO)_8]$$
$$\longrightarrow Mn_2Au(PPh_3)(\mu\text{-}PPh_2)(CO)_8 + [PPN]NO_3$$

Procedure

■ **Caution.** *Due to the toxic nature of CO, which may be released in reactions of metal carbonyls, the reaction must be carried out in a well-ventilated hood.*

A solution of $Au(PPh_3)NO_3$ (0.058 g, 0.111 mmol) is prepared by stirring $Au(PPh_3)Cl$ 0.055 g, 0.111 mmol) and $AgNO_3$ (0.025 g, 0.150 mmol) in 5 mL of N_2 saturated acetone under N_2 for 30 min. The solution is then filtered under N_2 from the excess $AgNO_3$ in a Schlenk tube containing a magnetically stirred solution of $[PPN][Mn_2(\mu\text{-}PPh_2)(CO)_8]$ (0.105 g, 0.10 mmol) in 5 mL of CH_2Cl_2 under N_2. Reaction commences immediately and is accompanied by a gradual darkening in color of the solution from orange to deep brown. Stirring is continued for $\sim$ 15 min until the IR spectrum shows no absorptions attributable to $[PPN][Mn_2(\mu\text{-}PPh_2)(CO)_8]$. The reaction mixture is then taken to dryness at 25 °C under high vacuum and the residue is redissolved in a minimum of hexane–CH_2Cl_2 (4:1) and applied to the top of a chromatography column (2 × 15 cm, silica gel 200–300 mesh, Crosfield). Elution with hexane–CH_2Cl_2 (4:1) gives only one major band which, on removal of solvent, affords $[Mn_2(\mu\text{-}AuPPh_3)(\mu\text{-}PPh_2)(CO)_8]$ as an orange-red microcrystalline solid (0.070 g, 71%).

Anal. Calcd. for $C_{38}H_{25}AuMn_2O_8P_2$: C, 46, 46.6; H, 2.6. Found: C, 46.9; H, 2.6.

2.   (μ-Chloromercurio)-(μ-diphenylphosphino)-bis(tetracarbonyl manganese) (Mn—Mn)

$$HgCl_2 + [PPN][Mn_2(\mu\text{-}PPh_2)(CO)_8]$$

$$\longrightarrow Mn_2(\mu\text{-}HgCl)(\mu\text{-}PPh_2)(CO)_8 + [PPN]Cl$$

Procedure

A solution of $[PPN][Mn_2(\mu\text{-}PPh_2)(CO)_8]$, (0.022 g, 0.021 mmol) in 5 mL of CH_2Cl_2 is added dropwise to a vigorously stirred solution of $HgCl_2$ (0.052 g, 0.2 mmol) in 10 mL of MeOH. A golden orange solution results together with a slight precipitate of the red pentametallic complex $[Hg\{Mn_2(\mu\text{-}PPh_2)(CO)_8\}_2]$.[4] This is filtered off, the solution taken to dryness, and the residue dissolved in the minimum quantity of CH_2Cl_2. Column chromatography on silica (70–230 mesh) using 10:1 CH_2Cl_2–hexane as eluant affords after removal of solvent (in order of decreasing R_f values) the red pentametallic complex (0.0015 g, 11%) and $Mn_2(\mu\text{-}HgCl)(\mu\text{-}PPh_2)(CO)_8$ (0.014 g, 89%) as an analytically pure orange powder.

Anal. Calcd. for $C_{20}H_{10}ClHgMn_2O_8P$: C, 39.2; H, 2.0. Found: C, 39.1; H, 1.9.

Properties

The trinuclear dimanganesegold and dimanganesemercury complexes show four intense bands in the carbonyl region of their IR spectra. These are at 2022, 1981, 1954, and 1927 cm^{-1} for the gold complex and at 2061, 2017, 1998, and 1976 for the mercury complex. The gold complex is soluble in most common organic solvents but the mercury complex is soluble only in polar solvents such as CH_2Cl_2. An X-ray study of the gold complex[3] reveals that the gold atom symmetrically bridges the Mn—Mn bond and the structure of the mercury complex is presumably similar. The mercury complex reacts with transition metal carbonyl anions of other metals to give heterometallic clusters such as $Mn_2(\mu\text{-}HgCo(CO)_4)(\mu\text{-}PPh_2)(CO)_8$. An X-ray study of this complex shows that the mercury atom is three-coordinate and symmetrically bridges the manganese atoms.[5]

References

1. M. L. H. Green and J. T. Moelwyn-Hughes, *Z. Naturforsch., Teil. B*, **17**, 783 (1962).
2. R. G. Hayter, *J. Am. Chem. Soc.*, **86**, 823 (1964).
3. J. A. Iggo, M. J. Mays, P. R. Raithby, and K. Henrick, *J. Chem. Soc. Dalton Trans.*, **1984**, 633.
4. J. A. Iggo and M. J. Mays, *J. Chem. Soc. Dalton Trans.*, **1984**, 643.

5. K. Henrick, M. McPartlin, A. D. Horton and M. J. Mays, *J. Chem. Soc., Dalton Trans.*, **1988**, 1083; A. C. Kemball, Ph.D. Thesis, University of Cambridge, 1985.
6. J. A. Iggo, A. C. Kemball, M. J. Mays, P. R. Raithby, and K. Henrick, *J. Chem. Soc., Dalton Trans.*, **1987**, 2669.
7. K. Henrick, J. A. Iggo, M. J. Mays, and P. R. Raithby, *J. Chem. Soc. Chem. Commun.*, **1984**, 209.
8. A. D. Horton, A. C. Kemball and M. J. Mays, *J. Chem. Soc., Dalton Trans.*, **1988**, 2953.

41. PREPARATION OF SOME BIMETALLIC μ-(η^1-C, O) ACETYL COMPLEXES

Submitted by T. C. FORSCHNER and A. R. CUTLER*
Checked by P. A. GOODSON,[†] C. P. CASEY,[†] W. BECK,[‡] and K. H. SÜNKEL[‡]

A number of bimetallic μ-(η^1-C, O) and μ-(η^2-C, O) acyl complexes have been prepared by reacting organometallic Lewis acids[1]—coordinatively unsaturated (i.e., formally 16 electron) and cationic metal electrophiles—with either η^1-alkyl or η^1-acyl compounds.[2,3] In the former reaction, the organometallic Lewis acid promotes the alkyl–CO insertion reaction and then binds to the acyl ligand. Extremely reactive electrophiles $[Fe(\eta^5\text{-}C_5H_5)(CO)_2]^+$ and $[Mo(\eta^5\text{-}C_5H_5)(CO)_3]^+$ thus have been useful in converting a number of organoiron and organo-molybdenum acyl compounds to their bimetallic acyl derivatives. These typically are orange to red salts (with $[BF_4]^-$, $[PF_6]^-$, and $[SbF_6]^-$ counterions) that are air stable and readily recrystallized from CH_2Cl_2–diethyl ether. Their solution stability, with respect to dissociation to starting acyl complex and solvated Lewis acid, generally runs: $CH_2Cl_2 > CH_3NO_2 \gg$ acetone.

Detailed experimental procedures are given for complexes derived from the two acetyl compounds $Fe(\eta^5\text{-}C_5H_5)(CO)_2COCH_3$ and $Fe(\eta^5\text{-}C_5H_5)(CO)PPh_3(COCH_3)$ and the two electrophiles $[Fe(\eta^5\text{-}C_5H_5)(CO)_2]^+$ and $[Mo(\eta^5\text{-}C_5H_5)(CO)_3]^+$. The labile Lewis acid precursors $[Fe(\eta^5\text{-}C_5H_5)(CO)_2(\overline{O(CH_2)_3CH_2})][PF_6]$ and $[Mo(\eta^5\text{-}C_5H_5)(CO)_3(CH_2Cl_2)][PF_6]$ are used in these reactions. Rosenblum and coworkers have reported experimental details for isolating the methallyl complex $Fe(\eta^5\text{-}C_5H_5)(CO)_2\{CH_2C(CH_3) = CH_2\}$ and for converting it to the isobutylene salt $[Fe(\eta^5\text{-}C_5H_5)(CO)_2\{CH_2 = C(CH_3)_2\}][BF_4]$.[4]

*Department of Chemistry, Rensselaer Polytechnic Institute, Troy, NY 12181.
†Department of Chemistry, University of Wisconsin, Madison, WI 53706.
‡Institut für Anorganische Chemie, Universität München, Meiserstrasse 1, 8000 München, Federal Republic of Germany.

All manipulations are carried out under a nitrogen atmosphere using standard syringe–septum and Schlenk-type bench-top techniques for moderately air-sensitive organometallics.[5a−e] Accordingly solutions, including those used for IR and NMR spectral studies, of $[Na][Fe(\eta^5\text{-}C_5H_5)(CO)_2]$ and all neutral complexes of $Fe(\eta^5\text{-}C_5H_5)(CO)L$ (L = CO or PPh_3) and must be made with solvents previously purged with nitrogen for ~ 20 min. The vacuum, following solvent removal in the Buchi rotary evaporator, is also refilled with nitrogen gas. Cationic organometallic compounds typically are not oxygen sensitive; they are, however, often moisture sensitive. Therefore precipitates of these salts are also filtered under nitrogen in Schlenk filter apparatus, in order to avoid condensing moisture as the last traces of solvent are evaporated. Tetrahydrofuran (THF) and diethyl ether are distilled under a nitrogen atmosphere from (purple) sodium benzophenone ketyl solution, and are transferred using a 50-mL syringe. The Ace model E6620 solvent still is recommended.* Dichloromethane is distilled under nitrogen from phosphorus pentoxide.

■ **Caution.** *The drying of the THF and diethyl ether may result in serious explosions under certain conditions.*[4d]

A. DICARBONYL(η^5-CYCLOPENTADIENYL)-(TETRAHYDROFURAN)IRON HEXAFLUOROPHOSPHATE

$$[Fe(\eta^5\text{-}C_5H_5)(CO)_2]_2 + Na(Hg) \longrightarrow 2[Na][Fe(\eta^5\text{-}C_5H_5)(CO)_2]$$

$$[Na][Fe(\eta^5\text{-}C_5H_5)(CO)_2] + ClCH_2CH(CH_3)=CH_2$$

$$\longrightarrow Fe(\eta^5\text{-}C_5H_5)(CO)_2\{CH_2CH(CH_3)=CH_2\} + NaCl$$

$$Fe(\eta^5\text{-}C_5H_5)(CO)_2\{CH_2CH(CH_3)=CH_2\} + H[PF_6]$$

$$\longrightarrow [Fe(\eta^5\text{-}C_5H_5)(CO)_2\{CH_2=C(CH_3)_2\}][PF_6]$$

$$[Fe(\eta^5\text{-}C_5H_5)(CO)_2\{CH_2=C(CH_3)_2\}][PF_6] + \overline{CH_2(CH_2)_3O}$$

$$\longrightarrow [Fe(\eta^5\text{-}C_5H_5)(CO)_2\{\overline{O(CH_2)_3\overset{\frown}{C}H_2}\}][PF_6] + CH_2=C(CH_3)_2$$

Materials

The starting dimer is prepared by the published procedure[5f,g] or obtained commercially (Strem). The crude dimer should be recrystallized from CH_2Cl_2–heptane as follows: It is first dissolved in a minimum volume of

*Ace Glass, Inc., P.O. Box 996, Vineland, NJ 08360.

CH_2Cl_2, filtered through a Celite® filter aid pad, and then diluted by one-half again the volume in heptane. The CH_2Cl_2 is removed on a Buchi rotary evaporator (25 °C, 30-torr vacuum) causing purple crystals of $[Fe(\eta^5$-$C_5H_5)$ $(CO)_2]_2$ to form. These are collected in a Schlenk filter, washed with pentane, and freed of solvent with a gentle flow of nitrogen.

Sodium amalgam (1%) is produced by adding Na chunks to the requisite amount of mercury, which is being stirred in a 500-mL three-necked amalgam-reduction flask.[6] This flask is equipped with a stopcock at the bottom to remove amalgam after completion of reaction.

▪ **Caution.** *Dissolution of sodium is extremely exothermic, and considerable mercury vapor is released.*

After cooling to room temperature, the amalgam (now having the consistency of liquid mercury) is drained through the bottom stopcock into a stoppered storage vessel.

Procedure

A three-necked amalgam reduction flask (500 mL) is fitted with an overhead stirrer and a nitrogen inlet adapter connected to a mineral oil gas bubbler, which is attached to a T-tube in the gas line between the nitrogen cylinder and the flask. This permits venting of excess pressure while maintaining a slightly positive gas pressure. The flask is thoroughly flushed with nitrogen, and excess 1% sodium amalgam is decanted in (~ 70 mL). Tetrahydrofuran (275 mL) and $[Fe(\eta^5$-$C_5H_5)(CO)_2]_2$ (17.50g, 0.049mol) are then added and the flask is sealed with a rubber septum. The purple solution in contact with sodium amalgam is stirred vigorously for 0.75 h yielding a dark orange-yellow supernatant solution of $[Na][FeCp(CO)_2]$. Little or no "mercury dust" or other insoluble residues should be evident if both recrystallized iron dimer and decanted sodium amalgam are used. The amalgam is carefully removed through the lower stopcock, and methallyl chloride (3-chloro-2-methyl-1-propene) (10.0 mL, 0.102 mol) is dripped in by syringe over a 15-min period.

▪ **Caution.** *Exothermic reaction.*

A yellow-brown suspension of NaCl results. After sitting for another 30 min, the THF suspension is filtered through an 8-cm Celite® pad, which is further washed with CH_2Cl_2 (5 × 25 mL).

The combined filtrates containing $FeCp(CO)_2\{CH_2C(CH_3)=CH_2\}$ should be protonated immediately; no attempt should be made to remove the solvent or otherwise to store this crude allylic complex.

The acid medium is prepared in a 2-L Erlenmeyer flask that is immersed in an ice–water slush. A nitrogen atmosphere is unnecessary. To 75 mL of cold acetic anhydride is added *dropwise* 15.0 mL of 65% aqueous HPF_6.

■ **Caution.**   *An extremely exothermic reaction occurs.*

This acid solution is further diluted with diethyl ether (600 mL) to which is slowly added the $THF-CH_2Cl_2$ filtrates [containing $FeCp(CO)_2\{CH_2C(CH_3)=CH_2\}$]. The resulting yellow precipitate is collected in a large Schlenk filter and washed with diethyl ether (3 × 40 mL). Purification is effected by washing the yellow solid through the frit with CH_2Cl_2 (6 × 50 mL), and reprecipitating the $[Fe(\eta^5-C_5H_5)(CO)_2\{CH_2=C(CH_3)_2\}][PF_6]$ by addition of diethyl ether (600 mL). The lemon yellow precipitate is filtered and washed with diethyl ether (3 × 40 mL). It is then dried, first under a flow of nitrogen, and then *in vacuo*. Yield: 31.76 g (84%).

Anal. Calcd. for $C_{11}H_{13}O_2FePF_6$: C, 35.45; H, 3.47. Found: C, 35.15; H, 3.56.

The preparation of $[Fe(\eta^5-C_5H_5)(CO)_2\{\overline{O(CH_2)_3C}H_2\}][PF_6]$ is a slight modification of the procedure reported by Rosenblum and Scheck.[7c] A three-necked flask (500 mL) is equipped with a magnetic stirring bar, a reflux condenser containing a gas inlet adapter, and an electric heating mantle. The flask is purged with nitrogen for several minutes and then charged with $[Fe(\eta^5-C_5H_5)(CO)_2\{\eta^5-CH_2=C(CH_3)_2\}][PF_6]$ (7.20 g, 19.0 mmol), and 150 mL each of CH_2Cl_2 and THF. The flask is then sealed with a rubber septum and heated at 40 °C for 8 h under a slightly positive nitrogen atmosphere. The initial dark yellow suspension turns dark red. After cooling at room temperature, the mixture is filtered through a 1 cm bed of Celite,® which is extracted with CH_2Cl_2 (6 × 30 mL) until the filtrates are colorless. The combined CH_2Cl_2 and THF filtrates are concentrated under reduced vacuum to ~ 150 mL, and they are added slowly to excess diethyl ether (600 mL) in a 1-L Erlenmeyer flask, with swirling or stirring. The product $[Fe(\eta^5-C_5H_5)(CO)_2\{\overline{O(CH_2)_3C}H_2\}][PF_6]$, precipitates as a dark red powder; it is filtered, washed with diethyl ether (3 × 30 mL), dried in a stream of nitrogen, and further dried under vacuum (10^{-2} torr, 45 min) (4.80 g, 61% yield).

Anal. Calcd. for $C_{11}H_{13}O_2FePF_6$: C, 33.53; H, 3.33. Found: C, 33.85; H, 3.17.

Properties

Both the iron η^1-methallyl and η^2-isobutylene complexes have been thoroughly characterized.[8] The complex $[Fe(\eta^5-C_5H_5)(CO)_2\{CH_2=C(CH_3)_2\}][PF_6]$ is an air-stable, lemon yellow salt that readily dissolves in CH_2Cl_2 and all dipolar aprotic solvents. It can be stored indefinitely in a tightly stoppered bottle in the refrigerator. The IR spectrum in CH_2Cl_2 solution exhibits two intense carbonyl absorptions, $\nu_{(CO)}$ at 2072 and 2033 cm^{-1}. The ^1H NMR spectrum (in CF_2CO_2H solution, with internal tetramethylsilane, TMS) shows three sharp resonances: δ 5.53 (s, 5H, Cp); 4.90 (s, 2H, $=CH_2$); 1.96 [s, 6H, $=C(CH_3)_2$]. The utility of this salt stems from the lability of isobutylene at 40 to 60 °C in CH_2Cl_2 or $ClCH_2CH_2Cl$.[8] As a result, a large number of $[Fe(\eta^5-C_5H_5)(CO)_2L]^+$ complexes (where L is a two electron donor ligand or Lewis base) are accessible through displacement of the isobutylene.

The tetrahydrofuranate $[PF_6]^-$ salt[7] is characterized by IR, $\nu_{(CO)}$ 2075, 2028 cm^{-1} (in CH_2Cl_2) and ^{1}H NMR (CF_3CO_2H solution): δ 5.36 (s, 5H, Cp); 3.50 (m, 4H α-CH_2); 1.86 (m, 4H, β-CH_2); (CD_3NO_2) δ 5.43, 3.50, and 1.80. The dark red salt is moisture sensitive and extremely labile in solution, although the solid can be stored under nitrogen in the refrigerator for extended periods of time.

B. (μ-ACETYL-*C:O*)-BIS(DICARBONYL-η⁵-CYCLOPENTADIENYLIRON)(1 +) HEXAFLUOROPHOSPHATE(1 −)

$$2Fe(\eta^5-C_5H_5)(CO)_2CH_3 + [Ph_3C][PF_6] + CO$$
$$\longrightarrow [Fe(\eta^5-C_5H_5)(CO)_2\{C(CH_3)O\}Fe(CO)_2(\eta^5-C_5H_5)][PF_6]$$
$$+ Ph_3CCH_3$$

Materials

Triphenylmethylium (trityl) hexafluorophosphate may be obtained commercially (Aldrich) or prepared by the following adaptation of Dauben's original synthesis.[9] A 1-L Erlenmeyer flask containing 200 mL of acetic anhydride is immersed in an ice-water bath, and with gentle swirling, (21 mL of 65%) aqueous $H[PF_6]$ is added dropwise.

- **Caution.** *The dehydration of the aqueous $H[PF_6]$ is extremely exothermic.*

Triphenylmethanol (45.0 g, 0.17 mol) is added with swirling, giving a bright yellow-orange suspension. The amount of precipitate is increased by adding ethyl acetate (600 mL). Care must be takent to avoid use of an ether in place of ethyl acetate in this step, as this substitution promotes spectacular decomposition of the product.

The crude $[Ph_3C][PF_6]$ is filtered in a Schlenk filter apparatus, washed with ethyl acetate (3 × 40 mL), dried in a flow of nitrogen, and then washed through the frit with CH_2Cl_2 (total 300 mL) into ethyl acetate (600 mL). The color of the product is lemon yellow (as a powder) or orange (crystalline), depending on the rate of addition of the CH_2Cl_2 filtrate to the ethyl acetate. The product is collected, washed with ethyl acetate ($3 \times 40\,mL$), freed of residual solvent in a stream of nitrogen, and dried *in vacuo* (10^{-2} torr, 22 °C, 2.0 h). Yield: 48.1 g (73%). Although stored under nitrogen at + 5 °C, trityl salts slowly decompose[10] (as evidenced by appearance of HF acid fumes), and this necessitates periodic reprecipitation from CH_2Cl_2–ethyl acetate and vacuum drying. When purified as above, $[Ph_3C][PF_6]$ is free of residual ethyl acetate: its IR spectra (0.20 mmol/1.0 mL of CH_2Cl_2), are free of the ethyl acetate carbonyl absorption ($1732\,cm^{-1}$), under conditions where at least 2% could be detected.

The methyl complex, $Fe(\eta^5\text{-}C_5H_5)(CO)_2CH_3$, used in this procedure can be prepared by following the published method[11] and purified as follows:

Caution.   *The complex* $Fe(\eta\text{-}C_5H_5)(CO)_2CH_3$, *is a volatile and potentially toxic metal carbonyl. All operations should be conducted in a well-ventilated hood.*

A quantity of alumina (activity I Camag, neutral, 45 g) is placed into a Schlenk filter (3.6-cm i.d.) to which is added a slurry of the iron methyl complex (15 g) in pentane. Extraction of the slurry with additional quantities of pentane (300 mL) without letting it go dry, continuously removes the desired yellow methyl complex, while leaving $[Fe(\eta\text{-}C_5H_5)(CO)_2]_2$ as a dark brown band at the top of the "column." The pentane extracts are concentrated to a volume of 10 mL on the rotary evaporator (30 torr, 20 °C); the vacuum is broken with nitrogen, and the remaining pentane is evaporated under a stream of nitrogen. Waxy yellow crystals of spectroscopically pure $Fe(\eta\text{-}C_5H_5)(CO)_2CH_3$ remain.

Procedure

Caution.   *This procedure employing toxic carbon monoxide must be conducted in a well-ventilated hood.*

A 100-mL sidearmed round-bottomed flask is equipped with a magnetic stirring bar and attached to a nitrogen line with a pressure-releasing oil bubbler. The flask is flame dried under nitrogen. Upon cooling, quantities of $Fe(\eta^5\text{-}C_5H_5)(CO)_2CH_3$ (0.600 g, 3.13 mmol), $[Ph_3C][PF_6]$ (0.602 g, 1.55 mmol) and CH_2Cl_2 (40 mL) are added. A standard taper adapter containing a sintered-glass gas dispersion frit is immediately fitted to the flask, and a gentle carbon monoxide purge through the yellow solution is started concurrent with vigorous stirring. The CO is vented through the oil bubbler–nitrogen line, as the nitrogen flow is stopped. No further change in the reddish-yellow solution (only a trace of sediment) is observed after stirring in the presence of a gentle stream of CO for 2 h. After a total of 4 h, the solution is filtered, and the filtrate is added dropwise to diethyl ether (200 mL) with stirring. The resulting red precipitate is filtered and washed with diethyl ether (3×20 mL). It is then dried, first under a flow of nitrogen gas and then under vacuum. Yield of $[Fe(\eta^5-C_5H_5)(CO)_2\{C(CH_3)O\}Fe(CO)_2(\eta^5-C_5H_5)][PF_6]$ is 0.26 to 0.64 g (32–79%).

Properties

The starting material, $Fe(\eta^5-C_5H_5)(CO)_2CH_3$, is a yellow, waxy solid (mp79 to 80 °C) (lit[11] 78–82 °C); it is air stable for short periods of time. Prolonged storage is best accomplished in sealed vessels, under nitrogen, and in a refrigerator. Its diagnostic solution spectral data are IR (CH_2Cl_2) 2003, 1948 cm^{-1}; ^1H NMR ($CDCl_3$) δ 4.70 (s, 5H, Cp), 0.15 (s, 3H, FeCH$_3$).

The bimetallic μ-acetyl salt $[Fe(\eta^5-C_5H_5)(CO)_2\{C(CH_3)O\}Fe(CO)_2(\eta^5-C_5H_5)][PF_6]$ (Refs. 2 and 3) is an air stable, red solid that is readily soluble in CH_2Cl_2, acetone, nitromethane, and other dipolar aprotic solvents. Acetone solutions, however, degrade slowly into the starting acetyl complex, $Fe(\eta^5-C_5H_5)(CO)_2COCH_3$ and the acetone solvate of the Lewis acid, $[Fe(\eta^5-C_5H_5)(CO)_2\{O=C(CH_3)_2\}][PF_6]$.

A convenient assay of the bimetallic salt is to react it with iodide: One equivalent of $[(butyl)_4N][I]$ quantitatively transforms $[Fe(\eta^5-C_5H_5)(CO)_2\{C(CH_3)O\}Fe(CO)_2(\eta^5-C_5H_5)][PF_6]$ (0.10 mmol/1.5 mL CH_2Cl_2) into $Fe(\eta^5-C_5H_5)(CO)_2COCH_3$, $Fe(\eta^5-C_5H_5)(CO)_2I$ and $[(butyl)_4N][PF_6]$ in ~5 min, (as judged by examination of IR spectra). Selected spectral data for $[Fe(\eta^5-C_5H_5)(CO)_2\{C(CH_3)O\}Fe(CO)_2(\eta^5-C_5H_5)][PF_6]$: IR ($CH_2Cl_2$) $\nu_{(CO)}$ 2062, 2040, 2018, 1992 cm^{-1} (CO); 1552 cm^{-1} (η-COCH$_3$). ^1H NMR (CD_3NO_2) δ 5.40 (s, 5H, CpFeO), 5.03 (s, 5H, CpFeC), 2.63 (s, 3H, CH$_3$).

Anal. Calcd. for $C_{16}H_{13}O_5Fe_2PF_6$: C, 35.46; H, 2.42. Found: C, 35.58; H, 2.64.

C. (μ-ACETYL-2κC^1:1κO)-TRICARBONYL-1κ^2C,2κC-BIS[1:2(η^5-CYCLOPENTADIENYL)](TRIPHENYL-PHOSPHINE-2κP)DIIRON(1 +) HEXAFLUOROPHOSPHATE(1 −)

$$Fe(\eta^5-C_5H_5)(CO)_2CH_3 + PPh_3$$

$$\longrightarrow Fe(\eta^5-C_5H_5)(CO)(PPh_3)COCH_3$$

$$[Fe(\eta^5-C_5H_5)(CO)_2\{\overline{O(CH_2)_3CH_2}\}][PF_6]$$

$$+ Fe(\eta^5-C_5H_5)(CO)(PPh_3)COCH_3$$

$$\longrightarrow [Fe(\eta^5-C_5H_5)(CO)(PPh_3)\{C(CH_3)O\}Fe$$

$$\cdot(CO)_2(\eta^5-C_5H_5)][PF_6] + THF$$

Procedure

This preparation represents a convenient, albeit inefficient, procedure, as the 3 h reflux time corresponds to the point of diminishing returns for acetyl formation *versus* decomposition. Other procedures using THF or acetonitrile in place of nitromethane have been reported, but the reaction times are substantially longer.[12]

A 250-mL three-necked flask is fitted with a magnetic stirring bar and a condenser attached to a nitrogen line with an oil bubbler pressure release. The flask is flushed with nitrogen for 15 min, and then $Fe(\eta^5\text{-}C_5H_5)(CO)_2CH_3$ (8.00 g, 41.8 mmol), triphenylphosphine (14.42 g, 55.0 mmol), and degassed nitromethane (130 mL, reagent grade) are added. The apparatus is sealed with a glass stopper, and the yellow-brown solution is refluxed for 3 h. After cooling to room temperature, the resulting orange-brown solution is evaporated on a rotary evaporator (30 torr, 35 °C. The solid is extracted with dichloromethane (9 × 25 mL) and the extracts are filtered through a 4-cm bed of alumina (neutral, activity 3) in a 3.6-cm diameter Schlenk filter apparatus until the filtrates are colorless. The combined filtrates are diluted with heptane (75 mL), and the CH_2Cl_2 is selectively distilled off on a rotary evaporator (30 mm, 22 °C). When almost all of the dichloromethane is removed, orange crystals are formed in the flask. These are collected, washed with pentane, and dried first under a nitrogen flow and then *in vacuo* (10^{-2} torr, 2 h). Yield: 12.32 g, (65%).

Recrystallization of $Fe(\eta^5\text{-}C_5H_5)(CO)(PPh_3)COCH_3$ from CH_2Cl_2– heptane eliminates excess phosphine—provided not quite all of the CH_2Cl_2 is removed ($\sim$ 5–10% of volume should remain). Analytical samples can be procured either by recrystallization from CH_2Cl_2–ethanol or by column chromatography on activity 3 alumina (neutral). The acetyl complex is put on the column in a minimum volume of CH_2Cl_2, eluted first with 3:1 pentane– CH_2Cl_2 to remove small amounts of organic (PPh_3) and starting material, and is then eluted with CH_2Cl_2 to cleanly remove the yellow band containing $(\eta^5\text{-}C_5H_5)(CO)(PPh_3)FeCOCH_3$.

Anal. Calcd. for $C_{27}H_{23}O_2FeP$: C, 68.74; H, 5.10. Found: 68.97; H, 5.37.

A 50-mL three-necked flask is fitted with a magnetic stirring bar, a thermometer, and a condenser connected to a nitrogen line–oil bubbler. The nitrogen assembly is used to purge the apparatus and then to maintain a positive nitrogen atmosphere once the flask is stopped. Quantities of $[Fe(\eta^5\text{-}C_5H_5)(CO)_2\{\overline{O(CH_2)_3CH_2}\}][PF_6]$ (0.500 g, 1.15 mmol), $Fe(\eta^5\text{-}C_5H_5)(CO)(PPh_3)(COCH_3)$ (0.630 g, 1.38 mmol), and CH_2Cl_2 (25 mL) are added giving a red-orange solution. This is heated to 40 °C (using a water bath) for 3 h, giving a purple-orange solution. The solution is filtered to remove traces of suspended matter and added dropwise to 150 mL of diethyl ether with stirring. The resulting red-purple crystals of $[Fe(\eta^5\text{-}C_5H_5)(CO)(PPh_3)\{C(CH_3)O\}Fe(CO)_2(\eta^5\text{-}C_5H_5)][PF_6]$ are filtered, washed with diethyl ether (3 × 20 mL), and freed of solvent under a stream of nitrogen (until they are no longer wet) and then under vacuum; yield 0.752 g (77%).

Anal. Calcd. for $C_{33}H_{28}O_4Fe_2P_2F_6$: C, 51.05; H, 3.61. Found: C, 51.47; H, 3.50.

Properties

The iron acetyl complex, $Fe(\eta^5\text{-}C_5H_5)(CO)(PPh_3)COCH_3$,[12] is an air stable and orange-yellow crystalline compound that is freely soluble in CH_2Cl_2, benzene, and acetone and other dipolar aprotic solvents, slightly soluble in diethyl ether, and insoluble in pentane–heptane. IR (CH_2Cl_2) 1952 cm^{-1} (CO), 1600 cm^{-1} (COCH$_3$); ^{1}H NMR (CDCl$_3$) δ 7.33 (s, 15H, PPh$_3$), 4.36 (s, 5H, Cp), 2.25 (s, 3H, CH$_3$); mp (N$_2$ filled capillary) 145–147 °C (lit 145 °C).[12]

The μ-acetyl salt $[Fe(\eta^5\text{-}C_5H_5)(CO)(PPh_3)\{C(CH_3)O\}Fe(CO)_2(\eta^5\text{-}C_5H_5)][PF_6]$ (Ref. 2) has solubility properties similar to $[Fe(\eta^5\text{-}C_5H_5)(CO)_2\{C(CH_3)O\}Fe(CO)_2(\eta^5\text{-}C_5H_5)][PF_6]$; acetone solutions, however, decompose only very slowly: ^{1}H NMR (acetone-d_6) δ7.55 (m, 15H, PPh$_3$), 5.36 (s, 5H, CpFeO), 4.65 ($d, J = 1.5$ Hz, 5H, CpFeC), 2.68 (s, 3H, CH$_3$). As a further example of its diminished reactivity with nucleophiles, the reaction with one equivalent of iodide [as described for $[Fe(\eta^5\text{-}C_5H_5)(CO)_2\{C(CH_3)O\}Fe(CO)_2(\eta^5\text{-}C_5H_5)][PF_6]$ takes at least 20 min for complete conversion to $Fe(\eta^5\text{-}C_5H_5)(CO)(PPh_3)COCH_3$ and $Fe(\eta^5\text{-}C_5H_5)(CO)_2I$. Spectral data for $[Fe(\eta^5\text{-}C_5H_5)(CO)(PPh_3)\{C(CH_3)O\}Fe(CO)_2(\eta^5\text{-}C_5H_5)][PF_6]$: IR (CH_2Cl_2) 2060, 2017, 1951 cm^{-1} (CO), 1475 cm^{-1} (μ-COCH$_3$); ^{1}H NMR (CD$_3$NO$_2$) 7.43 (m, 15H, PPh$_3$), 5.08 (s, 5H, CpFeO), 4.51 ($d, J = 1.5$ Hz, 5H, CpFeC), 2.61 (s, 3H, CH$_3$).

D. (μ-ACETYL-2 κC^1: 1 κO)-PENTACARBONYL-1 $\kappa^3 C$, 2 $\kappa^2 C$- BIS[1:2(η^5-CYCLOPENTADIENYL)] MOLYBDENUMIRON (1 +) HEXAFLUOROPHOSPHATE(1 −)

Preparation

$$[Fe(\eta^5\text{-}C_5H_5)(CO)_2]_2 + Na(Hg) \longrightarrow 2[Na][Fe(\eta^5\text{-}C_5H_5)(CO)_2]$$

$$[Na][Fe(\eta^5\text{-}C_5H_5)(CO)_2] + CH_3COCl$$

$$\longrightarrow Fe(\eta^5\text{-}C_5H_5)(CO)_2COCH_3 + NaCl$$

$$Mo(\eta^5\text{-}C_5H_5)(CO)_3H + [Ph_3C][PF_6]$$

$$\longrightarrow [Mo(\eta^5\text{-}C_5H_5)(CO)_3][PF_6] + Ph_3CH$$

$$[Mo(\eta^5\text{-}C_5H_5)(CO)_3][PF_6] + Fe(\eta^5\text{-}C_5H_5)(CO)_2COCH_3$$

$$\longrightarrow [Fe(\eta^5\text{-}C_5H_5)(CO)_2\{C(CH_3)O\}Mo(CO)_3(\eta^5C_5H_5)][PF_6]$$

Procedure

A solution of $[Na][Fe(\eta^5\text{-}C_5H_5)(CO)_2]$ is prepared (see the procedure in Section A) by sodium amalgam reduction of $[Fe(\eta^5\text{-}C_5H_5)(CO)_2]_2$ (20.00 g, 56.50 mmol) in the THF (275 mL). The excess amalgam is drained away leaving an orange-yellow solution. Acetyl chloride (12.0 mL, 168.0 mmol) is dripped in via syringe (15 min); the solution warms up due to the heat of reaction. After cooling to room temperature, the resulting yellow-brown suspension is transferred to a 500-mL round-bottomed flask, and solvent is removed by rotary evaporation. The solids are extracted with CH_2Cl_2 (8×30 mL) and the extracts filtered through Celite® (2-cm bed). The volume of the combined extracts is reduced to 125 mL, and heptane (125 mL) is added. The CH_2Cl_2 is selectively removed on a rotary evaporator (30 torr, 22 °C).

The flask is then prepared for low-temperature crystallization and washing of the product. A septum is placed on one neck of the flask through which is placed a 20 gauge syringe needle leading to a nitrogen line–oil bubbler. The septum is also pierced by one end of a 16 gauge, double tipped, stainless steel transfer tube (70 vm), which is fitted with a coarse porosity frit. The flask is then immersed in a Dry Ice–acetone slush (0.5 h), causing more $Fe(\eta^5\text{-}C_5H_5)(CO)_2COOH_3$ to crystallize out. During this cooling operation the end of the transfer needle fitted with the sintered-glass frit is kept above the solution level, while maintaining the nitrogen atmosphere. The supernatant is forced over by lowering the frit and applying nitrogen pressure. Additional heptane (2×40 mL) is syringed in, cooled (5 min) and swirled, and transferred out with nitrogen pressure. The flask is warmed to room temperature and residual solvent is evaporated in a flow of nitrogen; the resulting yellow-brown crystals of $Fe(\eta^5\text{-}C_5H_5)(CO)_2COCH_3$ are vacuum dried (0.10–2 torr, 1h). Yield: 20.51 g (82%).

Anal. Calcd. for $C_9H_8O_3Fe$: C, 49.13; H, 3.67. Found: C, 48.87; H, 3.42.

The starting molybdenum hydride, $Mo(\eta^5\text{-}C_5H_5)(CO)_3H$, is prepared according to published procedures.[13] As an alternative purification procedure, the molybdenum hydride can be crystallized from cold pentane using the method previously outlined for $Fe(C_5H_5)(CO)_2COCH_3$. The pale pink crystalline solid so obtained can be briefly transferred in air, but prolonged storage rigorously requires a nitrogen atmosphere and an all-glass storage vessel in the refrigerator.

A 100-mL round-bottomed Shlenck flask is equipped with a magnetic stirring bar and connected to a nitrogen line–oil bubbler. The flask is flamed and cooled in a flow of nitrogen. A yellow solution of $[Ph_3C][PF_6]$ (see the preparation in Section B) (1.55 g, 4.0 mmol) in CH_2Cl_2 (40 mL) is prepared and cooled to $-80\,°C$. Then, $Mo(\eta^5\text{-}C_5H_5)(CO)_3H$ (0.984 g, 4.0 mmol) is added with stirring; the resulting red solution is warmed to $-20\,°C$ (30 min) turning deep burgundy. The temperature is tracked through the solvent of the cooling bath (100 × 50-mm evaporating dish) as it is allowed to warm. At $-20\,°C$, $Fe(\eta^5\text{-}C_5H_5)(CO)_2(COCH_3)$ (0.896 g, 4.0 mmol) is added, yielding a precipitate. The mixture is warmed to room temperature (~ 40 min) and filtered. The deep red filtrate is then slowly added to diethyl ether (150 mL) with stirring. A reddish-brown powder is isolated by filtering, washing with ether (3 × 15 mL), and drying first in a flow of nitrogen and then *in vacuo* (10^{-2} torr). Yield: 1.73 g (71%).

Anal. Calcd. for $C_{17}H_{13}O_6FeMoPF_6$: C, 33.47; H, 2.15. Found: C, 33.25; H, 2.17.

E. (μ-ACETYL-2κC^1:1κO)-TETRACARBONYL-1κ^3C, 2κC-BIS[1:2(η^5-CYCLOPENTADIENYL)] (TRIPHENYLPHOSPHINE-1κP)-MOLYBDENUMIRON(1 +) HEXAFLUOROPHOSPHATE(1 −)

$$Mo(\eta^5\text{-}C_5H_5)(CO)_3H + [Ph_3C][PF_6]$$
$$\longrightarrow [Mo(\eta^5\text{-}C_5H_5)(CO)_3][PF_6] + Ph_3CH$$
$$[Mo(\eta^5\text{-}C_5H_5)(CO)_3][PF_6] + Fe(\eta^5\text{-}C_5H_5)(CO)(PPh_3)COCH_3$$
$$\longrightarrow [Fe(\eta^5\text{-}C_5H_5)(CO)(PPh_3)\{C(CH_3)O\}Mo(CO)_3(\eta^5\text{-}C_5H_5)][PF_6]$$

Procedure

A solution of $[Mo(\eta^5\text{-}C_5H_5)(CO)_3][PF_6]$ (4.0 mmol/30 mL CH_2Cl_2) is generated as seen in the preparation in Section D. To this solution at $-20\,°C$ is added $Fe(\eta^5\text{-}C_5H_5)(CO)(PPh_3)(COCH_3)$ (see the preparation is Section C) (1.816 g, 4.0 mmol). The resulting dark red solution is brough to room temperature (0.75 h) and the reaction is filtered to remove a trace of sediment. The product is precipitated with diethyl ether (150 mL, with stirring and scraping), filtered, rinsed with diethyl ether (3 × 20 mL), and dried under a nitrogen flow, followed by pumping under vacuum (10^{-2} torr). The product $[Fe(\eta^5\text{-}C_5H_5)(CO)(PPh_3)\{C(CH_3)O\}Mo(CO)_3(\eta^5\text{-}C_5H_5)][PF_6]$ is isolated as a brick red powder. Yield: 2.01 g (59%).

Anal. Calcd. for $C_{34}H_{28}O_5FeMoPF_6$: C, 48.36; H, 3.34. Found: C, 48.16: H, 3.59.

Properties

The iron acetyl starting material, $Fe(\eta^5\text{-}C_5H_5)(CO)_2COCH_3,^{14}$ is a yellow-brown, crystalline solid that is air stable for short periods of time. Prolonged storage is best accomplished in sealed vessels, under nitrogen, and in a refrigerator (mp 56–58 °C vs. lit[14] 56–57 °C). Its solution spectral data are diagnostic: IR (CH_2Cl_2) 2020, 1960 cm^{-1} (CO), 1645 cm^{-1} ($COCH_3$); ^1H NMR $(CDCl_3)$ δ 4.89 (s, 5H, Cp), 2.57 (s, 3H, Cp). Small quantities can be further purified by chromatography on activity 3 alumina (neutral): The sample is put in a minimum volume of CH_2Cl_2, and $[Fe(\eta^5\text{-}C_5H_5)(CO)_2]_2$ and then $Fe(\eta^5\text{-}C_5H_5)(CO)_2(COCH_3)$ are removed and separated (as brown and yellow bands, respectively) with 3:1 pentane–CH_2Cl_2.

The iron–molybdenum μ_2-acetyl salts $[(\eta^5\text{-}C_5H_5)(CO)(L)Fe\{C(CH_3)O\}Mo(CO)_3(\eta^5\text{-}C_5H_5)]$ $[PF_6]$ (L = CO or PPh_3) are air-stable reddish- and orange-brown solids, respectively, that are readily precipitated from CH_2Cl_2–diethyl ether. Both react quickly (5 min) and quantitatively with one equivalent of iodide, giving $Mo(\eta^5\text{-}C_5H_5)(CO)_3I$ and releasing the corresponding iron acetyl compound $Fe(\eta^5\text{-}C_5H_5)(CO)_2(COCH_3)$ or $Fe(\eta^5\text{-}C_5H_5)(CO)(PPh_3)(COCH_3)$. For $[Fe(\eta^5\text{-}C_5H_5)(CO)_2\{C(CH_3)O\}Mo(CO)_3(\eta^5\text{-}C_5H_5)][PF_6]$: IR (CH_2Cl_2) 2064, 2043, 1993 cm^{-1} (CO), 1511 cm^{-1} (μ-$COCH_3$); ^1H NMR (CD_3NO_2) δ 6.10 (s, 5H, CpMo), 5.08 (s, 5H, CpFe), 2.62 (s, 3H, CH_3). $[(\eta^5\text{-}C_5H_5)(CO)(PPh_3)Fe\{C(CH_3)O\}Mo(CO)_3(\eta^5\text{-}C_5H_5)][PF_6]$: IR (CH_2Cl_2) 2059, 1970 cm^{-1} (CO), 1430 cm^{-1} (μ-$COCH_3$); ^1H NMR (CD_3NO_2) δ 7.44 (m, 15H, PPh_3), 5.88 (s, 5H, CpMo), 4.66 (s, 5H, CpFe), 2.40 (s, 3H, CH_3).

References

1. (a) W. Beck and K. Schloter, *Z. Naturforsch.*, **33b**, 1214 (1978); (b) K. Sunkel, G. Urgan, and W. Beck, *J. Organometal. Chem.*, **252**, 187 (1983).

2. S. J. LaCroce and A. R. Cutler, *J. Am. Chem. Soc.*, **104**, 2312 (1982).

3. (a) J. Sunkel, K. Schloter, W. Beck, K. Ackermann, and U. Schubert, *J. Organometal. Chem.*, **241**, 332 (1983); (b) K. Sunkel, U. Nagel, and W. Beck, *J. Organomet. Chem.*, **251**, 227 (1983).

4. M. Rosenblum, W. P. Giering, and Sari-beth Samuels, *Inorg. Synth.*, **24**, 163 (1986).

5. (a) R. B. King, *Organometallic Syntheses*, Part I, Vol. 1, Academic Press, New York, 1965; (b) D. F. Shriver, *The Manipulation of Air-Sensitive Compounds*, Part II, McGraw-Hill, New

York, 1969; (c) H. C. Brown, *Organic Syntheses via Boranes*, Wiley, New York, 1975, Chapter 9. (d) J. J. Eisch, *Organometallic Syntheses*, Part I, Vol. 2, Academic Press, New York, 1981, (e) R. C. Moore, S. S. White, Jr., and H. C. Kelly, *Inorg. Synth.*, **12**, 111 (1970) and Safety Note, p. 317; G. J. Kubas and D. F. Shriver, *Inorg. Synth.*, **17**, 14 (1977).

(f) R. B. King, *Organometallic Syntheses*, Part I, Vol. 1, Academic Press, New York, 1965, p. 114; (g) R. B. King and F. G. A. Stone, *Inorg. Synth.*, **7**, 110 (1963).

6. (a) R. B. King, *Organometallic Syntheses*, Part I, Vol. 1, Academic Press, New York, 1965, pp. 149, 151; (b) B. D. Dombek and R. J. Angelici, *Inorg. Synth.*, **17**, 100 (1977).

7. (a) D. L. Reger and C. Coleman, *J. Organometal. Chem.*, **131**, 153 (1977); (b) E. K. G. Schmidt and C. H. Thiele, *J. Organometal. Chem.*, **209**, 373 (1981); (c) M. Rosenblum and D. Scheck, *Organometallics*, **1**, 397 (1982).

8. (a) W. P. Giering and M. Rosenblum, *J. Chem. Soc. Chem. Commun.*, **1971**, 441; (b) A. Cutler, D. Ehnthold, P. Lennon, K. Nicholas, D. F. Marten, M. Madhavarao, S. Raghu, A. Rosan, and M. Rosenblum, *J. Am. Chem. Soc.*, **97**, 3149 (1975).

9. H. J. Dauben, Jr., L. R. Honnen, and K. M. Harmon, *J. Org. Chem.*, **25**, 1442 (1960).

10. D. Lloyd, D. J. Walton, J. P. Declercq, G. Germain, and M. Van Meerssche, *J. Chem. Res. S*, **7**, 249 (1979).

11. (a) R. B. King, *Organometallic Syntheses*, Part I, Vol. 1, Academic Press, New York, 1965, p. 151; (b) T. S. Piper and G. Wilkinson, *J. Inorg. Nucl. Chem.*, **3**, 104 ((1956).

12. (a) J. B. Bibler and A. Wojcicki, *Inorg. Chem.*, **5**, 889 (1966); (b) I. S. Butler, F. Basolo, and R. G. Pearson, *Inorg. Chem.*, **6**, 2074 (1966); (c) M. Green and D. J. Westlake, *J. Chem. Soc. A*, 367 (1971); (d) D. L. Reger, D. J. Fauth, and M. D. Dukes, *Syn. React. Inorg. Metal Org. Chem.*, **7**, 151 (1977).

13. (a) R. B. King, Organometallic Syntheses, Part I, Vol. 1, Academic Press, New York, 1965, p. 156; (b) R. B. King and F. G. A. Stone, *Inorg. Synth.*, **7**, 107 (1963); (c) E. O. Fischer, *Inorg. Synth.*, **7**, 136 (1963).

14. R. B. King, *J. Am. Chem. Soc.*, **85**, 1918 (1963).

42. TRINUCLEAR METAL COMPLEXES

Submitted by L. MARKÓ and J. TAKÁCS*
Checked by KENTON H. WHITMIRE,[†] B. A. MATRANA,[‡] and H. D. KAESZ[‡]

All three members of the isoelectronic series $Co_nFe_{3-n}H_{2-n}(CO)_9S$ ($n = 0.1, 2$) are known.[1-3] They may be prepared by acidification of solutions containing the $[FeH(CO)_4]^-$ and the $[Co(CO)_4]^-$ anions in the presence of sodium sulfide.[2,3] The yields of this method are only moderate, however,

*Institute of Organic Chemistry, University of Chemical Engineering, H-8200 Veszprém, Hungary.
†Department of Chemistry, Rice University, P.O. Box 1892 Houston, Tx 77251.
‡Department of Chemistry and Biochemistry, University of California, Los Angeles, CA 90024-1569.

and in the case of the mixed-metal clusters the purification of the product is troublesome since the Co_2Fe and $CoFe_2$ clusters can hardly be separated by chromatography.

The methods described here are based on the thermal stability of such clusters under CO pressure.[4] In the case of $Co_2Fe(CO)_9S$, the complex itself is stable under "hydroformylation conditions" (130–180 °C, 50–150 bar CO and H_2) and is formed under such conditions from almost every sulfur-containing substance, iron, cobalt, and carbon monoxide.[1] Ethanethiol is used in the procedure as the sulfur source because of its reactivity. The dihydride cluster $Fe_3H_2(CO)_9S$ is not directly accessible by this method because it is a thermally much less stable species but the anion $Fe_3H(CO)_9S^-$ can be prepared in this way and its acidification yields the desired complex.

A.   NONACARBONYLDIHYDRIDO-μ_3-THIO-TRIIRON

$$3Fe(CO)_5 + HS^- \xrightarrow[\text{140 bar CO}]{\text{150 °C}} [Fe_3H(CO)_9S]^- + 6CO$$

$$[Fe_3H(CO)_9S]^- + H^+ \longrightarrow Fe_3H_2(CO)_9S$$

Procedure

■ **Caution.** *Carbon monoxide and pentacarbonyliron are poisons. All manipulations should be carried out in a well-ventilated hood.*

A rocking autoclave (10 mL) is charged with $Fe(CO)_5$ (30 mmol, 4.11 mL), $Na_2S \cdot 9H_2O$ (10 mmol, 2.40 g), NaOEt (50 mmol, 3.40 g), and ethanol (40 mL). The autoclave is flushed with CO and pressurized with CO at room temperature to 100 bar. The reaction is carried out at 150 °C for 3 h (the pressure rises to ~140 bar), after which the autoclave is cooled, and the carbon monoxide slowly vented into a well-ventilated hood. The reaction product, a red solution containing some precipitate, is poured under Ar into a Schlenk tube (300 mL). The product is evaporated to dryness under vacuum and the residue is dissolved under Ar in methanol (50 mL). Following this, water (20 mL) and hexane (80 mL) are added. The resulting two-phase mixture is acidified under vigorous stirring with 60 mL 20% cold HCl. Acidification and separation of the two phases should be done quickly because $Fe_3H_2(CO)_9S$ is sensitive to mineral acids.

The two phases are separated and the lower layer is extracted four times with 80-mL portions of hexane. The combined hexane solutions are washed with 40 mL of water and the hexane layer is separated and dried over Na_2SO_4, filtered under Ar, and cooled to -78 °C. The separating crystals usually do not need further purification, their purity can be checked by infrared (IR) spectroscopy. Minor amounts of $Fe_3(CO)_9S_2$ may be present as an impurity but its presence is easily detected by a band at 2062 cm^{-1} in the ν_{CO} spectrum in hexane. In the latter case, the product should be recrystallized from hexane. Yield: 1.8–2.8 g (40–61%).

Anal. Calcd. for $C_9H_2Fe_3O_9S$: C, 23.82; H, 0.44; Fe, 36.92; S, 7.06. Found: C, 23.5; H, 0.4; Fe, 36.5; S, 7.1.

Properties

The complex forms dark brown crystals that are sensitive to air. It is highly soluble in nonpolar organic solvents. Its IR spectrum in the ν_{CO} region (hexane, cm^{-1}): 2106 (m), 2069.5 (vs), 2050 (vs), 2040 (s), 2034.5 (m), 2013.5 (s), 2001 (m), 1992.5(w), 1985.5(w). Due to decomposition, $Fe_3H_2(CO)_9S$ cannot be chromatographed on silica gel. It dissociates one of its hydride ligands as a proton in polar (but not basic) organic solvents like ethanol; in basic solvents like 1-butanamine both hydride ligands are lost as protons. The IR spectra of these anionic species in the ν_{CO} region are the following: $[Fe_3H(CO)_9S]^-$ (ethanol, cm^{-1}): 2052 (w), 2009 (vs), 1982 (vs), 1965 (s), 1950(m, sh), 1914(w); $[Fe_3(CO)_9S]^{2-}$ (*n*-BuNH$_2$, cm^{-1}): 1999 (w), 1933 (vs), 1901 (s), 1880 (m, sh). The $[Et_4N^+]$ salt of the dianion has been prepared.[2]

B. NONACARBONYL-μ_3-THIO-DICOBALTIRON

$$Co_2(CO)_8 + Fe(CO)_5 + C_2H_5SH \xrightarrow[\text{200 bar CO + 100 bar H}_2]{150\,°C} Co_2Fe(CO)_9S + \cdots$$

Procedure

■ **Caution.** *Carbon monoxide, ethanethiol, and pentacarbonyliron are poisonous and ethanethiol has a very strong unpleasant odor. All manipulations should be carried out in a well-ventilated hood.*

A rocking autoclave (100 mL) is charged with $Co_2(CO)_8$ (5 mmol, 1.71 g), $Fe(CO)_5$ (5 mmol, 0.69 mL), C_2H_5SH (6 mmol, 0.44 mL), and hexane (50 mL). The autoclave is flushed with CO and pressurized with a CO–H$_2$ (2:1) gas mixture at room temperature to 210 bar. The reaction is carried out at 150 °C for 3 h (the pressure rises to about 300 bar) after which the autoclave is cooled, and the gas is slowly vented into a well-ventilated hood. The product, a brown solution containing some crystals is poured under Ar into a Schlenk tube (100 mL). The solvent is evaporated under vacuum and the solid residue dissolved in hexane (120 mL). The solution is filtered under Ar and chilled to -78 °C. Yield: 1.5 g (65%).

Anal. Calcd. for $C_9Co_2FeO_9S$: C, 23.53; Co, 25.74; Fe, 12.20; S, 7.01. Found: C, 23.7; Co, 25.5; Fe, 12.1; S, 7.0.

Properties

The complex forms shining black crystals that are moderately stable in air. It is highly soluble in organic solvents. Its IR spectrum in the ν_{CO} region (hexane, cm^{-1}): 2104 (w), 2066 (vs), 2053.5 (vs), 2041.5 (s), 2028.5 (m), 1984 (m).

References

1. S. A. Khattab, L. Markó, G. Bor, and B. Markó, *J. Organometal. Chem.*, **1**, 373 (1964).
2. L. Markó, J. Takács, S. Papp, and B. Markó-Monostory, *Inorg. Chim. Acta Lett.*, **45**, L189 (1980).
3. L. Markó, *J. Organometal. Chem.*, **213**, 271 (1981).
4. L. Markó, *Gazz. Chim. Ital.*, **109**, 247 (1979).

43.   BIS[μ-NITRIDO-BIS(TRIPHENYLPHOSPHORUS) (1 +)] μ₄-CARBIDO-DODECACARBONYL-TETRAFERRATE(2 −), [PPN]₂[Fe₄C(CO)₁₂]

Submitted by J. W. KOLIS,* M. A. DREZDZON,* and D. F. SHRIVER*
Checked by F. R. FURUYA† and W. L. GLADFELTER†

Introduction

Until recently, syntheses of carbide clusters have been limited to pyrolyses of carbonyl clusters or reactions of metal carbonyls with halocarbons.[1] Both of these procedures tend to give unpredictable products in variable yields. These products almost always contain interstitial carbides with high coordination numbers, and little or no reactivity at the carbide atom. Recently, clusters containing a four coordinate carbide ion situated in a butterfly array of iron atoms have been shown to undergo attack by protons,[2] electrophiles,[3] and nucleophiles,[4] as well as cluster building reactions.[5] The preparation of $[Fe_4C(CO)_{12}]^{2-}$ reported here is based on the reductive cleavage of $[Fe_4(CO)_{12}COR]^-$.[6-8] Previous methods for the synthesis of this type of cluster are the reaction of $[Fe_4(CO)_{13}]^{2-}$ with neat CF_3SO_2OH and the oxidative cleavage of iron vertices from a higher coordinate carbide cluster,

*Department of Chemistry, Northwestern University, Evanston, IL 60201.
†Department of Chemistry, University of Minnesota, Minneapolis, MN 55455.

which were previously synthesized by pyrolysis.[2,4] Although both of these preparations provide workable quantities of clusters, the reaction reported here[6,8] provides much higher overall yields by a rational pathway directly from a carbonyl cluster. Also, the carbonyl precursor can be easily enriched by stirring under ^{13}CO overnight, and therefore the carbide atom may be labeled for NMR and isotopic studies.[9]

■ **Caution.** *This preparation should be performed in an efficient fume hood in the event unforeseen decomposition reactions lead to evolution of CO or* $Fe(CO)_5$*, both highly toxic*

1. $[PPN]_2[Fe_4(CO)_{13}] + CH_3C(O)Cl$

$$\xrightarrow[25\,°C]{CH_2Cl_2} [PPN][Fe_4(CO)_{12}(COC\{O\}CH_3)]$$
$$+ [PPN]Cl$$

2. $[PPN][Fe_4(CO)_{12}(COC\{O\}CH_3)] + 2Na$

$$\xrightarrow[THF]{(C_6H_5)_2CO} [PPN]Na[Fe_4C(CO)_{12}]$$
$$+ NaOC(O)CH_3$$

3. $[PPN]Na[Fe_4C(CO)_{12}] + [PPN]Cl$

$$\longrightarrow [PPN]_2[Fe_4C(CO)_{12}] + NaCl$$

Procedure

All manipulations are carried out using standard Schlenk techniques.[10] It is especially important to use anhydrous solvent and thoroughly dried (100 °C) glassware in reaction 1. Solids are handled in a nitrogen purged dry box and liquids are transferred via syringe. Tetrahydrofuran (THF) is dried by distillation from sodium benzophenone ketyl and dichloromethane is distilled from phosphorus pentoxide and purged with nitrogen before use. Anhydrous diethyl ether and reagent grade methanol are purged with nitrogen before use. Acetyl chloride must be free from HCl and acetic acid. It is purified by distillation from phosphorus pentachloride under nitrogen followed by trap-to-trap vacuum distillation from quinoline on a vacuum line. Benzophenone is recrystallized from petroleum ether. μ-Nitrido-bistriphenylphosphorus(1 +) chloride, [PPN]Cl, is recrystallized from hot water and dried under vacuum. The starting material, $[PPN]_2[Fe_4(CO)_{13}]$, is prepared via a literature method.[11]

In a 100-mL oven-dried Schlenk flask containing a magnetic stirring bar and purged with dry nitrogen, 2.5 g (1.5 mol) of $[PPN]_2[Fe_4(CO)_{13}]$ is dissolved in 10 mL of dry CH_2Cl_2. To this is added 2 mL (23 mmol) $CH_3C(O)Cl$, using an oven-dried, gas-tight syringe, and the solution is stirred for 1 h. At this point the IR spectrum of the red-brown solution may be monitored. A single broad feature with a sharp minimum at 1998 cm^{-1} indicates a clean product. A broadened minimum extending from 1998 to 1988 cm^{-1} indicates the presence of $[HFe_4(CO)_{13}]^-$, and this material is unsatisfactory for the next step. Deoxygenated diethyl ether, 70 mL, is introduced by syringe, and the solution is stirred for 10 min and filtered under nitrogen, leaving behind white [PPN]Cl. The solvent is removed under vacuum and the solid is dried for 30 min under vacuum. In the meantime, a mixture of 2.0 g $(C_6H_5)_2CO$ and 0.5 g Na in 60 mL of THF is prepared in a 100-mL Schlenk flask under dry nitrogen and stirred vigorously to give a blue solution. The solid cluster compound from this acylation is dissolved in 6 mL of THF, and the deep blue $Na–(C_6H_5)_2CO–THF$ solution is added dropwise by syringe with repeated removal of aliquots to monitor the reaction by IR spectroscopy (an IR cell with $\sim$ 0.1-mm pathlength is appropriate). When the reactant band at 1998 cm^{-1} has disappeared and new bands at 1970 and 1943 cm^{-1} have grown in, the reaction is terminated (addition of excess sodium benzophenone ketyl must be avoided). The brown solution is filtered and THF is removed under vacuum. The resulting oily solid is swirled with 15 mL of diethyl ether and filtered under nitrogen in a Schlenk filter. A solution containing 1.5 g (2.6 mmol) [PPN]Cl in 25 mL of CH_2Cl_2 in a Schlenk flask under nitrogen is added to the black-brown solid, and the mixture is Schlenk filtered into a 100-mL flask. The filtrate is concentrated to 15 mL under vacuum, and 10 mL of CH_3OH is added followed by four 10-mL aliquots of diethyl ether while the solution is being swirled. Filtration followed by washing with MeOH (3 × 15 mL) and diethyl ether (2 × 10 mL) and drying for 10 min under vacuum, affords 2.0 g (1.2 mmol, 80%) of crystalline $[PPN]_2[Fe_4C(CO)_{12}]$.

Anal. Calcd. for $[PPN]_2[Fe_4C(CO)_{12}]$: C, 61.92; H, 3.67; N, 1.72. Found: C, 61.64; H, 3.87; N, 1.83.

Properties

The compound is a shiny brown-black crystalline solid. It is stable in air for short periods in the solid state but rapidly oxidizes in solution. The [PPN]$^+$ salt is soluble in CH_2Cl_2, CH_3CN, THF, and acetone to give intense brown solutions. The IR spectrum in CH_2Cl_2 displays strong bands at 1968 (s) and 1942 (vs) cm^{-1} and weaker bands at 2003 (w) and 1912 (sh) cm^{-1}. The IR spectrum of a Nujol mull contains iron–carbide stretching bands at 921 and

$667\,cm^{-1}$. The ^{13}C NMR spectrum of the enriched species in CD_2Cl_2 at $-90\,°C$ displays a downfield carbide resonance at 477.9 and terminal CO resonances at 222.8 and 220.0 in a ratio of 1:6:6. This anion is protonated first on the Fe—Fe skeleton and next on an Fe—C bond.[2] Alkylation occurs on the carbide,[8] and reaction with $Fe_2(CO)_9$ produces a five-iron carbide anion. In a strongly acidic medium, the carbide ligand is converted to CH_4.[7]

References

1. P. Chini, G. Longoni, and V. G. Albano, *Adv. Organomet. Chem.*, 14, 285 (1976).
2. M. Tachikawa and E. L. Muetterties, *J. Am. Chem. Soc.*, 102, 4541 (1980).
3. E. M. Holt, K. H. Whitmire, and D. F. Shriver, *J. Am. Chem. Soc.*, 104, 5621 (1982).
4. J. S. Bradley, G. B. Ansell, and E. W. Hill, *J. Am. Chem. Soc.*, 101, 7417 (1979).
5. M. Tachikawa, R. Geerts, and E. L. Muetterties, *J. Organomet. Chem.*, 213, 11 (1981).
6. J. W. Kolis, E. M. Holt, M. A. Drezdzon, K. H. Whitmire, and D. F. Shriver, *J. Am. Chem. Soc.*, 104, 6134 (1982).
7. E. M. Holt, K. W. Whitmire, and D. F. Shriver, *J. Organomet. Chem.*, 213, 125 (1981).
8. A. Ceriotti, P. Chini, G. Longoni, and G. Piro, *Gazz. Chim. Ital.*, 112, 353 (1982).
9. K. W. Whitmire, D. F. Shriver, and E. M. Holt, *J. Chem. Soc. Chem. Commun.*, 1980, 780.
10. D. F. Shriver and M. A. Drezdzon, *The Manipulation of Air Sensitive Compounds*, Wiley-Interscience, New York, 1986.
11. K. W. Whitmire, J. Ross, C. B. Cooper III, and D. F. Shriver, *Inorg. Synth.*, 21, 66 (1982).

44. DINUCLEAR RUTHENIUM(II) CARBOXYLATE COMPLEXES

Submitted by MICHEL O. ALBERS,* ERIC SINGLETON,* and JANET E. YATES*
Checked by FRED B. McCORMICK[†]

Carboxylate anions are versatile ligands that are found in a variety of coordination modes including unidentate, chelate, and a number of bis(monodentate) bridging modes.[1] Such versatility makes the chemistry of carboxylate complexes particularly interesting. It is now also apparent that the diverse catalytic activity shown by many metal carboxylates may be rationalized in terms of the chemistry of the carboxylato ligand.[2] In general though, few rational syntheses of metal carboxylates are known.[1] This

*National Chemical Research Laboratory, Council for Scientific and Industrial Research, P.O. Box 395, Pretoria 001, Republic of South Africa.
†3M Central Research, St. Paul., MN 55144.

hampers the investigation of catalytic and synthetic applications of metal carboxylates as few, well-characterized, closely related series of compounds are known. In this contribution we describe a high-yield route to a series of ruthenium(II) dimeric carboxylate complexes of general formula $[\{Ru(diolefin)(O_2CR)\}_2(\mu\text{-}O_2CR)_2(\mu\text{-}OH_2)]$ based on the protonation of $(\eta^4\text{-}$cycloocta-1, 5-diene)bis(η^3-2-propenyl)ruthenium(II)   and   (η^4-bicyclo [2.2.1]hepta-2, 5-diene)bis(η^3-2-propenyl)ruthenium(II)   complexes   by carboxylic acids.[3] These dimeric ruthenium carboxylates have been found to possess interesting catalytic properties[4] and to be precursors to a wide range of mononuclear   (carboxylato)(phosphine)-,   (amine)(carboxylato)-,   and (carbonyl)(carboxylato)ruthenium(II) complexes.[3]

The   (η^4-bicyclo[2.2.1]hepta-2, 5-diene)bis(η^3-2-propenyl)ruthenium(II) and (η^4-cycloocta-1, 5-diene)bis(η^3-2-propenyl)ruthenium(II) complexes have previously been reported,[5,6] but we present here scaled-up procedures that produce synthetically useful quantities of these compounds. The preparation of Grignard reagents is based upon the methods described by Eisch.[7]

General Procedure

Unless otherwise stated, the syntheses are all performed in standard Schlenk glassware and under a nitrogen atmosphere.[8] In the case of reactions involving Grignard reagents the glassware is also oven dried prior to use. Solvents are all, unless otherwise stated, of Analar Grade, and are dried and distilled under nitrogen.[9] The carboxylic acids are of Chemical Purity and used as purchased. Allyl bromide (3-bromo-1-propene) is purified by distillation.[9] Magnesium turnings are washed with dry diethyl ether and subsequently oven dried for 1 h. Manipulations are all routinely performed in a fume hood. Melting points are all determined in air and are uncorrected.

A.   DI-μ-CHLORO-(η^4-BICYCLO[2.2.1]HEPTA-2,5-DIENE)RUTHENIUM(II) POLYMER[10]

$$2RuCl_3 \cdot 3H_2O + 2C_7H_8 + CH_3CH_2OH$$
$$\longrightarrow 2[RuCl_2(\eta^4\text{-}C_7H_8)]_x + 2HCl + CH_3CHO$$

Procedure

■   **Caution.**   *In order to minimize the strong odor of norbornadiene, these procedures are best carried out in a well-ventilated fume hood.*

This reaction may be carried out in air. A 500-mL capacity Erlenmeyer flask containing a large Teflon-coated stirring bar is charged with a filtered solution of $RuCl_3 \cdot 3H_2O$ (10.0 g, ~ 0.04 mol) in absolute ethanol (400 mL), and bicyclo[2.2.1]hepta-2, 5-diene (norbornadiene) (20 mL, 0.24 mol). The mixture is vigorously stirred at room temperature for 24 h. During this time the brick red to brown product precipitates from solution. On completion of the reaction the suspension is filtered using a medium porosity glass filter frit and washed thoroughly with acetone (50 mL). Drying gives the analytically pure product. Yield: 10.1 g (95%).

Anal. Calcd. for $C_7H_8Cl_2Ru$: C, 31.84; H, 3.05; Cl, 26.85. Found: C, 31.62; H, 3.16; Cl, 26.43.

Properties

The complex is a brick red to brown, highly insoluble solid believed to have a polymeric, halogen-bridged structure.[10] The principle reaction pathways of this complex involve chloride-bridge cleavage leading to a range of ruthenium(II) products.[11]

B. (η^4-BICYCLO[2.2.1]HEPTA-2, 5-DIENE)BIS-(η^3-2-PROPENYL)RUTHENIUM(II)

$$[RuCl_2(\eta^4\text{-}C_7H_8)]_x + 2C_3H_5MgBr$$
$$\longrightarrow [Ru(\eta^3\text{-}C_3H_5)_2(\eta^4\text{-}C_7H_8)] + 2MgBrCl$$

Procedure

■ **Caution.** *Owing to the poisonous nature of 3-bromo-1-propene, the following procedure must be carried out in a well-ventilated fume hood. No further 3-bromo-1-propene should be added until it is certain that the reaction with magnesium turnings, as evidenced by a slightly milky appearance of the diethyl ether solution, has been successfully initiated.*

A 500-mL three-necked round-bottomed flask is equipped with a dropping funnel, a reflux condenser, and a nitrogen inlet. Stirring is achieved by using a large Teflon-coated stirring bar and a suitable magnetic stirring device. The flask is oven dried prior to use and flushed with dry nitrogen. Under a stream of nitrogen, the flask is charged with magnesium turnings (4.0 g, 200 mmol) and oxygen-free and peroxide-free diethyl ether (80 mL). Freshly distilled 3-bromo-1-propene (11.4 g, 100 mmol) is placed in the dropping funnel and a few drops added to the vigorously stirred magnesium suspension to initiate the reaction. The remainder of the 3-bromo-1-propene is diluted with oxygen-free and, peroxide-free diethyl ether (80 mL) and added dropwise over 1 h to the vigorously stirred suspension. On completion of the addition, the suspension is stirred for another hour. The solution of the Grignard reagent is separated from excess magnesium turnings using syringe techniques and stored in an oven-dried, nitrogen-purged 500-mL Schlenk flask. The remaining solid residue is washed with two portions of oxygen-free and peroxide-free diethyl ether (2×20 mL) and the washings added to the bulk of the solution of the Grignard reagent. Commercially available allylmagnesium bromide solutions (Aldrich) may also be used in this procedure.

Freshly synthesized $[RuCl_2(\eta^4\text{-}C_7H_8)]_x$ is necessary in the next step to obtain a high yield of product. The reported[5] low yields of the compound $[Ru(\eta^3\text{-}C_3H_5)_2(\eta^4\text{-}C_7H_8)]$ may be attributable to the use of old stocks of the polymer. Finely divided $[RuCl_2(\eta^4\text{-}C_7H_8)]_x$ enhances the efficiency of the reaction; grinding with a mortar and pestle is recommended.

Freshly prepared and finely ground $[RuCl_2(\eta^4\text{-}C_7H_8)]_x$ (5.0 g, 19.0 mmol) and oxygen-free and peroxide-free diethyl ether (150 mL) are placed in a 500-mL round-bottomed flask equipped with a dropping funnel. Stirring is achieved using a large Teflon-coated stirring bar and a magnetic stirring device. The solution of the Grignard reagent is placed in the dropping funnel and over 30 min, added slowly to the stirred suspension. On completion of the addition, the reaction mixture is stirred until all the $[RuCl_2(\eta^4\text{-}C_7H_8)]_x$ disappears (~ 12 h) and a yellow or yellow-orange solution is formed. The solution is filtered under nitrogen using a medium porosity filter and Celite® filter aid, and cooled to 0 °C.

A quantity (75 mL) of distilled water is nitrogen purged immediately prior to use to avoid product decomposition. Excess Grignard reagent is hydrolyzed using 5-mL portions of nitrogen-purged and cooled distilled water (5 °C, 75 mL total volume) while stirring the solution rapidly.

■ **Caution.** *If water is added too rapidly a vigorous exothermic reaction sometimes occurs that can lead to product loss.*

The diethyl ether layer is separated and dried over anhydrous $MgSO_4$ for 1 to 2 h. Filtration, using a medium porosity frit and Celite® filter aid, followed by solvent removal under reduced pressure gives an orange gum.

Subsequent to the Grignard reagent hydrolysis step, it is permissible to use reagent grade solvents that are purged with nitrogen immediately prior to use. The orange residue above is redissolved in pentane (10 mL) and filtered through a neutral alumina column (6% H_2O, 3×5 cm) using 6×20-mL portions of pentane as eluent. A pale yellow solution is obtained. The solvent is evaporated under reduced pressure giving the product as a pale yellow or white waxy solid. Yield: 3.8–4.0 g (73–80%).

Anal. Calcd. for $C_{13}H_{18}Ru$: C, 56.70; H, 6.59. Found: C, 56.88; H, 6.52, mp 108–110 °C (dec).

Due to its air sensitivity, the product is best used immediately upon isolation rather than attempting prolonged storage.

Properties

The complex $[Ru(\eta^3-C_3H_5)_2(\eta^4-C_7H_8)]$ is soluble in most common organic solvents. It is air sensitive in both the solid state and in solution. It may be stored under an inert atmosphere and at $-20\,^\circ C$, however, it is best to prepare it immediately prior to its use. The ^1H NMR spectrum recorded in C_6D_6 (90 MHz) shows the following resonances assigned to the bicyclo[2.2.1]hepta-2, 5-diene ligand: 4.10 ppm (m, 2H, olefin), 3.5 ppm (m, 2H, methine), 1.35 ppm (m, 2H, olefin), 1.20 ppm (t, 2H, methylene). Resonances assignable to the allyl ligands are 2.9–3.25 ppm (two overlapping m, 2H, central and 2H, anti), 3.75 ppm (m, 2H, syn), 1.7 ppm (m, 2H, syn′), -0.1 ppm (m, 2H, anti′).

C. DI-μ-CHLORO-(η^4-CYCLOOCTA-1,5-DIENE)RUTHENIUM(II)[12]

$$2RuCl_3 \cdot 3H_2O + 2C_8H_{12} + CH_3CH_2OH$$
$$\longrightarrow 2[RuCl_2(\eta^4-C_8H_{12})]_x + 2HCl + CH_3CHO$$

Procedure

■ **Caution.** *In order to minimize the strong, sometimes irritating odor of cycloocta-1, 5-diene, these procedures should be carried out in a well-ventilated fume hood.*

This reaction may be carried out in air. A 750-mL capacity Erlenmeyer flask containing a large Teflon-coated stirring bar is fitted with a reflux condenser. The flask is charged with a mixture of $RuCl_3 \cdot 3H_2O$ (50.0 g, ~ 0.2 mol) and cycloocta-1, 5-diene (50 mL, 0.4 mol) in absolute ethanol (400 mL) and heated under reflux with stirring for 24 h. During this time the brown product precipitates from solution. On completion of the reaction, the solution is cooled to room temperature and filtered. Washing with diethyl ether (50 mL) and drying gives the analytically pure product. Yield: 51–54 g (91–96%).

Anal. Calcd. for $C_8H_{12}Cl_2Ru$: C, 34.29; H, 4.29; Cl, 25.40. Found: C, 34.75; H, 4.40; Cl, 25.16.

Properties

The complex is a brown solid, highly insoluble in most organic solvents. It is thought to have a polymeric chloride-bridged structure.[12] It reacts via halogen-bridge cleavage and is a synthetic precursor to a variety of ruthenium(II) complexes.[11]

D. (η^4-CYCLOOCTA-1, 5-DIENE)BIS(η^3-2-PROPENYL)-RUTHENIUM(II)

$$[RuCl_2(\eta^4\text{-}C_8H_{12})]_x + 2C_3H_5MgBr$$
$$\longrightarrow [Ru(\eta^3\text{-}C_3H_5)_2(\eta^4\text{-}C_8H_{12})] + 2MgBrCl$$

Procedure

■ **Caution.** *Owing to the poisonous nature of 3-bromo-1-propene, the procedure must be carried out in a well-ventilated fume hood.*

The procedure and the amounts of the appropriate reagents used are analogous to those described before for the preparation of $[Ru(\eta^3\text{-}C_3H_5)_2(\eta^4\text{-}C_7H_8)]$. Freshly prepared and finely divided $[RuCl_2(\eta^4\text{-}C_8H_{12})]_x$ (ground by mortar and pestle) is recommended to ensure good yields of the product $[Ru(\eta^3\text{-}C_3H_5)_2(\eta^4\text{-}C_8H_{12})]$. Yield: 3.5–3.8 g (68–73%).

Anal. Calcd. for $C_{14}H_{22}Ru$: C, 57.71; H, 7.61. Found: C, 57.82; H, 7.62, mp 145–150 °C (dec).

Due to its air sensitivity, the product is best used immediately upon isolation rather than attempting prolonged storage.

Properties

The complex $[Ru(\eta^3\text{-}C_3H_5)_2(\eta^4\text{-}C_8H_{12})]$ is soluble in most common organic solvents. It is air sensitive, decomposing over a few hours in the solid state and rapidly in solution. The ^1H NMR spectrum recorded in C_6D_6 (90 MHz) shows four multiplet resonances assigned to the cycloocta-1, 5-diene ligand: 3.95 ppm (2H, olefin), 2.8 ppm (4H, methylene), 1.5–2.0 ppm (4H, methylene), 1.3 ppm (2H, olefin). Resonances assigned to the allyl ligands appear at 3.1–3.4 ppm (m, 2H, central), 3.65 ppm (dd, 2H, syn), 2.6 ppm (d, 2H, anti), 1.8 ppm (overlap, 2H, syn′), − 0.1 ppm (d, 2H, anti′).

E. μ-AQUA-BIS-(μ-TRIFLUOROACETATO)-BIS[(η^4-CYCLOOCTA-1, 5-DIENE)(TRIFLUOROACETATO)-RUTHENIUM(II)]

$$2[Ru(\eta^3\text{-}C_3H_5)_2(\eta^4\text{-}C_8H_{12})] + 4CF_3COOH + H_2O$$
$$\longrightarrow [\{Ru(\eta^4\text{-}C_8H_{12})(O_2CCF_3)\}_2(\mu\text{-}O_2CCF_3)_2(\mu\text{-}OH_2)]$$
$$+ 4C_3H_6$$

Procedure

A 100-mL Schlenk flask is purged with nitrogen and while under a nitrogen flow charged with freshly prepared $[Ru(\eta^3\text{-}C_3H_5)_2(\eta^4\text{-}C_8H_{12})]$ (2.0 g, 6.86 mmol), oxygen-free and peroxide-free diethyl ether (40 mL), and trifluoroacetic acid (2.7 mL, 28.1 mmol). The solution is stirred at 25 °C for 2 h. A color change from very pale yellow to orange is observed within a few minutes, with further darkening during the period of stirring. The solvent is removed under reduced pressure giving the product as an orange, crystalline solid. The solid is washed with chilled reagent grade pentane (0 °C, 5–10 mL) and removed by filtration and dried. Yield: 2.5–2.7 g (82–89%). The material so obtained is generally sufficiently pure for most synthetic uses. The product may, however, be further purified by recrystallization under a nitrogen atmosphere. Thus, 1.0 g of material is dissolved in pentane (100 mL) under nitrogen, and the solution is filtered. The volume of the filtrate is reduced under vacuum to ~ 10 mL. Cooling to − 78 °C for 3 to 5 h gives orange crystals of the product. The mother liquors are decanted and the crystals are washed with a minimum of chilled pentane (0 °C, 2–4 mL). Drying under vacuum gives 0.75 g of analytically pure product (yield 75%).

Anal. Calcd. for $C_{24}H_{26}O_9F_{12}Ru_2$: C, 32.44; H, 2.95. Found: C, 32.60; H, 2.96, mp 148–150 °C.

Properties

The complex $[\{Ru(\eta^4\text{-}C_8H_{12})(O_2CCF_3)\}_2(\mu\text{-}O_2CCF_3)_2(\mu\text{-}OH_2)]$ is an air-stable, orange crystalline solid soluble in both polar and nonpolar organic solvents. Solutions of the compound in air decompose slowly over a period of days. Reactions are, however, best carried out under an inert atmosphere. The infrared(IR) spectrum (Nujol mull) shows a broad band at 1680 to 1700 cm^{-1} assignable to $\nu_{(COO)}$. The ^{1}H NMR spectrum recorded in CDCl$_3$ (90 MHz) shows a broad multiplet resonance centered at 2.4 ppm (16H), and two broad singlet resonances at 4.7 ppm (4H) and 4.8 ppm (4H), assigned to the cycloocta-1, 5-diene ligand protons. A characteristic sharp singlet resonance at 13.0 ppm (2H) is assignable to the protons of the aqua ligand. Molecular weight determinations (benzene, found: 852 ± 17, calcd. for $C_{24}H_{26}O_9F_{12}Ru_2$: 888.6) indicate the dinuclear nature of the complex. The structure of this compound[3] has been determined, confirming the formulation as a dimer in which two trifluoroacetato ligands and an aqua ligand bridge the two ruthenium atoms.

F. μ-AQUA-BIS(μ-CHLOROACETATO)-BIS[(CHLOROACETATO)-(η⁴-CYCLOOCTA-1, 5-DIENE)RUTHENIUM(II)]

$$2[Ru(\eta^3\text{-}C_3H_5)_2(\eta^4\text{-}C_8H_{12})] + 4CH_2ClCOOH + H_2O$$
$$\longrightarrow [\{Ru(\eta^4\text{-}C_8H_{12})(O_2CCH_2Cl)\}_2(\mu\text{-}O_2CCH_2Cl)_2(\mu\text{-}OH_2)]$$
$$+ 4C_3H_6$$

Procedure

A 100-mL Schlenk flask is purged with nitrogen and under a flow of nitrogen gas is charged with freshly prepared $[Ru(\eta^3\text{-}C_3H_5)_2(\eta^4\text{-}C_8H_{12})]$ (2.0 g, 6.86 mmol) in oxygen-free and peroxide-free diethyl ether (40 mL). A solution of chloroacetic acid (2.0 g, 21.3 mmol) in oxygen-free and peroxide-free diethyl ether (10 mL) is prepared under nitrogen and then added. The initially colorless solution is stirred at 25 °C for 1 h, during which time the color darkens to yellow-orange. The solvent is removed under reduced pressure to give a yellow oil. Crystallization under nitrogen using acetone–hexane (1:2) gives the product as a yellow solid. Yield: 1.5–2.0 g (58–71%).

Anal. Calcd. for $C_{24}H_{34}O_9Cl_4Ru_2$: C, 35.56; H, 4.23; Cl, 17.50. Found: C, 35.91; H, 4.21; Cl, 17.53, mp 125–127 °C.

Properties

The complex $[\{Ru(\eta^4\text{-}C_8H_{12})(O_2CCH_2Cl)\}_2(\mu\text{-}O_2CCH_2Cl)_2(\mu\text{-}OH_2)]$ is a yellow, crystalline, air-stable compound, soluble in most organic solvents. The IR spectrum (Nujol mull) shows a strong $\nu_{(COO)}$ absorption at $\sim 1610 \, cm^{-1}$. The ¹H NMR spectrum recorded in $CDCl_3$ (90 MHz) shows three broad multiplet resonances centered at 2.4 ppm (16 H), 4.3 ppm (4 H), and 4.55 ppm (4H) assigned to the cyclooctadiene ligand, two singlets at 4.0 ppm (4H) and 4.2 ppm (4H) arising from the protons of the carboxylato ligands, and a sharp singlet at 13.8 ppm (2H) assignable to the protons of the aqua ligand.

G. μ-AQUA-BIS(μ-TRICHLOROACETATO)-BIS[(η⁴-BICYCLO[2.2.1]-HEPTA-2, 5-DIENE)(TRICHLOROACETATO)RUTHENIUM(II)]

$$[Ru(\eta^3\text{-}C_3H_5)_2(\eta^4\text{-}C_7H_8)] + 4CCl_3COOH + H_2O$$
$$\longrightarrow [\{Ru(\eta^4\text{-}C_7H_8)(O_2CCCl_3)\}_2(\mu\text{-}O_2CCCl_3)_2(\mu\text{-}OH_2)] + 4C_3H_6$$

Procedure

A 150-mL Schlenk flask is purged with nitrogen and under a flow of nitrogen gas is charged with freshly prepared $[Ru(\eta^3\text{-}C_3H_5)_2(\eta^4\text{-}C_7H_8)]$ (2.0 g, 7.26 mmol) and oxygen-free and peroxide-free tetrahydrofuran (THF) (50 mL). A solution of CCl_3COOH (3.68 g, 22.5 mmol) in oxygen-free and peroxide-free THF (20 mL) is prepared under nitrogen and then added. The reaction solution is stirred at 25 °C for 1 h. The initially colorless solution turns dark orange during this time. The solvent is removed under reduced pressure, leaving an orange-brown solid. This material is washed with chilled reagent grade pentane (0 °C, 5–8 mL). Filtration gives the product as an orange solid. Yield: 2.6 g (71%). Recrystallization is achieved from dichloromethane–hexane (1:7) under an atmosphere of nitrogen.

Anal. Calcd. for $C_{22}H_{18}O_9Cl_{12}Ru_2$: C, 25.07; H, 1.72; Cl, 40.36. Found: C, 25.50; H, 1.73; Cl, 39.88, mp 172–175 °C (dec).

Properties

The compound $[\{Ru(\eta^4\text{-}C_7H_8)(O_2CCCl_3)\}_2(\mu\text{-}O_2CCCl_3)_2(\mu\text{-}OH_2)]$ is a mustard yellow, air-stable solid. It is soluble in most common organic solvents. The IR spectrum (Nujol mull) shows an intense band at $\sim 1660\,cm^{-1}$ assignable to $v_{(COO)}$. The ^1H NMR spectrum recorded in $CDCl_3$ (90 MHz) shows a singlet resonance at 1.8 ppm (4H), two multiplet resonances centered at 4.0 ppm (2H) and 4.2 ppm (2H), and three triplets of doublets centered at 4.98 ppm (4H), 5.1 ppm (2H), and 5.3 ppm (2H). All these resonances are assignable to the norbornadiene ligand. A sharp singlet at 12.6 ppm (2H) is assignable to the protons of the aqua ligand.

Related Derivatives

The complexes $(\eta^4\text{-cycloocta-1,5-diene})bis(\eta^3\text{-2-methyl-1-propenyl})$-ruthenium(II), $[\{Ru(\eta^4\text{-}C_8H_{12})(O_2CCCl_3)\}_2(\mu\text{-}O_2CCCl_3)_2(\mu\text{-}OH_2)]$, and $[\{Ru(\eta^4\text{-}C_7H_8)(O_2CCF_3)\}_2(\mu\text{-}O_2CCCl_3)_2(\mu\text{-}OH_2)]$ may be prepared by procedures similar to those described for the related complexes in Sections D, E, and G.

Properties

$[Ru\{\eta^3\text{-}CH_2C(CH_3)CH_2\}_2(\eta^4\text{-}C_8H_{12})]$. This complex is soluble in most common organic solvents except alcohols in which it is only sparingly soluble. It is moderately air stable in the solid state but decomposes in solution. If stored for extended periods the compound should be kept under an inert atmosphere and at $-20\,°C$. The ^1H NMR spectrum recorded in $CDCl_3$ (90 MHz) shows the following characteristic resonances C_8H_{12}: 3.95 ppm (m, 2H, olefin), 2.7–3.05 ppm (m, 4H, methylene), 1.65–2.2 ppm (m, 4H, methylene), 1.05–1.35 ppm (m, 2H, olefin); $CH_2C(CH_3)CH_2$: 3.55 ppm (d, 2H, syn), 2.85 ppm (s, 2H, anti), 1.8 ppm (s, 6H, CH_3), 1.55 ppm (s, 2H, syn′), 0.15 ppm (s, 2H, anti′).

[{Ru(η⁴-C₈H₁₂)(O₂CCCl₃)}₂(μ-O₂CCCl₃)₂(μ-OH₂)]. This complex is a yellow crystalline solid. It is air stable. Although it has limited solubility in dichloromethane and chloroform, in general, it is insoluble in most common organic   solvents. The IR spectrum (Nujol mull) shows a broad band at ∼ 1670 cm⁻¹ assignable to $v_{(COO)}$. The ^1H NMR spectrum recorded in CD_2Cl_2 (90 MHz) shows two broad multiplet resonances centered at 2.4 ppm (16H) and 4.8 ppm (8H) for the cyclooctadiene ligand protons and a sharp singlet resonance at 13.2 ppm (2H) assignable to the protons of the aqua ligand.

[{Ru(η⁴-C₇H₈)(O₂CCF₃)}₂(μ-O₂CCF₃)₂(μ-OH₂)]. This compound is an air-stable, orange-brown crystalline solid possessing good solubility in most common organic solvents. The IR spectrum (Nujol mull) displays a strong $v_{(COO)}$ absorption at ∼ 1680 cm⁻¹. The ^1H NMR spectrum recorded in $CDCl_3$ (90 MHz) shows a singlet resonance at 1.8 ppm (4H), multiplet resonances centered at 3.9 ppm (2H) and 4.2 ppm (2H), two triplets of doublets centered at 5.0 ppm (2H) and 5.2 ppm (2H), and a pseudotriplet centered at 4.9 ppm (4H). All these resonances are assigned to the norbornadiene ligand. A sharp singlet at 12.3 ppm (2H) is assignable to the protons of the aqua ligand.

References

1. R. C. Mehrotra and R. Bohra, *Metal Carboxylates*, Academic Press, London, 1983.
2. A. Dobson and S. D. Robinson, *Inorg. Chem.*, **16**, 137 (1977).
3. M. O. Albers, D. C. Liles, E. Singleton, and J. E. Yates, *J. Organomet. Chem.*, **272**, C62 (1984).
4. M. O. Albers and E. Singleton, *J. Molecular Catal.*, **31**, 211 (1985); M. O. Albers, E. Singleton, and M. M. Viney, *J. Molecular Catal.*, **34**, 235 (1986).
5. J. Powell and B. L. Shaw, *J. Chem. Soc. A*, **1968**, 159.
6. R. R. Schrock, B. F. G. Johnson, and J. Lewis, *J. Chem. Soc. Dalton Trans.*, **1974**, 951.
7. J. J. Eisch, *Organometallic Syntheses*, Vol. 2, Academic Press, New York, 1981, p. 105.
8. D. F. Shriver, *The Manipulation of Air-Sensitive Compounds*, McGraw-Hill, New York, 1969.
9. D. D. Perrin, W. L. F. Armarego, and D. R. Perrin, *Purification of Laboratory Chemicals*, 2nd ed., Pergamon Press, Oxford, 1980.
10. E. W. Abel, M. A. Bennett, and G. Wilkinson, *J. Chem. Soc.*, **1959**, 3178.
11. M. A. Bennett, M. I. Bruce, and T. W. Matheson, in *Comprehensive Organometallic Chemistry*, Vol. 4, G. Wilkinson (ed.), Pergamon Press, Oxford, 1982, pp. 748–750.
12. M. A. Bennett and G. Wilkinson, *Chem. Ind.*, **1959**, 1516.

45. POLYNUCLEAR RUTHENIUM COMPLEXES

A. DODECACARBONYLTRIRUTHENIUM, $Ru_3(CO)_{12}$

Submitted by M. I. BRUCE,* C. M. JENSEN,[†] and N. L. JONES[‡]
Checked by GEORG SÜSS-FINK,[§] GERHARD HERRMANN,[§] and VERA DASE[§]

$$RuCl_3 \cdot nH_2O + CO \longrightarrow Ru_3(CO)_{12}$$

Dodecacarbonyltriruthenium can be prepared by several methods. Johnson and Lewis[1] have reported a procedure in which ruthenium trichloride hydrate is converted to tris(2,4-pentanedionato)ruthenium(III), which is turn is reacted with hydrogen and carbon monoxide. Reaction pressure and temperature are high (160 atm and 165 °C) and the yield is in the range from 50 to 55%.

James and coworkers[2] reported a method for the synthesis of dodecacarbonyltriruthenium from hexakis(μ-acetato)-trisaquooxotriruthenium(III) acetate, which requires only ambient pressures of carbon monoxide. The reaction time can be long and the yield is 59% based on the starting triruthenium complex.

Mantovani and Cenini[3] have also reported a two-step ambient pressure synthesis of dodecacarbonyltriruthenium starting with ruthenium trichloride hydrate resulting in a 50 to 60% yield but the product requires recrystallization.

We give here details of a one-step, high yield (70% or greater), medium pressure (65 atm) synthesis of $Ru_3(CO)_{12}$ from $RuCl_3 \cdot nH_2O$.[4] No solvent purification is necessary and this synthesis can be completed in 1 day.

Procedure

■ **Caution.** *All manipulations with carbon monoxide should be carried out in a well-ventilated area.*

A mixture of $RuCl_3 \cdot nH_2O$ (Strem or Aldrich) (25.4 g) and anhydrous methanol (Mallinckrodt, fresh bottle with no further drying or deaerating)

*Department of Physical and Inorganic Chemistry, University of Adelaide, Adelaide, South Australia 5000.
†Department of Chemistry, University of Hawaii, 2545 The Mall, Honolulu, HI 96822.
‡Department of Chemistry, La Salle University, Philadelphia, PA 19141.
§Laboratorium für Anorganische Chemie, Universität Bayreuth, Universitätstrasse 30, D-8580 Bayreuth, Federal Republic of Germany.

(300 mL) is pressurized to $\sim$ 1000 psi (65 atm) with carbon monoxide in a 1-L autoclave. The autoclave is heated at 250 °F (125 °C) with stirring. After 8 h of heating the autoclave is cooled and then vented in a well-ventilated hood.

■ **Caution.** *Highly toxic carbonyl chloride (phosgene) may be formed as a by-product and therefore use of an efficient fume hood is mandatory.*

The crude orange crystalline dodecacarbonyltriruthenium is separated by filtration in air on a Buchner funnel. The crude product is extracted into dichloromethane (3.5–4 L) leaving a blue-black solid behind. The amount of blue-black solid formed varies and based on Ru and Cl elemental analyses it is identified as RuO_2.

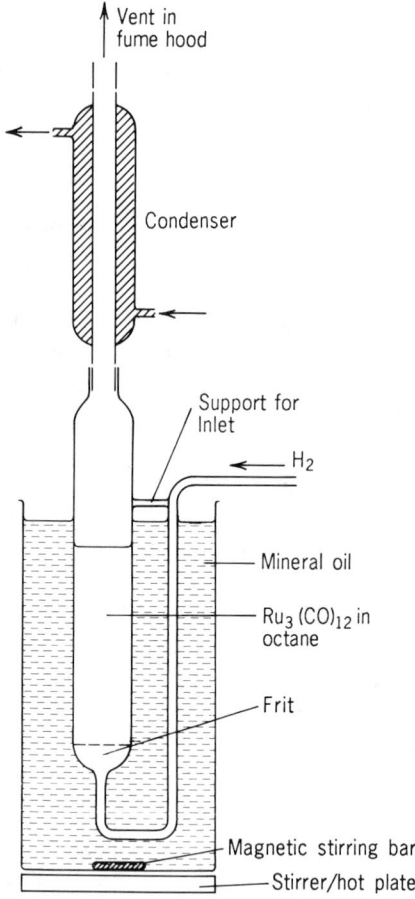

Fig. 1. Apparatus for preparation of $Ru_4H_4(CO)_{12}$.

Anal. Calcd. for RuO_2: Ru, 75.95%, Cl, 0.0%; Found: Ru, 76.30%, Cl < 0.1%.

The solution is concentrated by rotary evaporation at room temperature. Spectroscopically pure orange crystalline $Ru_3(CO)_{12}$ is isolated by filtration. Yields vary slightly from preparation to preparation and are typically $\sim 70\%$ (15.4g) but sometimes can be as high as 92%.

Anal. Calcd. for $Ru_3(CO)_{12}$: C, 22.54; Found: C, 22.50.

Properties

Dodecacarbonyltriruthenium is an orange, air- and light-stable crystalline solid. It is soluble in most organic solvents. Its infrared (IR) spectrum in hexane displays three bands attributable to terminal CO ligands: 2061 (vs), 2031 (s), and 2011 (m) cm^{-1}. No band assignable to a bridging carbonyl ligand is observed.

Note. The mother liquor can be recycled, an amount of $RuCl_3 \cdot nH_2O$ equal to the amount of $Ru_3(CO)_{12}$ formed in the previous preparation being added to the solution. We have successfully operated this process for up to four successive preparations, with essentially quantitative conversion of the added $RuCl_3 \cdot nH_2O$ to $Ru_3(CO)_{12}$.

In some instances, particularly when the ruthenium trichloride sample contains more than the usual amount of water (this may occur, e.g., with old samples or on long exposure to moist air), the isolated product may be a mixture of $Ru_3(CO)_{12}$ and $Ru_4(\mu\text{-}H)_4(CO)_{12}$ (as indicated by the IR $v_{(CO)}$ spectrum). In such cases, depending upon the final product required (a) the product may be used directly as in the synthesis of $Ru_4(\mu\text{-}H)_4(CO)_{12}$ described below, when conversion to the cluster carbonyl hydride is completed by reaction with H_2; or (b) treatment of the product with CO for 1 h while suspended in refluxing octane, using the apparatus depicted in Fig. 1, results in conversion of any $Ru_4(\mu\text{-}H)_4(CO)_{12}$ to $Ru_3(CO)_{12}$.

B. DODECACARBONYLTETRA-μ-HYDRIDO-
TETRARUTHENIUM, Ru$_4$(μ-H)$_4$(CO)$_{12}$

$$4Ru_3(CO)_{12} + 6H_2 \longrightarrow 3Ru_4(\mu\text{-H})_4(CO)_{12} + 12CO$$

Submitted by MICHAEL I. BRUCE* and MICHAEL L. WILLIAMS*
Checked by GUY LAVIGNE[†] and THÉRÈSE ARLIGUIE[†]

This tetranuclear ruthenium carbonyl hydride was described on several occasions,[5] but early preparations were usually contaminated with Ru$_3$(CO)$_{12}$, giving rise to suggestions of the existence of two isomeric forms. The situation was clarified by the work of Kaesz and coworkers,[6] who discovered the direct route from Ru$_3$(CO)$_{12}$ and hydrogen, which is described below. The compound is often obtained from reactions between Ru$_3$(CO)$_{12}$ and substrates containing hydrogen (hydrocarbons, ethers, alcohols, water, etc.) and by acidification of anionic ruthenium cluster carbonyls.[7]

■ **Caution.** *Due to the highly toxic nature of carbon monoxide and flammable nature of hydrogen gas, this procedure must be carried out in a well-ventilated hood; the autoclave room must also be well ventilated.*

Procedure

1. Ru$_4$(μ-H)$_4$(CO)$_{12}$ can be obtained from Ru$_3$(CO)$_{12}$ by the original method,[6] in which hydrogen is passed through a solution of Ru$_3$(CO)$_{12}$ in refluxing octane.

    Finely powdered Ru$_3$(CO)$_{12}$ (250 mg, 0.35 mmol) is suspended in octane (80 mL), which has been washed successively with conc. H$_2$SO$_4$ and water and distilled, contained in the apparatus shown in Figure 1. A gentle stream of H$_2$ is passed through the solution while it is being heated at the reflux point in the oil bath. After 1 h, or when the $v_{(CO)}$ band of Ru$_3$(CO)$_{12}$ at 2061 cm^{-1} is no longer present, the solution is filtered hot through a short (10 × 5 cm) column of silica gel. Reduction of volume to ~ 10 mL (rotary evaporator) results in deposition of Ru$_4$(μ-H)$_4$(CO)$_{12}$ as a yellow powder, which is filtered, washed with cold petroleum ether (2 × 2 mL), and dried. Yield: 180 mg (90%). The cluster hydride can be recrystallized from a dichloromethane–hexane mixture; however, it is generally pure enough for further reactions.

*Department of Physical and Inorganic Chemistry, University of Adelaide, South Australia 5000.
[†]CNRS Laboratoire de Chimie de Coordination, 31077 Toulouse, France.

Note. This reaction does not give high yields if more concentrated solutions of $Ru_3(CO)_{12}$ are used.

2. Alternatively, this hydrogenation may be carried out in a small laboratory autoclave [$Ru_3(CO)_{12}$ in cyclohexane solution ($1\,g\,50\,mL^{-1}$); H_2 at 25 atm, 120 °C, for 2 h].
   Finely powdered $Ru_3(CO)_{12}$ (1.0 g, 1.4 mmol) is added to octane (50 mL), purified as in (a), contained in the close-fitting glass linear of a 100-mL capacity stainless steel laboratory autoclave (Röth). After initial pressurization with hydrogen and venting, the autoclave is charged, with H_2 (25 atm) and heated at 120 °C for 2 h. After cooling and venting, the solution is removed from the autoclave and filtered; removal of solvent (rotary evaporator) gives $Ru_4(\mu\text{-}H)_4(CO)_{12}$. Yield: 0.70 g (90%).

Properties

The compound $Ru_4(\mu\text{-}H)_4(CO)_{12}$ is obtained as a yellow air-stable powder, which is soluble in most organic solvents, but insoluble in water. The IR spectrum contains $\nu_{(CO)}$ bands at 2081 (s), 2067 (vs), 2030 (m), 2024 (s), and 2009 (w) cm^{-1} (cyclohexane solution); the ^1H NMR spectrum has a resonance at $\delta -17.98$ (CDCl$_3$ solution). The molecular structure of $Ru_4(\mu\text{-}H)_4(CO)_{12}$ has been determined by X-ray diffraction: the four hydrogen atoms bridge the edges of the tetrahedral Ru_4 core in a D_{2d} arrangement, while three CO ligands are terminally bonded to each ruthenium.[8] The deuterated complex $Ru_4(\mu\text{-}D)_4(CO)_{12}$ can be prepared in the same way if D_2 is used in place of H_2.[6]

References

1. B. F. G. Johnson and J. Lewis, *Inorg. Synth.*, **13**, 92 (1972) and references cited therein.
2. B. R. James, G. L. Rempel, and W. K. Teo, *Inorg. Synth.*, **16**, 45 (1975).
3. A. Mantovani and S. Cenini, *Inorg. Synth.*, **16**, 47 (1975).
4. M. I. Bruce, J. G. Matisons, R. C. Wallis, J. M. Patrick, B. W. Skelton, and A. H. White, *J. Chem. Soc. Dalton Trans.*, **1983**, 2365.
5. J. W. S. Jamieson, J. V. Kingston, and G. Wilkinson, *Chem. Commun.*, **1966**, 569; B. F. G. Johnson, R. D. Johnston, J. Lewis, and B. H. Robinson, *Chem. Commun.*, **1966**, 851; H. Pichler, H. Meier zu Kocker, W. Gabler, R. Gartner, and D. Kioussis, *Brennstoff Chem.*, **48**, 266 (1967); F. Piacenti, M. Bianchi, P. Frediani, and E. Benedetti, *Inorg. Chem.*, **10**, 2759 (1971).
6. S. A. R. Knox, J. W. Koepke, M. A. Andrews, and H. D. Kaesz, *J. Am. Chem. Soc.*, **97**, 3942 (1975).
7. B. J. G. Johnson, R. D. Johnston, J. Lewis, B. H. Roninson, and G. Wilkinson, *J. Chem. Soc. A*, **1958**, 2856.
8. R. D. Wilson, S. M. Wu, R. A. Love, and R. Bau, *Inorg. Chem.*, **17**, 1271 (1978).

46. A PHOSPHINO BRIDGED RUTHENIUM CLUSTER: NONACARBONYL-μ-HYDRIDO-(μ-DIPHENYL-PHOSPHINO)TRIRUTHENIUM(0)

$$Ru_3(CO)_{12} + PHPh_2 \xrightarrow{Ph_2CO^-} Ru_3(CO)_{11}(HPPh_2) + CO$$

$$Ru_3(CO)_{11}(HPPh_2) \xrightarrow{\Delta} Ru_3(\mu\text{-}H)(CO)_{10}(\mu\text{-}PPh_2) + CO$$

$$Ru_3(\mu\text{-}H)(CO)_{10}(\mu\text{-}PPh_2) \xrightarrow{Me_3NO} Ru_3(\mu\text{-}H)(CO)_9(\mu\text{-}PPh_2) + CO_2$$

Submitted by D. NUCCIARONE, S. A. MACLAUGHLIN, and A. J. CARTY*
Checked by STEPHEN B. COLBRAN[†]

Dialkyl and diarylphosphino groups (PR_2) have attracted considerable interest in metal cluster chemistry as strongly bound yet flexible, supporting ligands potentially capable of maintaining the integrity of a polynuclear framework during chemical transformations.[1] Numerous phosphino bridged clusters have been synthesized in the last few years using a variety of methods, including those established in the early work of Hayter[2] and Issleib and Wenschuk[3] and Chatt and Davidson,[4] namely, (a) elimination of HX in the reaction of R_2PH with a halogen–metal compound; (b) halide displacement by the anion R_2P^-; (c) halide displacement from R_2PCl by a carbonyl anion; (d) P—P bond cleavage in the reaction of a diphosphine R_2PPR_2 with a transition metal compound or on thermolysis or photochemical activation of a monodentate diphosphine complex; (e) oxidative addition of R_2PH to a metal carbonyl.

For metal carbonyl clusters, methods (b),[1n,o] (c),[5] (d),[6] and (e),[1f,e,m] or variations of them have been successfully employed. Thus an effective strategy for mixed-metal cluster synthesis is chloride ion displacement from a coordinated R_2PCl ligand by a carbonyl metallate.[1j] Several new routes to μ-PR_2 clusters have been developed in recent years. Vahrenkamp and Keller[7] have successfully adapted a method first described by Benson et al.[8] involving elimination of propene in reactions of η^3-C_3H_5 complexes with secondary phosphines. Thus, treatment of η^3-C_3H_5 $Co(CO)_3$ with $Co(CO)_2(NO)(PMe_2H)$ affords several products including the unusual pentanuclear cluster $Co_5(CO)_{11}(PMe_2)_3$. Haines has prepared

*Guelph-Waterloo Centre for Graduate Work in Chemistry, Department of Chemistry, University of Waterloo, Waterloo, Ontario, N2L 3G1.
†University Chemical Laboratory, Lensfield Road Cambridge, CB2 1EW, United Kingdom.

$Rh_2Fe_2(CO)_8(PPh_2)_4$ from $[Rh(\eta^3\text{-}C_3H_5)_2Cl]_2$ and $Fe(CO)_4(PPh_2H)$ via a similar strategy.[1g] For mixed-metal systems, generation of intermediate, reactive metallophosphine species [e.g., $Co(CO)_4(PPh_2)^5$] via halide displacement from PPh_2Cl with carbonyl anion or $[Fe_2(\mu\text{-}PPh_2)_2(CO)_5(PPh_2)]^-$ (Ref. 1e) via deprotonation of a secondary phosphine complex] followed by carbonyl or halide displacement from a second metal has been successfully employed. The facile oxidative cleavage of a $P\text{—}C_{sp}$ bond in phosphinoalkynes $R_2PC{\equiv}CR'$, which occurs in thermal or photochemical reactions with metal carbonyls, has been used to generate a wide range of μ-phosphino, μ-acetylido clusters.[1c,9] Other routes to metal cluster phosphides include the use of metallated secondary phosphine complexes such as *cis*-$Mo(CO)_4(Me_2PLi)_2$ in displacement reactions with carbonyl halides,[10] elimination of hydrogen from carbonyls and secondary phosphines,[11] and thermolysis of tertiary phosphine complexes with $P\text{—}C$ bond cleavage.[1i,12]

In this report we describe a synthetic route to the highly unusual phosphino bridged cluster $Ru_3(\mu\text{-}H)(CO)_9(\mu\text{-}PPh_2)$, a molecule that is a member of the increasingly important class of unsaturated clusters. In this 46-electron species, a $P\text{—}Ph$ group of a $\mu\text{-}PPh_2$ moiety, blocks a vacant axial site on the unique ruthenium atom;[13] reactions with Lewis bases are rapid, leading to saturated 48-electron complexes. Such molecules are potentially useful synthons in organometallic and cluster chemistry. The procedure involves the generation of $Ru_3(CO)_{11}(PPh_2H)$ via the ketyl route of Bruce et al.[14] followed by oxidative addition of the coordinated diphenylphosphine to the cluster and finally CO displacement from $Ru_3(\mu\text{-}H)(CO)_{10}(\mu\text{-}PPh_2)$.

Procedure

All solvents used are heated at reflux over an appropriate drying agent and freshly distilled under nitrogen prior to use. Tetrahydrofuran (THF) is dried by heating at reflux over sodium benzophenone ketyl, and CH_2Cl_2 by distillation from P_4O_{10}. The compounds C_7H_8 and C_7H_{16} are dried by heating at reflux over $Li[AlH_4]$. Reactions are carried out on a standard Stock[15] line using Schlenk apparatus and syringe techniques. Triruthenium dodecacarbonyl is obtained from Strem, or synthesized according to a procedure appearing elsewhere in this volume.[16] The $PHPh_2$ is also obtained from Strem.

■ **Caution.** *Due to evolution of poisonous CO gas, and due to the poisonous nature of PHPh$_2$, all manipulations must be carried out in a well-ventilated hood using protective gloves.*

A solution of sodium benzophenone ketyl is prepared by adding benzophenone (Fisher Scientific) (10 g) to THF (100 mL) containing excess sodium.

A 100-mL two-necked round-bottomed flask is flame dried and maintained under a positive pressure of nitrogen. To this is added a magnetic stirring bar, $Ru_3(CO)_{12}$ (200 mg, 0.313 mmol), and THF (30 mL). Upon complete dissolution of the $Ru_3(CO)_{12}$, one equivalent of $PHPh_2$ (54 μL) is added. With rapid stirring, three to four drops of a previously prepared THF solution of purple sodium benzophenone ketyl is added. If moisture has not been rigorously excluded, the first few drops of ketyl will be destroyed; an additional three or four drops should be added. Reaction may be followed by monitoring the disappearance of $Ru_3(CO)_{12}$ (IR). Small amounts of residual $Ru_3(CO)_{12}$ should be tolerated: Any further addition of ketyl should then be avoided, due to competition of the monosubstituted product for the free phosphine. The solution is stirred for 10 min and then taken to dryness *in vacuo*.

The residue is taken up in heptane (5 mL) and placed at the top of a dry-packed column of Florisil (100–200 mesh, 20 × 2 cm) and eluted with heptane. The first band eluted is yellow and typically contains $Ru_3(CO)_{12}$ (5 mg) and $Ru_3(\mu$-$H)(CO)_{10}(\mu$-$PPh_2)$ (5 mg). The second band is deep red $Ru_3(CO)_{11}(PHPh_2)$.

This latter solution (in $\sim$ 20 mL of heptane) is transferred to a Schlenk tube equipped with nitrogen inlet and connected to an oil-bubbler venting into the hood. Under an atmosphere of nitrogen, the solution is heated at 50 to 55 °C for 12 h. The reaction mixture is allowed to cool, and its volume is reduced to half, by evacuation. The solution is then refrigerated ($\sim$ − 15 °C) giving reddish-orange crystals of $Ru_3(\mu$-$H)(CO)_{10}(\mu$-$PPh_2)$ overnight (123–130 mg). The mother liquor is then chromatographed as previously described with heptane as eluent. The first band is $Ru_3(CO)_{12}$ in trace quantities followed by a deeper yellow band containing $Ru_3(\mu$-$H)(CO)_{10}(\mu$-$PPh_2)$ (30–35 mg). This gives a total yield for $Ru_3(\mu$-$H)(CO)_{10}(\mu$-$PPh_2)$ of 160 to 170 mg [61–73% based on $Ru_3(CO)_{12}$].

A 50-mL two-necked round-bottomed flask is equipped with a magnetic stirring bar. Under an atmosphere of nitrogen is added $Ru_3(\mu$-$H)(CO)_{10}(\mu$-$PPh_2)$ (200 mg, 0.27 mmol) and CH_2Cl_2 (15 mL). To this solution is added one equivalent of trimethylamine *N*-oxide (dissolved in 1 mL of ethanol and diluted with 3 mL of CH_2Cl_2), via syringe, over 2 to 3 min. The solution is allowed to stir for 10 min and then taken to dryness *in vacuo*.

The residue is redissolved in toluene (2 mL) and the resulting solution is diluted with heptane (5 mL). Chromatography of this solution on Florisil (as previously described) gives first a deep yellow band of unchanged starting material ($\sim$ 10 mg), followed by an orange-red band of $Ru_3(\mu$-$H)(CO)_9(\mu$-$PPh_2)$. Resolution of these two bands is poor; however, pure product can be obtained by fractional crystallization from a concentrated solution of heptane–toluene (4:1 v/v) at − 15 °C. This gives a yield for $Ru_3(\mu$-$H)(CO)_9(\mu$-$PPh_2)$ of 115 to 120 mg (54–62%).

Anal. Calcd. for $Ru_3PC_{22}O_{10}H_{11}$: C, 34.36; H, 1.44; P, 4.02. Found: C, 34.73; H, 1.46; P, 4.35. *Anal.* Calcd. for $Ru_3PC_{21}O_9H_{11}$: C, 34.01; H, 1.50. Found: C, 34.22; H, 1.66.

Properties

All three cluster complexes are air stable for short periods of time in the solid state and are thermally unstable. They should be stored under nitrogen in a refrigerator. Infrared spectra (in cyclohexane solution) in the $v_{(CO)}$ region show the following bands: for $Ru_3(CO)_{11}(PHPh_2)$: 2096 (w), 2045 (vs), 2028 (s), 2014 (vs), 1993 (w), 1985 (w), 1975 (w, sh), 1959 (w, sh); for $Ru_3(\mu\text{-}H)(CO)_{10}(\mu\text{-}PPh_2)$ 2098 (w), 2054 (s), 2048 (m), 2020 (s), 2002 (w), 1996 (m), 1987 (w); for $Ru_3(\mu\text{-}H)(CO)_9(\mu\text{-}PPh_2)$: 2083 (s), 2054 (vs), 2029 (vs), 2014 (s), 1993 (m), 1986 (m). $\{^1H\}$ ^{31}P NMR data in C_6D_6 (85% H_3PO_4 reference) show singlets at 133.8 and 134.2 for $Ru_3(\mu\text{-}H)(CO)_{10}(\mu\text{-}PPh_2)$ and $Ru_3(\mu\text{-}H)(CO)_9(\mu\text{-}PPh_2)$, respectively. The ^{13}C $\{^1H\}$ NMR spectrum of $Ru_3(\mu\text{-}H)(CO)_9(\mu\text{-}PPh_2)$ in $(CD_3)_2CO$ at $-80\,^\circ C$ [δ wrt TMS, 202.3 (s), 201.6 (d) ($^2J_{pc} = 60.6$ Hz) 197.6 (d), ($^2J_{pc} = 4.5$ Hz), 195.0 (s), 190.6 (s) (CO); 143.6 (d) ($^1J_{pc} = 15.7$ Hz); 81.5 (d) ($^1J_{pc} = 33.3$ Hz) (C_i); 129.5 (d) ($^2J_{pc} = 9.0$ Hz), 129.2 (d) ($^2J_{pc} = 12.5$ Hz) (C_o); 138.0 (s), 128.0 (d) ($^3J_{pc} = 7.0$ Hz) (C_m); 136.8 (s), 130.5 (s) (Cp)] shows a very high-field doublet at 81.5 ppm due to the ipso carbon atom of the P—Ph ring, which interacts with the unique ruthenium atom.

The formally 46-electron $Ru_3(\mu\text{-}H)(CO)_9(\mu\text{-}PPh_2)$ reacts rapidly with CO regenerating its precursor $Ru_3(\mu\text{-}H)(CO)_{10}(\mu\text{-}PPh_2)$,[13] with acetylenes RC≡CR to give 50-electron adducts $Ru_3(\mu\text{-}H)(CO)_9(\mu\text{-}PPh_2)(RC≡CR)$ (Ref. 9) with phosphines forming 48-electron adducts $Ru_3(\mu\text{-}H)(CO)_9(\mu\text{-}PPh_2)(PR_3)$ (Ref. 17), and on pyrolysis affords polynuclear phosphinidene clusters including $Ru_4(CO)_{13}(\mu_3\text{-}PPh)$ (Ref. 13) and $Ru_7(CO)_{18}(\mu_4\text{-}PPh)_2$.[18] An iron analog has been described by Huttner and coworkers.[19]

References

1. For a selection of recent references to phosphido bridged complexes see, for example (a) A. J. Carty, S. A. MacLaughlin, and D. Nucciarone, in *Phosphorus-31 NMR Spectroscopy in Stereochemical Analysis: Organic Compounds and Metal Complexes*; J. G. Verkade, L. D. Quin (eds), VCH, New York, 1986, Chapter 16, pp. 559–619; (b) D. Nucciarone, S. A. MacLaughlin, N. J. Taylor, and A. J. Carty, *Organometallics*, 7, 106 (1988); (c) T. S. Targos, G. L. Geoffroy, and A. Rheingold, *Organometallics*, 5, 12 (1986); (d) S. Rosenberg, G. L. Geoffroy, and A. L. Rheingold, *Organometallics*, 4, 1184 (1985); (e) Y. F. Yu, A. Wojcicki, M. Calligaris, and G. Nardin, *Organometallics*, 5, 47 (1986); (f) S. K. Kang, T. A. Albright, T. C. Wright, and R. A. Jones, *Organometallics*, 4, 666 (1985); (g) S. B. Colbran, B. F. G. Johnson, J. Lewis, and S. M. Sorrel, *J. Organomet. Chem.*, 296, C1 (1985); (h) L. Chen, D. J. Kountz, and D. W. Meek, *Organometallics*, 4, 598 (1985); (i) H. Werner and R. Zolk, *Organometallics*,

4, 601 (1985); (j) J. Powell, J. F. Sawyer, and S. J. Smith, *J. Chem. Soc., Chem. Commun.*, **1985**, 1312; (k) Y. F. Yu, J. C. Galluci, and A. Wojcicki, *J. Chem. Soc., Chem. Commun.*, **1984**, 653; (l) K. Henrick, K. Iggo, M. J. Mays, and P. R. Raithby, *J. Chem. Soc. Chem. Commun.*, **1984**, 209; (m) R. Bender, P. Braunstein, B. Metz, and P. Lemoine, *Organometallics*, **3**, 381 (1984); (n) B. Deppisch, H. Schafer, D. Binder, and W. Leske, *Z. Anorg. Allg. Chem.*, **519**, 53 (1984); (o) J. S. McKennis and E. P. Kyba, *Organometallics*, **2**, 1249 (1983); (p) R. T. Baker, T. H. Tulip, and S. S. Wreford, *Inorg. Chem.*, **24**, 1379 (1985); (q) R. J. Haines, N. D. C. T. Steen, and R. B. English, *J. Chem. Soc. Chem. Commun.*, **1981**, 407; (r) E. Keller and H. Vahrenkamp, *Chem. Ber.*, **111**, 2347 (1979); (s) K. Fischer and H. Vahrenkamp, *Z. Anorg. Allg. Chem.*, **475**, 109 (1981); (t) K. Natarajan, L. Zsolnai, and G. Huttner, *J. Organomet. Chem.* **220**, 365 (1981).

2. R. G. Hayter, in *Preparative Inorganic Reactions*, W. L. Jolly (ed.), Interscience, New York, **1965**, p. 211.

3. K. Issleib and E. Wenschuk, *Z. Anorg. Allg. Chem.*, **15**, 305 (1960).

4. J. Chatt and J. M. Davidson, *J. Chem. Soc.*, **1964**, 2433.

5. A. D. Harley, G. J. Guskey, and G. L. Geoffroy, *Organometallics*, **2**, 53 (1983).

6. L. Staudacher and H. Vahrenkamp, *Chem. Ber.*, **109**, 218 (1976).

7. E. Keller and H. Vahrenkamp, *Angew Chem. Int. Ed. Engl.* **16**, 542 (1977).

8. B. C. Benson, R. Jackson, K. K. Joshi, and D. T. Thompson, *J. Chem. Soc. Chem. Commun.*, **1968**, 1506.

9. A. J. Carty, S. A. MacLaughlin, and N. J. Taylor, *Organometallics*, **2**, 1194 (1983).

10. (a) U. Stelzer and E. Unger, *Chem. Ber.*, **110**, 3430 (1970); (b) M. J. Breen, M. R. Duttera, G. L. Geoffroy, G. C. Novotnak, D. A. Roberts, D. M. Shulman, and G. R. Steinmetz, *Organometallics*, **1**, 1008 (1982).

11. H. Vahrenkamp and E. J. Wucherer, *Angew Chem. Int. Ed. Engl.*, **20**, 680 (1981).

12. N. J. Taylor, P. C. Chieh, and A. J. Carty, *J. Chem. Soc. Chem. Commun.*, **1975**, 448.

13. S. A. MacLaughlin, N. J. Taylor, and A. J. Carty, *Can. J. Chem.*, **60**, 87 (1982).

14. M. I. Bruce, D. C. Kehoe, J. G. Matisons, B. K. Nicholson, P. H. Rieger, and M. L. Williams, *J. Chem. Soc. Chem. Commun.*, **442** (1982).

15. D. F. Shriver and M. A. Drezdzon, *The Manipulation of Air-sensitive Compounds*, 2nd ed., McGraw-Hill, New York, 1986.

16. (a) M. I. Bruce, C. M. Jensen, and N. L. Jones, *Inorganic Syntheses*, **26**, 259 (1989); (b) M. I. Bruce, J. G. Matisons, R. G. Wallis, J. M. Patrick, B. W. Skelton, and A. H. White, *J. Chem. Soc., Dalton Trans.*, **1983**, 2365–2373.

17. F. van Gastel, S. A. MacLaughlin, M. Lynch, A. J. Carty, E. Sappa, A. Tiripicchio, and M. Tiripicchio-Camellini, *J. Organomet. Chem.*, **326**, C65 (1987).

18. F. van Gastel, N. J. Taylor, and A. J. Carty, *J. Chem. Soc. Chem. Commun.*, 1049 (1987).

19. K. Knoll, G. Huttner, L. Zsolnai, O. Orama, and M. Wasiucionek, *J. Organomet. Chem.*, **310**, 225 (1986).

47. [μ-NITRIDO-BIS(TRIPHENYLPHOSPHORUS)(1 +)]-[DECACARBONYL-1κ³C, 2κ³C, 3κ⁴C-μ-HYDRIDO-1: 2κ²H-BIS(TRIETHYLSILYL-1κSi, 2κSi)-*TRIANGULO*-TRIRUTHENATE(1 −)]

$$Ru_3(CO)_{12} + Na[BH_4] \longrightarrow Na[Ru_3H(CO)_{11}] + CO + \tfrac{1}{2}B_2H_6$$

$$Na[Ru_3H(CO)_{11}] + 2Et_3SiH \longrightarrow Na[Ru_3H(CO)_{10}(SiEt_3)_2]$$
$$+ CO + H_2$$

$$Na[Ru_3H(CO)_{10}(SiEt_3)_2] + [N(PPh_3)_2]Cl$$
$$\longrightarrow [N(PPh_3)_2][Ru_3H(CO)_{10}(SiEt_3)_2] + NaCl$$

Submitted by GEORG SÜSS-FINK*
Checked by YEA-JER CHEN† and HERBERT D. KAESZ†

The cluster anion $[Ru_3H(CO)_{10}(SiEt_3)_2]^-$ was first detected in catalytic reactions of triethylsilane using the anion $[Ru_3H(CO)_{11}]^-$ as the catalyst.[1] It is formed smoothly by reaction of $Na[Ru_3H(CO)_{11}]^{2,3}$ with Et_3SiH in tetrahydrofuran (THF) solution;[1,4] the anionic product can be isolated as the μ-nitrido-bis(triphenylphosphorus) salt by crystallization from methanol, giving yields up to 65%. In the same way the analogous anions $[Ru_3H(CO)_{10}(ER_2R')_2]^-$ (E = Si, R = R' = Ph; E = Si, R = Et, R' = Me; E = Ge, R = R' = Ph; E = Sn, R = R' = Ph) can be prepared and isolated.[4] The cluster anion $[Ru_3H(CO)_{10}(SiEt_3)_2]^-$ was found to act as a catalyst for the silacarbonylation[1] or hydroxylation[5] of olefins and the hydrosilylation of carbon dioxide; in particular, it catalyzes the silane-supported spirocylization of alkyl isocyanates providing access to a novel series of spiroheterocyclic compounds.[6]

General remarks

■ **Caution.** *Reactions with metal carbonyls involving evolution of highly toxic CO, and evolution of toxic and highly flammable* B_2H_6, *and manipulations of flammable and toxic triethylsilane should be carried out in a well-ventilated hood.*

The reaction can be conducted with standard Schlenk techniques.[7] All

*Laboratorium für Anorganische Chemiè der Universität Bayreuth, Universitätsstrasse 30, D-8580 Bayreuth, Federal Republic of Germany; present address: Institut de Chimie, Université de Neuchatel, Ave. de Bellevaux 51, CH-2000 Neuchatel, Switzerland.
†Department of Chemistry, University of California, Los Angeles, CA 90024-1569.

manipulations must be carried out under purified nitrogen, and all solvents must be distilled over drying agents and saturated with nitrogen prior to use.[7] Triruthenium dodecacarbonyl is commercially available or may be synthesized from ruthenium trichloride trihydrate (Alpha) by a procedure given elsewhere in this volume.[8] All other reagents are commercially available: Na[BH$_4$] (Fluka), triethylsilane (Fluka), [μ-nitrido-bis(triphenyl-phosphorus)(1 +)] chloride (Alpha).

Procedure

In a 250-mL Schlenk tube 640 mg (1 mmol) Ru$_3$(CO)$_{12}$ and 160 mg (4 mmol) (Na[BH$_4$]) are dissolved in 100 mL of THF. The solution, which turns dark red due to the formation of [Ru$_3$H(CO)$_{11}$]$^-$, is stirred at 25 °C for 30 min. Then the reaction mixtures is filtered through a layer of filter pulp using a 250-mL Schlenk frit. Triethylsilane (1 mL, 6.25 mmol) is then added to the filtrate and the solution is stirred overnight at 45 °C. After 15 h the orange-red reaction mixture is filtered through a 2-cm layer of filter pulp using a 250-mL Schlenk frit. The filtrate is evaporated to dryness in a 250-mL Schlenk tube using an oil pump vacuum. The residue is dissolved in methanol (60 mL) to which is added 1 g (1.75 mmol) of [N(PPh$_3$)$_2$]Cl dissolved in 10 mL of methanol. The solution is concentrated under reduced pressure and allowed to stand under vacuum until the product starts to crystallize. Then the mixture is cooled to −30 °C and allowed to stand for 2 days at this temperature. A red crystalline precipitate of analytically pure [N(PPh$_3$)$_2$][Ru$_3$H(CO)$_{10}$(SiEt$_3$)$_2$] is obtained and separated by decantation, washed with cold methanol (5 mL), and dried under vacuum (10^{-5} torr). A typical yield is 850 mg (63%).

Anal. Calcd. for C$_{58}$H$_{61}$NO$_{10}$P$_2$Ru$_3$Si$_2$: C, 59.99; H, 3.75; N, 0.85; P, 3.77; Si, 3.42. Found: C, 59.67; H, 3.82; N, 0.89; P, 3.79; Si 3.37.

Properties

The compound [N(PPh$_3$)$_2$][Ru$_3$H(CO)$_{10}$(SiEt$_3$)$_2$] is a red, crystalline material. The crystals are only slightly air sensitive and decompose above 120 °C over a broad temperature range. They dissolve in polar solvents such as THF, dichloromethane, trichloromethane, acetonitrile, methyl alcohol, or acetone. The red solutions are much more sensitive to oxygen than the solid. The infrared spectrum of a THF solution displays characteristic absorptions at 2070 (w), 2019 (m), 1992 (m), 1984 (vs), 1975 (w), 1965 (m), and 1925 (sh) cm^{-1} (Nicolet MX-1 Spectrometer). The structure of the compound, tautomerism of the CO groups, and electrochemical characteristics have been reported.[9]

References

1. G. Süss-Fink, *Angew. Chem.*, **94**, 72 (1982); *Angew. Chem. Int. Ed. Engl.*, **21**, 73 (1982); *Angew. Chem. Suppl.*, **1982**, 71.

2. B. F. G. Johnson, J. Lewis, P. R. Raithby, and G. Süss-Fink, *J. Chem. Soc. Dalton Trans.*, **1979**, 1356.

3. G. Süss-Fink, *Inorg. Synth.*, **24**, 168 (1986).

4. G. Süss-Fink, J. Ott, B. Schmidkoṅz, and K. Guldner, *Chem. Ber.*, **115**, 2487 (1982).

5. G. Süss-Fink and J. Reiner, *J. Organometal. Chem.*, **221**, C36 (1981).

6. G. Süss-Fink, G. Herrmann, and U. Thewalt, *Angew. Chem.*, **95**, 899 (1983); *Angew. Chem. Int. Ed. Engl.*, **22**, 880 (1983); *Angew. Chem. Suppl.*, **1983**, 1203.

7. S. Herzog and J. Dehnert, *Z. Chem.*, **4**, 1 (1964); english translation: S. Herzog, J. Dehnert, and K. Lübdes, *Technique of Inorganic Chemistry*, Vol. 8, Interscience, New York, 1968, p. 119.

8. M. I. Bruce, N. L. Jones, and C. M. Jensen, *Inorg. Synth.*, **26**, 259 (1989).

9. H.-P. Klein, U. Thewalt, G. Herrmann, G. Süss-Fink, and C. Moinet, *J. Organometal. Chem.*, **286**, 225 (1985).

48. TRI- AND TETRANUCLEAR CARBONYL–RUTHENIUM CLUSTER COMPLEXES CONTAINING ISOCYANIDE, TERTIARY PHOSPHINE, AND PHOSPHITE LIGANDS. RADICAL ION-INITIATED SUBSTITUTION OF METAL CLUSTER CARBONYL COMPLEXES UNDER MILD CONDITIONS

Submitted by MICHAEL I. BRUCE,* BRIAN K. NICHOLSON,[†] and MICHAEL L. WILLIAMS*
Checked by THÉRÈSE ARLIGUIE[‡] and GUY LAVIGNE[‡]

Replacement of CO by tertiary phosphines, arsines, and similar ligands in metal cluster carbonyls is a reaction that is highly dependent on the metal carbonyl. In some cases, such as $Co_4(CO)_{12}$ or $Ir_4(CO)_{12}$, the reaction proceeds readily at room temperature or on gentle warming, and the stoichiometric products can be obtained fairly easily.[1] In other cases, such as $Ru_3(CO)_{12}$, $Ru_4H_4(CO)_{12}$, or $Os_3(CO)_{12}$, the reactions proceed only under vigorous conditions, and usually result in (a) polysubstitution of the cluster carbonyl, or (b) the formation of a mixture of products, or (c) further reaction of the initial substitution product. Examples of these situations are the formation of $Ru_3(CO)_9(PPh_3)_3$ as the sole product when mixtures of $Ru_3(CO)_{12}$ and PPh_3 are heated in refluxing benzene, even when a deficiency

*Department of Physical and Inorganic Chemistry, University of Adelaide, South Australia, 5001.
†Department of Chemistry, University of Waikato, Hamilton, New Zealand.
‡Laboratoire de Chimie de Coordination, CNRS, 31077 Toulouse, France.

of the phosphine is present;[2] the formation of $Ru_4H_4(CO)_{12-n}[P(OMe)_3]_n$ ($n = 0$–4) on heating mixtures of $P(OMe)_3$ and $Ru_4H_4(CO)_{12}$, individual complexes being separated with difficulty and in low overall yield;[3] and the formation of P—C and C—H bond cleavage products among the nine complexes isolated after heating mixtures of $Os_3(CO)_{12}$ and PPh_3 are heated in xylene.[4]

The current interest in cluster complexes as possible catalysts has made desirable a method of synthesis of selectively substituted derivatives, which does not suffer the disadvantages just outlined. Cluster-bound molecules have high and often unique reactivity, and the introduction of tertiary phosphines containing functional groups is often difficult.

It was earlier observed that radical anions of binuclear metal carbonyls, generated electrochemically, were unusually susceptible to attack by nucleophilic ligands.[5] Specific substitution of CO ligands was thus achieved. This was followed by the demonstration of substitution reactions initiated by radical anions, such as sodium benzophenone ketyl.[6] A range of metal cluster carbonyls can be used in the reactions, which proceed under mild conditions and in high yields. Development of the method led to syntheses of specifically substituted compounds containing two or more different ligands.[7] It has resulted in the possibility of the syntheses of complexes that are chiral by virtue of having four different ligands on a tetrahedral M_4 core, such as $Ru_4H_4(CO)_9(PMe_2Ph)[P(OMe)_2Ph][P(OCH_2)_3CEt]$.[8]

Under these conditions, it is also possible to prepare (a) complexes in which bidentate ligands contain one cluster moiety attached to each donor atom, such as $[Ru_3(CO)_{11}]PPh_2(CH_2)_2PPh_2[Ru_3(CO)_{11}]$ (Ref. 9) or $[Ru_3(CO)_{11}]PPh_2C{\equiv}CPPh_2[Ru_3(CO)_{11}]$,[10] or (b) derivatives that readily rearrange thermally, such as $Ru_3(CO)_{10}(Ph_2PC_6H_4CH{=}CH_2)$, which is converted to $Ru_3H_2(CO)_8(Ph_2PC_6H_4C{\equiv}CH)$ at $40\,°C$,[11] or $Ru_3(CO)_{10}[(Ph_2P)_2CH_2]$;[10] the products obtained from the thermal reaction between $Ru_3(CO)_{12}$ and $CH_2(PPh_2)_2$ are $Ru_3(CO)_8[(Ph_2P)_2CH_2]_2$ (in xylene, at 80–$85\,°C$, 73%) and $Ru_3(CO)_7(\mu_3\text{-}PPh)(\mu\text{-}CHPPh_2)[(Ph_2P)_2CH_2]$ (in xylene, $> 130\,°C$, 42%).[12]

The following syntheses describe the method as applied to $Ru_3(CO)_{12}$ and $Ru_4H_4(CO)_{12}$, the experimental application of which is quite general. Brief accounts of similar reactions with $Co_3(\mu_3\text{-}CR)(CO)_9$, $Rh_6(CO)_{16}$, or $Os_3(CO)_{12}$ have been given.[13] Some of these substitution reactions may also be catalyzed by various $[(Ph_3P)_2N]^+$ salts.[14]

Reagents

Dodecacarbonyltriruthenium, $Ru_3(CO)_{12}$, is available commercially from Strem, Pressure Chemicals, or Aldrich; it may also be made by any of the methods described in previous volumes of this series,[15] or by the modification described in Ref. 16. The synthesis of $Ru_4(\mu\text{-}H)_4(CO)_{12}$ is also described elsewhere in this volume.[17]

The tertiary phosphines PMe_2Ph, $CH_2(PPh_2)_2$, and $C_2(PPh_2)_2$, the phosphite $P(OC_6H_4Me\text{-}p)_3$, and the isocyanide $CN\text{-}t\text{-}Bu$ are available from Strem or Pressure Chemicals.

A. SODIUM BENZOPHENONE KETYL SOLUTION

$$Ph_2CO + Na \longrightarrow Na[Ph_2CO]$$

▪ **Caution.** *Sodium metal is highly reactive. It should be handled under a covering layer of mineral oil and an atmosphere of dry nitrogen.*

Procedure

A 50-mL two-necked round-bottomed flask is removed from a hot oven and flushed with nitrogen. A sample of benzophenone (0.091 g, 0.5 mmol) together with 20 mL of freshly distilled and deoxygenated tetrahydrofuran (THF) and a stirring bar are placed in the flask. A small lump of sodium metal (~ 0.50 g) is cut into smaller pieces directly into the flask against the emergent stream of nitrogen. The second neck of the flask is then closed with a septum cap. The mixture is stirred for 0.5 to 2 h in the stoppered flask under nitrogen until the initial ultramarine blue is replaced by an intense blue-purple color. The resulting solution is ~ 0.025 mol L^{-1} in $Na[Ph_2CO]$. This solution may decolorize in 1 or 2 days' time but it can be regenerated by addition of more benzophenone.

Properties

A solution of sodium benzophenone ketyl is exceedingly reactive, and is immediately decolorized by air or traces of water, alcohols, and so on. The solution should therefore be kept under nitrogen at all times. Aliquots may be removed using a dry syringe via the septum cap. Under these conditions, the solution appears to be stable for ~ 2 days.

B. UNDECACARBONYL(DIMETHYL-PHENYLPHOSPHINE)TRIRUTHENIUM, $Ru_3(CO)_{11}(PMe_2Ph)$

$$Ru_3(CO)_{12} + PMe_2Ph \xrightarrow{\text{Na[Ph}_2\text{CO]}} Ru_3(CO)_{11}(PMe_2Ph) + CO$$

■ **Caution.** *Due to evolution of highly toxic carbon monoxide (a colorless and odorless gas), and also because of the toxic nature and bad odor of tertiary phosphine, phosphite, or isocyanide ligands, these reactions should be performed in a well-ventilated hood.*

■ **Caution.** *The procedure for obtaining accurately small amounts of PMe_2Ph (or similar toxic liquids with bad odor) is to weigh ~ 0.4 g of the ligand into a stoppered, nitrogen-flushed flask. Sufficient dry THF is then added to make a solution of 1.0 mmol mL^{-1}. A calibrated syringe can be used to transfer the required volume of this standard solution.*

Procedure

A 50-mL Schlenk tube, previously kept overnight in a hot oven, is flushed with nitrogen while cooling. After attaching a gas inlet to the side arm, the flask is charged with $Ru_3(CO)_{12}$ (0.20 g, 0.31 mmol), 25 mL of dry, deoxygenated THF, 0.33 mL of a solution of PMe_2Ph in THF (~ 1.0 mmol mL^{-1}) and a magnetic stirring bar, and is then closed with a rubber septum cap. The Schlenk tube is placed in a oil bath on a stirrer hot plate, and warmed to $\sim 40\,°C$ to dissolve the $Ru_3(CO)_{12}$. A dry 1-mL syringe is loaded with the $Na[Ph_2CO]$ solution, which is added dropwise to the reaction mixture through the septum cap until the solution rapidly darkens in color (typically 5–10 drops; see cautionary note immediately above). To check that reaction is complete, an aliquot of the reaction mixture is removed by a second syringe to an IR solution cell; the 2061 cm^{-1} $v_{(CO)}$ band of $Ru_3(CO)_{12}$ is monitored. If still present, several more drops of the $Na[Ph_2CO]$ are added until disappearance of the band. The reaction mixture is then transferred into a 100-mL round-bottomed flask, and solvent is removed (rotary evaporator), to leave an oily residue, which is extracted with two 15-mL portions of warm ($\sim 40\,°C$) petroleum ether (boiling range 40–60 °C). The combined extracts are filtered, concentrated to ~ 10 mL, and cooled overnight ($-10\,°C$) to give red-orange crystals, which are collected on a sintered-glass filter and air-dried. Yield: 0.18 g (78%).

Anal. Calcd. for $C_{19}H_{11}O_{11}PRu_3$: C, 30.46; H, 1.48. Found: C, 30.26; H, 1.23.

Properties

The complex $Ru_3(CO)_{11}(PMe_2Ph)$ is a reasonably air-stable red-orange crystalline solid (mp 104–106 °C), which dissolves readily in organic solvents. The IR spectrum (hexane) shows $v_{(CO)}$ bands at 2096(m), 2044(s), 2028(s),

2016(s), 2000(w,sh) and 1987 (w) cm^{-1}. The ^{1}H NMR spectrum (CDCl$_3$) shows a broad multiplet at δ 7.52 (Ph) and a doublet at δ 1.97 [$J_{(HP)}$ 10 Hz, Me].

C. DECACARBONYL(DIMETHYLPHENYLPHOSPHINE)-(2-ISOCYANO-2-METHYLPROPANE)TRIRUTHENIUM, Ru$_3$(CO)$_{10}$(CN-t-Bu)(PMe$_2$Ph)

$$Ru_3(CO)_{11}(PMe_2Ph) + CN\text{-}t\text{-}Bu \xrightarrow{\text{Na[Ph}_2\text{CO]}}$$

$$Ru_3(CO)_{10}(CN\text{-}t\text{-}Bu)(PMe_2Ph) + CO$$

■ *See cautionary notes, Section B*

Procedure

A dry nitrogen flushed 50-mL Schlenk tube is set up as described in Section B, and charged with 0.10 g (0.133 mmol) Ru$_3$(CO)$_{11}$(PMe$_2$Ph), 10-mL dry, deoxygenated THF, a stirring bar and 0.14 mL of a solution of 2-isocyano-2-methylpropane (CN-t-Bu, t-butyl isocyanide) in THF (1.0 mmol mL^{-1}). The tube is then sealed with a septum cap. A solution of Na[Ph$_2$CO] is added dropwise by syringe as before, $\sim$5 to 10 drops usually being required.

The amount of Na[Ph$_2$CO] solution required depends on how dry the reaction mixture is. When the reaction appears visually to be complete, a small aliquot should be examined (solution IR spectrum) to ensure that the precursor cluster complex has reacted completely [using the 2096 cm^{-1} $\nu_{(CO)}$ band of Ru$_3$(CO)$_{11}$(PMe$_2$Ph)]. Alternatively, the progress of the reaction may be monitored by TLC (thin layer chromatography) (silica gel, petroleum ether eluant).

After the reaction is determined to be complete, solvent is removed (rotary evaporator), and the resulting deep red oil is extracted with two 10-mL aliquots of warm ($\sim$40 °C) petroleum ether (boiling range, 40–60 °C). The extracts are filtered, concentrated to $\sim$5 mL under vacuum, and cooled at -10 °C overnight to give Ru$_3$(CO)$_{10}$(CN-t-Bu)(PMe$_2$Ph) (0.052 g, 42%) as deep red crystals.

Anal. Calcd. for C$_{23}$H$_{20}$NO$_{10}$PRu$_3$: C, 34.33; H, 2.51; N, 1.74. Found: C, 33.82; H, 2.37; N, 1.59.

Properties

The deep red crystals (mp 128–132 °C) of $Ru_3(CO)_{10}(PMe_2Ph)(CN-t-Bu)$ are reasonably air stable, although they darken in color after several days in air. An IR spectrum (hexane) shows a $v_{(CN)}$ band at 2168 cm^{-1}, and $v_{(CO)}$ bands at 2098 (m), 2068 (w), 2046 (m), 2029 (s), 2016 (s), 2004 (w), and 1988 (w) cm^{-1}. The ^1H NMR spectrum ($CDCl_3$) contains resonances at δ 7.42 (multiplet, Ph), 1.88 [$J_{(HP)}$ 9 Hz, PMe] and 1.49 (singlet, CMe_3).

D. DECACARBONYL[METHYLENEBIS(DIPHENYL-PHOSPHINE)]TRIRUTHENIUM, $Ru_3(CO)_{10}[(Ph_2P)_2CH_2]$

$$Ru_3(CO)_{12} + CH_2(PPh_2)_2 \xrightarrow{\text{Na[Ph}_2\text{CO]}} Ru_3(CO)_{10}[(Ph_2P)_2CH_2]$$
$$+ 2CO$$

■ *See cautionary notes, Section B*

Procedure

A dry, nitrogen flushed 100-mL Schlenk tube is charged with $Ru_3(CO)_{12}$ (1.00 g, 1.56 mmol), methylenebis(diphenylphosphine) (0.62 g, 1.61 mmol) and 70 mL of freshly distilled, deoxygenated THF. A magnetic stirring bar is added and the mixture is warmed to about 40 °C to dissolve the $Ru_3(CO)_{12}$. A solution of Na[Ph$_2$CO] is added dropwise as before, carbon monoxide being evolved and the solution rapidly darkening. An aliquot is withdrawn by syringe to ensure that the $Ru_3(CO)_{12}$ is consumed by monitoring the 2061 cm^{-1} $v_{(CO)}$ band in the IR spectrum. The reaction mixture is then transferred into a 100-mL round-bottomed flask and concentrated (rotary evaporator) to $\sim$ 3 mL. Addition of 40 mL of methanol and cooling in a freezer overnight give orange-red crystals, which are collected on a sintered-glass filter and air-dried. Yield: 1.37 g (91%).

Anal. Calcd. for $C_{35}H_{22}O_{10}P_2\dot{R}u_3$: C, 43.44; H, 2.29. Found: C, 43.76; H, 2.05.

Properties

The complex forms orange-red, air-stable crystals (mp 180–181 °C), which readily dissolve in acetone, dichloromethane, chloroform, THF, and benzene. It is sparingly soluble in cyclohexane, petroleum ether, and methanol. The IR spectrum (cyclohexane) shows $v_{(CO)}$ bands at 2086 (m), 2024 (sh), 2018 (s), 2005 (s), 1991 (w), 1968 (m), 1965 (m), and 1947 (w) cm^{-1}. The ^1H NMR spectrum ($CDCl_3$) shows a broad multiplet at δ 7.37 (Ph) and a triplet at δ 4.29 [$J_{(HP)}$, 10.5 Hz; CH_2].

E. [μ-ETHYNEDIYL BIS(DIPHENYLPHOSPHINE)]-BIS[UNDECACARBONYLTRIRUTHENIUM], $[Ru_3(CO)_{11}]_2[\mu\text{-}C_2(PPh_2)_2]$

$$2Ru_3(CO)_{12} + C_2(PPh_2)_2$$

$$\xrightarrow{\text{Na[Ph}_2\text{CO]}} [Ru_3(CO)_{11}]_2[\mu\text{-}C_2(PPh_2)_2] + 2CO$$

■ *See cautionary notes, Section B*

Procedure

A dry, nitrogen flushed 50-mL Schlenk tube is set up as described in Section B, and charged with $Ru_3(CO)_{12}$ (0.15 g, 0.235 mmol), and $C_2(PPh_2)_2$ (0.047 g, 0.12 mmol) and 20 mL of freshly distilled THF. A magnetic stirring bar is added and the mixture is warmed to $\sim 30\,^\circ$C to dissolve the $Ru_3(CO)_{12}$. A solution of Na[Ph$_2$CO] solution is added dropwise from a syringe as in Section B or C, ~ 5 to 10 drops being required. The progress of the reaction is assessed by monitoring the $v_{(CO)}$ band of $Ru_3(CO)_{12}$ at 2061 cm^{-1}. When the reaction is complete the solution is transferred to a 50-mL round-bottomed flask and solvent is removed (rotary evaporator). The residue is dissolved in 1 mL of dichloromethane, and 5 mL of hexane is added. Cooling in a freezer for 24 h gives the complex as an orange powder, which is collected on a sintered-glass filter and air-dried. Yield: 0.135 g (71%).

Anal. Calcd. for $C_{48}H_{20}O_{22}P_2Ru_6$: C, 35.65; H, 1.25. Found: C, 35.56; H, 1.21.

Properties

The complex is obtained as an orange microcrystalline solid [mp $> 150\,^\circ$C (dec)], which is readily soluble in acetone, dichloromethane, chloroform, and THF, but only sparingly soluble in light petroleum ether, cyclohexane, and methanol. The IR spectrum (cyclohexane) shows $v_{(CO)}$ bands at 2102 (m), 2068 (sh), 2052 (vs), 2034 (s), 2021 (vs), 2005 (sh), 1997 (w), 1987 (sh), and 1970 (w) cm^{-1}. A broad multiplet at δ 7.50 (Ph) is the only resonance in the ^{1}H NMR spectrum (CDCl$_3$).

F. UNDECACARBONYLTETRAHYDRIDO[TRIS(4-METHYLPHENYL) PHOSPHITE]TETRARUTHENIUM, $Ru_4H_4(CO)_{11}[P(OC_6H_4Me\text{-}p)_3]$

$$Ru_4H_4(CO)_{12} + P(OC_6H_4Me\text{-}p)_3$$

$$\xrightarrow{\text{Na[Ph}_2\text{CO]}} Ru_4H_4(CO)_{11}[P(OC_6H_4Me\text{-}p)_3] + CO$$

■   *See cautionary notes, Section B*

Procedure

A dry, nitrogen flushed 100-mL Schlenk tube is charged with $Ru_4H_4(CO)_{12}$ (0.21 g, 0.28 mmol), 20 mL of freshly distilled, deoxygenated THF, and 0.3 mL of a solution of tris(4-methylphenyl) phosphite in THF (1.0 mmol mL^{-1}). A stirring bar is added and the mixture is warmed to about 40 °C to dissolve the $Ru_4H_4(CO)_{12}$. A solution of $Na[Ph_2CO]$ is next added as described in Sections B or C. The reaction is monitored by following the disappearance of the 2081 cm^{-1} $v_{(CO)}$ band of the starting cluster complex. An additional 0.4 to 0.5 mL of the $Na[Ph_2CO]$ solution may be required for complete reaction.

   The reaction mixture is transferred to a 50-mL round-bottomed flask and solvent is removed (rotary evaporator). The oily residue is dissolved in 2 mL of benzene and 8 mL of absolute ethanol is added. The mixture is kept in a freezer for 24 h. The bright yellow crystals that form are collected on a glass sinter and vacuum dried (1 torr, 25 °C). Yield: 0.20 g (66%).

Anal. Calcd. for $C_{32}H_{25}O_{14}PRu_4$: C, 35.97; H, 2.35; P, 2.90. Found: C, 36.41; H, 2.31; P, 2.94.

Properties

The golden yellow crystals [mp 130–133 °C(dec)] of $Ru_4H_4(CO)_{11}$-$[P(OC_6H_4Me-p)_3]$ are apparently stable in air indefinitely. A hexane solution shows $v_{(CO)}$ bands at 2098 (m), 2071 (s), 2061 (s), 2037 (s), 2028 (w), 2016 (s), 2001 (w), and 1983 (w) cm^{-1}. The ^1H NMR spectrum $[(CD_3)_2CO]$ has peaks at δ 7.87 (multiplet, C_6H_4) and 2.33 (multiplet, CH_3), with the metal hydride resonance at δ − 17.7 [d, $J_{(HP)}$ 7 Hz].

G.   DECACARBONYL(DIMETHYLPHENYLPHOSPHINE)-TETRAHYDRIDO[TRIS(4-METHYLPHENYL)-PHOSPHITE]TETRARUTHENIUM, $Ru_4H_4(CO)_{10}(PMe_2Ph)[P(OC_6H_4Me-p)_3]$

$Ru_4H_4(CO)_{11}[P(OC_6H_4Me-p)_3] + PMe_2Ph$

$\xrightarrow{\text{Na[Ph}_2\text{CO]}} Ru_4H_4(CO)_{10}(PMe_2Ph)[P(OC_6H_4Me-p)_3] + CO$

■   *See cautionary notes, Section B*

Procedure

A dry, nitrogen flushed 100-mL Schlenk tube is loaded with $Ru_4H_4(CO)_{11}[P(OC_6H_4Me\text{-}p)_3]$ (0.18 g, 0.17 mmol), 0.17 mL of a solution of PMe_2Ph in THF (1.0 mmol mL^{-1}), and 8 mL of dry, deoxygenated THF as in Sections B or C. A solution of $Na[Ph_2CO]$ is next added dropwise from a syringe through the septum cap, again as described in Sections B or C. Typically, 5 to 10 drops are sufficient for complete reaction as evidenced by the disappearance of the 2097 cm^{-1} $v_{(CO)}$ band of the precursor cluster. The reaction mixture is evaporated under vacuum and the residue is dissolved in 1 mL of benzene. Addition of 3 mL of petroleum ether (boiling range 40–60 °C), and cooling in a freezer overnight gives the product as a yellow-brown powder (0.14 g, 64%), which is collected on a glass sinter and air-dried.

Anal. Calcd. for $C_{39}H_{36}O_{13}P_2Ru_4$: C, 39.74; H, 3.08. Found: C, 40.53; H, 2.96.

Properties

The compound $Ru_4H_4(CO)_{10}(PMe_2Ph)[P(OC_6H_4Me\text{-}p)_3]$ is a yellow-brown powder (mp 136–137 °C), which is stable in air and very soluble in organic solvents. The IR spectrum (hexane) has $v_{(CO)}$ bands at 2079 (m), 2059 (vs), 2043 (w), 2029 (s), 2017 (s), 2002 (m), 1998 (sh), and 1956 (w) cm^{-1}. The ^{1}H NMR spectrum contains singlet resonances at δ 7.43 (Ph), 7.11 (C_6H_4), and 2.30 (C_6H_4Me), a doublet at δ 1.87 [$J_{(HP)}$ 9 Hz, PMe], and a broad, poorly resolved triplet at δ −19 [$J_{(HP)}$ 10 Hz, RuH].

References

1. D. J. Darensbourg and M. J. Incorvia, *Inorg. Chem.*, **19**, 2585 (1980); D. Sonnenberger and J. D. Atwood, *Inorg. Chem.*, **20**, 3243 (1981).

2. M. I. Bruce, G. Shaw, and F. G. A. Stone, *J. Chem. Soc. Dalton Trans.*, **1972**, 2094, and references cited therein.

3. S. A. R. Knox and H. D. Kaesz, *J. Am. Chem. Soc.*, **93**, 4594 (1971).

4. C. W. Bradford, R. S. Nyholm, G. J. Gainsford, J. M. Guss, P. R. Ireland, and R. Mason, *J. Chem. Soc. Chem. Commun.*, **1972**, 87; G. J. Gainsford, J. M. Guss, P. R. Ireland, R. Mason, C. W. Bradford, and R. S. Nyholm, *J. Organomet. Chem.*, **40**, C70 (1972).

5. M. Arewgoda, B. H. Robinson, and J. Simpson, *J. Am. Chem. Soc.*, **105**, 1893 (1983).

6. M. I. Bruce, J. G. Matisons, and B. K. Nicholson, *J. Organomet. Chem.*, **247**, 321 (1983).

7. M. I. Bruce, J. G. Matisons, B. K. Nicholson, and M. L. Williams, *J. Organomet. Chem.*, **236**, C57 (1982).

8. M. I. Bruce, B. K. Nicholson, J. M. Patrick, and A. H. White, *J. Organomet. Chem.*, **254**, 361 (1983).

9. M. I. Bruce, T. W. Hambley, B. K. Nicholson, and M. R. Snow, *J. Organomet. Chem.*, **235**, 83 (1982).

10. M. I. Bruce, M. L. Williams, J. M. Patrick and A. H. White, *J. Chem. Soc. Dalton Trans.*, **1985**, 1229; M. I. Bruce, M. L. Williams, J. M. Patrick, B. W. Skelton and A. H. White, *J. Chem. Soc. Dalton Trans.*, **1986**, 2557.

11. M. I. Bruce, B. K. Nicholson, and M. L. Williams, *J. Organomet. Chem.*, **243**, 69 (1983).

12. G. Lavigne and J.-J. Bonnet, *Inorg. Chem.*, **20**, 2713 (1981).

13. M. I. Bruce, D. C. Kehoe, J. G. Matisons, B. K. Nicholson, P. H. Rieger, and M. L. Williams, *J. Chem. Soc. Chem. Commun.*, **1982**, 442.

14. G. Lavigne and H. D. Kaesz. *J. Am. Chem. Soc.*, **106**, 4647 (1984).

15. B. F. G. Johnson and J. Lewis, *Inorg. Synth.*, **13**, 92 (1972); B. R. James, G. L. Rempel, and W. K. Teo, *Inorg. Synth.*, **16**, 45 (1976); A. Mantovani and S. Cenini, *Inorg. Synth.*, **16**, 47 (1976).

16. (a) M. I. Bruce, J. G. Matisons, R. C. Wallis, J. M. Patrick, B. W. Skelton, and A. H. White, *J. Chem. Soc. Dalton Trans.*, **1983**, 2365; (b) M. I. Bruce, N. L. Jones, and C. M. Jensen, *Inorg. Synth.*, **26**, 259 (1989)

17. M. I. Bruce and M. L. Williams, *Inorg. Synth.*, **26**, 262 (1989)

49. CARBIDO-CARBONYL RUTHENIUM CLUSTER COMPLEXES

Submitted by J. N. NICHOLLS* and M. D. VARGAS*†
Checked by J. HRILJAC‡ and MICHAEL SAILOR‡

Interest in the high nuclearity carbonyl clusters of ruthenium has centered around the hexanuclear carbido-cluster $Ru_6C(CO)_{17}$ and its derivatives, $Ru_5C(CO)_{15}$ and, more recently, $[Ru_{10}C_2(CO)_{24}]^{2-}$.[1] The chemistry of both $Ru_6C(CO)_{17}$ and $Ru_5C(CO)_{15}$ has been investigated in some detail. Their syntheses are described below.

The cluster $Ru_6C(CO)_{17}$ was first prepared by pyrolysis of $Ru_3(CO)_{12}$ in nonane, a route which gave, at best, 19% yields.[2] A later preparation, of which a modified version appears here, involves pyrolysis of $Ru_3(CO)_{12}$ in hydrocarbon solvents under a moderate pressure of ethylene; this gives much higher yields.[3] The pentanuclear cluster $Ru_5C(CO)_{15}$ is prepared in > 90% yield by the carbonylation of $Ru_6C(CO)_{17}$.[4]

Although $Ru_6C(CO)_{17}$ and $Ru_5C(CO)_{15}$ are useful starting materials for reactions involving the addition of nucleophilic reagents to clusters, the dianion $[Ru_5C(CO)_{14}]^{2-}$ is useful in that it will react with electrophilic reagents. The reduction of $Ru_5C(CO)_{15}$ to $[Ru_5C(CO)_{14}]^{2-}$ is rather sensitive

*University Chemical Laboratory, Lensfield Road, Cambridge, CB2 1EW, United Kingdom.
†Address correspondence to Professor M. D. Vargas, Universidade Estadual de Campinas, Instituto de Quimica, Caixa Postal 6154, 13100, Campinas, S. Paulo, Brasil.
‡Department of Chemistry, Northwestern University, Evanston, IL 60201.

to the conditions and handling employed.[5] We therefore also describe it in detail here.

General Remarks

Preparations in Sections A and B involve the use of a 100-mL capacity magnetically stirred autoclave, fitted with a glass liner tube. For both reactions, it is crucial that the autoclave head and base should be completely free of chlorinated solvents and any other traces of halide or chalcogenide, which will reduce yields drastically. The authors keep a special autoclave for this preparation, which is never used for reactions involving halides or chalcogenides. The checkers cleaned their autoclave by soaking it in a 25% solution of nitric acid for 8 h followed by a soaking in distilled water for a day, periodically changing the water.

The heptane solvent should be dried over sodium wire and a few milliliters of heptane should be added to the autoclave outside the liner tube to improve thermal contact between the autoclave and the tube. Triruthenium dodecacarbonyl is available commercially (Alpha and Strem) or may be made from hydrated ruthenium trichloride by well-known procedures.[6]

■ **Caution.** *Metal carbonyls and carbon monoxide are highly toxic and should be handled in a well-ventilated hood. Care should also be taken to provide adequate shielding of all equipment used under pressure.*

A. μ_6-CARBIDO-HEPTADECACARBONYLHEXARUTHENIUM

$$Ru_3(CO)_{12} \xrightarrow[165\,°C]{C_2H_4\,30\,atm} Ru_6C(CO)_{17}$$

Procedure

Heptane (50 mL) is added to a 100-mL autoclave liner tube containing $Ru_3(CO)_{12}$ (400 mg, 0.626 mmol) and a Teflon-coated stirring bar. The autoclave is assembled and pressurized twice to 30 atm of ethylene. Then, the ethylene is vented to atmospheric pressure and the autoclave is finally repressurized with ethylene to 30 atm and sealed. The pressure will drop to 15 to 20 atm on stirring. The autoclave is then heated to 165 °C over a period of 0.75 h, with rapid stirring, and maintained at 165 °C for an additional 4.25 h (thermocouple measuring temperature at the interface between autoclave and heating mantle).

Note. Inefficient stirring will drastically reduce yields due to the low solubility of $Ru_3(CO)_{12}$.

The autoclave is allowed to cool overnight in the heating mantle, to ensure slow crystallization of the product. The pressure is then released, the liner tube is removed from the autoclave, and the yellow solution is decanted immediately and discarded. The checkers observed a red-orange solution at this point, which could be due to the presence of $Ru_6(ethylene)_n$ derivatives if the prescribed reaction time is exceeded.

The purple crystalline solid is rinsed in the liner, with three 10-mL portions of dry pentane, the washings being decanted and discarded. The $Ru_6C(CO)_{17}$ may then be removed from the liner as a pentane slurry, from which the liquid is decanted. The product is dried under vacuum. Yield: 214 mg (62%).

Anal. Calcd. for $Ru_6C_{18}O_{17}$; C, 19.74%. Found: C, 19.90%.

The product usually requires no further purification. Poor stirring or insufficient reaction time may lead to crystallization of unreacted $Ru_3(CO)_{12}$ whose orange crystals are best removed by hand sorting. Prolonged reaction of $Ru_6C(CO)_{17}$ with ethylene yields ethylene derivatives of the hexanuclear cluster, again reducing yields. These derivatives are removed in the yellow heptane solution. Purity of $Ru_6C(CO)_{17}$ may be checked by infrared (IR) and thin layer chromatography (TLC) using the following data.

Properties

Data for the desired product and its principal impurity, $Ru_3(CO)_{12}$, are given in Table I.

Over long periods, the hexanuclear cluster is light sensitive and it is best stored in the cold in the absence of light. It is soluble in polar solvents and slightly soluble in hydrocarbons, giving air-stable solutions.

It undergoes a range of reactions among which are reduction with methanolic KOH[3] to give $[Ru_6C(CO)_{16}]^{2-}$, substitution (for example, by PR_3)[7] to yield $Ru_6C(CO)_{17-n}(PR_3)_n$ ($n = 1$–4), and degradation by halogen atoms, X_2 (X = Cl, Br, I) and CO to the dinuclear species $Ru_2(CO)_6X_2$ and $Ru_5C(CO)_{15}$, respectively.[8]

TABLE I.   Carbonyl Absorptions and TLC Data.

Compound	IR Frequencies $\nu_{(CO)}$ (cm^{-1})(CH$_2$Cl$_2$)	R_f (20% CHCl$_3$–hexane)
$Ru_6C(CO)_{17}$	2066 (s), 2045 (s, br)	0.5
$Ru_3(CO)_{12}$	2059 (s), 2027 (m); 2007 (w);	0.75

B. μ_5-CARBIDO-PENTADECACARBONYLPENTARUTHENIUM

$$Ru_6C(CO)_{17} \longrightarrow Ru_5C(CO)_{15} + Ru(CO)_5$$

■ **Caution.** *Carbon monoxide must be carefully released in a well-ventilated hood, as indicated previously.*

Up to 250 mg of $Ru_6C(CO)_{17}$ can be converted to $Ru_5C(CO)_{15}$ under the following conditions.

Heptane (50 mL) and a Teflon-coated stirring bar are added to the autoclave liner containing the $Ru_6C(CO)_{17}$ (214 mg, 0.196 mmol) produced in Section A. The autoclave is assembled and pressurized twice to 30 atm with carbon monoxide. The gas is then vented to atmospheric pressure and repressurized with 80 atm of carbon monoxide and sealed. The autoclave is then heated to 90 °C with stirring over a period of 0.3 h and maintained at 90 °C for 3.7 additional hours. Again, efficient stirring is crucial for complete conversion of the hexanuclear cluster to the pentanuclear cluster.

The work-up may be performed as soon as the autoclave has cooled sufficiently to allow handling. The autoclave is vented to atmospheric pressure and the red solution is transferred to a 100-mL round-bottomed flask *in a well-ventilated fume hood.* The flask is wrapped in aluminum foil, and with magnetic stirring, the solution is evaporated to dryness on a vacuum line at ambient temperature.

■ **Caution.** *The Ru(CO)_5, collected in the trap along with the heptane solvent is volatile and highly toxic. The exhaust from the pump should therefore be released into a well-ventilated fume hood.*

When evaporation of the reaction solution is completed the trap should be removed to a well-ventilated fume hood. The $Ru(CO)_5$ in the trap is allowed to convert to $Ru_3(CO)_{12}$ under ambient illumination (the process is assisted by light).

Purification of the product is effected by extraction of the red solid with successive portions of boiling hexane (total volume ~ 80 mL) until all the solid is dissolved, or, in the presence of $Ru_3(CO)_{12}$ impurity, until the solution starts to appear orange signifying the presence of significant amounts of $Ru_3(CO)_{12}$. The extracts are filtered and combined into a single flask. The combined liquid volume at this point should not exceed 80 mL. The solution must be kept overnight in a -25 °C freezer to allow crystallization. The temperature is critical in obtaining the stated yields.

The red crystals of $Ru_5C(CO)_{15}$ are collected by filtration. Yield: 170 mg, 93%. A second crop of $Ru_5C(CO)_{15}$ may be obtained by reducing the volume of the mother liquor and cooling further. Purity is assessed by IR $\nu_{(CO)}$ (see properties section in this section). Impurity due to unreacted $Ru_6C(CO)_{17}$ is removed by recarbonylation under the same conditions as before. Impurity due to $Ru_3(CO)_{12}$ is dealt with by recrystallization from hexane.

Anal. Calcd. for $Ru_5C_{16}O_{15}$: C, 20.49%. Found: C, 20.68%.

Properties

Absorptions in the carbonyl stretching region of the IR for $Ru_5C(CO)_{15}$ are (hexane solution, cm^{-1}): 2067 (s), 2034 (m), 2015 (w). Absorptions of its principal impurities $Ru_6C(CO)_{17}$ and $Ru_3(CO)_{12}$ are given in Table I.

Crystalline $Ru_5C(CO)_{15}$ is somewhat light sensitive over periods of several weeks, and should be stored in the cold, in the absence of light. The compound may be handled in air. It is soluble in hydrocarbon and chlorinated solvents to give red solutions. When using THF as a solvent, care should be taken to use dry distilled solvent and to perform the operations under an inert atmosphere.

The compound reacts with a variety of nucleophiles to give addition or substitution products. Thus, yellow-orange adducts of the formula $Ru_5C(CO)_{15}L$ are formed upon reaction with L = MeOH, MeCN, or CO.[4] With PR_3, a range of substitution products, $Ru_5C(CO)_{15-n}(PR_3)_n$ ($n = 1$–4), is formed.[4] The compound $Ru_5C(CO)_{15}$ also reacts with halogens, pseudohalogens, or with HX and $AuPR_3X$ (X = Cl, Br, I, SR)[4,7,8] to give addition and substitution products, $HRu_5C(CO)_{15-n}X_n$ and $Au(PR_3)Ru_5C(CO)_{15-n}X_n$ ($n = 0, 1, 2$).

C. DISODIUM OR BIS[μ-NITRIDO-BIS(TRIPHENYL-PHOSPHORUS)(1 +)] SALTS OF [μ$_5$-CARBIDO-TETRADECA-CARBONYLPENTARUTHENATE (2 −)]

$$Ru_5C(CO)_{15} \xrightarrow[\text{MeOH}]{Na_2CO_3} [Na]_2[Ru_5C(CO)_{14}]$$

$$Na_2[Ru_5C(CO)_{14}] + 2[(Ph_3P)_2N][Cl]$$

$$\longrightarrow [(Ph_3P)_2N]_2[Ru_5C(CO)_{14}] + 2NaCl$$

Procedure

A suspension of sodium carbonate (200 mg, 19 mmol) in freshly distilled methanol (60 mL) is degassed by bubbling nitrogen through it, with stirring, for ~ 15 min. Solid $Ru_5C(CO)_{15}$ (170 mg, 0.18 mmol) is added and the suspension is stirred under nitrogen until all of the red solid has dissolved (~ 30 min), giving a bright orange solution.

Solutions of the sodium salt of $[Ru_5(CO)_{14}]^{2-}$ are air sensitive. Under an atmosphere of nitrogen, the reaction solution is filtered rapidly through a sinter containing Celite® into a 100-mL round-bottomed flask containing [PPN]Cl, [(μ-nitrido-bis(triphenylphosphorus)(1 +)] chloride, (Ventron, 115 mg, 0.2 mmol). The volume of solvent is reduced to about 5 mL by a fast stream of nitrogen, which will also result in cooling of the solution. The orange crystals are collected by filtration and washed with 1 mL of cold methanol (-25 °C) before drying under vacuum. Yield: 166 mg, 63%. Purity can be monitored by IR ν_{CO}, see properties, below.

Anal. Calcd. for $Ru_5C_{51}H_{15}O_{14}P_2$: C, 52.59; H, 3.02; N, 1.41%. Found: C, 52.35; H, 3.05; N, 1.35%.

Properties

The sodium salt of the dianion is soluble in a wide range of polar solvents to give air-sensitive solutions, which must be handled in inert atmosphere. IR (dichloromethane solution, cm^{-1}): 2033 (w), 1975 (s), 1965 (sh), 1919 (w), 1753 (w). The [PPN]$^+$ salt of the dianion reacts with $Au(PMe_2Ph)ClO_4$ to give the neutral compound $Ru_5C(CO)_{14}(AuPMe_2Ph)_2$.[9]

■ **Caution.** *Perchlorate complexes are extremely hazardous. Use of the salts of [Au(PMe$_2$Ph)]$^+$ with other weakly coordinating anions such as [PF$_6$]$^-$, [BF$_4$]$^-$, or [CF$_3$SO$_3$]$^-$ is advised.*

Crystalline samples of [PPN]$_2$[Ru$_5$(CO)$_{14}$] may be handled in air but are best stored under nitrogen in a freezer.

References

1. C. T. Hayward, J. R. Shapley, M. R. Churchill, C. Bueno, and A. L. Rheingold, *J. Am. Chem. Soc.*, **104**, 7347 (1982).

2. B. F. G. Johnson, R. D. Johnson, and J. Lewis, *J. Chem. Soc. A*, **1968**, 2865.

3. B. F. G. Johnson, J. Lewis, S. W. Sankey, K. Wong, M. McPartlin, and W. J. H. Nelson, *J. Organomet. Chem.*, **191**, C3 (1980).

4. B. F. G. Johnson, J. Lewis, J. N. Nicholls, J. Puga, P. R. Raithby, M. J. Rosales, M. McPartlin, and W. Clegg, *J. Chem. Soc. Dalton Trans.*, **1983**, 787.

5. B. F. G. Johnson, J. Lewis, W. J. H. Nelson, J. N. Nicholls, J. Puga, P. R. Raithby, M. J. Rosales, M. Schroder, and M. D. Vargas, *J. Chem. Soc. Dalton Trans.*, **1983**, 2447.

6. (a) A. Mantovani and S. Cenini, *Inorg. Synth.*, **16**, 47 (1976); (b) M. I. Bruce N. L. Jones, and C. M. Jensen, *Inorg. Synth.*, **26**, 259 (1989).

7. B. F. G. Johnson, J. Lewis, J. N. Nicholls, J. Puga, and K. H. Whitmire, *J. Chem. Soc. Dalton Trans.*, **1983**, 787.

8. A. G. Cowie, B. F. G. Johnson, J. Lewis, J. N. Nicholls, P. R. Raithby, and M. J. Rosales, *J. Chem. Soc. Dalton Trans.*, **1983**, 2311.

9. A. G. Cowie, B. F. G. Johnson, J. Lewis, and P. R. Raithby, *J. Chem. Soc. Chem. Commun.*, **1984**, 1710.

50. NITRIDO-RUTHENIUM CLUSTER COMPLEXES*

Submitted by MARGARET L. BLOHM† and WAYNE L. GLADFELTER†
Checked by A. G. COWIE‡

The inherent difficulty in understanding mechanisms of heterogeneous catalysis has created recent interest in low-valent transition metal clusters. These homogeneous systems provide molecular models for the interaction of metal surfaces with atoms and small molecules such as N, C, NO, and CO. For this reason the synthesis of clusters with interstitial carbon or nitrogen atoms has been actively pursued. Previously, the only source of the nitrogen atoms for nitrido clusters was nitric oxide or nitrosyl salts. These syntheses, however, often proceed in low yield, since the high reactivity and oxidizing properties of the nitrosyl cation tend to produce several undesired products. We have recently reported an alternative method for the preparation of nitrido clusters that selectively forms $[Ru_6N(CO)_{16}]^-$ in high yield.[1,2] The reaction of $PPN[N_3]$ with $Ru_3(CO)_{12}$ generates the isocyanates, $PPN[Ru_3(NCO)(CO)_{11}]$ and $PPN[Ru_3(NCO)(CO)_{10}]$, which when pyrolyzed in the presence of excess $Ru_3(CO)_{12}$, cleanly form $PPN[Ru_6N(CO)_{16}]$ in high yield.

General procedure

■ **Caution.** *Reactions with metal carbonyls requiring the use of highly toxic CO or of its evolution from reaction mixtures should be performed in a well-ventilated hood, as also the use of benzene as a solvent due to its toxic nature.*

Tetrahydrofuran (THF) and diethyl ether are freshly distilled from sodium benzophenone ketyl under nitrogen. Hexane is freshly distilled from sodium metal under nitrogen. Triruthenium dodecacarbonyl is available commercially or it can be synthesized from ruthenium trichloride trihydrate (Alfa) using the procedure of Mantovani and Cenini,[3a] or alternately, by the procedure given in this volume.[3b] The [μ-nitrido-bis(triphenylphosphorus)(1 +)] chloride can be purchased from Strem or synthesized according to the literature procedure[4]

A. μ-NITRIDO-BIS(TRIPHENYLPHOSPHORUS)(1 +) AZIDE, PPN[N₃]

$$PPN[Cl] + NaN_3 \longrightarrow PPN[N_3] + NaCl$$

*PPN = μ-nitrido-bis(triphenylphosphorous)(+ 1), [(Ph₃P)₂N].⁺
†Department of Chemistry, University of Minnesota, Minneapolis, MN 55455.
‡Department of Chemistry, University Chemical Laboratory, Lensfield Rd., Cambridge CB2 1EW, United Kingdom.

After 20 mL of deionized water is heated to 40 °C on a hot plate, 0.5 g (0.93 mmol) of PPN[Cl] is added, and stirred until all of the solid has dissolved. A solution of 0.12 g of NaN_3 (1.86 mmol) in 5 mL of deionized water is added dropwise to the rapidly stirring aqueous PPN[Cl] solution, immediately precipitating PPN[N_3]. After addition is complete, the solution is removed from the hot plate, cooled to room temperature, and filtered. The white solid is then redissolved in 75 mL of hot water (75 °C). Excess NaN_3 (241.4 mg, 3.714 mmol) dissolved in 5 mL of water is added dropwise to the warm aqueous PPN[N_3] solution, resulting in the immediate precipitation of white crystals. The suspension is cooled very slowly to room temperature; rapid cooling will lead to formation of a yellow oil and discoloration of the final product. After the white crystals are isolated by filtration, they are dissolved in 20 mL of acetone; the solution is filtered to remove any remaining sodium salts, and the filtrate is dried over magnesium sulfate. After 30 min, the drying agent is removed by filtration and the volume of the acetone filtrate reduced to 10 mL on a warm hot plate. The addition of 45 mL of diethyl ether results in the crystallization of white PPN[N_3]. After cooling the reaction solution in an ice bath, 0.31 g of PPN[N_3] are isolated by filtration (62% from PPN[Cl]).[5]

Properties

The compound PPN[N_3] contains some acetone of crystallization. It is an air stable, white crystalline solid (mp 195–196 °C). A chloride analysis of $< 0.2\%$ indicates metathesis from the chloride to the azide salt is complete. It is very soluble in dichloromethane, acetone, and acetonitrile, and only slightly soluble in refluxing THF or hot water. It is insoluble in hydrocarbon solvents and diethyl ether.

B. μ-NITRIDO-BIS(TRIPHENYLPHOSPHORUS)(1 +) HEXADECACARBONYLNITRIDOHEXARUTHENATE(1 −), PPN[$Ru_5N(CO)_{16}$]

$$PPN[N_3] + 2Ru_3(CO)_{12} \xrightarrow{\text{THF}} PPN[Ru_6N(CO)_{16}]$$

A 50-mL three-necked round-bottomed flask is assembled with a serum cap, a gas inlet tube, and a condenser equipped with a gas inlet tube, and is charged with 100.0 mg (0.156 mmol) of $Ru_3(CO)_{12}$, 45.4 mg (0.0782 mmol) of PPN[N_3], and a magnetic stirring bar. The reaction vessel is evacuated and filled with nitrogen three times to remove oxygen and a slow, continuous purge of nitrogen is begun. The solvent (15 mL of THF) is added by syringe. Evolution of gases occurs immediately and the solution becomes deep red. The reaction mixture is stirred at room temperature for 10 min, after which time the flask is lowered into an oil bath at 80 °C. The solution becomes red-brown after 15 min, and reflux is continued for 12 h. The oil bath is lowered and, while the solution is still warm, the solvent is distilled under vacuum into a cold trap cooled with liquid nitrogen. The resulting red-brown oil is extracted with six 10-mL portions of diethyl ether. The deep red, combined extracts are filtered and then reduced in volume to 10 mL, to which 15 mL of hexane is then added. Rapid mixing, followed by slow removal of diethyl ether under vacuum leads to the formation of brick red microcrystals that are isolated by filtration. The product can be further purified by recrystallization from diethyl ether–hexane under nitrogen.

Anal. Calcd. for $[(C_6H_5)_3P)_2N][Ru_6N(CO)_{16}]$: C, 38.86; H, 1.88; N, 1.74. Found: C, 38.68; H, 2.00; N, 1.63. Yield of $PPN[Ru_6N(CO)_{16}]$: 103 mg (82%).

Properties

The compound $PPN[Ru_6N(CO)_{16}]$ is obtained as brick red crystals. The solid is stable in air for several hours, but it should be stored in an inert atmosphere. It is unstable in solution over extended periods of time, even under an inert atmosphere, decomposing to form $PPN[Ru_5N(CO)_{14}]$. The crystals are soluble in THF, diethylether, dichloromethane, and acetone and, insoluble in hexane and water. The IR spectrum of a THF solution contains the following carbonyl stretching absorptions: 2010 (vs), 1965 (w sh), 1839 (w) cm^{-1}.

C. μ-NITRIDO-BIS(TRIPHENYLPHOSPHORUS)(1 +)-TETRADECACARBONYLNITRIDOPENTARUTHENATE(1 −), PPN[Ru$_5$N(CO)$_{14}$]

$$PPN[Ru_6N(CO)_{16}] + 2CO$$

$$\xrightarrow{\text{THF}} PPN[Ru_5N(CO)_{14}] + \tfrac{1}{3}Ru_3(CO)_{12}$$

■ **Caution.** *Procedures requiring the use or the evolution of highly toxic CO should be carried out in a well-ventilated hood.*

A 20-mL Schlenk tube equipped with a magnetic stirring bar is charged with 50.0 mg (0.0311 mmol) of $PPN[Ru_6N(CO)_{16}]$. The reaction flask is evacuated and filled with nitrogen, and 10 mL of THF is added by syringe. Carbon monoxide is gently bubbled through the deep red solution for 15 m, while the color gradually lightens to a bright orange-red. The solvent is removed *in vacuo*, and the resulting orange-red oil is washed with three 10-mL portions of hexane, decanting off the hexane solutions of the $Ru_3(CO)_{12}$ by-product.

The remaining oil is extracted with two 10 mL portions of diethyl ether. The combined extracts are filtered, reduced in volume to 5 mL, and 5 mL of hexane is then added under rapid stirring to form orange-red crystals of the product.

Anal. Calcd. for $[(C_6H_5)_3P)_2N][Ru_5N(CO)_{14}]$: C, 41.41; H, 2.08; N, 1.93. Found: C, 41.38; H, 2.17; N, 1.91. Yield: 42.8 mg (95%).

The compound $PPN[Ru_5N(CO)_{14}]$ is obtained as orange-red crystals. The solid is stable in air for short periods, but it should be stored under an inert atmosphere. It is soluble in diethyl ether, THF, acetone, and dichloromethane, and insoluble in hexane and water. The IR spectrum of a THF solution exhibits the following carbonyl stretching absorptions: 2060 (w), 2013 (vs), 1999 (s), 1960 (m), 1820 (m) cm^{-1}. The square pyramidal geometry of ruthenium atoms has been established by X-ray crystallography.[2]

References

1. M. L. Blohm, D. E. Fjare, and W. L. Gladfelter, *Inorg. Chem.*, **22**, 1004 (1983).
2. M. L. Blohm and W. L. Gladfelter, *Organometallics*, **4**, 45 (1985).
3. (a) A. Montovani and S. Cenini, *Inorg. Synth.*, **16**, 47 (1976); (b) M. I. Bruce, N. L. Jones, and C. L. Jensen, *Inorg. Synth.* **26**, 259 (1989).
4. J. K. Ruff and W. J. Schlientz, *Inorg. Synth.*, **15**, 84 (1974).
5. Solvent volumes suggested by checker.

51. SOME USEFUL DERIVATIVES OF DODECACARBONYLTRIOSMIUM

Submitted by J. N. NICHOLLS* and M. D. VARGAS*[†]
Checked by A. J. DEEMING[†] and S. E. KABIR[‡]

The preparation of substituted derivatives of $Os_3(CO)_{12}$ is hampered by the reluctance of this cluster complex to undergo replacement of carbonyl ligands by other donor ligands.[1] There is, however, an alternative route to such species via the derivatives $Os_3(CO)_{11}(CH_3CN)$, **1**, and $Os_3(CO)_{10}(CH_3CN)_2$, **2**, the

*University Chemical Laboratory, Lensfield Road, Cambridge, CB2 1EW, United Kingdom.
[†]Address correspondence to Professor M. D. Vargas, Universidade Estadual de Campinas, Instituto de Quimica, Caixa Postal 6154, 13100, Campinas, S. Paulo, Brasil.
[‡]Department of Chemistry, University College, London, WC1H OAJ, United Kingdom.

acetonitrile ligands of which are readily displaced by other nucleophiles such as PR_3, C_5H_5N, HX (X = Cl, Br, I), C_2H_4, and so on.[2,3] Much milder conditions are needed to replace the acetonitrile ligands in the cluster complexes **1** or **2** than the CO ligands in $Os_3(CO)_{12}$. Simple substitution products such as $Os_3(CO)_{11}(C_2H_4)$, and $Os_3(CO)_{11}(C_5H_5N)$, for instance, can thus be prepared only from the monoacetonitrile derivative, **1**. By contrast, the direct reaction of $Os_3(CO)_{12}$ with C_2H_4 or C_5H_5N proceeds to give $HOs_3(CO)_{10}(C_2H_3)$ or $HOs_3(CO)_{10}(C_5H_4N)$, respectively.[2]

The preparation of complexes **1** or **2** involves the reaction of $Os_3(CO)_{12}$ in acetonitrile with trimethylamine oxide, which removes CO ligands as CO_2; the vacant site(s) on the cluster are then filled by acetonitrile rather than the resulting trimethylamine, which is a poorer coordinating ligand. Complexes **1** or **2** are easily isolated in high yields. Their syntheses are reported in detail here, together with the synthesis of a typical derivative $Os_3(CO)_{11}(C_5H_5N)$.

A.   (ACETONITRILE)UNDECACARBONYLTRIOSMIUM

$$Os_3(CO)_{12} + Me_3NO \xrightarrow{CH_3CN} Os_3(CO)_{11}CH_3CN + Me_3N + CO_2$$

■ **Caution 1.**   *The metal carbonyls described in this synthesis are toxic by absorption through the skin and should therefore be handled with care. All procedures should be carried out in a well-ventilated fume hood.*

■ **Caution 2.**   *Dry trimethylamine oxide can be explosive; it should be handled and stored only in small quantities.*

Trimethylamine oxide (Aldrich) must be freshly sublimed for the procedure that follows. Typically, 2 g of trimethylamine oxide is sublimed in a cold finger apparatus (60 °C, 0.05 torr, 16 h). Dry trimethylamine oxide is very hygroscopic. It is important to handle it in a dry atmosphere using suitably dried solvents.

In a 1-L three-necked round-bottomed flask and under an atmosphere of nitrogen, a suspension of $Os_3(CO)_{12}$ (0.5 g, 0.55 mmol)[4] is prepared in acetonitrile (750 mL), freshly distilled from a slurry of phosphorus pentoxide under nitrogen. To ensure dissolution of the maximum amount of $Os_3(CO)_{12}$, the suspension is heated under reflux for 1 h. The solution is then allowed to cool to room temperature, and under a nitrogen atmosphere, a solution of freshly sublimed, dry trimethylamine oxide (42 mg, 0.55 mmol) in acetonitrile (200 mL) is added to the stirred suspension over a period of 4 h using a pressure-equalized dropping funnel.

After being stirred for a total of 12 h, the mixture is filtered through silica to remove any excess amine oxide. The solvent is then removed at room temperature and the yellow solid is taken up in dichloromethane (120 mL) from which it is crystallized at 0 °C. Yield: 406 mg (80%). Checkers noted that the product may alternatively be recrystallized from acetonitrile (15 mL) at 0 °C.

Properties

The air-stable yellow product $Os_3(CO)_{11}(CH_3CN)$ is soluble in a variety of organic solvents apart from hydrocarbons in which it is only sparingly soluble. Its purity can be checked by IR spectrum: (CH_2Cl_2 solution) $v_{(CO)}$, 2107 (w), 2054 (vs), 2040 (vs), 2017 (s, sh), 2008 (vs), 1981 (m) cm^{-1}.

It seems to be virtually impossible to obtain 100% pure $Os_3(CO)_{11}(MeCN)$ via this route. We generally prefer to use exactly one equivalent of amine oxide in the preparation. Although traces of $Os_3(CO)_{12}$ may then be present in the product, this compound is much less reactive and therefore less likely to interfere with subsequent reactions than $Os_3(CO)_{10}(MeCN)_2$, the impurity found if excess amine oxide is used. Possible impurities are $v_{(CO)}$ in CH_2Cl_2, cm^{-1}: $Os_3(CO)_{12}$: 2066 (s), 2032 (s), 2011 (w), 1997 (w); $Os_3(CO)_{10}(CH_3CN)_2$: 2079 (w), 2025 (s, sh), 2021 (vs), 1983 (m), 1960 (w).

Anal. Calcd. for $C_{13}H_3NO_{11}Os_3$: C, 16.95; H, 0.35; N, 1.50. Found: C, 17.2; H, 0.5; N, 1.2.

A typical acetonitrile replacement reaction on $Os_3(CO)_{11}(CH_3CN)$ is that with pyridine, described in Section B.

B.   UNDECACARBONYL(PYRIDINE)TRIOSMIUM

$$Os_3(CO)_{11}(CH_3CN) + C_5H_5N$$

$$\xrightarrow{\text{cyclohexane}} Os_3(CO)_{11}(C_5H_5N) + CH_3CN$$

Procedure

■   *See cautionary note (1) in Section A. Pyridine must be handled in a well-ventilated fume hood.*

In a 1.5-L three-necked round-bottomed flask and under a nitrogen atmosphere, a quantity of $Os_3(CO)_{11}(CH_3CN)$ (700 mg, 0.76 mmol) is dissolved in degassed cyclohexane (1 L). Excess pyridine (0.5 mL, 6.2 mmol) is then added. The solution is stirred at 60 °C for 15 min, after which time the reaction is complete. The progress of the reaction should be monitored by IR spectroscopy so that heating is not continued after completion. This could lead to orthometallation of the pyridine derivative to $HOs_3(CO)_{10}(C_5H_4N)$.[2]

The solvent and excess pyridine are removed by a fast stream of nitrogen (in a well-ventilated fume hood), and the yellow-green residue is crystallized from dichloromethane–diethyl ether (x:y) at $-10\,°C$. The product is obtained as yellow-orange crystals. A green powder might precipitate. This is somewhat less pure $Os_3(CO)_{11}(C_5H_5N)$. Yield: 537 mg (74%).

Anal. Calcd. for $[C_{16}H_5NO_{11}Os_3]CH_2Cl_2$: C, 20.60; H, 0.60; N, 1.35, Found: C, 20.55; H, 0.60; N, 0.90.

Properties

The compound $Os_3(CO)_{11}(C_5H_5N)$ is very soluble in acetonitrile, dichloromethane, acetone, and methanol, and its solutions are stable to air. It is sparingly soluble in hydrocarbons. Its purity can be checked by IR spectroscopy (dichloromethane solution): $v_{(CO)}\,cm^{-1}$: 2106 (w), 2052 (s), 2035 (vs), 2008 (s), 1976 (m). Other physical properties have been reported.[2] The vacuum pyrolysis of this compound provides a high-yield route to the carbido-dianion $[Os_{10}C(CO)_{24}]^{2-}$.[5]

C. BIS(ACETONITRILE)DECACARBONYLTRIOSMIUM

$$Os_3(CO)_{12} + 2(CH_3)_3NO$$

$$\xrightarrow{CH_3CN} Os_3(CO)_{10}(CH_3CN)_2 + 2Me_3N + 2CO_2$$

■ *See cautionary notes in Section A.*

Using apparatus and sublimation procedure for trimethylamine oxide described in Section A a suspension of $Os_3(CO_{12})$ (0.5 g, 0.55 mmol) is prepared in acetonitrile, (700 mL) freshly distilled from phosphorous pentoxide under nitrogen. In order to dissolve the maximum amount of $Os_3(CO)_{12}$, the suspension is heated under reflux for 1 h. The solution is then cooled to $40\,°C$. Slightly more than two equivalents of trimethylamine oxide (100 mg, 1.31 mmol) in acetonitrile (200 mL) is added under nitrogen over a period of 2 h. The excess Me_3NO does not react further with $Os_3(CO)_{10}(CH_3CN)_2$ under these conditions. The mixture is left stirring at this temperature for a further 2 h. The dark yellow solution is filtered through silica to remove excess trimethylamine oxide; the solvent is then removed under vacuum at room temperature to yield a brown-yellow solid.

This solid is suspended in 2 mL of acetonitrile and transferred by pipette to a 25-mL round-bottomed flask. The supernatant liquid is returned to the original flask and the procedure repeated as many times as is necessary to transfer all the solid. At this stage, evaporation of the solvent gives slightly impure $Os_3(CO)_{10}(CH_3CN)_2$ in yields between 85 and 95%. The trace brown impurity does not impair the reactivity of the compound even in sensitive reactions. We therefore generally use it without going through the purification process described next.

Purification of the product can be achieved by washing it with acetonitrile (2 mL), leaving the suspension overnight at $-25\,°C$, and then twice decanting the brown solution. Pure $Os_3(CO)_{10}(CH_3CN)_2$ is thus obtained in $\sim 60\%$ yield. The supernatant of the brown solution may be purified by thin layer chromatography, using Merck plates precoated with silica gel 60F-254, and 50% $CHCl_3$, 45% hexane and 5% MeCN as eluent. (R_f, 0.4). Additional product is obtained by this procedure.

Anal. Calcd. for $C_{14}H_6N_2O_{10}Os_3$: C, 18.0; H, 0.6; N, 3.0. Found: C, 18.2; H, 0.6; N, 3.2.

Properties

Infrared absorptions in the carbonyl region for $Os_3(CO)_{10}(CH_3CN)_2$ in CH_2Cl_2 are 2079(w), 2025(sh), 2021(vs), 1983(m), 1960(w) cm^{-1}. The acetonitrile ligands are readily displaced by other donor ligands; See Ref. 2.

References

1. B. F. G. Johnson and J. Lewis, *Gazz. Chim. Ital.*, **109**, 271 (1979).
2. B. F. G. Johnson, J. Lewis, and D. A. Pippard, *J. Chem. Soc. Dalton Trans.*, **1981**, 407.
3. Johnson et al., *J. Chem. Soc. Dalton Trans.*, **1981**, 407.
4. B. F. G. Johnson and J. Lewis, *Inorg. Synth.*, **13**, 93 (1972).
5. P. F. Jackson, B. F. G. Johnson, J. Lewis, W. J. H. Nelson, and M. McPartlin, *J. Chem. Soc. Dalton Trans.*, **1982**, 2099.

52. DODECACARBONYLTETRA-µ-HYDRIDO--*TETRAHEDRO*-TETRAOSMIUM

Submitted by CAMILO ZUCCARO*
Checked by G. PAMPLONI and F. CALDERAZZO†

$$4Os_3(CO)_{12} + 6H_2 \rightleftarrows 3Os_4H_4(CO)_{12} + 12CO$$

The complex $Os_4H_4(CO)_{12}$ was first synthesized in low yield by acidification of the anions formed in the reaction of methanolic potassium hydroxide and

*Departamento de Química, Facultad de Ciencias, Universidad de Los Andes, Mérida 5101, Venezuela.
†Departimento di Chimica e Chimica Industriale, Via Risorgimento, 35-56100 Pisa, Italy.

$Os_3(CO)_{12}$.[1] In 1975, Kaesz and coworkers synthesized $Os_4H_4(CO)_{12}$ in $\sim 30\%$ yield using atmospheric hydrogenation of $Os_3(CO)_{12}$ in refluxing octane over a period of 50 h.[2] In 1978, Lewis and coworkers synthesized $Os_4H_4(CO)_{12}$ in the range from 60 to 70% yield using 24 h of reaction at 120 atm of hydrogen and 100 °C.[3] Bau and coworkers[4] found that in the direct hydrogenation procedure of Kaesz[2] the yield could be doubled by raising the temperature from that of refluxing octane (120 °C) to that of a mixture of refluxing xylenes (137–144 °C). The compound $Os_4H_4(CO)_{12}$ so obtained is found to be contaminated with many uncharacterized impurities, which make its purification tedious. A high-yield synthetic method is reported here to facilitate further study of the chemistry of $Os_4H_4(CO)_{12}$.

Procedure

A suspension of $Os_3(CO)_{12}$ (2 g) in hexane (60 mL) is placed into a stirred stainless steel autoclave (450-mL total capacity), under a nitrogen atmosphere.[5] After the autoclave is sealed, the gases are evacuated and the autoclave is pressurized to 25 atm of hydrogen. The autoclave is then heated with stirring to 140 °C for 24 h. After cooling, the gas is slowly vented. After coming to atmospheric pressure the autoclave is flushed with an atmosphere of nitrogen, and opened.

■ **Caution.**   *The vented gases may contain CO; ventilation of hydrogen gas into the atmosphere must be carried out with care in a well-ventilated hood.*

Pale yellow crystals of $Os_4H_4(CO)_{12}$ are separated by decantation of the supernatant solution. The sample thus obtained is sufficiently pure for most purposes. The yield of the $Os_4H_4(CO)_{12}$ is usually in the range from 80 to 85% (1.8 g). Checkers report a yield of 45% for a smaller scale reaction, namely, 1 g $Os_3(CO)_{12}$ in 30 mL of solvent in an autoclave of 125-mL total capacity.

Further purification may be carried out by extracting the solids with chloroform (200 mL) in a Soxhlet extractor.

Properties

The complex $Os_4H_4(CO)_{12}$ is a pale yellow solid, insoluble in common organic solvents. The mass spectrum shows the parent ion peak, and other fragments arising from successive loss of twelve CO groups. The IR spectrum (in cyclohexane solution) in the carbonyl region consists of four peaks 2085 (m), 2036 (s), 2020 (s), 1997 (w) cm^{-1}. The proton NMR spectrum in CD_2Cl_2 shows a sharp singlet at $-20.5\,\delta$ throughout the temperature range from 30 to -50 °C, consistent with the fact that the hydrogen atoms are magnetically equivalent as is shown in the X-ray structure.[6]

Anal. Calcd.: C, 13.00%; H, 0.46%.

References

1. B. F. G. Johnson, J. Lewis, and P. A. Kilty, *J. Chem. Soc. A*, **1968**, 2889.

2. S. A. R. Knox, J. W. Keople, M. A. Andrews, and H. D. Kaesz. *J. Am. Soc.*, **97**, 3942 (1975).

3. B. F. G. Johnson, J. Lewis, P. R. Raithby, G. M. Sheldrick, K. Wong, and M. McPartlin. *J. Chem. Soc. Dalton Trans.*, **1978**, 673.

4. C.-Y. Wey, L. Garlaschelli, R. Bau, and T. F. Koetzel. *J. Organomet. Chem.*, **213**, 63 (1981).

5. J. R. Norton, J. P. Collman, G. Dolcetti, and W. T. Robinson. *Inorg. Chem.* **11**, 382 (1972); $Os_3(CO)_{12}$ may also be purchased from Strem Chemicals.

6. B. F. G. Johnson, J. Lewis, P. R. Raithby, and C. Zuccaro. *Acta Cryst.*, **B34**, 1728 (1981).

53. HIGH NUCLEARITY CARBONYL CLUSTER COMPOUNDS OF OSMIUM

Submitted by J. N. NICHOLLS* and M. D. VARGAS*[†]
Checked by R. D. ADAMS[†] and K. NATARAJAN[‡]

Studies in the chemistry of high nuclearity osmium carbonyl clusters have centered around the hexanuclear species $Os_6(CO)_{18}$.[1,2] This is largely due to the existence of a high yield, one-step synthesis of this cluster via the pyrolysis of $Os_3(CO)_{12}$ and also to the fact that $Os_6(CO)_{18}$ has proved to be a good starting material for the syntheses of both higher and lower nuclearity clusters, for example, $[Os_8H(CO)_{22}]^-$ and $[Os_5(CO)_{15}]^{2-}$. We therefore describe in detail the preparation of $Os_6(CO)_{18}$. This hexanuclear cluster undergoes reduction reversibly to give $[Os_6(CO)_{18}]^{2-}$, and irreversibly to give $[Os_5(CO)_{15}]^{2-}$, both useful starting materials for reactions with electrophiles. In addition to the preparation of these dianions, we also describe in detail the synthesis of the dihydride $Os_6H_2(CO)_{18}$.

A. OCTADECACARBONYLHEXAOSMIUM

$$Os_3(CO)_{12} \xrightarrow[\text{vacuum}]{\Delta} Os_6(CO)_{18} + \text{by-products} + CO$$

The pyrolysis of $Os_3(CO)_{12}$ to higher nuclearity clusters is the subject of continuous investigation. This is because although $Os_6(CO)_{18}$ is the major

*University Chemical Laboratory, Cambridge University, Lensfield Road, Cambridge CB2 1EW, United Kingdom.
[†]Address correspondence to Professor M. D. Vargas, Universidade Estadual de Campinas, Instituto de Quimica, Caixa Postal 6154, 13100, Campinas, S. Paulo, Brasil.
[‡]Department of Chemistry, University of South Carolina, Columbia, SC 29208.

product, the pyrolysis also yields a large number of other clusters,[1,3] the proportions of which are very sensitive to the conditions employed. Thus far, clusters of nuclearity 5 to 11 have been isolated from this reaction. The conditions described below are optimum for formation of $Os_6(CO)_{18}$. The advantage of this route is that it is a one-step synthesis even though separation of the $Os_6(CO)_{18}$ from the reaction mixture is effectively achieved only by fractional crystallization over a period of weeks.

Procedure

■ **Caution.** *Due to the toxicity of metal carbonyls, the heating oven used in the following procedure should be located in a well-ventilated hood. The Carius tube should be constructed of thick walled (min 2 mm) glass tubing. Care should be taken when sealing the tube to form a glass seal of uniform thickness, to reduce likelihood of explosion. After sealing, the tube should be placed in a heavy iron pipe as a shield against possible implosion or explosion under heating, see Fig. 1.*

The Carius tube used in the procedure is previously washed with acetone and dried for 12 h at 130 °C. It *must* be free of water since the pyrolysis of

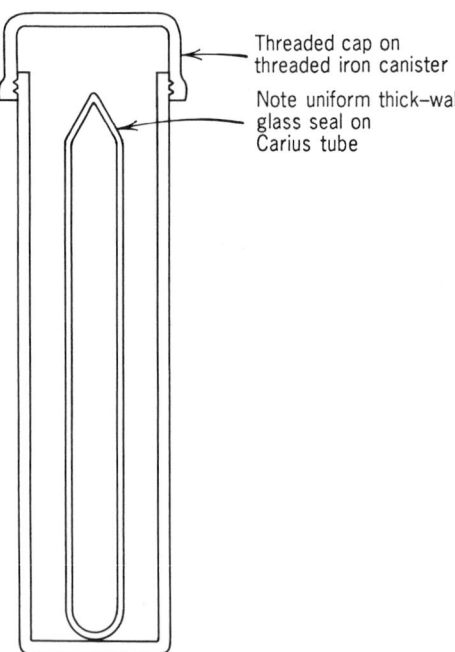

Threaded cap on threaded iron canister

Note uniform thick–wall glass seal on Carius tube

Fig. 1.   Details of the sealed heavy-walled Carius tube and iron pipe assembly.

$Os_3(CO)_{12}$ in the presence of water gives hydrido clusters instead of the binary carbonyl clusters. A tube of 165-mL volume is used in the following procedure. The charge to volume ratio is $12.7 \, g \, L^{-1}$ and should be maintained if the scale of the reaction is altered.

Finely powdered $Os_3(CO)_{12}$ (2.07 g, 2.28 mmol)[4] is placed in a Carius tube of 163-mL capacity, previously washed with acetone and thoroughly dried as explained previously. The tube is then evacuated to 0.015 torr for 1 h and sealed under vacuum making sure that the glass seal is uniform with the thickness of the tube, see Fig. 1. The tube is then placed in an iron pipe and heated in an oven equipped with a fan for approximately 60 h at 260 °C.

■ **Caution.** *Opening of the Carius tube after completion of the heating must be carried out with suitable precaution against explosion, using heavy-duty safety gloves and protected by an explosion shield. The entire assembly should be placed in a well-ventilated hood.*

Upon cooling, the tube is removed from the lead pipe container using suitable heavy-duty gloves behind an explosion shield. The Carius tube is placed in a small Dewar flask such that the upper portion is available for opening. Liquid nitrogen is placed in the Dewar flask to cool the lowest portion of the Carius tube to $-196 \, °C$, reducing any gas pressure that may have developed in the tube. A scratch is made near the uppermost end of the tube with a standard tungsten–carbide tool or iron file. The Carius tube is then cracked open by applying the heated end of a Pyrex rod to the tip of the scratch, the tip of the Pyrex rod having been heated close to its softening point. A stream of nitrogen should be directed around the scratch on the Carius tube to minimize condensation of moisture when the Carius tube is cracked in the previous procedure. After opening, the Carius tube is quickly removed from the liquid nitrogen bath and the bottom of the tube warmed rapidly to room temperature by immersion in an acetone bath.

The contents of the Carius tube should appear as a brown solid. This is extracted with 50-mL portions of ethyl acetate each brought to boil in the Carius tube by heating it in an oil bath under stirring. When the solvent is pale pink in color, all the neutral clusters will have been extracted. The residue contains a mixture of high nuclearity osmium anions of which $[Os_{10}C(CO)_{24}]^{2-}$ and $[Os_{11}C(CO)_{27}]^{2-}$ have been fully characterized.[3] These can be extracted with a methanol–acetone (1:1) solution of [PPN]Cl, (μ-nitrido-bis(triphenylphosphorus)(1 +)] chloride (Ventron) and separated by fractional crystallization from acetone–methanol.

The combined fractions are heated under reflux for 30 min. The volume of ethyl acetate is then reduced to 300 mL and the solution is decanted and filtered through Celite®. The products are separated by fractional crystallization over a period of 3 weeks at $-5 \, °C$.

The first compound to crystallize out after ~ 3 days is unreacted

$Os_3(CO)_{12}$ (yellow crystals). Reduction of the solvent volume by small portions lead to crystallization of the products in the following order: small amounts of $Os_5(CO)_{16}$ (brown), $Os_5C(CO)_{15}$ (dark red crystals), $Os_6(CO)_{18}$ (dark brown crystals 0.95–1.12 g, yield 50–60%), $Os_7(CO)_{21}$ (brown yellow crystals), $Os_8(CO)_{23}$ (dark brown crystals), and $Os_8C(CO)_{21}$ (purple-brown powder). The compound $Os_5C(CO)_{15}$ is not always formed in this reaction. Appearance of this compound may be related to the presence of air and/or water in the reaction tube.

Anal. Calcd. for $Os_6C_{18}O_{18}$: C, 13.13%. Found: C, 13.31%.

Properties

The compound $Os_6(CO)_{18}$ is stable indefinitely to air in the solid state. In solution it is stable in noncoordinating solvents. In coordinating solvents adducts are formed,[5] which can lead in the case of nitriles and amines to reduction of the cluster to give $[Os_6(CO)_{18}]^{2-}$ or $[HOs_5(CO)_{15}]^{-}$. Purity can be checked by thin layer chromatography (TLC) using 2% ethyl acetate–hexane as eluent and by carbonyl absorptions in the IR (see Table I). Possible impurities are $Os_3(CO)_{12}$, $Os_5(CO)_{16}$, $Os_5C(CO)_{15}$, $Os_7(CO)_{21}$, $Os_8(CO)_{23}$, and $Os_8C(CO)_{21}$. Mass spectral characterisation of these clusters has been described elsewhere.[1]

The octadecacarbonylhexaosmium cluster has been used as a starting

TABLE I. Properties of Osmium Cluster Complexes.

Cluster	Color	IR Frequenceis $\nu_{(CO)}$ (cm^{-1}) CH$_2$Cl$_2$ solution	TLC $R_f{}^a$
$Os_3(CO)_{12}$	Yellow	2066 (vs), 2032 (s), 2011 (w) 1997 (w)	0.73
$Os_5(CO)_{16}$	Brown	2065 (vs), 2052 (s), 2039 (s, sh)	0.70
$Os_6(CO)_{18}$	Brown	2108 (w), 2077 (vs), 2063 (vs), 2040 (vs), 2007 (w), 1955 (w)	0.63
$Os_7(CO)_{21}$	Orange	2112 (vw), 2076 (m), 2060 (vs), 2038 (w), 2002 (w)	0.57
$Os_8(CO)_{23}$	Brown	2113 (vw), 2083 (m), 2071 (s), 2061 (vs), 2045 (w), 2038 (w), 2029 (w), 2012 (vw), 2000 (vw)	0.53
$Os_8(CO)_{21}$	Purple	2120 (w), 2091 (m), 2073 (s), 2039 (w), 2030 (m), 2021 (m)	0.35

^{a}Thin layer chromatography (TLC) using Merck precoated silica gel 60F-254 plates; solvent 15:85 CH$_2$Cl$_2$–hexane.

material for the syntheses of lower and higher nuclearity compounds; for example, carbonylation yields $Os_5(CO)_{19}$ (Ref. 6) and thermolysis in isobutyl alcohol gives $[Os_8H(CO)_{22}]^-$.[7] The cluster reacts with nucleophiles, but a more satisfactory route to its substituted derivatives is via $Os_6CO_{17}(CH_3CN)$, which is prepared by the amine oxide route. Thus, $Os_6(CO)_{17}(C_5H_5N)$ and $Os_7H_2(CO)_{20}$ (Ref. 8) have been prepared by the action of pyridine and $H_2Os(CO)_4$, respectively. The reduction of $Os_6(CO)_{18}$ is dealt with in detail in Section B.

Reductions of Octadecacarbonylhexaosmium

The reduction of $Os_6(CO)_{18}$ can follow one of two routes depending on the reducing agent, the types of solvents, and the concentration of the reagents used.

Under drastic conditions, that is, using high concentration of methanolic potassium hydroxide, $Os_6(CO)_{18}$ is "decapped" to give the dianion $[Os_5(CO)_{15}]^{2-}$ in high yields.[9] Under milder conditions (high dilution and a weaker base such as iodide), $[Os_6(CO)_{18}]^{2-}$ (Ref. 10) is the product of the reaction.

B. BIS[µ-NITRIDO-BIS(TRIPHENYLPHOSPHORUS)(1+)] PENTADECACARBONYLPENTAOSMATE (2−)

$$Os_6(CO)_{18} + KOH \xrightarrow{CH_3OH} K_2[Os_5(CO)_{15}] + K_2CO_3$$

$$K_2[Os_5(CO)_{15}] \xrightarrow[(2)[PPN]Cl/CH_3OH]{(1)CH_2Cl_2} [PPN]_2[Os_5(CO)_{15}] + KCl$$

- **Caution** *Due to the toxicity of metal carbonyls, this procedure should be carried out in a well-ventilated hood.*

Procedure

A degassed solution of potassium hydroxide (570 mg 9.62 mmol) in freshly distilled methanol (8 mL) is prepared under nitrogen in a 50 mL round-bottomed flask equipped with a magnetic stirring bar. Solid $Os_6(CO)_{18}$ (150 mg 0.09 mmol) is added and the mixture is stirred at room temperature for 30 min. After addition of [PPN]Cl (Ventron) (330 mg, 0.57 mmol), the solution is cooled to −25 °C for 16 h. A brown powder is formed that is removed by filtration and washed with ice-cold methanol (10 mL). Yield: 190 mg (85%). The compound does not need to be purified further.

Anal. Calcd. for $Os_5C_{87}H_{60}N_2O_{15}P_4$: C, 42.63; H, 2.45; N, 1.14%. Found: C, 42.98; H, 2.58; N, 1.07%.

Properties

The compound $[PPN]_2[Os_5(CO)_{15}]$ is a dark brown air stable solid, soluble in a range of dry, polar solvents to give air stable solutions. Infrared (dichloromethane solution, cm^{-1}) 2006 (vs), 1963 (vs), 1921 (w), 1875 (vw). It is readily protonated to give the salt of the monohydride $[PPN][Os_5H(CO)_{15}]$ whose characteristic absorptions in the IR are (dichloromethane solution, cm^{-1}) 2035 (vs), 2023 (s), 2007 (m br), 1958 (w br). The dihydride $Os_5H_2(CO)_{15}$ may be obtained upon further protonation with conc. sulfuric acid in CH_3CN. Infrared (hexane solution, cm^{-1}): 2076 (s), 2069 (vs), 2048 (s), 2037 (m), 2029 (w), 2015 (w), 2007 (w), 1996 (w). Other spectroscopic data on $[PPN][Os_5H(CO)_{15}]$ and $Os_5H_2(CO)_{15}$ have been reported elsewhere.[9]

C.   BIS[μ-NITRIDO-BIS(TRIPHENYLPHOSPHORUS)](1+) OCTACARBONYLHEXAOSMATE(2−)

$$Os_6(CO)_{18} + [(C_4H_9)_4N]I \xrightarrow{\text{THF}} [(C_4H_9)_4N]_2[Os_6(CO)_{18}] + I_2$$

$$[(C_4H_9)_4N]_2[Os_6(CO)_{18}] \xrightarrow[\text{(2)[PPN]Cl/CH}_3\text{OH}]{\text{(1)CH}_2\text{Cl}_2} [PPN]_2[Os_6(CO)_{18}]$$

$$+ 2[(C_4H_9)_4N]Cl$$

■   *See cautionary note, Section B.*

Procedure

A suspension of $[(C_4H_9)_4N]I$ (74 mg, 0.200 mmol) (Fisons) is prepared in freshly distilled and degassed tetrahydrofuran (THF, 20 mL). This is added under nitrogen atmosphere to a solution of $Os_6(CO)_{18}$ (150 mg, 0.091 mmol) in THF (150 mL), in a 250-mL two-necked flask equipped with a magnetic stirring bar. The mixture is stirred for 3 h after which time the reaction is complete.

The red solution is evaporated to dryness by a fast nitrogen flow and the residue dissolved in a minimum of dichloromethane (2 mL). Addition of a solution of [PPN]Cl (157 mg, 0.273 mmol) (Ventron) in methanol (4 mL) followed by cooling at −25 °C overnight, gives dark orange microcrystals of $[PPN]_2[Os_6(CO)_{18}]$ that are filtered off, washed with ice-cold methanol (2 mL), and dried under vacuum. Yield: 160 mg (65%).

Anal. Calcd. for $Os_6C_{90}H_{60}N_2O_{18}P_4$: C, 49.45; H, 2.75; N, 1.28%. Found: C, 49.17; H, 2.98, N, 1.25.

Properties

The compound $[PPN]_2[Os_6(CO)_{18}]$ is insoluble in hydrocarbons, sparingly soluble in methanol, and very soluble in acetone, acetonitrile, dichloromethane, and chloroform to give solutions that are stable indefinitely at room temperature. Its IR spectrum in CH_2Cl_2 exhibits an intense $\nu_{(CO)}$ absorption at 1991 and very weak absorptions at 1964, 1938, and 1910 cm^{-1}. The compound is easily oxidized back to $Os_6(CO)_{18}$ by treatment of a CH_2Cl_2 solution with iodine I_2.[11]

D. OCTADECACARBONYLDIHYDRIDOHEXAOSMIUM

$$[PPN]_2[Os_6(CO)_{18}] + HCl \xrightarrow{CH_2Cl_2} Os_6H_2(CO)_{18} + 2[PPN]Cl$$

The compound $Os_6H_2(CO)_{18}$ is obtained by protonation of a suitable salt of the dianion $[Os_6(CO)_{18}]^{2-}$. Depending on the conditions used, however, one of two isomers of the dihydride is formed. Protonation of the dianion, whose metal framework is octahedral, with aqueous sulfuric acid in acetonitrile gives the capped, square based pyramidal isomer,[11] whereas gaseous hydrogen chloride bubbled through a dichloromethane solution of the salt of the dianion gives a different isomer.[12] This latter isomer is proposed to have the same octahedral structure as $Ru_6H_2(CO)_{18}$, on the basis of the similarity of their IR spectra.[13]

Although both preparations give $Os_6H_2(CO)_{18}$ in yields of $\sim 90\%$, we find the second route more convenient since the hydrochloric acid is more easily removed from the crystalline product, and therefore, the cluster can be obtained in a higher state of purity. Conversion of the isomer produced via this route to the capped square base pyramidal $Os_6H_2(CO)_{18}$ occurs readily on standing in solution.

Procedure

▪ **Caution.** *Due to the toxic nature of metal carbonyl and the lachrymatory properties of gaseous hydrogen chloride, this procedure must be carried out in a well-ventilated hood.*

In a 25 mL two-necked flask a solution of $[PPN]_2[Os_6CO_{18}]$ (110 mg, 0.050 mmol) in dichloromethane (2 mL) is prepared. Gaseous hydrogen chloride is bubbled through the solution for 10 to 20 min, during which time an orange-red solid $Os_6H_2(CO)_{18}$, precipitates. Solvent is evaporated under reduced pressure to a volume of 1 mL, the flask is stoppered and left overnight in a refrigerator at $-5\,^\circ$C. The solids are removed by filtration, followed by washing with cold methanol (2 × 3-mL portions) to remove [PPN]Cl. Yield of $Os_6H_2(CO)_{18}$: 77 mg (90%). No further purification is necessary.

Anal. Calcd. for $Os_6C_{18}H_2O_{18}$: C, 13.11%. Found: C, 13.34%.

Properties

The dihydride $Os_6H_2(CO)_{18}$ is a microcrystalline air-stable orange solid, sparingly soluble in hydrocarbons and soluble in solvents such as dichloromethane, acetone, acetonitrile, and THF. It is characterized by IR absorptions, (dichloromethane solution, cm^{-1}): 2081 (w), 2067 (vs), 2056 (m), 2036 (w), 2022 (w), 2004 (w).

On standing in the above-mentioned polar solvents, the octahedral $Os_6H_2(CO)_{18}$ rearranges to give the square base pyramidal isomer (in dichloromethane, 30 min), which exhibits the following IR absorptions: (dichloromethane solution, cm^{-1}) 2112 (w), 2081 (m), 2073 (vs), 2045 (m), 2018 (w), 2002 (w). Additional spectroscopic data on $Os_6H_2(CO)_{18}$ have been reported.[11]

References

1. C. R. Eady, B. F. G. Johnson, and J. Lewis, *J. Chem. Soc. Dalton Trans.*, **1975**, 2606.

2. R. Mason, K. M. Tomas, and D. M. P. Mingos, *J. Am. Chem. Soc.*, **95**, 3802 (1973).

3. P. F. Jackson, B. F. G. Johnson, J. Lewis, W. J. H. Nelson, and M. McPartlin, *J. Chem. Dalton Trans.*, **1982**, 2099; D. Braga, B. F. G. Johnson, J. Lewis, M. McPartlin, W. J. H. Nelson, A. Sironi, and M. D. Vargas, *J. Chem. Soc. Chem. Commun.*, **1983**, 1131.

4. B. F. G. Johnson and J. Lewis, *Inorg. Synth.*, **13**, 93 (1972).

5. G. R. John, B. F. G. Johnson, and J. Lewis, *J. Organomet. Chem.*, **181**, 143 (1979).

6. D. H. Farrar, B. F. G. Johnson, J. Lewis, J. N. Nicholls, P. R. Raithby, and M. J. Rosales, *J. Chem. Soc. Chem. Commun.*, **1981**, 273.

7. D. Braga, K. Kendrick, B. F. G. Johnson, J. Lewis, M. McPartlin, W. J. H. Nelson, and M. D. Vargas, *J. Chem. Soc. Chem. Commun.*, **1982**, 419.

8. E. J. Ditzel, H. D. Holden, B. F. G. Johnson, J. Lewis, A. Saunders, and M. J. Taylor, *J. Chem. Soc. Chem. Commun.*, **1982**, 1373.

9. C. R. Eady, J. J. Guy, B. F. G. Johnson, J. Lewis, M. C. Malatesta, and G. M. Sheldrick, *J. Chem. Soc. Chem. Commun.*, **1976**, 807.

10. G. R. John, B. F. G. Johnson, J. Lewis, and A. L. Mann, *J. Organomet. Chem.*, **171**, C9 (1979).

11. C. R. Eady, B. F. G. Johnson, and J. Lewis, *J. Chem. Soc. Chem. Commun.*, **1976**, 302.

12. D. H. Farrar, B. F. G. Johnson, and J. Lewis, unpublished results.

13. C. R. Eady, P. F. Jackson, B. F. G. Johnson, J. Lewis, M. C. Malatesta, M. McPartlin, and W. J. H. Nelson, *J. Chem. Soc. Dalton Trans.*, **1980**, 383.

54. THIOOSMIUM CLUSTERS

Submitted by RICHARD D. ADAMS* and ISTVAN T. HORVATH*
Checked by JEAN-JACQUES BONNET,[††] CHRISTIAN BERGOUNHOU,[†] and
JULIAN P. ATTARD[‡]

Recent studies have shown that sulfido ligands can play an important role in the syntheses and chemistry of transition metal cluster compounds.[1-3] Their value is derived from their abilities to form relatively strong bonds to nearly all the transition elements, and to serve as bridging ligands involving variable degrees of electron donations. This, combined with the tendency of the third-row transition elements to form strong metal–metal bonds, has led to the syntheses of a large number of new and unusual thioosmium carbonyl cluster compounds.[2,3]

The detailed synthesis of $Os_3(CO)_{10}(\mu\text{-}H)(\mu\text{-}SC_6H_5)$, **1**,[4] $Os_3(CO)_9(\mu_3\text{-}CO)(\mu_3\text{-}S)$, **2**,[5] $Os_3(CO)_9(\mu_3\text{-}S)_2$, **3**,[6-8] $Os_4(CO)_{13}(\mu\text{-}S)_2$, **4**,[3,9] and $Os_4(CO)_{12}(\mu\text{-}S)_2$, **5** is given here.[3,9]

*Department of Chemistry, University South Carolina, Columbia, SC 29208.
†Laboratories de Chimie de Coordination, CNRS, 205 Route de Narbonne 31400 Toulouse, France.
‡University Chemical Laboratory, Lensfield Road, Cambridge, CB2 1EW, United Kingdom.

A. (μ-BENZENETHIOLATO)DECACARBONYL-μ-HYDRIDO-TRIOSMIUM, (1)[4]

$$Os_3(CO)_{12} + HSC_6H_5 \longrightarrow Os_3(CO)_{10}(\mu\text{-}H)(\mu\text{-}SC_6H_5) + 2CO$$

Procedure

■ **Caution.** *The reaction should be carried out in a well-ventilated hood, because benzenethiol(thiophenol) and carbon monoxide are toxic. The latter is evolved during the reaction. Any excess thiophenol can be destroyed by treatment with a mild oxidant such as chlorine water (5% NaOCl solution, known commercially as Clorox).*

Dedecarbonyltriosmium is commercially available (Strem) or may be prepared according to a procedure appearing in these volumes.[10]

A 200-mL two-necked round-bottomed flask is fitted with a reflux condenser connected to a nitrogen bubbler. The flask is charged with 0.36 g (0.40 mmol) of dodecacarbonyltriosmium and 100 mL of octane. Under nitrogen 0.073 mL (0.70 mmol) of thiophenol is added and the mixture is heated to reflux with stirring for 3 h. The mixture is then cooled and the solvent is evaporated *in vacuo*. The yellow residue is dissolved in 15 mL hexane (or heptane) and chromatographed on neutral alumina (Brockman Activity 1, 80-200 mesh, Fischer) using hexane or heptane as eluent. A column 12 in. long with a 1-in. diameter is an adequate size. The analytically pure sample can be obtained by recrystallization from hexane. Yield: 0.27–0.29 g (70–76%), mp 131 °C.

Anal. Calcd. for $C_{16}H_6O_{10}Os_3S$: C, 19.99; H, 0.63. Found: C, 20.26; H, 0.71.

Properties

Compound **1** is an air-stable, yellow, crystalline solid. It is readily soluble in organic solvents such as hexane, benzene, and dichloromethane. Solutions of **1** in these solvents are stable in air for several weeks. The IR spectrum of **1** contains the following CO absorptions (hexane): 2109 (w), 2068 (s), 2059 (m), 2024 (vs), 2018 (m), 2002 (m), 1990 (w), and 1985 (w) cm^{-1}. The NMR spectrum (CDCl$_3$) shows a singlet at -17.0 ppm due to the hydride ligand and a multiplet at 7.26 ppm due to the phenyl ring.

Chemical Properties. When subjected to UV irradiation[5] or temperatures $> 150\,°C$,[7] compound **1** eliminates 1 mol of benzene and is transformed into a variety of new thioosmium carbonyl cluster compounds.

B. μ_3-CARBONYL-NONACARBONYL-μ_3-THIO-TRIOSMIUM, 2^5

$$Os_3(CO)_{10}(\mu\text{-H})(\mu\text{-SC}_6H_5) \xrightarrow[CO]{hv} Os_3(CO)_9(\mu_3\text{-CO})(\mu_3\text{-S}) + C_6H_6$$

Procedure

■ **Caution.** *The reaction should be carried out in a well-ventilated hood because carbon monoxide—a highly toxic, colorless, and odorless gas—is used in large amounts in this reaction; and benzene—toxic and also possibly carcinogenic—is formed in the reaction.*

A 500-mL two-necked round-bottomed Pyrex flask is fitted with a gas dispersion tube and a reflux condenser connected to a bubbler. The flask is charged with 0.191 g (0.20 mmol) of (μ-benzenethiolato)-decarbonyl-μ-hydrido-triosmium and 250 mL of cyclohexane. The solution is purged with carbon monoxide (flow rate: ~ 50 mL min^{-1}) for 15 min. Then the flow rate is increased to ~ 100 mL min^{-1} and the solution is irradiated for 2.0 to 2.5 h by using an *external* high-pressure mercury lamp (e.g., Philips HPK 125 W). The lamp should be maintained at least 12 in. from the reaction flask to minimize heating of the solution. Heating decreases the yield of **2**.

■ **Caution.** *Exposure of the eyes to UV light should be avoided at all times.*

If the radiation time is exceeded the following compounds also will be formed: $Os_3(CO)_9(\mu_3\text{-S})_2$, **(3)** ($R_f = 0.52$); $Os_4(CO)_{13}(\mu_3\text{-S})^7$ (an orange band at $R_f = 0.30$ described in the following separation).

After removal of solvent *in vacuo*, the yellow residue is chromatographed on silica thin layer chromatography (TLC) plates using hexane–CH_2Cl_2(85:15) as eluent. The major product, **2**, is a nearly colorless band observed at $R_f = 0.17$. This band can be seen better by using an UV lamp and UV sensitized TLC plates (precoated TLC plates, silica gel 60 F-254, layer thickness 0.25 mm, Merck). This band is separated from the starting material ($R_f = 0.47$) and several minor bands and the products are eluted from the silica gel with CH_2Cl_2. An analytically pure sample can be obtained by recrystallization from hexane at $-20\,°C$. Yield: 0.053–0.070 g (30–40%).

Anal. Calcd. for $C_{10}O_{10}Os_3S$: C, 13.61. Found: C, 13.53.

Properties

Compound **2** is an air-stable, pale yellow, crystalline solid. It is readily soluble in organic solvents such as hexane, benzene, and dichloromethane. Solutions of **2** in these solvents are stable in air for several days. The IR spectrum of **2** contains the following CO absorptions (hexane): 2109 (vw), 2075 (s), 2028(s), 2012 (w), 1685 (w) cm^{-1}. The crystal structure of **2** has been reported.[5]

Chemical Properties. Compound **2** is readily decarbonylated upon exposure to UV irradiation.[5] Irradiated solutions of **2** readily yield addition products of sulfur containing small molecules such as COS, CS$_2$, and H$_2$S. In the absence of reagents it will form the hexanuclear compound Os$_6$(CO)$_{17}$(μ_4-S)$_2$. It reacts with other metal complexes to form higher nuclearity osmium clusters and heteronuclear metal cluster compounds.[5,11,12]

C.   NONACARBONYLDI-μ_3-THIO-TRIOSMIUM, 3[6-8]

$$2Os_3(CO)_{10}(\mu\text{-H})(\mu\text{-SC}_6H_5) + 4CO$$

$$\longrightarrow Os_3(CO)_9(\mu_3\text{-S})_2 + 3Os(CO)_5 + 2C_6H_6$$

Procedure

■   **Caution.**   *The synthesis uses carbon monoxide—a highly toxic, color-less, and odorless gas—therefore the preparation should be carried out in a well-ventilated room. Carbon monoxide detection systems are available from Devco. In the reaction benzene and Os(CO)$_5$ are formed, the benzene vapors are noxious and probably carcinogenic. The work-up should be performed in a well-ventilated hood.*

(μ-Benzenethiolato)decacarbonyl-μ-hydrido-triosmium (0.23 g, 0.24 mmol) is dissolved in octane (50 mL) and the solution is placed in a high-pressure apparatus with CO at 2000 psi (minimum volume 200 mL). The apparatus is pressurized and then heated to 160 °C for 10 h. After removal from the high-pressure apparatus, it can be shown by IR that the reaction mixture consists of two major products: Os$_3$(CO)$_9$(μ_3-S)$_2$, **3**, and Os(CO)$_5$ (IR: 2043 (s), 1991 (s) cm^{-1}). The solvent and Os(CO)$_5$ is collected by distillation at 0.1 torr/25 °C into a nitrogen trap (this can be used as Os(CO)$_5$ "stock" solution for later syntheses). The yellow residue is chromatographed on silica TLC plates using hexane as eluent. The major product, **3**, is a yellow band observed at $R_f = 0.35$. This band is separated from several minor bands and the product is eluted from the silica gel with CH$_2$Cl$_2$. An analytically pure sample can be obtained by recrystallization from hexane at -20 °C. Yield: 0.098 g (92%).

Properties

Compound **3** is an air-stable, yellow, crystalline solid. It is soluble in common organic solvents such as hexane, benzene, and dichloromethane. Solutions of **3** in these solvents are stable in air for several weeks. The IR spectrum of the compound contains the following CO absorptions (hexane): 2079 (s), 2059 (s), 2019(s), 2014 (sh) cm^{-1}. The crystal structure of **3** has been reported.[8]

D.   TRIDECACARBONYLDI-μ_3-THIO-TETRAOSMIUM, 4[3,9]

$$Os_3(CO)_9(\mu_3\text{-}S)_2 + Os(CO)_5 \xrightarrow{h\nu} Os_4(CO)_{13}(\mu_3\text{-}S)_2 + CO$$

Procedure

- **Caution.** *The cautions for use of CO and Os(CO)$_5$ were given earlier.*

A 200-mL three-necked round-bottomed Pyrex flask is fitted with a pressure-equalizing dropping funnel covered with aluminum foil and a reflux condenser connected to a bubbler. The flask is charged with nonacarbonyldi-μ_3-thiotriosmium (0.044 g, 0.05 mmol) and octane (50 mL) and the dropping funnel is charged with a solution of Os(CO)$_5$ (0.066g, 0.20 mmol) in octane (50 mL). Under nitrogen, the Os(CO)$_5$ solution is added dropwise and the mixture is irradiated for 2 h using an external high-pressure mercury lamp. Suitable amounts of the Os(CO)$_5$ solution collected in Section C may be used at this time.

- **Caution.** *Exposure of the eyes to UV light should be avoided at all times.*

After removal of solvent *in vacuo*, the orange-red residue is chromatographed on silica TLC plates using hexane as eluent. In addition to some small fractions, two main bands are observed. At $R_f = 0.35$, a yellow band (unreacted starting material, 0.01 g) and at $R_f = 0.17$, an orange band. This band is separated and the product is eluted from the silica gel with CH$_2$Cl$_2$. An analytically pure sample can be obtained by recrystallization from hexane at − 20 °C. Yield: 0.023 g (40%).

Properties

Compound **4** is an air-stable, orange-red, crystalline solid. It is soluble in common organic solvents such as hexane, benzene and dichloromethane. Solutions of **4** in these solvents are stable in air for several days. The IR spectrum of the compound contains the following CO absorptions (hexane): 2090 (s), 2065 (s), 2028 (vs), 2022 (m), 1995 (w), 1983 (w), 1936 (w) cm^{-1}. The crystal structure of **4** has been reported.[9]

E.   DODECACARBONYLDI-μ_3-THIO-TETRAOSMIUM, 5[3,9]

$$Os_4(CO)_{13}(\mu_3\text{-}S)_2 \xrightarrow{\Delta} Os_4(CO)_{12}(\mu_3\text{-}S)_2 + CO$$

Procedure

- **Caution.**   *The reaction should be carried out in a well-ventilated hood, as carbon monoxide—a highly poisonous, colorless, and odorless gas—is evolved.* A 25-mL three-necked round-bottomed flask, is equipped with a magnetic stirring bar and a reflux condenser connected to a nitrogen bubbler. To this is added tridecacarbonyldi-μ_3-thio-tetraosmium (0.0245 g, 0.020 mmol) and nonane (10 mL). Under nitrogen, the mixture is heated to reflux with stirring for 20 min. The mixture is cooled and the solvent is removed *in vacuo*. The red residue is crystallized from hexane at 0 °C. Yield: 0.024 g (100%).

Properties

Compound **5** is an air-stable, red, crystalline solid. It is soluble in common organic solvents such as hexane, benzene, and dichloromethane. Solutions of **5** in these solvents are stable in air for several weeks. The IR spectrum of the compound contains the following CO absorptions (hexane): 2074 (vs), 2066 (s), 2059 (vs), 2024 (m), 2019 (m), 2001 (s) cm^{-1}. The crystal structure of **5** has been reported.[9]

F.   ANALOGOUS SELENIDO CLUSTERS

The compounds, $Os_3(CO)_{10}(\mu\text{-H})(\mu\text{-SeC}_6H_5)$, $Os_3(CO)_9(\mu_3\text{-CO})(\mu_3\text{-Se})$, $Os_3(CO)_9(\mu_3\text{-Se})_2$, $Os_4(CO)_{12}(\mu_3\text{-Se})_2$ may be prepared similarly.[10]

References

1. H. Vahrenkamp, *Angew. Chem. Int. Ed. Engl.*, **14**, 322 (1975); L. Markó, *Gazz. Chim. Ital.*, **109**, 247 (1979).

2. R. D. Adams, *Polyhedron*, **4**, 2003 (1985).

3. R. D. Adams, I. T. Horváth, P. Mathur, B. E. Segmüller, and L. W. Yang, *Organometallics*, **2**, 1078 (1983).

4. G. R. Crooks, B. F. G. Johnson, J. Lewis, and I. G. Williams, *J. Chem. Soc. A*, **1969**, 797.

5. R. D. Adams, I. T. Horváth, and H. S. Kim, *Organometallics*, **3**, 548 (1984). The authors have recently discovered that $Os_3(CO)_9(\mu_3\text{-CO})(\mu_3\text{-S})$ can be obtained more conveniently (~ 30% yield) from the reaction of $(CH_2)_2S$ with $Os_3(CO)_{10}(NCMe)_2$. The ruthenium homolog $Ru_3(CO)_9(\mu_3\text{-CO})(\mu_3\text{-S})$ has also been prepared in good yield by the reaction of $(CH_2)_2S$ with $Ru_3(CO)_{12}$ see R. D. Adams, J. E. Babin, and M. Tasi, *Inorg. Chem.*, **25**, 4514 (1986).

6. B. F. G. Johnson, J. Lewis, P. G. Lodge, and P. R. Raithby, *J. Chem. Soc. Chem. Commun.*, **1979**, 719.

7. R. D. Adams, I. T. Horváth, B. E. Segmüller, and L. W. Yang, *Organometallics*, **2**, 1301 (1983).

8. R. D. Adams, I. T. Horváth, B. E. Segmüller, and L. W. Yang, *Organometallics*, **2**, 144 (1983).

9. R. D. Adams, and L. W. Yang, *J. Am. Chem. Soc.*, **105**, 235 (1983).

10. B. F. G. Johnson and J. Lewis, *Inorg. Synth.*, **13**, 93 (1972).
11. R. D. Adams and T. S. A. Hor, *Inorg. Chem.*, **23**, 4723 (1984).
12. R. D. Adams, T. S. A. Hor, and P. Mathur, *Organometallics*, **3**, 634 (1984).

55. TRIS(η^5-CYCLOPENTADIENYL)BIS-(μ_3-PHENYLMETHYLIDYNE)-TRICOBALT

$$3(\eta^5\text{-Cp})\text{Co(CO)}_2 + \text{PhC}\equiv\text{CPh} \longrightarrow \text{Co}_3\text{PhC}(\eta^5\text{-Cp})_3\text{CPh} + 6\text{CO}$$

Submitted by S. B. COLBRAN,* B. H. ROBINSON,* and J. SIMPSON*
Checked by J. R. SHAPLEY[†] and WENN-YANN YEH[†]

Tris(η^5-cyclopentadienyl)-bis(μ_3-phenylmethylidyne)-tricobalt is a member of an important class of metal cluster complexes in which the metal framework is capped by a nonmetal group,[1] in this case formally a carbyne group $\equiv$CR. Vollhardt and coworkers were the first to report a general synthesis of the bicapped clusters $\text{Co}_3\text{CR}(\eta^5\text{-Cp})_3\text{CR}'$ from cleavage of an alkyne by $\text{Co}(\eta^5\text{-Cp})(\text{CO})_2$,[2] and have subsequently developed the organic chemistry of these compounds.[3] The temperature of the reaction mixture and the ratio of $\text{Co}(\eta^5\text{-Cp})(\text{CO})_2$ to alkyne are critical in reducing the number of products produced. The bis(carbyne) cluster is the only organometallic cobalt product in the procedure given, and this route can also be used to prepare other derivatives.[4] Yields decrease with the steric bulk of the alkyne substituent.

Procedure

■ **Caution.** *This reaction should be carried out in a well-ventilated fume hood as $\text{Co}(\eta^5\text{-C}_5\text{H}_5)(\text{CO})_2$ and the evolved CO are toxic.*

All operations should be carried out under an atmosphere of nitrogen in a thoroughly dried apparatus. The hydrocarbon solvents, dodecane, tridecane, and hexane, must be dried over sodium; dichloromethane should be distilled from CaH_2 (or dried over 4A molecular sieves).

Dicarbonyl(η^5-cyclopentadienyl)cobalt may be synthesized from $\text{Co}_2(\text{CO})_8$ (Ref. 5) or purchased from Strem or Pressure Chemicals. Diphenylacetylene (Farchan, commercial grade) can be used without purification.

A dry, 100-mL three-necked flask is equipped with a reflux condenser connected to a mineral oil bubbler for the release of gases, a 50-mL pressure-

*Department of Chemistry, University of Otago, P.O. Box 56, Dunedin, New Zealand.
[†]Department of Chemistry, University of Illinois, Urbana, IL 61801.

equalizing dropping funnel, an N_2 inlet, and a magnetic stirrer. The flask is charged with a dodecane–tridecane (2:1 mixture, 20 mL) and purged with N_2. Diphenylacetylene (0.42 g, 2.36 mmol) and dicarbonyl(η^5-cyclopentadienyl)cobalt (1.24 g, 7 mmol) in degassed dodecane–tridecane (2:1 mixture, 35 mL) is added to the dropping funnel. The stirred solvent mixture is heated (with the minimum external heating to minimize charring) to reflux temperature and the reactants are added dropwise from the dropping funnel at a rate such that the red color of $Co(\eta^5\text{-Cp})(CO)_2$ vapor is just maintained in the flask (this takes from 1 to 2 h). A slow N_2 flush of $\sim 2 \, cm^3 \, s^{-1}$ can be maintained to aid the removal of evolved CO. Under this treatment the solution quickly becomes purple-brown. Once the reactants have been added, the solution is refluxed for a further 30 min and then cooled to room temperature. Insoluble materials are removed by filtration and the residue is washed with CH_2Cl_2 until the washings show no purple color. The washings are combined with the filtrate and the CH_2Cl_2 is then evaporated from the combined filtrate and washings on a rotary evaporator. At this point, a thin layer chromatographic (TLC) analysis (Merck PF60 254, elutant hexane) should show only one purple spot, $R_f \sim 0.1$. Purple flakes of the product precipitate from the solution on standing at 0 °C overnight. These are separated by filtration and washed with three 3-mL portions of cold pentane to remove high-boiling hydrocarbon contaminants. Alternatively, the high boiling solvents are removed (and can be recovered) as follows. The solution obtained after removal of the dichloromethane is chromatographed on a silica gel column (400 × 60 mm containing 0.063-mm mesh Fluka AG reagent grade silicic acid). Elution with a 4:1 hexane–CH_2Cl_2 solvent mixture removes the alkane solvents, followed by the purple band of the product. Evaporation of the purple band and crystallization of the purple residue from hexane gives the pure product. Yield: 0.58–0.79 g (62–85% based on alkyne). The product can be recrystallized from dichloromethane–hexane (1:1) mp 255–256 °C.

Anal. Calcd. for $C_6H_5C\{(\eta^5\text{-}C_5H_5)Co\}_3CC_6H_5$: C, 63.27; H, 4.55. Found: C, 63.32; H, 4.50.

Properties

Tris(η^5-cyclopentadienyl)-bis(μ_3-phenylmethylidyne)-tricobalt is an air-stable, purple, crystalline solid. It is soluble in organic solvents to give intensely purple solutions that are stable in air for many hours. For prolonged storage, the compound should be kept under nitrogen. Its ^1H NMR spectrum (in CCl_4) consists of a singlet at δ 4.47 (15H, Cp), and multiplets at δ 7.40 (6H), and 8.20 (4H). The ^{13}C NMR spectrum [$CDCl_3$ solution, tetramethylsilane

(TMS) reference] consists of a singlet at δ 85.2 (15C, Cp), resonances due to the phenyl groups at δ 125.3, 126.7, 127.4, and 167.3, and a broad carbyne resonance at δ 334.6. The mass spectrum shows the parent ion at m/e 550 (100%) and major fragmentation peaks at 373 (27%, $Co_3Cp_3^+$), 189 (96%, $CoCp_2^+$), and 124 (30%, $CoCp^+$). Three bands are seen in the visible spectrum at (nm): 730 (sh) ($\varepsilon = 370$), 522 ($\varepsilon = 3480$), and 430 (sh) ($\varepsilon = 1890$). Although the structure of this cluster has not been determined those of $Co_3(\eta^5\text{-}Cp)_3(CSiME_3)_2$, (Ref. 6) $Co_3(\eta^5\text{-}Cp)_3[C(O)OMe](CMe)$ (Ref. 7), and $Co_3(\eta^5\text{-}Cp)_3[C\text{-}\eta^5\text{-}C_5H_4)Fe(\eta^5\text{-}C_5H_5)](CH)$,[4] have been reported. This compound undergoes a reversible one-electron oxidation to the monocation at 0.56 V (vs Ag–AgCl) in acetone. The cation can be isolated as a brown $[PF_6^-]$ salt by stoichiometric oxidation with $Ag[PF_6]$ in CH_2Cl_2.[4] Nucleophiles such as $(MeO)^-$ and PPh_3 (in air) react with the cation to regenerate the neutral complex. In contrast, analogous $Co_3(\eta^5\text{-}CP)_3(CR)(CH)$ complexes undergo electrophilic transformations at the carbyne group [involving Ag(I) in some cases] through prior attack across a Co—Co bond.[3]

References

1. H. Vahrenkamp, *Angew. Chem. Int. Ed. Engl.*, **17**, 379 (1978).

2. J. R. Fritch and K. P. C. Vollhardt, *Angew. Chem. Int. Ed. Engl.*, **19**, 559 (1980).

3. D. E. Van Horn and K. P. C. Vollhardt, *J. Chem. Soc. Chem. Commun.*, **1982**, 203; K. P. C. Vollhardt, and E. C. Walborsky, *J. Am. Chem. Soc.*, **105**, 5507 (1983) and references cited therein.

4. S. B. Colbran, B. H. Robinson, and J. Simpson, *Organometallics*, **3**, 1344 (1984).

5. R. B. King, *Organometallic Synthesis*, Vol. 1. Academic Press, New York, 1965.

6. J. R. Fritch, K. P. C. Vollhardt, M. R. Thompson, and V. W. Day, *J. Am. Chem. Soc.*, **101**, 2768 (1979).

7. H. Yamazaki, Y. Wakatsuki, and K. Aoki, *Chem. Lett.*, **1979**, 1041.

56. BIS(TETRAMETHYLAMMONIUM) HEXA-η-CARBONYL-HEXACARBONYLHEXANICKELATE(2 −), $[N(CH_3)_4]_2[Ni_6(CO)_6(\mu\text{-}CO)_6]$

$$5Ni(CO)_4 + 2NaOH \longrightarrow Na_2[Ni_5(CO)_{12}] + H_2O + CO_2 + 7CO$$

$$3Na_2[Ni_5(CO)_{12}] + 2H_2O \longrightarrow 2Na_2[Ni_6(CO)_{12}] + 3Ni(CO)_4$$
$$+ H_2 + 2NaOH$$

$$Na_2[Ni_6(CO)_{12}] + 2[N(CH_3)_4]Cl \longrightarrow [N(CH_3)_4]_2[Ni_6(CO)_{12}] + 2NaCl$$

Submitted by A. CERIOTTI, G. LONGONI, and G. PIVA*
Checked by JACKY ROSÉ and PIERRE BRAUNSTEIN†

Salts of the dianion $[Ni_6(CO)_{12}]^{2-}$ have been prepared by reduction of tetracarbonyl nickel under a variety of experimental conditions with yields seldom exceeding 40 to 50%.[1,2]

The procedure given here represents a high-yield reproducible synthesis for salts of $[Ni_6(CO)_{12}]^{2-}$ with several counterions, and affords fairly pure products. It involves the reduction of $Ni(CO)_4$ with alkali hydroxide in N,N-dimethylformamide (DMF) as solvent and the use of a low-temperature condenser to prevent $Ni(CO)_4$ loss due to the stripping action of the evolving carbon monoxide.

Initial attack of hydroxide ions onto $Ni(CO)_4$ gives as yet uncharacterized species,[1] which readily condense with unreacted $Ni(CO)_4$ to give variable mixtures of $[Ni_5(CO)_{12}]^{2-}$ and $[Ni_6(CO)_{12}]^{2-}$. Hydrolysis of $[Ni_5(CO)_{12}]^{2-}$ in water produces $[Ni_6(CO)_{12}]^{2-}$, according to the above reaction. The dianion $[Ni_6(CO)_{12}]^{2-}$ is isolated as the tetramethylammonium salt in a crystalline state in $\sim 70\%$ yield. The $[Ni_6(CO)_{12}]^{2-}$ dianion is a starting material for the synthesis of several other homo- and heterometallic nickel carbonyl cluster complexes.[1-9]

Procedure

■ **Caution.** *The reaction must be run in a well-ventilated hood owing to the high toxicity of Ni(CO)₄ and CO.*

This synthesis is performed in standard Schlenk equipment[10] under a

*Dipartimento di Chimica Inorganica e Metallorganica e Centro del CNR, via G. Venezian 21, 20133 Milano, Italy and Dipartimento de Chimica Fisica e Inorganica, via Risorgimento 4, 40136 Bologna, Italy.
†Laboratoire de Chimie de Coordination, Université Louis Pasteur, 4 rue Blaise Pascal, 67070 Strasbourg Cédex, France.

nitrogen atmosphere. Liquids are transferred using calibrated hypodermic glass syringes equipped with Luer lock valves[10] to prevent spilling. Commercial Ni(CO)$_4$, (available from Strem) is distilled before use. All solvents are laboratory grade, and are distilled and stored under an inert atmosphere.

A 250-mL three-necked round-bottomed, flask is equipped with a magnetic stirring bar, a serum-bottle cap, a stopcock attached to an inlet tube for nitrogen purge or vacuum, and a coil condenser carrying a stopcock and a bubbler at the top. The coil condenser is connected to a cryostat and cooled with circulating fluid at − 20 °C. Alternatively, the coil condenser and cryostat can be substituted by a Dewar condenser filled with a cooling bath of ethylene glycol and Dry Ice.

The apparatus is evacuated and filled with nitrogen several times. Under a brisk stream of nitrogen, the serum-bottle cap is lifted off, and DMF (25 mL) and NaOH in pellets (3 g) is placed into the flask. After recapping, the mixture is stirred for ∼ 20 min. A quantity of Ni(CO)$_4$ (4.2 mL, 32.5 mmol) is introduced into the flask through the serum-bottle cap with a calibrated syring equipped with a Luer lock valve. The reaction mixture is rapidly stirred at room temperature while a slow stream of nitrogen (10–15 bubbles per min) flushes through the reaction vessel, escaping through the bubbler at the top of the coil condenser. Within the first hour of reaction the solution undergoes several color changes. Evolution of carbon monoxide and Ni(CO)$_4$ reflux cease after ∼ 2 h.

After stirring for 6 h the reaction is completed and the solution (still containing unreacted NaOH) is brown. The coil condenser is replaced with a dropping funnel containing deaerated water (100 mL). This is rapidly introduced into the flask to quench the reaction. Upon standing overnight the reaction mixture turns red, owing to the transformation of $[Ni_5(CO)_{12}]^{2-}$ into $[Ni_6(CO)_{12}]^{2-}$. An air-free solution of tetramethylammonium chloride (2.5 g) in water (40 mL) is placed into the dropping funnel and slowly introduced into the rapidly stirred reaction mixture.

A red microcrystalline precipitate separates and is collected on a Schlenk frit. The solid product is washed with five 20-mL portions of air-free water and dried under vacuum (0.01 torr) overnight. Yield: 2.9–3.2 g [64–73% based on Ni(CO)$_4$]. This precipitate is sufficiently pure for most synthetic purposes. Further purification is obtained by extraction in acetone (20 mL) and crystallization by dropwise addition of 2-propanol (70 mL) under stirring.

Anal. Calcd. for $[N(CH_3)_4]_2$ $[Ni_6(CO)_{12}]\cdot CH_3C(O)CH_3$: Ni, 39.4%. Found: Ni, 39.0%. The product tightly retains solvent of crystallization, as shown by proton NMR.

Salts of $[Ni_6(CO)_{12}]^{2-}$ with other counterions such as $[N(C_2H_5)_4]^+$, $[N(CH_3)_3CH_2C_6H_5]^+$, $[N(n\text{-}C_4H_9)_4]^+$, or $[P(C_6H_5)_4]^+$, are similarly prepared in comparable yield.

Properties

Bis(tetramethylammonium) hexa-μ-carbonyl-hexacarbonylhexanickelate(2−) is obtained as shiny red crystals. The compound is air sensitive both in solution and in the solid state.

■ **Caution.** *The compound must be handled with care and under a hood both in the solid state and in solution because highly poisonous Ni(CO)$_4$ is set free upon decomposition.*

The compound is soluble in acetone, acetonitrile, DMF, and dimethyl sulfoxide; sparingly soluble in alcohols and tetrahydrofuran (THF); insoluble in nonpolar solvents and water. It decomposes in halogenated solvents such as $CHCl_3$ and CH_2Cl_2. The corresponding $[N(C_2H_5)_4]^+$ and $[N(C_4H_9)_4]^+$ salts are fairly soluble in THF. Solid samples can be stored for long periods of time under a nitrogen atmosphere and at room temperature.

The IR spectrum in acetonitrile shows carbonyl absorptions at 1981 (s), 1808 (m), and 1783 (ms) cm^{-1}. The ^{13}C NMR spectrum and a single-crystal X-ray structure have been reported.[3,4]

The dianion $[Ni_6(CO)_{12}]^{2-}$ is stable to hydrolysis in alkaline aqueous solution; however, in acidic conditions (pH = 3–6) it is converted quantitatively into the interstitial hydride complex $[Ni_{12}(CO)_{21}H_{4-n}]^{n-}$ (n = 2, 3).[5,6] The complex readily reacts in solution with carbon monoxide according to the following degradation–condensation equilibrium:[1]

$$[Ni_6(CO)_{12}]^{2-} + 4CO \rightleftharpoons [Ni_5(CO)_{12}]^{2-} + Ni(CO)_4$$

Condensation of M(CO)$_5$·THF (M = Cr, Mo, W) onto $[Ni_6(CO)_{12}]^{2-}$ readily results in the corresponding $[MNi_6(CO)_{17}]^{2-}$ derivatives.[2] Condensation with Ni(CO)$_4$ is slow, and affords $[Ni_7(CO)_{15}]^{2-}$ and $[Ni_9(CO)_{18}]^{2-}$, depending on the experimental conditions.[3,7] The latter is more conveniently obtained by controlled oxidation with Ni(II) and Fe(III) salts.[7,8] Reaction with Pt(II) salts, on the contrary, results in the recently characterized $[Ni_9Pt_3(CO)_{21}H_{4-n}]^{n-}$ (n = 2, 3, 4) and $[Ni_{38}Pt_6(CO)_{48}H_{6-n}]^{n-}$ (n = 3–6).[9–11] Condensation of $[Ni_6(CO)_{12}]^{2-}$ with Co$_3$(CO)$_9$CCl and [Rh(CO)$_2$Cl]$_2$ gives bimetallic Co—Ni and Ni—Rh clusters.[12–15] Reaction with stoichiometric amounts of CCl$_4$ affords the $[Ni_9(CO)_{17}C]^{2-}$ and $[Ni_8(CO)_{16}C]^{2-}$ interstitial carbides;[16] the corresponding reaction with perhalo hydrocarbons such as C_2Cl_4, C_2Cl_6, and C_3Cl_6 results in the direct synthesis of polycarbide clusters containing up to six interstitial carbon atoms.[17–20] The reaction of $[Ni_6(CO)_{12}]^{2-}$ with ERCl$_2$ [E = P, As; R = Me, Ph, and CH(SiMe$_3$)$_2$] gives Ni$_8$(CO)$_8$ (μ_4-PPh)$_6$, $[Ni_9(CO)_{15}$ (μ_5-AsPh)$_3]^{2-}$, and $[Ni_{10}(CO)_{18}(\mu_5$-AsMe)$_2]^{2-}$, as well as the diphosphene

$Ni_5(CO)_6[(Me_3Si)_2CHP\!=\!PCH\;(SiMe_3)_2]$ cluster.[21-23] Finally, the polymerization of acetylene in the presence of $[N(CH_3)_4]_2[Ni_6(CO)_{12}]$ has been reported.[24]

References

1. G. Longoni, P. Chini, and A. Cavalieri, *Inorg. Chem.*, **15**, 3025 (1976).
2. T. L. Hall and J. K. Ruff, *Inorg. Chem.*, **20**, 4444 (1981).
3. G. Longoni, B. T. Heaton, and P. Chini, *J. Chem. Soc. Dalton Trans.*, **1980**, 1537.
4. J. C. Calabrese, L. F. Dahl, A. Cavalieri, P. Chini, G. Longoni, and S. Martinengo, *J. Am. Chem. Soc.*, **96**, 2616 (1974).
5. R. W. Broach, L. F. Dahl, G. Longoni, P. Chini, A. J. Schultz, and J. M. Williams, *Adv. Chem. Ser.*, **167**, 93 (1978).
6. A. Ceriotti, P. Chini, R. Della Pergola, and G. Longoni, *Inorg. Chem.*, **22**, 1595 (1983).
7. G. Longoni and P. Chini, *Inorg. Chem.*, **15**, 3029 (1976).
8. D. A. Nagaki, L. D. Lower, G. Longoni, P. Chini, and L. F. Dahl, *Organometallics*, **5**, 1764 (1986).
9. A. Ceriotti, F. Demartin, G. Longoni, M. Manassero, M. Marchionna, G. Piva, and M. Sansoni, *Angew. Chem. Int. Ed. Engl.*, **24**, 697 (1986).
10. D. F. Shriver, *The Manipulation of Air-Sensitive Compounds*, McGraw-Hill, New York, 1969.
11. A. Ceriotti, F. Demartin, G. Longoni, M. Manassero, G. Piva, G. Piro, and M. Sansoni, *J. Organomet. Chem.*, **301**, C5 (1986).
12. A. Arrigoni, A. Ceriotti, R. Della Pergola, G. Longoni, M. Manassero, N. Masciocchi, and M. Sansoni, *Angew. Chem. Int. Ed. Engl.*, **23**, 322 (1984).
13. A. Ceriotti, R. Della Pergola, G. Longoni, M. Manassero, and M. Sansoni, *J. Chem. Soc. Dalton Trans.*, **1984**, 1181.
14. A. Ceriotti, R. Della Pergola, G. Longoni, M. Manassero, N. Masciocchi, and M. Sansoni, *J. Organomet. Chem.*, **330**, 237 (1987).
15. D. Nagaki, J. V. Badding, A. M. Stacey, and L. F. Dahl, *J. Am. Chem. Soc.*, **108**, 3825 (1986).
16. A. Ceriotti, G. Longoni, M. Manassero, M. Perego, and M. Sansoni, *Inorg. Chem.*, **24**, 117 (1985).
17. A. Ceriotti, G. Longoni, M. Manassero, N. Masciocchi, L. Resconi, and M. Sansoni, *J.Chem. Soc. Chem. Commun.*, 1985, 181.
18. A. Ceriotti, G. Longoni, M. Manassero, N. Masciocchi, G. Piro, L. Resconi, and M. Sansoni, *J. Chem. Soc. Chem. Commun.*, 1985, 1402.
19. A. Ceriotti, A. Fait, G. Longoni, G. Piro, L. Resconi, F. Demartin, M. Manassero, N. Masciocchi, and M. Sansoni, *J. Am. Chem. Soc.*, **108**, 5370 (1986).
20. A. Ceriotti, A. Fait, G. Longoni, G. Piro, F. Demartin, M. Manassero, N. Masciocchi, and M. Sansoni, *J. Am. Chem. Soc.*, **108**, 8091 (1986).
21. L. D. Lower and L. F. Dahl, *J. Am. Chem. Soc.*, **98**, 5046 (1976).
22. D. F. Rieck, R. A. Montag, T. S. McKechnie, and L. F. Dahl, *J. Am. Chem. Soc.*, **108**, 1330 (1986).
23. M. M. Olmstead and P. P. Power, *J. Am. Chem. Soc.*, **106**, 1495 (1984).
24. A. Ceriotti, G. Longoni, and P. Chini, *J. Organomet. Chem.*, **174**, C 27 (1979).

57. BIS(TETRABUTYLAMMONIUM) HEXA-μ-CARBONYL-HEXACARBONYLHEXAPLATINATE(2 −), $[N(C_4H_9)_4]_2[Pt_6(CO)_6(\mu\text{-}CO)_6]$

$$30Na_2[PtCl_6]\cdot 6H_2O + 121CO + 122CH_3COONa\cdot 3H_2O$$

$$\longrightarrow Na_2[Pt_{30}(CO)_{60}] + 180NaCl + 61CO_2 + 122CH_3COOH + 485H_2O$$

$$Na_2[Pt_{30}(CO)_{60}] + 8NaOH + 4CO \longrightarrow 5Na_2[Pt_6(CO)_{12}] + 4CO_2 + 4H_2O$$

$$Na_2[Pt_6(CO)_{12}] + 2[N(C_4H_9)_4]Br \longrightarrow [N(C_4H_9)_4]_2[Pt_6(CO)_{12}] + 2NaBr$$

Submitted by A. CERIOTTI, G. LONGONI, and M. MARCHIONNA* *
Checked by ANNE DEGREMONT, JACKY ROSE, and PIERRE BRAUNSTEIN[†]

Reductive carbonylation in alkaline methanol solution of $Na_2[PtCl_6]\cdot 6H_2O$ gives the series of dianions $[Pt_3(CO)_6]_n^{2-}$ ($n = 3$–10); the value of n depends on the alkaline reagent and the experimental conditions.[1] The lowest term of this series of inorganic oligomers, which has been isolated in the solid state, is the $[Pt_6(CO)_{12}]^{2-}$ dianion. Its reported synthesis is based on the reduction with lithium metal in tetrahydrofuran (THF) of $Pt(CO)_2Cl_2$, or of preformed $[Pt_3(CO)_6]_n^{2-}$ ($n > 3$).[1,2]

The two-step high-yield synthesis of $[Pt_6(CO)_{12}]^{2-}$ salts described here represents a slight modification of the original procedure, and employs readily available starting materials. The first step consists of the reductive carbonylation of $Na_2[PtCl_6]\cdot 6H_2O$ in methanol to a sparingly soluble material of approximate stoichiometry $Na_2[Pt_{30}(CO)_{60}]$, as previously described.[1] To avoid losses in case some more soluble green $Na_2[Pt_3(CO)_6]_n$ ($n < 10$) is present, precipitation is completed by addition of tetrabutylammonium bromide.

In the second step, the material is reduced with NaOH in N,N-dimethylformamide (DMF) and under a carbon monoxide atmosphere to $[Pt_6(CO)_{12}]^{2-}$. The resulting $[Pt_6(CO)_{12}]^{2-}$ dianion is isolated in the solid state as the tetrabutylammonium salt in yield > 80%.

The $[Pt_6(CO)_{12}]^{2-}$ dianion can be used as starting material for the synthesis of other platinum carbonyl cluster complexes.[1,3]

*Dipartimento di Chimica Inorganica e Metallorganica e Centro del CNR, via G. Venezian 21, 20133 Milano, Italy.
†Laboratoire de Chimi e de Coordination Université Louis Pasteur, 4 rue Blaise Pascal, 67070 Strasbourg Cédex, France.

Procedure

- **Caution.** *The reaction must be run in a well-ventilated hood owing to the toxicity of carbon monoxide.*

The synthesis is performed in standard Schlenk equipment[4] under a carbon monoxide atmosphere, if not otherwise stated. Transfer of liquid is carried out with hypodermic glass syringes equipped with Luer lock valves. All solvents are laboratory grade, distilled and stored under inert atmosphere.

A quantity of $Na_2[PtCl_6]\cdot 6H_2O$ (4.9 g; Pt, 34.24%) and methanol (60 mL) are placed into a 250-mL two-necked round-bottomed flask equipped with a magnetic stirring bar, a stopcock attached to an inlet tube for carbon monoxide purge or vacuum, and a stopper. The apparatus is evacuated and filled with carbon monoxide two times. While a brisk stream of carbon monoxide is flowing, a quantity of $CH_3COONa\cdot 3H_2O$ (4.9 g) is introduced into the solution and a stopcock adapter connected to a bubbler filled to a 30 mm height of mercury is fitted onto the flask. The solution is rapidly stirred at room temperature for ~ 24 h under a slow stream (~ 10 bubbles per minute) of carbon monoxide. Over this period of time a silky brown precipitate separates. Tetrabutylammonium bromide (0.5 g) in deaerated methanol (10 mL) is added to complete the precipitation. The solid is collected on a Schlenk frit, washed with five portions of methanol (20 mL), five portions of deaerated water (20 mL), and dried in vacuum (0.01 torr) overnight.

Most of the product is transferred under a carbon monoxide atmosphere into a Schlenk tube (250 mL); the last traces of solid are recovered from the frit with the help of five 10-mL portions of DMF previously saturated with NaOH. A magnetic stirring bar and NaOH pellets (2 g) are introduced into the Schlenk tube, and the suspension is stirred at room temperature for ~ 10 h while maintaining a slight positive pressure of carbon monoxide, which must contain < 2 ppm of oxygen.

Over this period of time the solution undergoes several color changes from green to red. The final orange-red solution is filtered from unreacted NaOH and the filtrate collected in a second Schlenk tube of similar volume. An air-free solution of tetrabutylammonium bromide (4.5 g) in deaerated water (30 mL) is placed into a dropping funnel and slowly added into the rapidly stirred reaction solution. Precipitation is completed by dropwise addition of water (50 mL).

The resulting red-brown precipitate is collected on a Schlenk frit, washed with five 20-mL portions of water, followed by a 1:1 mixture of 2-propanol and water (20 mL). The solid is dried in vacuum. This precipitate is sufficiently pure for most synthetic purposes. Yield·2.33–2.89 g (70–83% based on starting platinum).

Anal. Calcd. for $[N(C_4H_9)_4]_2[Pt_6(CO)_{12}]$: Pt, 58.8%; $[N(C_4H_9)_4]$, 24.31%. Found: Pt, 59.5%; $[N(C_4H_9)_4]$, 24.51%. Further purification is achieved by crystallization in either THF–heptane or acetone–2-propanol mixture.

Other salts of the $[Pt_6(CO)_{12}]^{2-}$ dianion ($[N(CH_3)_4]^+$, $[N(C_2H_5)_4]^+$, $[N(CH_3)_3CH_2C_6H_5]^+$, or $[P(C_6H_5)_4]^+$) can be similarly prepared in comparable yield.

Note Regarding Analytical Procedures. The analyses were performed by decomposition at $-70°C$ of the salts in methanol with a slow stream of chlorine to obtain NR_4^+ and $PtCl_6^{2-}$. After switching to water as solvent, the analysis of platinum was made by atomic absorption, whereas the analyses of NR_4^+ are conventional gravimetric analyses by precipitation with a solution of $Na[BPh_4]$. In general the most significant data for these compounds are the NR_4^+/M ratios and their percentages, rather than C, H, N conventional analyses, which are often misleading due to the presence of clathrated solvent in the sample.

Properties

Bis(tetrabutylammonium) hexa-μ-carbonyl-hexacarbonylhexaplatinate(2−) is obtained as dark red crystals. These show a weak Curie-type paramagnetism.[5] The compound is air sensitive both in the solid state and in solution. Aereal oxidation in solution is shown by change in color from orange-red to violet-red or green, owing to subsequent formation of the higher dianions, $[Pt_9(CO)_{18}]^{2-}$ and $[Pt_{12}(CO)_{24}]^{2-}$. The compound is soluble in THF, acetone, acetonitrile, and DMF; sparingly soluble or insoluble in 2-propanol, water and nonpolar organic solvents. It decomposes in halogenated solvents. Solid samples can be stored for a long period of time in an inert atmosphere and at room temperature.

The IR spectrum in THF shows carbonyl absorptions at 2003 (s), 1824 (m), 1803 (ms), and 1785 (sh, w) cm^{-1}. The ^{13}C and ^{195}Pt NMR spectra have been reported.[3,6] UV/Vis (THF, $\lambda_{max}(\varepsilon$, L mol^{-1} cm^{-1}): 468 nm (1.3×10^4), 422 nm (1.0×10^4), 305 nm (2.1×10^4), 256 nm (sh, 2.1×10^4).[7] The solid state structure has been established by X-ray studies of a single crystal of $[P(C_6H_5)_4]_2[Pt_6(CO)_{12}]$.[2] An X-ray photoemission study has also been reported.[8]

Controlled reaction of salts of $[Pt_6(CO)_{12}]^{2-}$ with several oxidizing agents may be used to synthesize the other members of the $[Pt_3(CO)_6]_n^{2-}$ series, $n = 2-5$. Reduction with Na–K alloy in THF affords the very unstable $[Pt_3(CO)_6]^{2-}$ dianion.[1] Intermolecular exchange of $M_3(CO)_3(\mu$-CO)$_3$ units between $[Pt_6(CO)_{12}]^{2-}$ and $[Ni_6(CO)_{12}]^{2-}$ affords the bimetallic derivative $[Ni_3Pt_3(CO)_{12}]^{2-}$.[3]

The complex $[N(C_4H_9)_4]_2[Pt_6(CO)_{12}]$ has also been used to prepare highly dispersed supported platinum aggregates.[9,10]

References

1. G. Longoni and P. Chini, *J. Am. Chem. Soc.*, **98**, 7225 (1976).
2. J. C. Calabrese, L. F. Dahl, P. Chini, G. Longoni, and S. Martinengo, *J. Am. Chem. Soc.*, **96**, 2614 (1974).
3. (a) C. Brown, B. T. Heaton, A. D. C. Towl, P. Chini, A. Fumagalli, and G. Longoni, *J. Organomet. Chem.*, **181**, 233 (1979); (b) M. J. D'Aniello, C. J. Carr, and M. G. Zammit, see procedures in the synthesis immediately following (No. 58).
4. D. F. Shriver and M. A. Drezdzon, *The Manipulation of Air Sensitive Compounds*, Wiley-Interscience, New York, 1986.
5. B. K. Teo, F. J. DiSalvo, J. V. Waszczak, G. Longoni, and A Ceriotti, *Inorg. Chem.*, **25**, 2262 (1986).
6. C. Brown, B. T. Heaton, P. Chini, A. Fumagalli, and G. Longoni, *J. Chem. Soc. Chem. Commun.*, **1977**, 309.
7. Data provided by M. J. D'Aniello, see procedures immediately following (No. 58).
8. G. Apai, S. T. Lee, M. G. Mason, L. J. Gerenser, and S. A. Gardner, *J. Am. Chem. Soc.*, **101**, 6880 (1979).
9. M. Ichikawa, *Chem. Lett. Jpn.*, **1976**, 335.
10. M. Ichikawa, *J. Chem. Soc. Chem. Commun.*, **1976**, 11.

58. PREPARATION OF THE CARBONYL PLATINUM ANIONS, $[Pt_3(CO)_6]_n^{2-}$, $(n = 3-5)$

Submitted by MICHAEL J. D'ANIELLO, JR., CONSTANCE J. CARR, and MICHAEL G. ZAMMIT*
Checked by ANNE DEGREMONT, JACKY ROSÉ, and PIERRE BRAUNSTEIN[†]

The carbonyl platinum anions, $[Pt_3(CO)_6]_n^{2-}$, $(n = 1-6, 10)$ were first synthesized and characterized by Chini and coworkers[1-3]. They obtained these compounds by reaction of Pt(IV) or Pt(II) salts at room temperature with bases such as sodium hydroxide or sodium acetate under a carbon monoxide atmosphere. The product composition is quite sensitive to the Pt–base ratio, reaction time, and reaction conditions. As a consequence of this sensitivity, product mixtures with $\Delta n = 1$ are usually obtained, which are separable only with difficulty by fractional crystallization. Interest in this series of compounds for (a) their unique redox solution chemistry, (b) their use as precursors for higher nuclearity carbonyl platinum anions,[4] and (c) their use as precursors for novel supported Pt catalysts[5-8] prompted efforts to develop

*Physical Chemistry Department, General Motors Research Laboratories, Warren, MI, 48090.
[†]Laboratoire de Chimie de Coordination, Université Louis Pasteur, 4 rue Blaise Pascal, 67070 Strasbourg, France.

improved, more consistent syntheses of these compounds, which do not involve the separation of mixtures. The preparations reported here are based primarily on the rapid, quantitative redox reactions that occur between nonconsecutive members of this series of compounds.

A. BIS(TETRAETHYLAMMONIUM) PENTAKIS[TRI-μ-CARBONYL-TRICARBONYLTRIPLATINATE(2−)], $[N(C_2H_5)_4]_2[Pt_3(CO)_6]_5$

$$15[PtCl_6]^{2-} + 61CO + 62OH^-$$

$$\longrightarrow [Pt_3(CO)_6]_5^{2-} + 90Cl^- + 31CO_2 + 31H_2O$$

Procedure

■ **Caution.** *Carbon monoxide is a colorless, odorless, very toxic gas. The following reaction must be conducted in an efficient, well-ventilated hood.*

The solvents used in this and the following preparations were purified by distillation under argon from the following reagents: methanol–$Mg(OCH_3)_2$; acetone–4A molecular sieves; tetrahydrofuran (THF) and *t*-butyl methyl ether–sodium benzophenone ketyl.

A 500-mL three-necked round-bottomed flask is equipped with a gas dispersion tube connected by a length of rubber tubing to a cylinder (preferably aluminum) of 99.99% CO. Alternatively, a lower grade of CO may be used if treated to remove impurities such as water, oxygen and, in particular, $Fe(CO)_5$ (a serious problem in old steel cylinders of CO). Another neck of the flask is connected to a standard Schlenk system, which allows alternate evacuation and backfilling (with argon or N_2) of the flask. The flask is then charged through the third neck with $Na_2PtCl_6 \cdot 6H_2O$ (15.0 g, 26.7 mmol, Strem), $NaC_2H_3O_2 \cdot 3H_2O$ (24.0 g, 176 mmol) and a stirring bar. The third neck is then sealed and the flask and associated dispersion tube and tubing are evacuated and refilled twice with argon. Methanol (175 mL) is introduced and a slow (~ 1 bubble per second) CO flow is established through the stirred, homogeneous orange solution, passing out through the Schlenk line bubbler (alternatively, a mineral oil bubbler can be connected at the third neck). The reaction is then allowed to proceed for 24 h, during which the solution changes color slowly from orange to dark green and a fine crystalline precipitate (presumably NaCl) appears.

The reaction mixture is worked up after 24 h using standard Schlenk or inert atmosphere box techniques. The reaction mixture is filtered with a medium porosity filter and the residue washed with methanol (2 × 10 mL). The filtrate and washes are combined and treated with $N(C_2H_5)_4Cl$ (4.5 g) in methanol (15 mL), which produces a violet precipitate. The mixture is stirred for 30 min after which the crude product is collected on a medium porosity filter and dried in vacuum. The dry, crude product is extracted on the filter with acetone (4 × 15 mL), each aliquot being drawn through the filter leaving NaCl and some Pt metal on the filter. The combined extracts are condensed to 30 mL under vacuum and then treated (slowly, with stirring) with *t*-butyl methyl ether (used in our laboratory as a diethyl ether substitute) and allowed to stir for 30 min. The product is then collected on a filter, washed with *t*-butyl methyl ether, and dried in vacuum. Yield: 4.2 g (60%).

Anal. Calcd. for $C_{46}H_{40}O_{30}N_2Pt_{15}$: Pt, 72.67. Found: Pt (by ignition), 72.81, 72.54. This material may occasionally contain up to 10% $[N(C_2H_5)_4]_2[Pt_3(CO)_6]_4$, which can be removed by a second precipitation with some loss in yield.

Additional material may be recovered by evaporation of the filtrate and recrystallization of the residue, as above. The second crop is often contaminated with $[N(C_2H_5)_4]_2[Pt_3(CO)_6]_4$.

Properties

The pentamer (yellow-green in solution, violet in the solid) is somewhat air and moisture sensitive. It begins to decompose $\sim 60\,°C$, but can be stored at room temperature indefinitely in an inert atmosphere. The compound is soluble in acetone, methanol, acetonitrile, and THF and insoluble in hydrocarbons and aliphatic ethers. It does not lose CO at an appreciable rate in solution at room temperature, but does so at elevated temperatures with the formation of larger clusters.[4] The product purity is best assayed by UV–Vis spectroscopy, where traces of $[Pt_3(CO)_6]_4^{2-}$ appear as a shoulder near 620 nm on the 706 nm peak of $[Pt_3(CO)_6]_5^{2-}$. IR (THF, cm^{-1}): 2057(s), 1896(w), 1872(m), 1845(w), 1831(w). UV–Vis [THF, λ_{max}, (ε, $L\,mol^{-1}\,cm^{-1}$)]: 706 nm (59,000), 410 nm(57,000), 342 nm (sh, 24,000), 271 nm (sh, 42,000).

B. BIS(TETRAETHYLAMMONIUM) TETRAKIS[TRI-μ-CARBONYL-TRICARBONYLTRIPLATINATE(2−)], $[N(C_2H_5)_4]_2[Pt_3(CO)_6]_4$

$$2[Pt_3(CO)_6]_5^{2-} + [Pt_3(CO)_6]_2^{2-} \longrightarrow 3[Pt_3(CO)_6]_4^{2-}$$

Procedure

The salt $[N(C_2H_5)_4]_2[Pt_3(CO)_6]_5$ is obtained as in the procedure in Section A. The salt $[N(C_2H_5)_4]_2[Pt_3(CO)_6]_2$ is obtained by the method of

Ceriotti, Longoni and Marchionna.[8] The solvents are purified as indicated in the procedure in Section A; care should be taken here since the dimer is very sensitive to oxygen and moisture.

A portion of $[N(C_2H_5)_4]_2[Pt_3(CO)_6]_5$ (1.000 g, 0.248 mmol) is dissolved in THF (60 mL) under an argon atmosphere. The resulting green solution is treated with $[N(C_2H_5)_4]_2[Pt_3(CO)_6]_2$ (0.220 g, 0.124 mmol); the solution immediately turns blue-green and is left to stir for 30 min. After this period, the THF is removed under vacuum and the residue dissolved in a minimum ($\sim$ 10 mL) of acetone. The product is precipitated by the slow addition, with agitation, of 50 mL of *t*-butyl methyl ether, collected on a filter, and dried in vacuum. Yield: 1.0–1.1 g (76–83%).

Anal. Calcd. for $C_{40}H_{40}O_{24}N_2Pt_{12}$: Pt, 71.51. Found: Pt (by ignition), 71.66, 71.48.

The reaction can be followed by UV–Vis or IR spectroscopy and adjustments in the product distribution can be made by adding small amounts of either the dimer or pentamer. This is generally not necessary unless the starting pentamer contains significant ($> 10\%$) amounts of the hexamer or tetramer, leading to the formation of pentamer or trimer, respectively. The trimer can be removed by the addition of more pentamer, while the pentamer can be removed by the addition of more dimer.

Properties

The properties of the tetramer (blue-green in solution) are similar to those of the pentamer. The purity of the product is best ascertained by UV–Vis spectroscopy. IR (THF, cm^{-1}): 2045 (s), 1880 (m), 1860 (s), 1825 (m). UV–Vis [THF, λ_{max}, (ε, L mol^{-1} cm^{-1})]: 620 nm (42,000), 513 nm (7,400), 394 nm (42,000).

C. BIS(TETRAETHYLAMMONIUM) TRIS[TRI-μ-CARBONYL-TRICARBONYLTRIPLATINATE(2−)], $[N(C_2H_5)_4]_2[Pt_3(CO)_6]_3$

$$[Pt_3(CO)_6]_5^{2-} + 2[Pt_3(CO)_6]_2^{2-} \longrightarrow 3[Pt_3(CO)_6]_3^{2-}$$

Procedure

This reaction is conducted in a manner very similar to that in the procedure in Section B. A portion of $[N(C_2H_5)_4]_2[Pt_3(CO)_6]_5$ (0.750 g, 0.186 mmol) is dissolved in THF (50 mL) under an argon atmosphere. The green solution is treated with $[N(C_2H_5)_4]_2[Pt_3(CO)_6]_2$ (0.660 g, 0.374 mmol), whereupon the solution turns violet-red. After stirring for 30 min, the THF is evaporated in vacuum and the residue dissolved in a minimum (~ 15 mL) of methanol. The product is precipitated by slow addition, with agitation, of *t*-butyl methyl ether (75 mL), collected on a filter, and dried in vacuum. Yield: 1.05 g (75%).

Anal. Calcd. for $C_{34}H_{40}O_{18}N_2Pt_9$: Pt, 69.65. Found: Pt (by ignition), 70.00, 69.66.

As noted in the preparation in Section B, the reaction can be followed by UV–Vis or IR spectroscopy. If necessary, small amounts of the dimer can be added to remove traces of tetramer while the pentamer can be added to remove traces of dimer.

Properties

The properties of the trimer (red-violet in solution) are similar to those of the other members of this series of compounds. The purity of the product is best ascertained by IR or UV–Vis spectroscopy. IR (THF, cm^{-1}): 2030 (s), 1862 (m), 1845 (ms), 1832 (sh), 1812 (w). UV–Vis [THF, λ_{max}, (ε, $L\,mol^{-1}\,cm^{-1}$)]: 562 nm (25,000), 506 nm (20,000), 422 nm (16,000), 367 nm (40,000), 247 nm (40,000).

References

1. G. Longoni, and P. Chini, *J. Am. Chem. Soc.*, **98**, 7225 (1976).
2. J. C. Calabrese, et al., *J. Am. Chem. Soc.*, **96**, 2614 (1974).
3. C. Brown, et al., *J. Organometal. Chem.*, **181**, 233 (1979).
4. D. M. Washecheck, et al., *J. Am. Chem. Soc.*, **101**, 6110 (1979).
5. M. Ichikawa, *Bull. Chem. Soc. Jpn.*, **51**, 2268 (1978).
6. A. F. Simpson, and R. Whyman, *J. Organometal. Chem.*, **213**, 157 (1981).
7. K. -i. Machida, et al., *J. Chem. Soc. Chem. Commun.*, **1987**, 1486.
8. A. Ceriotti, G. Longoni, and Marchionna, *Inorg. Synth.*, **26**, 316 (1989).

59. SYNTHESIS OF GOLD-CONTAINING MIXED-METAL CLUSTER COMPLEXES

Submitted by M. I. BRUCE,* B. K. NICHOLSON,[†] and O. BIN SHAWKATALY*
Checked by J. R. SHAPLEY[‡] and T. HENLY[‡]

Recently, there has been much interest in the syntheses and chemistry of mixed-metal compounds containing $Au(PR_3)$ moieties, and to a lesser extent, the silver and copper analogs. Although the first detailed studies of these compounds were reported in 1964,[1] recent work has largely resulted from the suggestion that the $Au(PR_3)$ group is isolobal with H, and that structural investigations of gold-containing complexes might give information about the corresponding hydrido clusters.[2] Since that time, several groups have synthesized a variety of cluster complexes containing gold, and it has become apparent that the isolobal relationship breaks down when more than one $Au(PR_3)$ group is present, as a result of the overriding tendency for the formation of Au—Au bonds.

Approaches to syntheses have generally employed reactions between cluster anions and $AuCl(PR_3)$, often in the presence of $Tl[PF_6]$, which increases conversion by precipitating $TlCl$;[3] the elimination of CH_4 in reactions between metal hydrides and $AuMe(PPh_3)$;[4] and the reaction of cluster anions with the μ_3-oxo-trigold(1+) salt, $[O\{Au(PPh_3)\}_3][BF_4]$.[5] Whereas the first two routes appear to be limited to the introduction of up to the same number of $Au(PR_3)$ units as there are charges or hydrides, respectively, the μ_3-oxo-trigold(1+) reagent can introduce three $Au(PPh_3)$ units directly, even in reactions with monoanions, with concomitant displacement of a two-electron ligand, usually CO. This reagent, which was first described by Russian workers,[6] was used by them to replace reactive hydrogen atoms in organic molecules, such as $CH_2(CN)_2$, PhC_2H, C_5HPh_4, and ferrocene, by $Au(PPh_3)$ groups. In the examples cited, the products were $Au[CH(CN)_2](PPh_3)$, $Au(C_2Ph)(PPh_3)$, $Au(C_5Ph_4)(PPh_3)$ and $[Au(PPh_3)]_3C_5Ph_4$, and $[Fe(\eta\text{-}C_5H_5)\{\eta\text{-}C_5H_4Au_2(PPh_3)_2\}][BF_4]$.[7] We have found that this reagent reacts with a range of anionic metal cluster complexes, to give mixed-metal clusters containing gold, many of which have unusual core geometries. This account describes the synthesis of $AuCl(PPh_3)$, its conversion to $[Au_3(\mu_3\text{-}O)(PPh_3)_3]$, and an example of the use of this reagent to make the cobalt–gold–ruthenium cluster,

*Department of Physical and Inorganic Chemistry, University of Adelaide, Adelaide, South Australia 5000.
[†]Department of Chemistry, University of Waikato, Hamilton, New Zealand.
[‡]Department of Chemistry, University of Illinois, Urbana, IL 61801.

$Au_3CoRu_3(CO)_{12}(PPh_3)_3$. The first two preparations are minor modifications of those reported in the literature.[6,8]

A. CHLORO(TRIPHENYLPHOSPHINE)GOLD, AuCl(PPh₃)

$$H[AuCl_4] + 2PPh_3 + H_2O \longrightarrow AuCl(PPh_3) + Ph_3PO + 3HCl$$

Sources

Hydrogen tetrachloroaurate(1 −), $H[AuCl_4] \cdot nH_2O$, or "brown gold chloride," was obtained from Johnson Matthey; this compound is quite hygroscopic, and may plate out gold on metal spatulas. Triphenylphosphine was obtained from BDH Chemicals.

Procedure

Triphenylphosphine (3.08 g, 11.74 mmol) is dissolved in 50 mL of absolute (i.e., 100%) ethanol contained in a 100-mL conical flask, warming if necessary to complete dissolution. A solution of 2.0 g (5.87 mmol) of hydrogen tetrachloroaurate(1 −), $H[AuCl_4] \cdot nH_2O$, in 10 mL of ethanol is filtered, to remove any insoluble gold-containing material, into the stirred solution of triphenylphosphine. This gives a white precipitate. After stirring for 15 min, the white microcrystalline solid is collected on a fritted glass filter (Corning porosity 4), washed with ethanol (2 × 5 mL), and dried in vacuum (0.1 torr) to give pure AuCl(PPh₃). Yield: 1.94–2.32 g (67–80%).

Anal. Calcd. for $C_{18}H_{15}AuClP$: C, 43.6; H, 3.1. Found: C, 43.5; H, 3.2.

Properties

Chloro(triphenylphosphine)gold forms white crystals (mp 242–243 °C), which are stable indefinitely in air, and are soluble in acetone, benzene, and dichloromethane. The crystals are insoluble in ethanol. The molecular structure has been determined, and shows the compound to contain linear two-coordinate gold.[9] This compound is a useful starting material for the preparation of gold(I) complexes, such as $AuMe(PPh_3)$,[10] $Au(C_2Ph)(PPh_3)$,[11] and $Au_2(\mu\text{-S})(PPh_3)_2$.[12] An alternative route to this compound is by displacement of Me_2S from $AuCl(SMe_2)$ by the tertiary phosphine.[13] In our hands, this reaction gives a lower overall yield of $AuCl(PPh_3)$, although we find it to be the method of choice for other tertiary phosphine complexes. The reason for this probably lies in the unique solubility properties of PPh_3 complexes, a feature that has been noted previously in the preparation of platinum metal derivatives.[14]

B. μ_3-OXO-[TRIS(TRIPHENYLPHOSPHINE)GOLD](1+) TETRAFLUOROBORATE(1−), [O{Au(PPh$_3$)}$_3$][BF$_4$]

$$3AuCl(PPh_3) + Ag_2O + NaBF_4 \longrightarrow [O\{Au(PPh_3)\}_3][BF_4]$$
$$+ 2AgCl + NaCl$$

Procedure

Silver oxide is prepared by adding a solution of 0.45 g (11.2 mmol) of sodium hydroxide in 10 mL of water to a solution of 1.88 g (11.0 mmol) of silver nitrate in 10 mL of water. The brown precipitate is removed by filtration on a glass frit, washed with water (2 × 5 mL), ethanol (2 × 5 mL), and acetone (2 × 5 mL), and air-dried. The freshly prepared solid silver oxide is added to a 250-mL round-bottomed, flask containing a solution of 1.5 g (3.0 mmol) of AuCl(PPh$_3$) in 100 mL of acetone, together with a magnetic stirring bar. This is followed by addition of 1.88 g (17.1 mmol) of sodium tetrafluoroborate. The mixture is stirred rapidly for 1 h. After this time, the acetone is removed (rotary evaporator), and the solid residue is extracted with chloroform (3 × 15 mL). The combined extracts are filtered into 150 mL of freshly distilled dry diethyl ether to precipitate the μ_3-oxo-trigold(1+) compound, which is removed by filtration on a glass frit and air-dried. Yield: 1.14–1.20 g (76–80%). The product is sufficiently pure to be used in further reactions, but may have a slight purple tinge. If necessary, the compound can be recrystallized by dissolving in the minimum amount of chloroform and adding ~ 1.5-fold volume of acetone.

Anal. Calcd. for $C_{54}H_{45}Au_3BF_4OP_3$: C, 43.8; H, 3.1; Found: C, 43.7; H, 2.9.

Properties

The pure μ_3-oxo-trigold(1+) compound forms white crystals [mp 207–208 °C (dec)], which are soluble in chloroform and dichloromethane but insoluble in hexane and diethyl ether. The IR spectrum (Nujol) contains a broad band from 1040 to 1070 cm^{-1} (BF$_4$). The structure of the cation has been described: It is dimeric, the two pyramidal [Au$_3$(μ_3-O)(PPh$_3$)$_3$] units interacting via gold–gold bonds involving two of the three gold atoms in each unit, with Au—Au′ separations of ~ 3.16 Å. The resulting six-membered Au$_2$OAu$_2$O heterocycle has a chair conformation.[6]

C. DODECACARBONYLTRIS(TRIPHENYLPHOSPHINE)-COBALTTRIGOLDTRIRUTHENIUM, $Au_3CoRu_3(CO)_{12}(PPh_3)_3$

$$[N(PPh_3)_2][CoRu_3(CO)_{13}] + [Au_3(\mu_3\text{-}O)(PPh_3)_3]$$
$$\longrightarrow Au_3CoRu_3(CO)_{12}(PPh_3)_3 + [N(PPh_3)_2][BF_4] + CO_2$$

■ **Caution.** *Due to toxicity of metal carbonyls and possible evolution of toxic carbon monoxide, this procedure should be carried out in a well-ventilated hood.*

Procedure

The reaction is carried out in a 100-mL Schlenk tube under a nitrogen atmosphere, using standard techniques for handling oxygen- and moisture-sensitive materials.[15] The Schlenk tube, which contains a magnetic stirring bar, is charged with 0.05 g (0.04 mmol) $[N(PPh_3)_2][Ru_3Co(CO)_{13}]$,[16] 0.054 g (0.036 mmol) $[Au_3(\mu_3\text{-}O)(PPh_3)_3][BF_4]$, and 10 mL of dry deoxygenated freshly distilled tetrahydrofuran. The mixture is stirred for 2 h at room temperature, after which time the mixture is evaporated to dryness (*in vacuo* at room temperature). The residue is extracted with 5 mL of benzene and 10 mL of petroleum ether (bp 40–60 °C) is carefully added to the filtered extract. After cooling for 18 h at − 30 °C, the black crystals, which have deposited, are filtered on a glass sinter, washed with 2-mL cold petroleum ether, and dried in vacuum (0.1 torr) to give the cluster complex as a benzene solvate. Yield: 0.047 g (55%).

Anal. Calcd. for $C_{66}H_{45}Au_3CoO_{12}P_3Ru_3 \cdot C_6H_6$: C, 40.15; H, 2.40. Found: C, 40.02; H, 2.54.

Properties

The complex forms black crystals, which are stable in air. It is soluble in polar organic solvents (acetone, dichloromethane), giving intensely colored brown solutions, which slowly decompose on prolonged exposure to air. The IR spectrum (CH_2Cl_2) contains $\nu_{(CO)}$ bands at 2058 (s), 2014 (s), 2000 (s), 1984 (m), 1956 (m), 1840 (w), and 1810 (w) cm^{-1}. The molecular structure has been determined. The Au_3CoRu_3 cluster has a bicapped trigonal pyramidal geometry, with two gold atoms capping the AuCoRu and $AuRu_2$ faces of the $AuCoRu_3$ core.[17]

References

1. C. E. Coffey, J. Lewis, and R. S. Nyholm, *J. Chem. Soc.*, **1964**, 1741.

2. J. W. Lauher and K. Wald, *J. Am. Chem. Soc.*, **103**, 7648 (1981).

3. J. Lewis and B. F. G. Johnson, *Pure Appl. Chem.*, **54**, 97 (1982); K. Burgess, B. F. G. Johnson, J. Lewis, and P. R. Raithby, *J. Chem. Soc. Dalton Trans.*, **1983**, 1661; and references cited therein.

4. L. J. Farrugia, J. A. K. Howard, P. Mitrprachachon, J. L. Spencer, F. G. A. Stone, and P. Woodward, *J. Chem. Soc. Chem. Commun.*, **1978**, 260; I. D. Salter and F. G. A. Stone, *J. Organomet. Chem.*, **260**, C71 (1984); and references cited therein.

5. M. I. Bruce and B. K. Nicholson, *J. Chem. Soc. Chem. Commun.*, **1982**, 1141; *J. Organomet. Chem.*, **252**, 243 (1983).

6. A. N. Nesmeyanov, E. G. Perevalova, Yu. T. Struchkov, M. Y. Antipin, K. I. Grandberg, and V. P. Dyadchenko, *J. Organomet. Chem.*, **201**, 343 (1980).

7. T. V. Baukova, Yu. L. Slovokhotov, and Yu. T. Struchkov, *J. Organomet. Chem.*, **220**, 125 (1981); **221**, 375 (1981).

8. C. Kowala and J. M. Swan, *Aust. J. Chem.*, **19**, 547 (1966).

9. N. C. Baenziger, W. E. Bennett, and D. M. Soboroff, *Acta Crystallogr.*, **B32**, 926 (1976).

10. G. E. Coates and C. Parkin, *J. Chem. Soc.*, **1963**, 421.

11. G. E. Coates and C. Parkin, *J. Chem. Soc.*, **1962**, 3220.

12. C. Lensch, P. G. Jones, and G. M. Sheldrick, *Z. Naturforsch*, **37b**, 944 (1980).

13. F. Bonati and G. Minghetti, *Gazz. Chim. Ital.*, **103**, 373 (1973); A. K. Al-Sa'ady, C. A. McAuliffe, R. V. Parish and J. A. Sandbank, *Inorg. Synth.*, **23**, 191 (1985).

14. N. Ahmad, J. J. Levison, S. D. Robinson, and M. F. Uttley, *Inorg. Synth.*, **15**, 45 (1974).

15. D. F. Shriver, *The Manipulation of Air-sensitive Compounds*, McGraw-Hill, New York, 1969.

16. G. L. Geoffroy, J. R. Fox, E. Burkhardt, H. C. Foley, A. D. Harley, and R. Rosen, *Inorg. Synth.*, **21**, 61 (1982).

17. M. I. Bruce and B. K. Nicholson, *Organometallics*, **3**, 101 (1984).

60. MERCURY-BRIDGED TRANSITION METAL CLUSTER DERIVATIVES AND THEIR PRECURSOR

Submitted by EDWARD ROSENBERG AND BRUCE NOVAK*
Checked by W. EDWARD LINDSELL[†]

Transition metal complexes with mercury directly bound to one or two transition metal centers are well known.[1] They have found use as relatively air-stable sources of carbonyl metallate anions in polar solvents and as intermediates in the syntheses of bimetallic transition metal complexes. They are probably the key intermediates in the many transition metal promoted

*Department of Chemistry, California State University, Northridge, Northridge, CA 91330.
†Department of Chemistry, Heriot-Watt University, Riccarton, Edinburgh EH14 4AS, United Kingdom.

coupling reactions that utilize organomercury compounds. Until recently, this class of compounds consisted of two-centre two-electron mercury–transition metal bonds.

There are three important routes to the formation of the mercury–transition metal bond: (a) displacement of halogen or pseudohalogen from mercury(II) salts with carbonyl metallate anions; (b) reaction of a halophenylmercury compound with a transition metal hydride; and (c) oxidative addition of a mercury halide to neutral zerovalent metals.[1] We report here the syntheses of three compounds containing three-centre, two-electron, mercury–ruthenium bonds utilizing trinuclear cluster anions and mercury(II) halides.[2–4]

The reactivities of these compounds have yet to be compared in detail with those of the mononuclear analogs, but some preliminary results from our laboratories are summarized here. Substitution of up to three carbonyl ligands by triphenylphosphine can be accomplished by simple thermolysis with $[Ru_3(\mu_3\text{-}\kappa^2:\eta^2:\eta^2\text{-}C_2\text{-}t\text{-}Bu)\,(CO)_9]_2\,(\mu_4\text{-}Hg)$. Substitution of phosphine for carbonyl in $[Ru_3(\mu_3\text{-}\kappa^2:\eta:\eta^2\text{-}C_2\text{-}t\text{-}Bu)\,(CO)_9]\,(\mu\text{-}Hg)X$ cannot be accomplished owing to the facile thermal redistribution process these compounds undergo at elevated temperatures and in the presence of electron-rich phosphine substitution promotors $(Pt[P(C_6H_5)_3]_4, (C_6H_5)_2\!-\!C\!=\!O^-Na^+)$, giving $[Ru_3(\mu_3\text{-}\kappa^2:\eta^2:\eta^2\text{-}C_2\text{-}t\text{-}Bu)(CO)_9]_2Hg$ and HgX_2.[3] With reducing agents [e.g., $(C_6H_5)_2C\!=\!O^-$], $[Ru_3(\mu_3\text{-}\kappa^2:\eta^2:\eta^2\text{-}C_2\text{-}t\text{-}Bu)(CO)_9]_2Hg$ acts as a two-electron sink, fragmenting the cluster into mercury metal and two stable triruthenium anions.

The procedures for the syntheses presented here are representative and are general for a variety of transition metal cluster and mononuclear anions.

A.   NONACARBONYL-μ_3-(κ^2:η^2:η^2-3, 3-DIMETHYL-1-BUTYNYL)-μ-HYDRIDO-*TRIANGULO*-TRIRUTHENIUM

$$Ru_3(CO)_{12} + (CH_3)_3C\!\equiv\!CH$$
$$\xrightarrow[\text{reflux}]{\text{heptane}} Ru_3(\mu\text{-}H)(\mu_3\text{-}\kappa^2:\eta^2:\eta^2\text{-}C_2\text{-}t\text{-}Bu)(CO)_9 + 3CO$$

■   **Caution**   *Due to evolution of highly toxic Carbon monoxide gas, this reaction should be carried out in a well-ventilated hood.*

Starting Materials and Solvents

Dodecacarbonyltriruthenium is available from several commercial sources (Strem, Aldrich, or Pressure Chemicals), or may be synthesized from ruthenium trichloride hydrate according to well-known procedures.[5] The compound 3, 3-dimethyl-1-butyne (*tert*-butylacetylene) is available from Farchan.* All other compounds or solvents are those available from the usual commercial vendors. A 1-L three-necked flask is equipped with water-cooled reflux condenser connected to a nitrogen gas line through a T-tube leading to a mineral oil bubbler. The flask is flamed under dry nitrogen and then charged with dodecacarbonyltriruthenium (1.41 g, 2.21 mmol) and a magnetic stirring bar. Heptane (400 mL, dried over molecular sieves) is added to the flask and deaerated with nitrogen gas. To this resulting solution, 3, 3-dimethyl-1-butyne (0.88 mL, 7.2 mmol) is added by syringe and the mixture, with magnetic stirring, is heated to reflux using a heating mantle. The red-orange solution is allowed to reflux until the color changes to yellow (checkers advise 10–15 min to avoid further reactions). The color change must be monitored closely, as continued heating past the onset of a yellow color leads to formation of a green-brown $Ru_3H_2(CO)_8\text{-}(C_2\text{-}t\text{-}Bu)_2$ complex and other polymeric products. The reaction mixture is then cooled to room temperature, and the solvent is removed under reduced pressure (rotary evaporator). The resulting yellow crystals are washed with 10 mL of dry diethyl ether and recrystallized from hexane at $-20\,°C$. Yield: 1.3–1.4 g of the desired product [92–99% yield based on $Ru_3(CO)_{12}$].

Properties

The compound $Ru_3H(CO)_9(C_2\text{-}t\text{-}Bu)$ is an air-stable, yellow crystalline solid [mp 156–157 °C (dec)]. It is soluble in common organic solvents. The IR spectrum of this compound in hexane solution shows six maxima in the terminal carbonyl region: 2097 (m), 2070 (vs), 2054 (vs), 2023 (vs), 2017 (vs), and 1992 (m) cm^{-1}. The proton NMR spectrum in $CDCl_3$ shows two peaks: a singlet at 1.4 ppm assigned to the nine *t*-butyl hydrogen atoms, and a singlet at -21.8 ppm, assigned to the one bridging hydrogen atom.

B.1.   NONACARBONYL-μ_3-(κ^2:η^2:η^2-3, 3-DIMETHYL-1-BUTYNYL)-μ-IODOMERCURIO-*TRIANGULO*-TRIRUTHENIUM

$Ru_3(\mu\text{-}H)(\mu_3\text{-}\kappa^2:\eta^2:\eta^2\text{-}t\text{-}Bu)(CO)_9$

$$\xrightarrow[\text{(2) HgI}_2]{\text{(1) KOH}} Ru_3HgI(\mu_3\text{-}\kappa^2:^2\text{-}C_2\text{-}t\text{-}Bu)(CO)_9 + KX + H_2O$$

---

*New address: 4702 E. 355 St., Willoughby, OH 44094.

- **Caution.** *Due to the toxicity of mercury compounds, both starting materials and products must be handled with care to avoid any skin contact. Due to the toxicity of carbon monoxide gas, this reaction must also be carried out in a well-ventilated hood.*

Procedure 1

A 250-mL three-necked flask is equipped with a 25-mL pressure-equalizing dropping funnel, a magnetic stirring bar, a glass inlet tube (to be used for bubbling carbon monoxide through the reaction mixture), and a gas adaptor attached to a mineral oil bubbler. The reaction vessel is flame-dried under nitrogen, cooled to room temperature, and then charged with $Ru_3H(CO)_9(C_2$-t-Bu) (0.891 g, 1.39 mmol). All further operations are done under a carbon monoxide atmosphere. Tetrahydrofuran (THF, 40 mL, freshly distilled from sodium benzophenone ketyl) is thoroughly saturated with carbon monoxide and added to the reaction mixture by syringe. To this solution, CO-saturated alcoholic potassium hydroxide (10.13 mL of 0.1587 M solution, 1.608 mmol) is added dropwise through the addition funnel with continuous stirring. The resulting amber solution is allowed to stir for $\frac{1}{2}$ h to ensure complete conversion to the $Ru_3(CO)_9(C_2$-t-Bu)$^-$ anion. Mercury(II) iodide (3.31 g, 7.29 mmol) dissolved in dry THF (20 mL, freshly distilled from sodium benzophenone ketyl, degassed and saturated with carbon monoxide) is then added to the reaction mixture by syringe. The reaction mixture is allowed to stir for 1 h in a carbon monoxide atmosphere and then evaporated to dryness (rotary evaporator). The resulting red-orange residue is extracted with hot hexane until no further residue dissolves (four 100-mL portions). The hexane extract is then concentrated to 100 mL and cooled to $-20\,°C$. The resulting crystals of product are isolated and dried under vacuum. Yield: 0.79–0.89 g [59–66% yield based on $Ru_3H(CO)_9$-$(C_2$-t-Bu)].

Anal. Calcd. for $Ru_3HgC_{15}HgO_9I$:C, 18.70; H, 0.90. Observed: C, 18.68; H, 0.95.

B.2. μ-(BROMOMERCURIO)-NONACARBONYL-μ$_3$-(κ:η^{2}η^2-3, 3-DIMETHYL-1-BUTYNYL)-*TRIANGULO*-TRIRUTHENIUM

$$Ru_3(\mu\text{-}H)(\mu_3\text{-}\kappa:\eta^2:\eta^2\text{-}t\text{-}Bu)(CO)_9$$

$$\xrightarrow[\text{(2) HgBr}_2]{\text{(1) KOH}} Ru_3HgBr(\mu_3\text{-}\kappa:\eta^2:\eta^2\text{-}t\text{-}Bu)(CO)_9 + KX + H_2O$$

Procedure 2

The bromo analog of the triruthenium–mercury cluster can be synthesized by following the identical procedure outlined in Section (B.1), by substituting mercury(II) bromide for mercury(II) iodide. Thin layer chromatography (TLC) must be used in the work-up of the bromo derivative to separate the desired product from $[Ru_3(CO)_9(C_2-t-Bu)]_2Hg$, which is formed to some extent.

The compound $Ru_3H(CO)_9(C_2-t-Bu)$ (0.713 g, 1.12 mmol) is treated with alcoholic potassium hydroxide (13.1 mL of 0.094 M, 1.23 mmol) followed by mercury(II) bromide (2.19 g, 6.07 mmol) as outlined in the procedure in Section (B,1). The solvent is evaporated under reduced pressure, and the residue is extracted with hot hexane until the solvent portions are no longer colored, ($\sim$ three 100-mL portions). The hexane extract is then evaporated to dryness under reduced pressure, and the resulting residue is redissolved in the minimum amount of dry dichloromethane. Thin layer chromatography (PF-254G silica gel with 1:4 dichloromethane–hexane as eluant) is then used to purify this mixture. Checkers used Silica Gel 60F254 plates on which the yellow (third) band gives an R_f value of 0.3, but extraction and isolation of product afforded only 0.20 g (20% yield) on first (and only) attempt. On Silica Gel PF-254G plates, the yellow (third) band ($R_f = 0.34$) is obtained and the compound extracted from it is recrystallized from hexane. Yield: 0.28 g of product [27% yield based on $Ru_3H(CO)_9(C_2-t-Bu)$].

Anal. Calcd. for $Ru_3HgC_{15}H_9O_9Br$: C, 19.64; H, 1.00. Observed: C, 19.80; H,1.12.

Properties

The compound $Ru_3HgI(CO)_9(C_2-t-Bu)$ is a yellow crystalline solid. It is moderately air stable but turns black on long exposure to light. The IR spectrum of this compound shows eight peaks in the terminal carbonyl region in hexane: 2098 (m), 2095 (m), 2071 (s), 2067 (s), 2055 (s), 2020 (s), 1997 (w), and 1991 (m) cm^{-1}. The proton NMR spectrum in CDCl$_3$ shows one singlet at 1.4 ppm due to the nine *t*-butyl protons.

The compound $RuHgBr(CO)_9(C_2-t-Bu)$ is a yellow-orange cyrstalline solid with chemical and physical properties similar to those of the iodo analog. It is moderately air stable but decolorizes on long exposure to light. It is soluble in most common organic solvents. The IR spectrum of this compound in hexane shows six peaks in the terminal carbonyl region: 2094 (m), 2066 (s), 2056 (s), 2024 (s), 2017 (sh), and 1997 (m) cm^{-1}. The proton NMR spectrum of this compound in CDCl$_3$ shows a singlet at 1.4 ppm.

C. μ_4-MERCURIO-BIS[NONACARBONYL-μ_3-(κ:η^2:η^2-3, 3-DIMETHYL-1-BUTYNYL)-*TRIANGULO*-TRIRUTHENIUM]

$$Ru_3(\mu\text{-}H)(\mu_3\text{-}\kappa\text{:}\eta^2\text{:}\eta^2\text{-}t\text{-}Bu)(CO)_9$$

$$\xrightarrow[\text{(2)HgI}_2]{\text{(1)KOH}} Hg[Ru_3(\mu_3\text{-}\kappa\text{:}\eta^2\text{:}\eta^2\text{-}C_2\text{-}t\text{-}Bu)(CO)_9]_2 + KI + H_2O$$

- *See cautionary note in the procedure in Section B*

The symmetrical hexaruthenium–mercury cluster is synthesized using the procedure presented in Section B.1, varying only the amount of mercury(II) iodide added.

The compound $Ru_3H(CO)_9(C_2\text{-}t\text{-}Bu)$ (1.243 g, 1.950 mmol) is treated with alcoholic potassium hydroxide (17.5 mL of 0.128 M, 2.24 mmol) followed by mercury(II) iodide (0.444 g, 0.978 mmol) as outlined in the procedure in Section (B.1). The resulting solution is allowed to stir under a carbon monoxide atmosphere until a bright yellow precipitate forms. The reaction time is typically 30 to 45 min. Upon precipitation, the reaction mixture is evaporated to dryness under reduced pressure. The resulting residue is washed first with 50 mL of absolute ethanol and then with 25 mL of dichloromethane. The remaining yellow residue is recrystallized from hot THF. Yield: 1.16–1.20 [81–84% yield based on $Ru_3H(CO)_9(C_2\text{-}t\text{-}Bu)$].

Anal. Calcd. for $Ru_3H(CO)_9(C_2\text{-}t\text{-}Bu$: C, 24.40; H, 1.22. Found: C, 24.63; H, 1.53.

Properties

The compound $Hg[Ru_3(CO)_9(C_2\text{-}t\text{-}Bu)]_2$ is a bright yellow crystalline solid with low overall solubility. It is slightly soluble in THF, dichloromethane, and chloroform, and has a very low solubility in typical hydrocarbon solvents. The solid shows moderate air stability but turns black on long exposure to light. The IR spectrum of this compound in CH_2Cl_2 shows four peaks in the terminal carbonyl region: 2090 (w), 2078 (m), 2055 (s), and 2002 (m, br) cm^{-1}. The proton NMR spectrum in $CDCl_3$ shows a singlet at 1.38 ppm.

D. NONACARBONYL-μ_3(κ:η^2:η^2-3, 3-DIMETHYL-1-BUTYNYL)-μ-{[(TRICARBONYL(η^5-CYCLOPENTADIENYL)MOLYBDIO]-MERCURIO}-*TRIANGULO*-TRIRUTHENIUM

$$Ru_3HgI(\mu_3\text{-}\kappa\text{:}\eta^2\text{:}\eta^2\text{-}C_2\text{-}t\text{-}Bu)(CO)_9 + Na^+[Mo(\eta^5\text{-}C_5H_5)(CO)_3]^-$$

$$\xrightarrow[\text{THF}]{} MoHgRu_3(CO)_{12}(\eta^5\text{-}C_5H_5)(\mu_3\text{-}\kappa\text{:}\eta^2\text{:}\eta^2\text{-}C_2\text{-}t\text{-}Bu) + KI$$

■ *Observe cautionary note in the procedure in Section B.*

For the procedure that follows, $Na[Mo(\eta^5\text{-}C_5H_5)(CO)_3]$ should be prepared in advance by reaction of $Na(\eta^5\text{-}C_5H_5)$ with $Mo(CO)_6$,[6] or by sodium amalgam reduction of $[Mo(\eta^5\text{-}C_5H_5)(CO)_3]_2$ (Strem).

A 250-mL two-necked flask is equipped with a magnetic stirring bar, a rubber septum, and a gas inlet fitting connected to a T-tube, one side of which is connected to a nitrogen line and the other side to a mineral oil bubbler. The flask is flame-dried under nitrogen, cooled to room temperature, and then charged with $Ru_3HgI(CO)_9(C_2\text{-}t\text{-}Bu)$ (0.399 g, 0.414 mmol). Tetrahydrofuran (40 mL, freshly distilled from sodium benzophenone ketyl) is added by syringe. The resulting solution is deaerated with nitrogen gas. All further operations are carried out under a nitrogen atmosphere. To this solution, $Na([Mo(C_5H_5)(CO)_3]$(19. mL of a $0.0232 M$ THF solution, 0.455 mmol) is added by syringe. It is important to keep the $Na[Mo(C_5H_5)(CO)_3]$ concentration around $0.02 M$ to ensure good yields. The reaction is considered to be complete after 45 min.

The solution is evaporated to dryness under reduced pressure and the residue is purified by thin layer chromatography (PF-254G silica gel with 1:4 dichloromethane–hexane eluant). Three yellow bands are eluted. The second band is removed and product extracted from it is recrystallized from hexane. Yield: 0.12–0.15 g (30–33%) of the desired product. Checkers used column chromatography (35 × 3-cm column, Silica Gel 60, 70–230 mesh, Merck) to give the product as the second band when eluting with 1:4 dichloromethane–hexane.

Anal. Calcd.: C, 25.54; H, 1.31. Found: C, 25.87; H, 1.74.

Properties

The compound $MoHgRu_3(CO)_{12}(\eta^5\text{-}C_5H_5)(C_2\text{-}t\text{-}Bu)$ is a moderately air-stable yellow crystalline solid. It is soluble in most common organic solvents. The IR spectrum of this compound in CH_2Cl_2 shows seven bands in the terminal carbonyl region: 2084 (m), 2054 (s), 2046 (sh), 2010 (s), 1982 (s), 1910 (m) and 1884 (s) cm^{-1}. The proton NMR spectrum of this compound in $CDCl_3$ shows two peaks: a singlet at 1.40 ppm (nine *t*-butyl protons) and a singlet at 5.29 ppm (five cyclopentadienyl protons).

References

1. M. J. Mays *International Review of Science, Part 2* Vol. 6 Butterworths, London, 1972.
2. E. Rosenberg, R. Fahmy, K. King, A. Tiripicchio, and M. T. Camellini, *J. Am. Chem. Soc.*, **102**, 3626 (1980).

3. E. Rosenberg, K. King, S. Ermer, K. I. Hardcastle, and A. Tiripicchio, *Inorg. Chem*, **23**, 1339 (1983).
4. Since our initial report we have noted three other transition metal mercury complexes in which the mercury bridges two or more metal atoms, which have been reported as products from sodium amalgam reductions. (a) D. Duffy, K. Mackay, and B. Nicholson, *J. Chem. Soc. Dalton Trans.*, **1981**, 381; (b) R. Jones, F. Real, and G. Wilkinson, *J. Chem. Soc. Dalton Trans.*, **1981**, 126; (c) J. Deutsche, S. Fadel, and M. Ziegler, *Angew Chem. Int. Ed. Engl.* **16**, 704 (1977).
5. (a) A. Mantovani and S. Cenini, *Inorg. Synth.*, **16**, 47 (1976); (b) M. I. Bruce, M. L. Jones, and C. M. Jensen, *Inorg. Synth.*, **26**, 259 (1989).
6. R. B. King and F. G A. Stone, *Inorg. Synth.*, **7**, 107 (1963).

61. HETEROBINUCLEAR NONACARBONYL COMPLEXES AND HYDRIDE COMPLEXES OF IRON–CHROMIUM, IRON–MOLYBDENUM, AND IRON–TUNGSTEN

Submitted by LARRY W. ARNDT, CHRISTOPHER J. BISCHOFF, and MARCETTA Y. DARENSBOURG*
Checked by JOHN E. ELLIS[†]

A chemical characteristic of metal hydride complexes is their ability to attach to other metal centers, both in the absence or in the presence of additional supporting bonds, as in $Re_3(\mu\text{-}H)(CO)_{14}$,[1] $[M_2(\mu\text{-}H)(CO)_{10}]^-$ (M = Cr, Mo, W),[2] $[Fe_2(\mu\text{-}H)(CO)_8]^-$,[3] or a variety of other complexes.[4]

Not unexpectedly, the $[FeH(CO)_4]^-$ anion aggregates with photochemically generated $[M(CO)_5]$ fragments (M = Cr, Mo, or W) to give the $[FeMH(CO)_9]^-$ anions.[5] These anions have interesting structures and reactivities, and can serve as useful precursors to the heterobimetallic anions $[FeM(CO)_9]^{2-}$,[5]. A preparation of the latter dianions, based on the reaction of $Na_2[Cr(CO)_5]$ and $Fe_2(CO)_9$, is available.[6] However, problems of separation and purification are considerably alleviated in the following procedures.

General Procedure

■ **Caution.** *Diethyl ether used as a solvent should be free of peroxides before further drying and distillation from lithium tetrahydroaluminate(1-) is attempted. During distillation, caution must be exercised never to permit the still pot to run dry.*

*Department of Chemistry, Texas A & M University, College Station, TX 77843-3255.
[†]Department of Chemistry, University of Minnesota, Minneapolis, MN 55455.

All operations are carried out under nitrogen using standard Schlenk techniques,[7] or in an argon atmosphere glove box. Tetrahydrofuran (THF) and hexane are distilled under nitrogen from sodium benzophenone ketyl. Acetonitrile is dried over calcium hydride, distilled three times from P_4O_{10}, and stored under nitrogen, over 3-Å molecular sieves. Diethyl ether is freshly distilled and degassed by a N_2 purge prior to use. All reagents were purchased from standard vendors as reagent grade or better, and used without further purification.

A.   μ-NITRIDO-BIS(TRIPHENYLPHOSPHORUS)(1+)
TETRACARBONYLHYDRIDOFERRATE(1−)[8]

$$Fe(CO)_5 + KOH + [PPN][Cl] \rightarrow [PPN][FeH(CO)_4]$$

Procedure

▪ **Caution.** *Metal carbonyls are volatile and toxic. The photolysis of metal carbonyls in THF solvent produces toxic carbon monoxide. Hence all reactions should be carried out in a well-ventilated hood.*

To a 100-mL Schlenk flask 1.0 g of KOH is dissolved in 10 mL of MeOH and 2 mL of water, which has been degassed for 20 min. This is allowed to stir for 1 h, after which 0.50 mL of $Fe(CO)_5$ is added, and the flask wrapped in aluminium foil to prevent exposure to light. A 2.10-g sample of [PPN][Cl] is dissolved in 5 mL of MeOH and then cannulated into the reaction mixture. After 30 min a creamy white precipitate can be seen, and an additional 15 mL of MeOH is added. The mixture is allowed to stir for 4 h, in the absence of light. A medium frit filter tube is fitted to the flask, and the solid collected by filtration. The solid is then dried *in vacuo*, and stored protected from light and air. The [PPN][FeH(CO)_4] exhibits IR absorptions at 1998 (w), 1906 (m), and 1877 (s) cm^{-1} in THF solution. The ^1H NMR resonance for FeH(CO)$_4^-$ is at -8.78 ppm in THF-d_8.

B.   μ-NITRIDO-BIS(TRIPHENYLPHOSPHORUS)(1+)
NONACARBONYLHYDRIDOFERRATE TUNGSTATE(1−)

$$W(CO)_6 + THF \xrightarrow{h\nu} W(CO)_5 \cdot THF + CO$$

$$W(CO)_5THF + [PPN][FeH(CO)_4] \longrightarrow [PPN][FeWH(CO)_9]$$

▪ **Caution.** *Metal carbonyls are volatile and toxic. The photolysis of metal* *carbonyls in THF solvent produces toxic carbon monoxide. Hence all reactions should be carried out in a well-ventilated fume hood.*

To a 100-mL water-jacketed photochemical reaction vessel is added 1.00 g (2.84 mmol) of $W(CO)_6$ (purchased from Strem). The inlets of the photochemical reactor are fitted with wired down rubber septa, and the reactor is evacuated and backfilled with nitrogen. Nitrogen is bubbled continuously through the solution during photolysis (1 h with a 450-W mercury vapor lamp). The conversion of $W(CO)_6$ to $W(CO)_5 \cdot THF$ is monitored by optimization of its $v_{(CO)}$IR bands at 2075 (w), 1936 (s), and 1895 (m) cm^{-1}. The yellow solution of $W(CO)_5 \cdot THF$ ($\sim 95\%$ conversion or better) is then transferred via cannula into a 100-mL Schlenk flask containing 2.01 g (2.84 mmol) of $[PPN][FeH(CO)_4]$ and a stirring bar. This mixture is stirred until a red-orange solution is obtained (~ 10 min). The solution is concentrated to 15 mL under evacuation accompanied by rapid stirring. Hexane is slowly added (~ 50 mL) under rapid stirring to precipitate a brown-orange solid. The supernatant liquid is removed via cannula, and the remaining solid washed twice with hexane to remove any excess $W(CO)_6$ or $W(CO)_5 \cdot THF$. The solid is dried under vacuum at room temperature. Under a steady stream of nitrogen, the flask is refitted with a medium frit filterstick, which is connected to a 100-mL Schlenk flask via a cannula.[9] The solid is extracted with 50 mL of diethyl ether, and the suspension is filtered leaving behind undissolved $[PPN][FeH(CO)_4]$. This solid is washed twice with diethyl ether. Under strong nitrogen flow, the filterstick is removed and the flask, containing the filtrate and washings is sealed with a straight stopcock connector. The solution is then concentrated to 10 mL and hexane is added slowly to precipitate a yellow solid. The solid is washed twice with two 15-mL portions of hexane. Time is allowed after each washing for the solid to settle out before the mother liquid is removed via cannula. The product weighs 2.70 g (92% yield).

Anal. Calcd. for $C_{45}H_{31}N_1O_9P_2Fe_1W_1$: C, 52.40; H, 3.03; N, 1.36. Found: C, 52.26; H, 3.06; N, 1.44.

Crystals of the dinuclear hydride may be obtained by layering hexane onto a concentrated methanol solution of hydride complex.

This procedure can be used to obtain salts containing other counter-cations such as Et_4N^+ and Ph_4P^+ using as starting materials the appropriate salt of $[FeH(CO)_4]^-$. The iron–chromium and iron–molybdenum analogs are prepared in precisely the same manner with similarly high yields of pure product. The yield is limited by the percentage conversion of $M(CO)_6$ to $M(CO)_5 \cdot THF$. Conversion of the $M(CO)_6$ to the $M(CO)_5 \cdot THF$ adducts is monitored by IR. For M = Cr: 2074 (w), 1938 (s), 1895 (m) and for M = Mo: 2078 (w), 1939 (s), 1895 (m) cm^{-1}.

Properties

The binuclear hydride salts are air sensitive, soluble in THF, acetone, CH_3CN, MeOH, and diethyl ether, and insoluble in hydrocarbon solvents. They can be stored cold under an inert gas or *in vacuo* for several months. The salts tend to slowly decompose into $W(CO)_6$ and $[PPN][FeH(CO)_4]$. Carbon monoxide rapidly (within minutes) degrades the dimer into the same products.[5] The salts react rapidly with CH_3COOD or stronger deuterated acids to form the H–D exchanged products, $[FeCrD(CO)_9]^-$. The hydrides also act as catalysts in olefin isomerization.[10]

Spectroscopic characterizations of the salts are listed in Table I. Despite the high-field chemical shift of the proton resonances as compared to the monomeric analogs and known bridging hydrides {i.e., $[FeH(CO)_4]^- = -8.78\,ppm$; $[WH(CO)_5]^- = -4.02\,ppm$; $[W_2(\mu\text{-}H)(CO)_{10}]^- = -12.6\,ppm$ $(J_{W-H} = 42.0\,Hz)$}, the small J_{W-H}, the single bond distance of Fe—W, $2.989(2)\,Å$,[5b] and theoretical calculations **B** imply the hydride to be isolated on Fe in terminal position, that is, structure **A** rather than **B**.

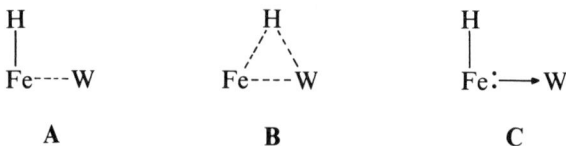

A	**B**	**C**

Further evidence comes from structural analyses of $[Ph_3PAuFeW(CO)_9]^-$ and $MeFeW(CO)_9^-$,[11,12] which find Ph_3PAu and Me groups, isolobal analogs to H, to be located solely on Fe. The Fe—W bond is regarded as a metallodonor–metalloacceptor interaction, Structure **C**.[13,14]

TABLE I.   Spectral Properties of $[PPN][FeMH(CO)_9]$.[5]

M	Color	^{1}H NMR Values (ppm)	IR Frequencies $\nu_{(CO)}\,(cm^{-1})^a$
Cr	Yellow	−14.6	2057 (vw), 2010 (w), 2000 (m, sh) 1942 (s), 1911 (m, sh), 1880 (m, br)
Mo	Yellow	−11.5	2062 (vw), 2009 (m, sh), 1951 (s), 1926 (m), 1874 (m)
W^{b,c}	Yellow-orange	−11.8	2063 (vw), 2009 (m, sh), 1940 (s) 1911 (m, sh), 1870 (m, br)

^{a}THF solution, 0.1 mm CaF_2 cells, Perkin-Elmer 283B spectrophotometer referenced to CO_2.
^{b}^{13}C NMR spectra (Varian HL200) in THF (2 °C) gives three resonances, 218.1, 203.9, and 203.0 ppm of approximate intensities of 4:1:4. These resonances correspond to $Fe(CO)_4$, $W(CO)$-*trans*, and $W(CO)$-*cis*, respectively. The $J_{(W-C)} = 125.6\,Hz$.
^{c}$J_{(W-H)} = 15.0\,Hz$.

C. BIS[μ-NITRIDO-BIS(TRIPHENYLPHOSPHORUS)(1 +)] NONACARBONYLCHROMATEFERRATE(1 −)

$$[PPN][FeCrH(CO)_9] + LiBu + PPNCl \longrightarrow LiCl + C_4H_{10}$$
$$+ [PPN]_2[FeCr(CO)_9]$$

Procedure

■ **Caution.** *Butyllithium reacts violently with water.*

A 100-mL Schlenk flask equipped with stirring bar and straight stopcock adapter is evacuated, flame dried under vacuum, and backfilled with nitrogen. To this flask is added 500 mg (1.11 mmol) of $[PPN][FeCrH(CO)_9]$ and 638 mg (1.11 mmol) of PPNCl. One can also use the $[PPN][FeCrH(CO)_9]$ solution generated *in situ* as described previously. Tetrahydrofuran (25 mL) is added via cannula into the flask with stirring, producing an orange heterogeneous solution. Butyllithium (− 0.5 mL of 1.6 *M* LiBu in hexane) is then added dropwise via syringe until the solution is a dark red color and a red-brown precipitate begins to form. The solution is stirred for 10 min and the solid is allowed to settle before filtering. The vacuum adapter is replaced with a medium frit filter tube connected to a 100-mL Schlenk flask, under a steady stream of nitrogen. The solution is filtered and the red-brown solid is washed twice with THF. The red-brown solid is dried in the frit *in vacuo* at room temperature, and is then transferred into a 40-mL Schlenk flask, and 25 mL of CH_3CN is added by cannula to dissolve the solid. The mixture is filtered through a medium-frit filterstick connected to a 50-mL Schlenk flask and washed twice with 10-mL portions of CH_3CN. The 50-mL flask is refitted under nitrogen with a straight stopcock connector. The solution is concentrated to 10 mL and diethyl ether is added to precipitate an orange solid. The product is washed twice with 10-mL portions of diethyl ether and dried *in vacuo* at room temperature. The dried solid weighs 410 mg (51% yield).

Anal. Calcd. for $C_{81}H_{60}P_4N_2O_9Fe_1W_1$: C, 67, 70; H, 4.21. Found: C, 67, 36; H, 4.44.

The iron–molybdenum and iron–tungsten analogs are prepared in precisely the same manner with comparative yields of pure product. Well-formed crystals of $[PPN]^+$ salts of the dinuclear dianion may be obtained by layering diethyl ether onto a concentrated CH_3CN solution of dianion complex.

An alternative route to the heteronuclear dianion with similar high yields is the aggregation of the $[Fe(CO)_4]^{2-}$ (generated from the deprotonation of

TABLE II.   Spectral Properties of [PPN]$_2$[Fe(CO)$_9$].

M	Color	IR Frequencies $\nu_{(CO)}$ (cm^{-1})^a
Cr	Orange	2009 (w), 1907 (w, sh), 1892 (s), 1814 (m)
Mo	Red	2028 (w), 1907 (s), 1809 (m)
W^b	Dark red	2026 (w), 1899 (s), 1808 (m)

^{a} CH$_3$CN solution.

^{b} ^{13}C NMR spectra in CH$_3$CN (Varian HL 200) gives two resonances at 23 °C, 225.6 and 214.4 ppm corresponding to Fe(CO)$_4$ and W(CO)$_5$, respectively. The $J_{(W-C)}$ = 133.4 Hz. *Note*: the axial and equatorial CO groups on W are either fluxional or their ^{13}C resonances overlap.

[FeH(CO$_4$)]$^-$ with LiBu or the reduction of Fe(CO)$_5$ with LiBu or the reduction of Fe(CO)$_5$ with Na–Hg) with photochemically generated Cr(CO)$_5$·THF.

The spectroscopic data for the salts of the dianions are listed in Table II. The compounds are air and moisture sensitive. The complexes are soluble in acetonitrile and CH$_2$Cl$_2$, and are insoluble in THF, diethyl ether, and hydrocarbon solvents. They can be stored under nitrogen or *in vacuo* for several months, and refrigeration is recommended. The dianions may be protonated in THF solution by stoichiometric amounts of MeOH or H$_2$O, regenerating the hydrides. The dianions have the ability to add a variety of bridging ligands. For example, the reaction of MeI with dianions gives bridging iodo species. Bridging phosphide compounds are obtained via the reaction of Ph$_2$PCl and these salts of the dianions.

References

1. (a) D. K. Huggins, W. Fellmann, J. M. Smith, and H. D. Kaesz, *J. Am. Chem. Soc.*, **86**, 4841 (1964); Compare HRe$_2$Mn(CO)$_{14}$, H. D. Kaesz, R. Bau, and M. R. Churchill, *J. Am. Chem. Soc.*, **89**, 2775 (1969).

2. R. G. Hayter, *J. Am. Chem. Soc.*, **88**, 4376 (1966).

3. J. P. Collman, R. G. Finke, P. L. Matlock, R. Wahren, R. G. Komoto, and J. I. Brauman, *J. Am. Chem. Soc.*, **100**, 1119 (1978).

4. H. Lehner, D. Matt, A. Togni, R. Thouvenot, L. M. Venanzi, and A. Albinati, *Inorg. Chem.*, **23**, 4254 (1984) and references cited therein.

5. (a) L. W. Arndt, T. Delord, and M. Y. Darensbourg, *J. Am. Chem. Soc.*, **106**, 456 (1984); (b) L. W. Arndt, M. Y. Darensbourg, T. Delord, and B. T. Bancroft, *J. Am. Chem. Soc.*, **108**, 2617 (1986).

6. E. Linder, H. Behrens, and D. Uhlig, *Z. Naturforsch.*, **28b**, 276 (1974).

7. D. F. Shriver and M. A. Drezdon, '*The Manipulation of Air-Sensitive Compounds*', Wiley, New York, 1987.

8. M. Y. Darensbourg, D. J. Darensbourg, and H. L. C. Barros, *Inorg. Chem.*, **17**, 297 (1978).

9. J. P. McNally, V. S. Leong, and N. J. Cooper, *Experimental Organometallic Chemistry* A. L. Wayda and M. Y. Darensbourg, (eds.) Symposium Series 357, American Chemical Society, Washington, DC, 1987, Chapter 2.

10. P. A. Tooley, L. W. Arndt, and M. Y. Darensbourg, *J. Am. Chem. Soc.*, **107**, 2422 (1985).

11. L. W. Arndt, M. Y. Darensbourg, J. P. Fackler, R. J. Lusk, D. O. Marler, and K. A. Youngdahl, *J. Am. Chem. Soc.*, **107**, 7218 (1985).

12. L. W. Arndt, B. T. Bancroft, M. Y. Darensbourg, C. P. Janzen, C. Kim, K. E. Varner, and K. A. Youngdahl, *Organometallics*, **7**, 1302 (1988).

13. M. B. Hall and C. Halpin, *J. Am. Chem. Soc.*, **108**, 1695 (1986).

14. M. Y. Darensbourg and C. E. Ash, *Adv. Organomet. Chem.*, **27**, 1 (1987).

62. CYCLOPENTADIENYLSODIUM AND SOME MONO-, TRI-, AND TETRANUCLEAR METAL CARBONYL DERIVATIVES AND CLUSTER COMPLEXES

Submitted by PIERRE BRAUNSTEIN,* ROBERT BENDER,* and JEANMARC JUD*
Checked by H. VAHRENKAMP,† GLEN C. VOGEL,‡ and GREGORY L. GEOFFROY§

Cyclopentadienylsodium is widely used for the preparation of main group and transition group metal complexes containing a σ- or π-bonded cyclopentadienyl ligand(s) and often, but not always, carbonyl ligands.[1] This reagent is usually prepared and used *in situ*, since it is pyrophoric in the solid state,[2] thus making it difficult to store over long periods of time. This presents certain inconvenience in synthetic chemistry. In contrast, the 1, 2-dimethoxyethane solvate of cyclopentadienylsodium is less air sensitive, and easier to isolate and to store in the solid state, Section A. The use of this reagent greatly facilitates synthetic work. Procedures are described below for the preparation of solid cyclopentadienylsodium and sodium carbonylmetallates of chromium, molybdenum, and tungsten, which are all easier to manipulate in the solid state when solvated with 1, 2-dimethoxyethane, $CH_3O-CH_2-CH_2-OCH_3$. Section B. The carbonylmetalates, $[M(CO)_3(\eta\text{-}C_5H_5)]^-$, (M = Cr,

*Département de Chimie, Université Louis Pasteur, 4 rue Blaise Pascal, F-67070 Strasbourg Cédex, France.

†Institut für Anorganische und Analytische Chemie, Universität Freiburg, Albertstrasse 21, D-7800 Freiburg, Federal Republic of Germany.

‡Department of Chemistry, Ithaca College, Ithaca NY 14850.

§Department of Chemistry, The Pennsylvania State University, University Park, PA 16802.

Mo, W) are very useful reagents in organometallic chemistry, in particular for the formation of metal–metal bonded complexes. This is illustrated below by the high-yield synthesis of a heterotrinuclear complex having a linear M— Pt—M core, Section C. The trinuclear complexes M—M′—M react with tertiary phosphines[3,4] to give heterotetranuclear clusters $M_2M_2'(\eta\text{-}C_5H_5)_2$ $(CO)_6(PR_3)_2$ and details of the synthesis of two of these (M = Mo; M′ = Pd, Pt) are presented in Section D.

A.   CYCLOPENTADIENYLSODIUM-1, 2-DIMETHOXYETHANE

$$Na + C_5H_6 \xrightarrow{\text{DME}} NaC_5H_5 \cdot DME + \tfrac{1}{2}H_2$$

Although solutions of cyclopentadienylsodium can be prepared according to an earlier synthesis,[2] an improved procedure makes use of the solvent 1, 2-dimethoxyethane (DME), which solvates NaC_5H_5 and stabilizes it in the solid.[5,6]

Procedure

▪ **Caution.** *H_2 is evolved during this reaction and should be released into a properly vented hood.*

All operations are performed in an atmosphere of nitrogen dried over molecular sieves (4 Å) (Merck 5708) and deoxygenated on catalyst BASF R3-11 (Imhoff and Stahl). A 250-mL round-bottomed Schlenk flask fitted with a reflux condenser and a Teflon encased magnetic stirring bar (45 mm long, 10-mm diameter) is evacuated and backfilled with purified nitrogen three times. Sodium metal (2.3 g, 0.1 mol) and toluene (100 mL) freshly distilled from sodium under nitrogen, are introduced into the flask, which is half-immersed in an oil bath placed on the heating plate of a magnetic stirrer. The mixture is heated to the boiling point of toluene. The molten sodium is vigorously stirred (~ 500 rpm) to produce sodium sand. After stirring is stopped, the mixture is immediately allowed to cool to room temperature. Under slow nitrogen flow, the toluene is removed with a syringe, pipette, or cannula, the sodium sand is washed twice with DME (50 mL) distilled from sodium benzophenone ketyl under N_2, and DME (75 mL) is added to the sodium sand.

Meanwhile, cyclopentadiene is prepared from commercial dicyclopentadiene by heating in a distillation apparatus at the boiling point.[2] Monomeric cyclopentadiene distills over at $\sim 42\,°C$ as a colorless liquid, which should be used within 1 h.

The reflux condenser on the round-bottomed flask containing the sodium sand is replaced by a pressure-equalizing dropping funnel. The whole apparatus is then flushed with a nitrogen flow, and freshly distilled cyclopentadiene (15 mL, 0.18 mol) is transferred into the funnel. The cyclopentadiene is added to the sodium sand over a period of 1 h with stirring, and the stirring is continued overnight. If some of the sodium fails to dissolve, more freshly distilled cyclopentadiene may be added and the stirring continued. Cyclopentadienylsodium-1,2-dimethoxyethane (11.1 g, 62.3% yield based on Na) precipitates as a white powder, which is removed by filtration, washed twice with purified DME (30 mL), and three times with 50-mL portions of pentane distilled from sodium under N_2, and dried under reduced pressure. When the pentane combines with the DME solution, further precipitation of $NaC_5H_5 \cdot DME$ occurs, this second crop is also collected (2.3 g, total yield 75% based on Na).

If the reaction mixture contains some impurities, the $NaC_5H_5 \cdot DME$ may be slightly pink.

Properties

The DME adduct of cyclopentadienylsodium is a white powder soluble in tetrahydrofuran (THF) and insoluble in toluene or diethyl ether. It darkens when exposed to air, but is stable enough to be easily transferred under N_2 from one Schlenk tube to another. This solid reagent can easily be weighted and used stoichiometrically: formula wt $[NaC_5H_5 \cdot DME] = 178.2$.

B. SODIUM TRICARBONYL(η^5-CYCLOPENTADIENYL)-METALLATES, DME SOLVATES Na[M(CO)$_3$(η^5-C$_5$H$_5$)]·2 DME (M = Cr, Mo, W)

$$NaC_5H_5 \cdot DME + Cr(CO)_6 \xrightarrow{DME} Na[Cr(CO)_3(\eta^5\text{-}C_5H_5)] \cdot 2DME + 3CO$$

The carbonylmetallate anions are extremely useful organometallic reagents, particularly for the syntheses of heteropolymetallic clusters.[3,4,7] However, they have rarely been isolated in the solid state and their solutions are obtained either by reduction of their dimers or by reaction of NaC_5H_5 with the parent metal carbonyls,[2] and immediately used. The first method is not very economical since the dimers are usually themselves obtained from the metallate anions. The second method has been improved as detailed below to allow the easy isolation of the pure Na[M(CO)$_3$(η^5-C$_5$H$_5$)] (M = Cr, Mo, W) solids as DME solvates.[4] The nonsolvated Na[M(CO)$_3$(η^5-C$_5$H$_5$)] salts can also be easily prepared from $M(CO)_3(CH_3CN)_3$ and NaC_5H_5 but are less stable in the solid state for long periods than the previously described DME solvates.[8]

Procedure

- **Caution.** *Carbon monoxide is evolved during this reaction and should be carefully released in a well-ventilated hood.*

A detailed procedure is given for M = Cr and can be similarly followed for M = Mo and W.

All operations are performed under an atmosphere of nitrogen dried and deoxygenated as described in the procedure in Section A, using standard Schlenk techniques and a vacuum nitrogen line.[9] 1,2-Dimethoxyethane solvent is distilled from sodium benzophenone ketyl, under nitrogen atmosphere.

A 1-L round-bottomed Schlenk flask equipped with a magnetic stirring bar is fitted with a reflux condenser topped with a Hg check valve. It is evacuated and backfilled with purified nitrogen. The compounds $Cr(CO)_6$ (6.40 g, 29.10 mmol, Ventron commercial grade), $NaC_5H_5 \cdot DME$ (5.19 g, 29.1 mmol) and DME (120 mL) are transferred into the round-bottomed flask.

The whole apparatus is flushed by a slow stream of nitrogen for 10 min. The mixture is refluxed for 24 h under constant stirring. The disappearance of $Cr(CO)_6$ can be followed by IR spectroscopy.

To precipitate the product, 400 mL of pentane, dried over sodium wire and distilled under N_2, is added to the solution at room temperature and the mixture cooled in Dry-Ice. The compound $Na[Cr(CO)_3(\eta^5\text{-}C_5H_5)]\cdot 2DME$ separates as a pale yellow powder, which is removed by filtration, washed with three portions (50 mL) of distilled pentane, and dried *in vacuo* to give 10.1 g of pure $Na[Cr(CO)_3(\eta^5\text{-}C_5H_5)]\cdot 2DME$ [86% yield based on $Cr(CO)_6$].

The same procedure as for M = Cr can be applied to M = Mo, using $Mo(CO)_6$ (7.66 g, 29 mmol). One obtains $Na[Mo(CO)_3(\eta^5\text{-}C_5H_5)]\cdot 2DME$ (90% yield based on Mo).

Similarly, $W(CO)_6$ (10.20 g, 29 mmol) leads to $Na[W(CO)_3(\eta^5\text{-}C_5H_5)]\cdot 2DME$ (90% yield, based on W).[4]

Properties

Sodium tricarbonyl (η^5-cyclopentadienyl)chromate(1−)-bis(1,2-dimethoxyethane), $Na[Cr(CO)_3(\eta^5\text{-}C_5H_5)]\cdot 2DME$, is a pale yellow powder, stable under N_2 but decomposes in air. It can be stored for several weeks under nitrogen at −20 °C. It is insoluble in nonpolar solvents such as hexane and soluble in polar organic solvents such as THF. The compound is best characterized by its carbonyl IR absorptions in THF solution: $\nu_{(CO)}$ 1898 (s), 1794 (s), 1744 (s) cm^{-1}.

The compounds $Na[Mo(CO)_3(\eta^5\text{-}C_5H_5)]\cdot 2DME$ and $Na[W(CO)_3(\eta^5\text{-}C_5H_5)]\cdot 2DME$ are white powders of properties similar to those of their chromium analog.

M = Mo   IR (THF) $\nu_{(CO)}$ 1900 (s), 1796 (s), 1745 (m) cm^{-1}
M = W   IR (THF) $\nu_{(CO)}$ 1895 (s), 1792 (s), 1742 (m) cm^{-1}

C. A LINEAR DIMETALLIOPLATINUM COMPLEX, *trans*-$Pt[Mo(CO)_3(\eta^5\text{-}C_5H_5)]_2(PhCN)_2 (Ph = C_6H_5)$

1. $PtCl_2 + 2PhCN \longrightarrow PtCl_2(PhCN)_2$

2. $PtCl_2(PhCN)_2 + 2Na[Mo(CO)_3(\eta^5\text{-}C_5H_5)]\cdot 2DME$

$\longrightarrow$ *trans*-$Pt[Mo(CO)_3(\eta^5\text{-}C_5H_5)]_2(PhCN)_2 + 2NaCl + 4DME$

The complex $PtCl_2$ is an extremely useful precursor in platinum chemistry.[10b] The labile benzonitrile ligands can be easily and quantatively replaced by neutral ligands (e.g., amines, phosphines, etc.). On the other hand, the chloride ligands can be substituted by nucleophiles of which an example is given below.

The trinuclear complex *trans*-$Pt[Mo(CO)_3(\eta^5\text{-}C_5H_5)]_2(PhCN)_2$ is prepared by the reaction of two equivalents of sodium carbonylmetallate with the square planar platinum(II) complex $PtCl_2(PhCN)_2$.[3] Both cis and trans isomers afford the trans trinuclear chain complex. The lability of the benzonitrile ligand in the latter can be used advantageously for cluster synthesis; this will be described in the next section.

Similar trinuclear complexes containing an isocyanide ligand RNC (e.g., *trans*-$Pt[Mo(CO)_3(\eta^5\text{-}C_5H_5)]_2(C_6H_{11}NC)_2$ have been prepared[11] and evaluated in the catalytic homogeneous hydrogenation of carbon–carbon multiple bonds.[12]

The homologous derivatives *trans*-$Pt[Cr(CO)_3(\eta^5\text{-}C_5H_5)]_2(PhCN)_2$ and *trans*-$Pt[W(CO)_3(\eta^5\text{-}C_5H_5)]_2(PhCN)_2$ can be similarly prepared in comparable yields and have been described in the literature.[3]

Bis(benzonitrile)dichloroplatinum

Preparation

A suspension of platinum dichloride (10 g, 37.6 mmol) in benzonitrile (commercial grade) (250 mL) is heated at 110 °C in a round-bottomed flask equipped with a magnetic stirring bar and a reflux condenser. When all the solid has dissolved, the yellow solution is filtered while hot and then cooled in an ice bath. The complex $PtCl_2(PhCN)_2$ precipitates as a yellow air-stable, crystalline powder that is removed by filtration, washed with three portions of hexane (commercial grade) (50 mL), and dried under reduced pressure. Addition of 300 mL of hexane to the filtrate affords a second crop of bis(benzonitrile)dichloroplatinum. Yield of $PtCl_2(PhCN)_2$: 16.3 g, (92%).

A Linear Dimetallic Platinum Complex

Preparation

All of the following operations must be performed under an atmosphere of nitrogen dried and deoxygenated as indicated in Section A, using standard Schlenk techniques, and a nitrogen vacuum line.[9] Bis(benzonitrile)-dichloroplatinum (5.903 g, 12.5 mmol) is introduced into a 250-mL round-bottomed flask equipped with a stopcock side arm and fitted with a magnetic stirring bar. The round-bottomed flask is evacuated and backfilled twice with purified nitrogen. Tetrahydrofuran, freshly distilled under N_2 over sodium benzophenone ketyl (160 mL), is added and the mixture is cooled in an acetone bath maintained at $-40\,°C$ with small additions of Dry Ice. Solid $Na[Mo(CO)_3(\eta^5\text{-}C_5H_5)]\cdot 2DME$ (11.21 g, 25 mmol) is added under N_2 in small portions and, under constant agitation, the temperature is slowly raised to $20\,°C$ in 8 h. The resulting red-brown mixture is filtered, and the solid is dried *in vacuo*. The solid is then washed with three portions of water (50 mL each), previously deoxygenated by boiling under a purge of N_2. The solid is then further washed by three portions of ethanol (30 mL each), previously deoxygenated by a purge of N_2, and the solid is finally dried under reduced pressure. The complex *trans*-$Pt[Mo(CO)_3(\eta^5\text{-}C_5H_5)]_2(PhCN)_2$ is isolated as an analytically pure red-brown powder. Yield: 8.91 g (80%).

Anal. Calcd. for $C_{30}H_{20}Mo_2N_2O_6Pt$: C, 40.41; H, 2.26; N, 3.14. Found: C, 40.40; H, 2.17; N, 3.33, MP $> 130\,°C$ (dec).

Properties

The complex *trans*-$Pt[Mo(CO)_3(\eta^5\text{-}C_5H_5)]_2(PhCN)_2$ is light sensitive. It can be stored in the dark under N_2 at $-20\,°C$ for months and is sufficiently stable toward oxygen that it can be rapidly weighted in air. The same is true for its Cr or W homologs.[3] All three complexes *trans*-$Pt[M(CO)_3(\eta^5\text{-}C_5H_5)]_2(PhCN)_2$ have a limited solubility in THF and dichloromethane but are insoluble in pentane. They are best characterized by carbonyl IR absorptions (KBr pellet):[3] M = Cr, 1884 (s), 1840 (s), 1808 (s); Mo, 2054 (w), 1900 (s), 1862 (sh), 1834 (s, br); W, 2060 (vw), 1900 (s), 1861 (sh), 1832 (s).

These complexes are convenient precursors for the high-yield synthesis of tetranuclear clusters, detailed in the following section for the Pt_2Mo_2 and the Pd_2Mo_2 derivatives.

D. PLANAR TETRANUCLEAR CLUSTERS WITH MIXED-METAL CORES, $M'_2Mo_2(\eta^5-C_5H_5)_2(CO)_6(PPh_3)_2$, $M' = Pd$ or Pt

$$2trans\text{-}M'[Mo(CO)_3(\eta^5\text{-}C_5H_5)]_2(PhCN)_2 + 2PPh_3$$

$$\longrightarrow M'_2Mo_2(\eta^5\text{-}C_5H_5)_2(CO)_6(PPh_3)_2 + [Mo(CO)_3(\eta^5\text{-}C_5H_5)]_2$$
$$+ 4PhCN$$

$$M' = Pd \text{ or } Pt$$

The method of choice is given here for the preparation of these mixed-metal cluster complexes and it can be extended to other phosphine ligands.[3, 4] A second route to these complexes starting with $M'Cl_2(PR_3)_2$ and $Na[Mo(CO)_3(\eta^5\text{-}C_5H_5)]\cdot 2DME$ is also known,[3, 4] but it is often of lesser value giving generally lower yields

1. Hexacarbonylbis(η^5-cyclopentadienyl)bis(triphenylphosphine)-dimolybdenumdiplatinum, $[Pt_2Mo_2(\eta^5-C_5H_5)_2(CO)_6(PPh_3)_2]$

Procedure

All operations are performed under an atmosphere of nitrogen dried and deoxygenated as indicated in the procedure in Section A, using standard Schlenk techniques and a vacuum nitrogen line.[9]

A 150-mL Schlenk tube fitted with a magnetic stirring bar is evacuated and backfilled with purified nitrogen. The solid complex *trans*-$Pt[Mo(CO)_3(\eta^5\text{-}C_5H_5)]_2(PhCN)_2$ (0.862 g, 0.97 mmol) is placed into the Schlenk tube, which is then again evacuated and backfilled twice with N_2. The 50 mL of THF, freshly distilled from sodium benzophenone ketyl under N_2, is added to the Schlenk tube. A solution of triphenylphosphine (0.254 g, 0.97 mmol) in purified THF (20 mL) is slowly added under N_2 at room temperature with a syringe, pipette, or cannula. The mixture is heated at reflux under continuous stirring for 3 h and filtered after cooling to room temperature. The solid is washed with 100 mL of a (1:9) THF–pentane mixture, and with diethyl ether (60 mL), distilled from sodium benzophenone ketyl. The solid is then dried *in vacuo*. Yield of $Pt_2Mo_2(\eta^5$-$C_5H_5)_2(CO)_6(PPh_3)_2$:   0.437 g   64%   based   on $Pt[Mo(CO)_3(\eta^5$-$C_5H_5)]_2(PhCN)_2$. The THF–pentane filtrate and the diethyl ether solution can be combined and evaporated to dryness. The residue from these combined solvents is recrystallized by slow diffusion of pentane into a toluene solution. Additional yield of $Pt_2Mo_2(\eta^5$-$C_5H_5)_2(CO)_6(PPh_3)_2$: 0.154 g [23% based on $Pt[Mo(CO)_3(\eta^5$-$C_5H_5)]_2(PhCN)_2$) and $[Mo(CO)_3Cp]_2$] 0.20 g (42% recovered Mo).

Anal. Calcd. for $C_{52}H_{40}Mo_2O_6P_2Pt_2$: C, 44.46; H, 2.87. Found: C, 44.48; H, 2.92, MP > 200 °C, formula wt.: 1404.9.

The related cluster $Pt_2W_2(\eta^5$-$C_5H_5)_2(CO)_6(PPh_3)_2$ can be prepared in a similar way using *trans*-$Pt[W(CO)_3(\eta^5$-$C_5H_5)]_2(PhCN)_2$ as a precursor.[3]

Properties

Hexacarbonylbis(η^5-cyclopentadienyl)bis(triphenylphosphine)dimolybdenumdiplatinum, $Pt_2Mo_2(\eta^5$-$C_5H_5)_2(CO)_6(PPh_3)_2$ is a dark brown or black powder, stable for months under N_2, which can be handled in air for a short time. It is soluble in dichloromethane, toluene, or THF, but insoluble in hexane. The compound is best characterized by its carbonyl IR absorptions (KBr pellet): $v_{(CO)}$ 1831 (s), 1798 (m sh), 1753 (sh), 1739 (s) cm^{-1}. ^1H, ^{31}P, ^{13}C NMR data have also been collected[3] and the crystal structure of the related compound $Pt_2Mo_2(\eta^5$-$C_5H_5)_2(CO)_6(PEt_3)_2$ has been reported.[3,13] The molecule has a planar core with five metal–metal bonds, its center of symmetry is located at the middle of the Pt–Pt bond. The complex $Pt_2W_2(\eta^5$-$C_5H_5)_2(CO)_6(PPh_3)_2$ has properties similar to those of its molybdenum analog[3] and the crystal structure of $Pt_2W_2(\eta^5$-$C_5H_5)_2(CO)_6(PEt_3)_2$ has been shown to be identical to that of its molybdenum analog.[3]

2.   Hexacarbonylbis(η^5-cyclopentadienyl)-
   bis(triphenylphosphine)-
   dimolybdenumdipalladium, $Pd_2Mo_2(\eta^5$-$C_5H_5)_2(CO)_6(PPh_3)_2$

1. $PdCl_2 + 2PhCN \longrightarrow PdCl_2(PhCN)_2$

2. $PdCl_2(PhCN)_2 + 2Na[Mo(CO)_3(\eta^5$-$C_5H_5)]\cdot 2DME$

   $\longrightarrow$ *trans*-$Pd[Mo(CO)_3(\eta^5$-$C_5H_5)]_2(PhCN)_2 + 2NaCl + 4DME$

   2*trans*-$Pd[Mo(CO)_3(\eta^5$-$C_5H_5)]_2(PhCN)_2 + 2PPh_3$

$$\longrightarrow Pd_2Mo_2(\eta^5\text{-}C_5H_5)_2(CO)_6(PPh_3)_2 + [Mo(CO)_3(\eta^5\text{-}C_5H_5)]_2 + 4PhCN$$

This method is the one of choice for the mixed-metal products.[4] This applies to a wide range of phosphines,[4] in a more economical way than an alternate one-step procedure from $PdCl_2L_2$ and $Na[Mo(CO)_3(\eta^5\text{-}C_5H_5)]\cdot2DME$.[4] The trinuclear Mo—Pd—Mo complex obtained in the second step must be prepared *in situ* and kept at low temperature, as detailed below. Procedures for the homologous complexes $Pd_2M_2(\eta^5\text{-}C_5H_5)_2(CO)_6(PPh_3)_2$, M = Cr or W, are also known and follow similar preparations.[4]

Procedure

Preparation of $PdCl_2(PhCN)_2$. The synthesis of bis(benzonitrile-dichloropalladium has been described previously.[10] It can be performed in a manner analogous to that given in Section C for $PtCl_2(PhCN)_2$, affording yields of 95%, based on the starting $PdCl_2$.

Preparation of $Pd_2Mo_2(\eta^5\text{-}C_5H_5)_2(CO)_6(PPh_3)_2$. All further operations must be performed in the absence of air under an atmosphere of nitrogen dried and deoxygenated as described in the procedure in Section A, and using standard Schlenk techniques and a vacuum nitrogen line.[9]

A 500-mL two-necked round-bottomed flask equipped with a stopcock side arm is fitted with a pressure-equalizing dropping funnel and a magnetic stirring bar. The compound $PdCl_2(PhCN)_2$ (0.384 g, 1 mmol) is introduced and the entire apparatus is evacuated and backfilled with purified nitrogen three times. Tetrahydrofuran (50 mL) dried over sodium benzophenone ketyl, is transferred into the round-bottomed flask and the mixture is cooled in an acetone–Dry Ice bath at $-78\,^\circ\mathrm{C}$. A quantity of $Na[Mo(CO)_3(\eta^5\text{-}C_5H_5)]\cdot2DME$ (0.897 g, 2 mmol) and THF (75 mL) are transferred into the dropping funnel and added dropwise to the stirred suspension of $PdCl_2(PhCN)_2$ in THF. After stirring for 15 min, PPh_3 (0.262 g, 1 mmol) and diethyl ether (30 mL) distilled from sodium benzophenone ketyl under N_2 are transferred into the emptied dropping funnel and added dropwise to the stirred red-violet solution of *trans*-$Pd[Mo(CO)_3(\eta^5\text{-}C_5H_5)]_2(PhCN)_2$ maintained at $-78\,^\circ\mathrm{C}$. Under constant agitation, the temperature is slowly raised to $20\,^\circ\mathrm{C}$ in 3.5 h. The resulting mixture is evaporated to a third of its volume, and pentane (200 mL) distilled from sodium wire under N_2 is added. A gray-green solid is removed by filtration, washed three times with 30 mL portions of each of the following solvents deoxygenated by a N_2 purge, diethyl ether, water (to remove NaCl), and ethanol and dried *in vacuo*. Recrystallization of the precipitate by slow diffusion of pentane into a dichloromethane solution (purification of solvents as indicated earlier) at $-20\,°C$ affords $Pd_2Mo_2(\eta^5\text{-}C_5H_5)_2(CO)_6(PPh_3)_2$. Yield: 0.549 g (89%) based on $PdCl_2(PhCN)_2$.

Anal. Calcd. for $C_{52}H_{40}Mo_2O_6P_2Pd_2$: C, 50.88; H, 3.28. Found: C, 51.02; H, 3.05, MP > 170 °C (dec), Formula wt. 1227.5.

Properties

The complex is a gray-green powder, soluble in dichloromethane, toluene, or THF and insoluble in pentane or hexane. It is stable for months under a nitrogen atmosphere and can be handled in air for a short time. It is best characterized by its IR spectrum in dichloromethane solution: $\nu_{(CO)}$: 1912 (m), 1851 (vs), 1830 (sh), 1785 (s) cm^{-1}. The NMR data for ^1H, ^{13}C and ^{31}P have been reported for $Pd_2Mo_2(\eta^5\text{-}C_5H_5)_2(CO)_6(PPh_3)_2$ as well as the crystal structure of the related $Pd_2Mo_2(\eta^5\text{-}C_5H_5)_2(CO)_6(PEt_3)_2$:[4] the molecule has a planar core with five metal–metal bonds, its center of symmetry is located at the middle of the Pd–Pd bond. This cluster has been used to prepare a cluster-derived heterogeneous catalyst for the selective carbonylation of aromatic nitro compounds $ArNO_2$ into isocyanates ArNCO.[14]

References

1. F. A. Cotton and G. Wilkinson, *Advanced Inorganic Chemistry*, 4th ed., Wiley, New York 1980.

2. R. B. King and F. G. A. Stone, *Inorg. Synth.*, **7**, 99 (1963).

3. R. Bender, P. Braunstein, J. M. Jud, and Y. Dusausoy, *Inorg. Chem.*, **23**, 4489 (1984).

4. R. Bender, P. Braunstein, J. M. Jud, and Y. Dusausoy, *Inorg. Chem.*, **22**, 3394 (1983).

5. D. S. Ginley, C. R. Bock, and M. S. Wrighton, *Inorg. Chim. Acta*, **23**, 85 (1977).

6. J. Smart and C. J. Curtis, *Inorg. Chem.*, **16**, 1788 (1977).

7. W. G. Gladfelter and G. L. Geoffroy, *Adv. Organometal. Chem.*, **18**, 207 (1980).

8. U. Behrens and F. Edelmann, *J. Organomet. Chem.*, **263**, 179 (1984).

9. D. F. Shriver, *The Manipulation of Air Sensitive Compounds*, McGraw-Hill, New York, 1969.

10. (a) M. S. Kharasch, R. C. Seyler, and F. R. Mayo, *J. Am. Chem. Soc.*, **60**, 882 (1938); (b) F. R. Hartley, *The Chemistry of Platinum and Palladium*, Wiley, New York, 1973, p. 462; (c) J. R. Doyle, P. E. Slade, and H. B. Jonassen, *Inorg. Synth.*, **6**, 218 (1960).

11. J. P. Barbier and P. Braunstein, *J. Chem. Res. Synop.*, 1978, 412; *J. Chem. Res., Miniprint*, 1978, 5029.

12. A. Fusi, R. Ugo, R. Psaro, P. Braunstein, and J. Dehand, *Phil. Trans. R. Soc. London*, **A308**, 125 (1982); *J. Mol. Catal.*, **16**, 217 (1982).

13. R. Bender, P. Braunstein, Y. Dusausoy, and J. Protas, *J. Organomet. Chem.*, **172**, C51 (1979).

14. P. Braunstein, R. Bender, and J. Kervennal, *Organometallics*, **1**, 1236 (1982).

63. MIXED TRINUCLEAR DICOBALTIRON AND DICOBALTRUTHENIUM CLUSTER COMPLEXES

Submitted by H. VAHRENKAMP*
Checked by DAQIANG XU,[†] HERBERT D. KAESZ,[†] EDWARD
ROSENBERG,[‡] and BRUCE NOVAK[‡]

The chemistry of organometallic clusters has moved from the exploratory stage into the stage of systematic reactivity studies.[1] This is due to the availability, in reasonable quantities, of clusters that at the same time have quite reactive metal–metal bonds, yet are rather inert towards total fragmentation. The first of these properties is natural for mixed-metal clusters[2]; the second is introduced by bridging main group element units.[3] We describe herein the syntheses of four such clusters, which we believe will be of general use in the further development of cluster chemistry. One of them, $FeCo_2(CO)_9S$, was first described more than 20 years ago,[4] while the others, $RuCo_2(CO)_9S$ (Ref. 5), $FeCo_2(CO)_9(PC_6H_5)$ (Ref. 6), and $RuCo_2(CO)_{11}$ (Ref. 7) are rather new.

General Procedures and Techniques

■ **Caution.** *High-pressure work is dangerous and should be performed by experienced personnel only. Hydrogen and carbon monoxide are poisonous and very easily inflammable gases. The compounds C_2H_5SH and $C_6H_5PH_2$ are volatile, toxic, and of highly obnoxious odor. Inhalation of the volatile metal carbonyls $Fe(CO)_5$ and $Co_2(CO)_8$ must be avoided by all means, and the compounds must be handled in a well-ventilated hood. Due to health hazards now recognized for benzene, toluene should be substituted for benzene in all procedures; as a consequence, there may be some minor variations in the reported yields.*

All materials are to be handled under an inert gas atmosphere in Schlenk-type vessels.[8a] All solvents must be dry, degassed, and distilled from natrium wire. The preparation of $FeCo_2(CO)_9(PC_6H_5)$ requires a medium scale chromatographic separation (silica gel Macherey & Nagel 0.06–0.2 mm, dried for

*Institut für Anorganische Chemie der Universität Freiburg, Albertstr. 21, D-7800 Freiburg, Federal Republic of Germany. These procedures are based on thesis work by F. Richter, M. Müller, E. Roland, and a private communication by L. Markó.

†Department of Chemistry and Biochemistry, University of California, Los Angeles, CA 90024-1569.

‡Department of Chemistry, California State University, Northridge, CA 91330.

12 h at 180 °C under 10^{-3} torr vacuum). The preparations of $FeCo_2(CO)_9S$ and $RuCo_2(CO)_9S$ require a 0.5-L high-pressure rotating autoclave equipped with temperature control, and supplies of hydrogen and carbon monoxide.

The starting materials $Fe(CO)_5$, $Co_2(CO)_8$, $RuCl_3 \cdot 3H_2O$, $C_6H_5PCl_2$, and C_2H_5SH are commercially available.[9] The metal compounds can be converted to $Fe_2(CO)_9$,[8a] $K[Co(CO)_4]$ (Ref. 10), $Ru_3(CO)_{12}$,[11] and $[Ru(CO)_3Cl_2]_2$ (Ref. 12) by well-described procedures. The compound $C_6H_5PH_2$ is prepared from $C_6H_5PCl_2$ and $Li[AlH_4]$.[13] Most of these refined starting materials are, however, also commercially available. The elemental analyses for all four clusters described here are given in the original publications.[4-7]

A. NONACARBONYL-μ_3-THIO-DICOBALTIRON, $FeCo_2(CO)_9S$

$$Co_2(CO)_8 + Fe(CO)_5 + C_2H_5SH \longrightarrow FeCo_2(CO)_9S + 4CO + C_2H_6$$

To 150 mL of hexane in a glass liner inside a 0.5-L autoclave $Co_2(CO)_8$ (13.3 g, 38.9 mmol), $Fe(CO)_5$ (7.6 g, 5.1 mL, 39 mmol), and ethanethiol (C_2H_5SH) (2.8 g, 3.3 mL, 45 mmol) are added under argon. Upon dissolution of the reactants vigorous evolution of CO gas occurs. The autoclave is closed and 70 bar of H_2 pressure and further 130 bar of CO pressure are applied (total pressure 200 bar). The autoclave is rotated and heated to 160 °C for 4 h, with care to avoid temperatures above 160 °C. After cooling to room temperature overnight and releasing the pressure, the liner is taken out of the autoclave and the mother liquor decanted. The solid residue (16.0 g, 90%) dried under vacuum is chemically pure $FeCo_2(CO)_9S$.

Properties

The compound $FeCo_2(CO)_9S$ thus prepared forms large black crystals. In this form it is virtually air stable (not so in solution). Its purity can be checked by IR spectroscopy [ν_{CO} bands in cyclohexane solution at 2105 (w), 2069 (vs), 2055 (vs), 2044 (s), 2030 (m), 1983 (m), 1951 (vw) cm^{-1}]. It is moderately soluble in all organic solvents to give reddish-brown solutions. It has been used for various ligand substitution and metal exchange reactions whereby the first optically active clusters were obtained.[14, 15]

B. NONACARBONYL-μ_3-THIO-DICOBALTRUTHENIUM, $RuCo_2(CO)_9S$

$$Co_2(CO)_8 + \tfrac{1}{3}Ru_3(CO)_{12} + C_2H_5SH \longrightarrow RuCo_2(CO)_9S + 3CO + C_2H_6$$

To 45 mL of hexane in a glass liner inside a 0.5-L autoclave $Ru_3(CO)_{12}$ (2.0 g, 3.1 mmol), $Co_2(CO)_8$ (3.2 g, 9.4 mmol), and ethanethiol (C_2H_5SH) (0.63 g, 0.75 mL, 10.1 mmol) are added under argon. Upon dissolution of the reactants, evolution of CO gas occurs. The autoclave is closed and 60 bar of H_2 pressure and 140 bar of CO pressure are applied (total pressure 200 bar). The autoclave is rotated and heated to 150 °C for 4 h. After cooling to room temperature overnight and releasing the pressure, the liner is taken out of the autoclave and the mother liquor decanted. After washing with 5 mL of pentane the solid residue (3.5 g, 75%), which is dried under vacuum, is analytically pure $RuCo_2(CO)_9S$.

Properties

The complex $RuCo_2(CO)_9S$ thus prepared forms black crystals. It is air stable in the solid state (not so in solution). In cyclohexane solution it shows v_{CO} bands at 2103 (vw), 2082 (vw), 2062 (vs), 2036 (s), and 1996 (w) cm^{-1}, being quite different from $FeCo_2(CO)_9S$ in this respect. It is moderately soluble in all organic solvents giving red-brown solutions. Its reactivity[5] is similar to that of $FeCo_2(CO)_9S$.

C. NONACARBONYL-μ_3-(PHENYLPHOSPHINIDENE)-DICOBALTIRON, $FeCo_2(CO)_9(PC_6H_5)$

$$\tfrac{1}{2}Fe_2(CO)_9 + C_6H_5PH_2 \longrightarrow Fe(CO)_4(PH_2C_6H_5) + \tfrac{1}{2}CO$$

$$Fe(CO)_4(PH_2C_6H_5) + Co_2(CO)_8 \longrightarrow FeCo_2(CO)_9(PC_6H_5) + H_2 + 3CO$$

A quantity of $Fe_2(CO)_9$ (12.8 g 35.2 mmol) is combined with $C_6H_5PH_2$ (2.0 g, 2.0 mL, 18.0 mmol) in 100 mL of toluene in a Schlenk flask wrapped with aluminum foil. The flask is connected to a mercury bubbler and the contents are stirred for 3 days at room temperature. The resulting solution containing $Fe(CO)_4(PH_2C_6H_5)$ is filtered into a 250-mL Schlenk flask. To this is added $Co_2(CO)_8$ (6.2 g, 18.0 m mol) (freshly recrystallized from petroleum ether at -78 °C) and the solution stirred for 24 h with irradiation by a normal 60-W bulb placed next to the flask. Then all volatile components are removed on a vacuum manifold equipped with a solvent trap, using oil pump vacuum. The residue is washed with benzene–hexane (1:10) onto a 4×60-cm chromatography column filled with silica gel (see above). Chromatography with this eluent first yields a green fraction containing 0.1 g of $Co_3(CO)_9(PC_6H_5)$. The second, purple main fraction is evacuated to dryness and recrystallized from 50 mL of benzene–hexane (1:1) at -25 °C yielding black crystalline $FeCo_2(CO)_9(PC_6H_5)$ (7.2 g, 75%).

Properties

The complex $FeCo_2(CO)_9(PC_6H_5)$ is black in the crystalline state and purple in solution. It is moderately air sensitive. It has a low solubility in aliphatic hydrocarbons and a moderately good solubility in aromatic hydrocarbons. In cyclohexane solution it shows ν_{CO} bands at 2101 (vs), 2059 (vs), 2048 (vs), 2039 (vs), 1981 (w), 1969 (w) cm^{-1}. It has been used for various types of basic metal cluster reactions.[16]

D.   UNDECACARBONYLDICOBALTRUTHENIUM, $RuCo_2(CO)_{11}$

$$\tfrac{1}{2}Co_2(CO)_8 + Na-K \longrightarrow K[Co(CO)_4]$$

$$2K[Co(CO)_4] + \tfrac{1}{2}[Ru(CO)_3Cl_2]_2 \longrightarrow RuCo_2(CO)_{11} + 2KCl$$

■   **Caution.**   *Kalium metal and Na–K are inflammable in air. These must be handled under inert atmosphere in a well-ventilated hood. Excess reagents must be carefully passivated by reaction with absolute t-butyl alcohol, under an inert atmosphere, in a well-ventilated hood.*

Preparation of Na–K alloy

Natrium (5.3 g, 0.23 mol) and kalium (25.6 g, 0.65 mol) are freshly cut, washed with petroleum ether, wiped dry with tissue paper, and placed in a 100-mL flask under nitrogen. Melting of the metals and concomitant formation of the alloy are brought about by careful warming with a heat gun. The Na–K must be stored under an inert atmosphere.

Preparation of K[Co(CO)$_4$][10]

A quantity of $Co_2(CO)_8$ (freshly recrystallized from petroleum ether at $-78\,°C$) (0.46 g, 1.33 mmol) is dissolved in 50 mL of dry THF. To this solution, 0.30 mL of Na–K alloy is syringed in and the solution stirred for 25 min. The excess alloy is removed by decanting under nitrogen and the solvent is then removed by a trap-to-trap distillation. The resulting white residue is kept under dynamic vacuum overnight to insure complete removal of the solvent.

Preparation RuCo$_2$(CO)$_{11}$

A quantity of solvent free kalium tetracarbonylcobaltate $(1-)$ (2.0 g, 9.5 mmol) is dissolved in a 250-mL Schlenk flask in 100 mL of water, which has previously been degassed and redistilled under inert atmosphere. To this mixture powdered hexacarbonyltetrachlorodiruthenium (1.2 g, 2.3 mmol) is added at once. The flask is closed with a stopcock and the mixture is stirred vigorously for 1 h at room temperature during which time the colorless starting material dissolves and the black product precipitates. After filtration the residue is dried in vacuum on the filter. It is then redissolved through the filter in 60 mL of hexane in three portions. The solution is cooled to $-35\,°C$ and the black crystalline $RuCo_2(CO)_{11}$ is obtained by quick filtration. The yield (1.3–1.9 g, 53–77%) depends critically on the purity of the starting materials and on the working conditions.

Properties

The complex $RuCo_2(CO)_{11}$ is black in the crystalline state and dark red in solution. It is air sensitive and must be stored in a refrigerator. It can be handled over extended periods of time only in hydrocarbon solvents in which it is moderately soluble. In cyclohexane it shows ν_{CO} bands at 2126 (vw), 2069 (s), 2056 (sh), 2050 (vs), 2028 (w), 2005 (w), 1820 (w) cm^{-1}. Controlled thermal decomposition above room temperature yields $Ru_2Co_2(CO)_{13}$.[15] It is very labile and especially suited for reactions that introduce μ_3 capping units.[17]

References

1. H. Vahrenkamp, *Adv. Organomet. Chem.*, **22**, 169 (1983).
2. W. L. Gladfelter and G. L. Geoffroy, *Adv. Organomet. Chem.*, **18**, 207 (1980).
3. H. Vahrenkamp, in A. Müller and E. Diemann (eds.), *Transition Metal Chemistry*, Verlag Chemie, Weinheim 1981, p. 35.
4. S. Khattab, L. Markó, G. Bor, and L. Markó, *J. Organomet. Chem.*, **1**, 373 (1964).
5. E. Roland and H. Vahrenkamp, *Chem. Ber.*, **117**, 1039 (1984).
6. J. C. Burt and G. Schmid, *J. Chem. Soc. Dalton Trans.*, **1978**, 1385.
7. E. Roland and H. Vahrenkamp, *Angew. Chem.*, **93**, 714 (1981); *Angew. Chem. Int. Ed. Engl.*, **20**, 679 (1981).
8. (a) R. B. King, in *Organometallic Syntheses*, Vol. 1, J. J. Eisch and R. B. King (ed.), Academic Press, New York 1965, p. 93; (b) D. F. Shriver and M. A. Drezdzon, *The Manipulation of Air-Sensitive Compounds*, 2nd ed., McGraw-Hill, New York, 1986.
9. For the synthesis of $Co_2(CO)_8$, see G. Brauer, *Handbuch der Präparativen Anorganischen Chemie*, Vol. 3, Enke, Stuttgart 1981, p. 1833.
10. J. E. Ellis and E. A. Flom, *J. Organomet. Chem.*, **99**, 263 (1975).
11. B. F. G. Johnson and J. Lewis, *Inorg. Synth.*, **13**, 92 (1972).
12. A. Mantovani and S. Cenini, *Inorg. Synth.*, **16**, 51 (1976).
13. F. Pass and H. Schindlbauer, *Monatsh. Chem.*, **90**, 148 (1959); L. Maier, in *Organic Phosphorus Compounds*, G. B. Kosolapoff and L. Maier (eds.), Wiley, New York 1972, p. 1.

14. R. Rossetti, G. Gervasio, and P. L. Stanghellini, *J. Chem. Soc. Dalton Trans.*, **1978**, 222.

15. F. Richter and H. Vahrenkamp, *Chem. Ber.*, **115**, 3224, 3243 (1982).

16. M. Müller and H. Vahrenkamp, *Chem. Ber.*, **116**, 2748, 2765 (1983).

17. E. Roland and H. Vahrenkamp, *Organometallics*, **2**, 1048 (1983).

64. STEPWISE SYNTHESES OF RUTHENIUM MIXED-METAL CLUSTER COMPLEXES

Submitted by PIERRE BRAUNSTEIN* and JACKY ROSE*
Checked by DAVID C. BUSBY[†]

Several synthetic methods are now available for the preparation of mixed-metal clusters.[1] However, when particular clusters are desired, two main points of concern for their synthesis often remain: the availability and price of the precursors, and the yield of the reaction. Mixed-metal clusters containing ruthenium have attracted considerable interest mainly because of the variety of structural and bonding types encountered, and of their potential for homogeneous and heterogeneous catalysis.[2]

"Commercial $RuCl_3 \cdot xH_2O$" is an obvious precursor in ruthenium chemistry. However, neither its percentage of metal content nor the ruthenium oxidation state are constant or well defined from one batch to another. For this reason, we report herein details of the high-yield synthesis of $(Et_4N)[RuCl_4(CH_3CN)_2]$, an easily accessible Ru (III) complex having a wide range of applicability.

We describe herein its conversion into the tetranuclear cluster anion $[Co_3Ru(CO)_{12}]^-$ itself leading to the pentanuclear species $Co_3RuCu(CH_3CN)(CO)_{12}$ in good yield.

A. TETRAETHYLAMMONIUM BIS(ACETONITRILE)-TETRACHLORORUTHENATE(1−), $(Et_4N)[RuCl_4(CH_3CN)_2]$

$$RuCl_3 \cdot xH_2O + [Et_4N]Cl \xrightarrow{HCl,Hg} \tfrac{1}{2}(Et_4N)_2[Ru_2Cl_7(OH)_3] \cdot HCl$$

$$\tfrac{1}{2}(Et_4N)_2[Ru_2Cl_7(OH)_3] \cdot HCl \xrightarrow{CH_3CN} (Et_4N)[RuCl_4(CH_3CN)_2] + xH_2O$$

*Institut Le Bel, Université Louis Pasteur, F-67070 Strasbourg, France.
[†]Union Carbide Corporation, P.O. Box 8361, S. Charleston, WV 25303.

Procedure

All the steps of this preparation are carried out in air; the procedure requires ~ 3 days. A solution of commercial $RuCl_3, xH_2O$ (4.0 g, 38% Ru, ~ 15 mmol) in HCl 12 N (80 mL) is placed in a 250-mL round-bottomed flask equipped with a magnetic stirring bar, and a reflux condenser. The solution is heated to 90 °C for 10 h, using an oil bath.

- **Caution.** *Concentrated hydrochloric acid should be manipulated carefully and handled in a well-ventilated hood. Mercury metal is poisonous and should be handled in a well-ventilated hood.*

The solution is allowed to cool and commercial $[Et_4N]Cl, xH_2O$ (3.7 g, 30% H_2O 15.5 mmol) dissolved in H_2O (40 mL) is added, together with few drops of mercury (1 mL). The mixture is shaken at ambient temperature until a blue-green solution is obtained (~ 3 h). The solution is decanted and filtered to eliminate the mercury salt, and then evaporated to dryness at 50 °C. The green-black solid residue is suspended in CH_3CN (80 mL), then vigorously stirred and heated under reflux until the solution becomes light yellow and a yellow product precipitates (~ 10 h). Diethyl ether (100 mL) is added and the mixture cooled to 0 °C for 3 h. The microcrystalline solid is filtered, washed with diethyl ether, and dried *in vacuo* to give 6.5 g (95% yield) of product. It can generally be used without further purification (see Section B). However, an analytically pure sample is easily obtained by suspending the product in 80 mL of CH_2Cl_2–CH_3CN (1 : 1 vv), with stirring for 1 h. The solid purified product is filtered and dried *in vacuo* (10^{-4} torr). Diethyl ether (40 mL) is added to the filtrate, which is cooled to $- 10$ °C and held at that temperature for 1 h, affording a second crop of $(Et_4N)[RuCl_4(CH_3CN)_2]$ (total yield 85–90%).

Anal. Calcd. for $C_{12}H_{26}N_3Cl_4Ru$: C, 31.66; H, 5.76; N, 9.23. Found: C, 31.4; H, 5.8; N, 9.5.

Properties

The salt $(Et_4N)[RuCl_4(CH_3CN)_2]$ is an air-stable, yellow solid, [mp 220–225 °C (dec)]. It is insoluble in nonpolar organic solvents such as hexane, and sparingly soluble at room temperature in more polar organic solvents such as dichloromethane and acetonitrile. The salt $(Et_4N)[RuCl_4(CH_3CN)_2]$ is stable in dichloromethane solution but is decomposed in methanol.[3] The compound is characterized by its IR spectrum, which shows a band due to $\nu_{(CN)}$ at 2298 cm^{-1} (KBr pellet). UV–Vis, magnetic moment, and conductivity data have also been reported for this compound.[3]

B. TETRAETHYLAMMONIUM DODECACARBONYLTRICO-
BALTATERUTHENATE(1 −), $(Et_4N)[Co_3Ru(CO)_{12}]$

$$(Et_4N)[RuCl_4(CH_3CN)_2] + \tfrac{5}{2}Co_2(CO)_8$$

$$\xrightarrow{\text{THF}} (Et_4N)[Co_3Ru(CO)_{12}] \cdot \tfrac{1}{3}THF$$

$$+ 8CO + 2CoCl_2 + 2CH_3CN$$

Procedure

■ **Caution.** *Carbon monoxide (poisonous) is evolved during this reaction, which must be carried out in a well-ventilated hood.*

This preparation requires ∼ 2 days. All manipulations are performed in Schlenk-type flasks under nitrogen dried over molecular sieves (4 Å) and deoxygenated on catalyst BASF R3-11 (Imhoff & Stahl). Tetrahydrofuran (THF) is dried over sodium benzophenone ketyl under N_2. The complex $Co_2(CO)_8$ was purchased from Strem.

A suspension of $(Et_4N)[RuCl_4(CH_3CN)_2]$ (1.50 g, 3.29 mmol) in THF (30 mL) and a solution of pure $Co_2(CO)_8$ (2.456 g, 7.18 mmol) in THF (50 mL) are placed in a 250-mL Schlenk flask with a gas inlet, equipped with a magnetic stirring bar and a reflux condenser topped with a gas inlet adapter. The solution is refluxed for 3 h and becomes deep red during the reaction. The reaction is monitored by following the intensity of the IR band at 1820 cm^{-1}. When this band has stopped growing, the reaction can be considered to be finished. There should no longer be any $[Co(CO)_4]^-$ present (IR band at 1889 cm^{-1}) but if so, small quantities of the ruthenium complex should be added to consume it.

The solution is kept overnight at ∼ − 40 °C and blue crystals of $CoCl_2$ deposit. The solution is filtered and evaporated to dryness under reduced pressure. Distilled and degassed H_2O (50 mL) is added to dissolve the remaining cobalt salts. The suspension is filtered under N_2 and the red solid residue is dried *in vacuo* (10^{-4} torr). The residue is then dissolved in THF (40 mL), filtered, and hexane (80 mL, previously dried over Na) is added. A red-purple microcrystalline solid is precipitated. This is filtered and washed with hexane (2 portions, 20 mL). Cooling the filtrate to − 20 °C affords a second crop of the THF solvated product $(Et_4N)[RuCo_3(CO)_{12}] \cdot \tfrac{1}{3}$ THF (total yield 1.9 g, 75% based on Ru).

Anal. Calcd. for $C_{20}H_{20}NCo_3O_{12}Ru$, $\tfrac{1}{3}$ THF: C, 33.35; H, 2.97; N, 1.82. Found: C, 33.4; H, 2.7; N, 1.9.

Properties

The complex $(Et_4N)[Co_3Ru(CO)_{12}]\cdot\frac{1}{3}$ THF is a moderately air-stable, red-purple crystalline solid (mp 195–198 °C) (becoming gray above 100 °C). It is best stored in crystalline form under N_2. It is insoluble in nonpolar organic solvents such as hexane, but highly soluble in polar organic solvents such as dichloromethane, THF, or acetone. Carbonyl absorptions (IR): $v_{(CO)(THF)}$: 2064 (w), 1998 (vs), 2012 (vs), 1968 (w), 1819 (w) cm^{-1}. Its reactivity towards alkynes[4] and towards methanol in the presence of $CO-H_2$,[5] has been investigated. The crystal structure of the analog (PPN) $[Co_3Ru(CO)_{12}]$ has been determined by X-ray diffraction.[5]

C. (ACETONITRILE)DODECACARBONYLTRICOBALTRUTHENIUMCOPPER $Co_3RuCu(CH_3CN)(CO)_{12}$

$$(Et_4N)[Co_3Ru(CO)_{12}]\cdot\tfrac{1}{3}THF + [Cu(CH_3CN)_4][BF_4]$$

$$\longrightarrow Co_3RuCu(CH_3CN)(CO)_{12} + (Et_4N)[BF_4] + 3CH_3CN$$

This preparation requires ~ 1.5 days. All manipulations are performed in Schlenk-type flasks under nitrogen. Solvents are dried and distilled under N_2 before use: acetone over $MgSO_4$, toluene over Na.

Solutions of $(Et_4N)[Co_3Ru(CO)_{12}]\cdot\frac{1}{3}$ THF (0.332 g, 0.43 mmol) in acetone (30 mL) and of $[Cu(CH_3CN)_4][BF_4]$ (Ref. 6) (0.280 g, 0.89 mmol) in acetone (20 mL) are mixed in a 100-mL Schlenk flask equipped with a magnetic stirring bar and a gas inlet. The solution is stirred for 1 h at room temperature. The deep red solution is then evaporated to dryness under reduced pressure. Extraction with 5×20-mL portions of hot toluene (60–70 °C) is continued until the toluene extract is almost colorless. While hot, these extracts are filtered, combined in a 100-mL Schlenk tube, and the volume is reduced *in vacuo* to 10 mL. Hexane (40 mL) is then layered over the toluene solution. Purple microcrystals of $Co_3RuCu(CH_3CN)(CO)_{12}$ are formed after overnight crystallization in Dry Ice. They are collected by filtration, washed with hexane (40 mL), and dried *in vacuo* to give 0.220 g (70.8% yield based on Ru) of product.

Anal. Calcd. for $C_{14}H_3Co_3CuNO_{12}Ru$: C, 23.40; H, 0.42; N, 1.95. Found: C, 23.5; H, 0.8; N, 1.7.

Properties

The complex $Co_3RuCu(CH_3CN)(CO)_{12}$ is a moderately air stable, purple, microcrystalline solid [mp 170–175 °C (dec)]. It can be stored under N_2 for weeks. It is insoluble in nonpolar organic solvents such as hexane, and very soluble in more polar solvents such as dichloromethane. It is not stable in donor solvents like THF. Its carbonyl absorptions (IR) are $v_{(CO)}(CH_2Cl_2$ solution): 2083 (m), 2018 (vs), 1983 (m), and 1852 (m) cm^{-1}.

References

1. W. L. Gladfelter and G. L. Geoffroy, *Adv. Organometal. Chem.*, **18**, 207 (1980).
2. P. Braunstein and J. Rosé, in *Stereochemistry of Organometallic and Inorganic Compounds*, Vol. 3, I. Bernal (ed.); Elsevier, Amsterdam, 1989 and references cited therein.
3. J. Dehand and J. Rosé, *J. Inorg. Chim. Acta*, **37**, 249 (1979).
4. P. Braunstein, J. Rosé, and O. Bars, *J. Organomet. Chem.*, **252**, C101 (1983).
5. M. Hidai, M. Orisaku, M. Ue, Y. Koyasu, T. Kodama, and Y. Uchida, *Organometallics*, **2**, 292 (1983).
6. G. J. Kubas, *Inorg. Synth.*, **19**, 90 (1979).

65. MIXED-METAL CLUSTER COMPLEXES OF NICKEL WITH RUTHENIUM OR WITH OSMIUM

Submitted by C. BERGOUNHOU,* J.-J. BONNET,* G. LAVIGNE* and
F. PAPAGEORGIOU*
Checked by S. G. SHORE† and U. SIRIWARDANE†

Several methods have been described that allow the directed syntheses of mixed-metal clusters.[1] One of these procedures involves the addition of coordinatively unsaturated monomeric complexes to $Os_3(\mu\text{-H})_2(CO)_{10}$, as illustrated below for the synthesis of $Os_3Pt(\mu\text{-H})_2(PR_3)(CO)_{10}$ (R = C_6H_{11} or Ph), eq. (1)[2]

1. $Pt(\eta^2\text{-}C_2H_4)_2(PR_3) + Os_3(\mu\text{-H})_2(CO)_{10}$

$\longrightarrow Os_3Pt(\mu\text{-H})_2(PR_3)(CO)_{10}$

Such a reaction occurs because the Os=Os double bond of $Os_3(\mu\text{-H})_2(CO)_{10}$ is well known to be reduced upon addition of a variety of nucleophiles including those that are metal centered.[3] This di-μ-hydrido triosmium

*Laboratoire de Chimie de Coordination du CNRS, associé à l'Université Paul Sabatier, 205 route de Narbonne, 31400 Toulouse, France.
†Department of Chemistry, Ohio State University, Columbus, OH 43210.

cluster is obtained in good yield by hydrogenation at atmospheric pressure of $Os_3(CO)_{12}$ at 125 °C, but extended treatment leads to $Os_4(\mu\text{-}H)_4(CO)_{12}$.[4] Under the same experimental conditions, hydrogenation of $Ru_3(CO)_{12}$ gives $Ru_4(\mu\text{-}H)_4(CO)_{12}$ as the only isolable product. If a dihydrido triruthenium intermediate is generated prior to cluster growth, then the chemistry of $Ru_3(CO)_{12}$ in the presence of molecular hydrogen could duplicate in some cases that of $Os_3(\mu\text{-}H)_2(CO)_{10}$.

Simple syntheses of new mixed-metal clusters through hydrogen-assisted condensations of polynuclear carbonyl complexes of the second- and third-row transition metals with first-row transition metal complexes can be performed. Our laboratory[5] and others[6] have simultaneously reported the synthesis of $Os_3Ni(\mu\text{-}H)_3(\eta^5\text{-}C_5H_5)(CO)_9$ via a hydrogen-assisted condensation. This complex is obtained in 66 to 74% yield [together with *tris*(Ni-η^5-C_5H_5)nonacarbonyl triosmium in 19–24% yield] in the reaction of Ni(-η^5-C_5H_5)(CO)$_2$ with $Os_3(\mu\text{-}H)_2(CO)_{10}$ under hydrogen.[7] In the present route, the unsaturated 46-electron cluster $Os_3(\mu\text{-}H)_2(CO)_{10}$ may be generated *in situ*. We have also been able to extend the reaction to the case of ruthenium[5] for which the trinuclear hydrido analog is unknown. The same reaction procedure is used for the preparation of $NiRu_3$ or $NiOs_3$ complexes, starting from $[Ni(\eta^5\text{-}C_5H_5)(CO)]_2$ and the trinuclear carbonyls $M_3(CO)_{12}$. All of the hydrido mixed-metal clusters are potential hydrogenation catalysts, for example, $Os_3Ni(\mu\text{-}H)_3(\eta^5\text{-}C_5H_5)(CO)_9$, supported on γ-alumina, has been shown to be useful for the high-efficiency methanation of CO and CO_2 at atmospheric pressure.[8]

General Procedure

■ **Caution.** *Owing to evolution of toxic carbon monoxide and explosive hydrogen, the gases should be carefully released in a well-ventilated hood.*

The syntheses of both complexes are performed using a 100-mL two-necked round-bottomed glass flask equipped with a reflux condenser, a bubbler, and a magnetic stirring bar. The flask is connected to a vacuum line. Reagent grade solvents may be used without further purification; they should be deaerated by several freeze–thaw cycles. The compound $Os_3(CO)_{12}$ may be purchased from Johnson Matthey or prepared according to the literature.[9] The complex $[Ni(\eta^5\text{-}C_5H_5)(CO)]_2$ is available from Pressure Chemicals. The compound $Ru_3(CO)_{12}$ is easily prepared from $RuCl_3 \cdot xH_2O$ (Johnson Matthey) by a published procedure,[10] or is also available commercially (Pressure Chemicals, Strem).

We have found that the amounts of starting materials and solvents for the synthesis in Ref. 10a can be tripled without any decrease in yield.

A. NONACARBONYL(η^5-CYCLOPENTADIENYL)-TRI-μ-HYDRIDO-NICKELOSMIUM, $Os_3Ni(\mu\text{-}H)_3(\eta^5\text{-}C_5H_5)(CO)_9$

$$Os_3(CO)_{12} + \tfrac{1}{2}[Ni(\eta^5\text{-}C_5H_5)(CO)]_2$$

$$\xrightarrow{H_2} NiOs_3(\mu\text{-}H)_3(\eta^5\text{-}C_5H_5)(CO)_9 + 4CO$$

Hydrogen is bubbled through an octane solution (50 mL) containing $Os_3(CO)_{12}$ (194 mg, 0.214 mmol) and an excess of $[Ni(\eta^5\text{-}C_5H_5)(CO)]_2$ (130 mg, 0.428 mmol). The solution is refluxed for 2 h during which time a metallic nickel mirror is deposited on the glass vessel. After cooling, the solution is reduced to ~ 2 mL under vacuum. Silica Gel 60 (Merck, 230–400 mesh, ~ 1.5 g) is added, and the solution is evaporated to dryness. The solid residue is placed on a 3×40-cm column packed with the same silica gel. Elution with pentane gives two violet bands: (a) $Os_3(\mu\text{-}H)_2(CO)_{10}$. Yield: 18.2 mg ($\sim 10\%$). (b) $NiOs_3(\mu\text{-}H)_3(\eta^5\text{-}C_5H_5)(CO)_9$. The mixed-metal complex is recrystallized from acetone as violet-black prismatic needles. Yield: 136–145 mg (67–71.4%), based on osmium. Further elution with acetone affords several unidentified products. Brown, untractable material is retained on the top of the column.

Anal. Calcd. for $C_{14}H_8O_9NiOs_3$: C, 17.71; H, 0.85; Ni, 6.18. Found: C, 17.85; H, 0.68; Ni, 5.95.*

Note. It is extremely difficult to determine the percentage of hydrogen in complexes containing osmium owing to the formation of highly volatile OsO_4. For a thorough discussion of the problem, see Ref. 5.

Properties

The complex $NiOs_3(\mu\text{-}H)_3(\eta^5\text{-}C_5H_5)(CO)_9$ is an air-stable, crystalline solid that can be handled in solution without undue precautions. It is soluble in many of the usual organic solvents. The compound is best characterized by its carbonyl IR absorptions, $\nu_{(CO)}$ (cyclohexane solution): 2090 (w), 2062 (s), 2004 (s), 1990 (w) cm^{-1}. The ^1H NMR spectrum 90 MHz, C_6D_6, room temperature, $\delta_{(CH_3)_4Si} = 0.00$ ppm shows a singlet at δ +5.73 ppm (5H, cyclopentadienyl) and a singlet at δ −17.8 ppm (3H, hydride). The mass spectrum of the compound at $\sim 75\,°C$ (70 eV) exhibits a parent peak centered at m/e (rel. int.) = 950 with a highly characteristic isotope pattern, the chief

*Service central de Microanalyse du CNRS à Vernaison 69390 France.

peaks being at $m/e = 945$ (4.0%), 946 (6.6%), 947 (7.9%), 948 (11.4%), 949 (10.5%), 950 (13.5%), 951 (9.7%), 952 (12.6%), 953 (4.7%), 954 (8.3%). Successive ions corresponding to loss of one through nine carbonyl groups are also observed, as is the ion for intact metal core Os_3Ni^+ ($m/e \sim 629$).

This complex proved to be an efficient catalyst precursor in the hydrogenation of olefins.[8]

The crystal structure of $NiOs_3(\mu\text{-}H)_3(\eta^5\text{-}C_5H_5)(CO)_9 \cdot \frac{1}{2}Os_3(\mu\text{-}H)_2(CO)_{10}$ (obtained by slow evaporation of the reaction mixture in octane, without chromatographic work-up) has been solved.[5] The molecular structure of the $NiOs_3(\mu\text{-}H)_3(\eta^5\text{-}C_5H_5)(CO)$, species is shown below. The crystal structure of the compound $NiOs_3(\mu\text{-}H)_3(\eta^5\text{-}C_5H_5)(CO)_9$, after chromatographic work-up has been reported by Shore et al.[6] who prepared this mixed-metal cluster from $Os_3(\mu\text{-}H)_2(CO)_{10}$.

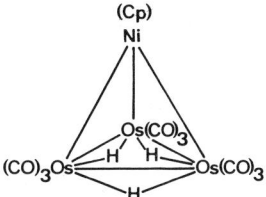

The monoanion of the complex has been prepared and used *in situ* for the synthesis of $NiOs_3(\mu\text{-}H)_2(CO)_9(\eta^5\text{-}C_5H_5)(\mu\text{-}MPPh_3)$ (M = Cu, Au)[11] and of $NiOs_3(\mu\text{-}H)_2(CO)_9(\eta^5\text{-}C_5\text{-}H_5)(\mu\text{-}HgBr)$.[12] Owing to its very high overall stability, CO substitution on the cluster may be achieved only in the presence of Me_3NO; monosubstituted and bisubstituted derivatives with PR_2R', PhCN, and diphenylphosphinoethane can be obtained.[13,14] The monosubstitution product plus PPh_2H easily yields the monoanion, which reacts further to yield $NiOs_3(\mu\text{-}H)_2(CO)_8(\eta^5\text{-}C_5H_5)(PPh_2H)(\mu\text{-}HgBr)$.[12]

The complex has been found to be an efficient hydrogenation catalyst in homogeneous conditions[15] and when supported on alumina[16] or on gas-chromatographic materials.[17,18]

B. NONACARBONYL(η^5-CYCLOPENTADIENYL)-TRI-μ-HYDRIDO-NICKELTRIRUTHENIUM, $Ru_3Ni(\mu\text{-}H)_3(\eta^5\text{-}C_5H_5)(CO)_9$

$$Ru_3(CO)_{12} + \tfrac{1}{2}[Ni(\eta^5\text{-}C_5H_5)(CO)]_2$$

$$\xrightarrow{\text{H}_2} NiRu_3(\mu\text{-}H)_3(\eta^5\text{-}C_5H_5)(CO)_9 + 4CO$$

■ *See cautionary note, Section A.*

Procedure

General remarks preceding Section A apply here as well.

Hydrogen is bubbled through an ethoxyethanol solution (50 mL) containing $Ru_3(CO)_{12}$ (780 mg, 1.22 mmol) and $[Ni(\eta^5\text{-}C_5H_5)(CO)]_2$ (370 mg, 1.22 mmol) at 80 to 85 °C for 8 h. After cooling and solvent removal under reduced pressure, chromatographic work-up with pentane in the same manner as in the preparation of $NiOs_3(\mu\text{-}H)_3(\eta^5\text{-}C_5H_5)(CO)_9$ described previously affords two bands: (a) a broad yellow band, $Ru_4(\mu\text{-}H)_4(CO)_{12}$ (traces), (b) a dark green band $NiRu_3(\mu\text{-}H)_3(\eta^5\text{-}C_5H_5)(CO)_9$. The mixed-metal cluster is recrystallized from acetone as prismatic needles. Yield: 575–626 mg (68–74%) based on ruthenium. Traces of other unidentified products are also eluted with toluene. In all cases, black, untractable material is retained at the top of the column.

Anal. Calcd. for $C_{14}H_8O_9NiRu_3$: C, 24.65; H, 1.18; Ni, 8.61. Found: C, 24.75; H, 1.19; Ni, 8.38.

Properties

The compound $NiRu_3(\mu\text{-}H)_3(\eta^5\text{-}C_5H_5)(CO)_9$ is an air-stable, dark green crystalline solid at room temperature. It can be handled in solution without undue precautions. It is soluble in many of the common organic solvents. Characterization is best accomplished by its carbonyl IR absorptions, $v_{(CO)}$ (cyclohexane solution): 2088 (w), 2060 (s), 2015 (s), 2000 (m) cm^{-1}. The ^{1}H NMR spectrum (250 MHz, CD$_2$Cl$_2$, room temperature) shows a singlet at $\delta + 5.71$ ppm (5H, cyclopentadienyl) and a singlet at $\delta - 16.07$ ppm (3H, hydride). Mass spectrum (65 °C, 70 eV): M$^+$ ($m/e = 683$) and Ru$_3$Ni$^+$ ($m/e = 362$). All nine carbonyl groups are lost successively, and there is an isotope pattern characteristic of the Ru$_3$Ni cluster. The crystal structure of this Ru—Ni mixed-metal cluster has not been solved, but the compound is clearly isomorphic with $NiOs_3(\mu\text{-}H)_3(\eta^5\text{-}C_5H_5)(CO)_9$.[5]

This complex has been shown to be an efficient hydrogenation catalyst in test reactions involving cyclohexene.[19]

References

1. D. A. Roberts and G. L. Geoffroy, *Comp. Organometal. Chem.*, **6**, 763–877 (1982) and references cited therein.
2. L. J. Farrugia, J. A. K. Howard, P. Mitrprachachon, J. L. Spencer, F. G. A. Stone, and P. Woodward, *J. Chem. Soc., Chem. Commun.*, 260 (1978).
3. A. J. Deeming, in *Transition Metal Clusters*, B. F. G. Johnson (ed.), Wiley, New York, 1980, p. 391 and references, cited therein.

4. S. A. R. Knox, J. W. Koepke, M. A. Andrews, and H. D. Kaesz, *J. Am. Chem. Soc.*, **97**, 3942 (1975).

5. G. Lavigne, F. Papageorgiou, C. Bergounhou, and J.-J. Bonnet, *Inorg. Chem*, **22**, 2485 (1983).

6. S. G. Shore, W. L. Hsu, C. R. Weisenberger, M. L. Caste, M. R. Churchill, and C. Bueno, *Organometallics*, **1**, 1405 (1982).

7. E. Sappa and M. Valle, *Inorg. Synth.*, **26**, 000 (1989).

8. P. Moggi, G. Albanesi, G. Predieri, and E. Sappa, *J. Organometal. Chem.*, **252**, C89 (1983).

9. B. F. G. Johnson and J. Lewis, *Inorg. Synth.*, **13**, 93 (1972).

10. (a) A. Mantovani and S. Cenini, *Inorg. Synth.*, **16**, 47 (1976); (b) M. I. Bruce, J. G. Matisons, R. G. Wallis, J. M. Patrick, B. W. Skelton, and A. H. White, *J. Chem. Soc., Dalton Trans.*, 2365 (1983); (c) M. I. Bruce, C. M. Jensen and N. L. Jones, *Inorg. Synth.*, **26**, 000 (1989).

11. P. Braunstein, J. Rose, A. M. Manotti Lanfredi, A. Tiripicchio, and E. Sappa, *J. Chem. Soc. Dalton Trans.*, 1843 (1984).

12. E. Sappa, G. Predieri, A. Tiripicchio, C. Vignali, and P. Braunstein, *J. Chem. Soc. Dalton Trans.*, 1135 (1986).

13. E. Sappa, M. Valle, G. Predieri, and A. Tiripicchio, *Inorg. Chim. Acta*, **88**, L23 (1984).

14. E. Sappa, G. Predieri, A. Tiripicchio, and M. Tiripicchio Camellini, *J. Organomet. Chem.*, **297**, 103 (1985).

15. M. Castiglioni, R. Giordano, E. Sappa, A. Tiripicchio, and M. Tiripicchio Camellini, *J. Chem. Soc. Dalton Trans.*, 23 (1986).

16. G. Albanesi, P. Moggi, R. Bernardi, G. Predieri, and E. Sappa, *Gazz. Chim. Ital.*, **116**, 385 (1986).

17. M. Castiglioni, R. Giordano, E. Sappa, G. Predieri, and A. Tiripicchio, *J. Organomet. Chem.*, **270**, C7 (1984).

18. M. Castiglioni, R. Giordano, E. Sappa, and P. Volpe, *J. Chromatogr.*, **349**, 173 (1985).

19. P. Fompeyrine, C. Bergounhou, and J.-J. Bonnet, unpublished results.

66. TRIOSMIUM AND TRINICKELTRIOSMIUM CLUSTER COMPLEXES

Submitted by ENRICO SAPPA* and MARIO VALLE*
Checked by SHELDON G. SHORE[†] and UPALI SIRIWARDANE[†]

A. NONACARBONYLTRIS(η^5-CYCLOPENTADIENYL)-TRINICKELTRIOSMIUM

$$3[Ni(CO)(\eta^5\text{-}C_5H_5)]_2 + 2Os_3(CO)_{12}$$

$$\longrightarrow 2Ni_3Os_3(CO)_9(\eta^5\text{-}C_5H_5)_3 + 12CO$$

*Istituto di Chimica Generale ed Inorganica, Universita' di Torino. Corso Massimo d'Azeglio 48, 10125 Torino, Italy.
†Department of Chemistry, Ohio State University, Columbus, OH 43210.

Procedure

■ **Caution.** *Because of the evolution of toxic CO and the possibility of generating highly toxic Ni(CO)₄ the reactions here described must be run in a well-ventilated hood.*

The reaction is run in a 250-mL flat-bottomed three-necked flask containing a magnetic stirrer and equipped with a gas inlet tube and a reflux condenser connected to a mercury check-valve to ventilate evolved gases. Octane is dried over sodium and freshly distilled before use. Triosmium-dodecacarbonyl and cyclopentadienylnickelcarbonyl dimer are commercially available (Strem). Alternatively, $Os_3(CO)_{12}$ may be synthesized from commercially available materials by a tested procedure.[1]

A suspension of $Os_3(CO)_{12}$ (0.304 g, 0.335 mmol) and $[Ni(CO)(\eta^5-C_5H_5)]_2$ (0.302 g, 0.994 mmol) in 100 mL of octane is stirred at room temperature for 10 min, under a stream of nitrogen bubbled through the solvent. The suspension is then refluxed for 40 min and allowed to cool under nitrogen. The initial red color turns dark brown and a thin metal mirror is deposited on the walls of the flask.

The cool solution is filtered under nitrogen on a Whatman paper disc giving a reaction mixture filtrate.

■ **Caution.** *The metal powder obtained when filtering the reaction solutions may be pyrophoric when dry.*

The metal powder remaining on the filter may still contain some unreacted $Os_3(CO)_{12}$. This can be recovered by Soxhlet extraction with cyclohexane. The metal powder remaining in the Soxhlet extraction thimble should be washed with acetone and then dissolved in warm nitric acid (37%) in order to deactivate it.*

The filtrate obtained from this reaction mixture is reduced to a volume of 15.0 mL in a rotating evaporator and purified by means of thin layer chromatography (TLC) preparative plates of 1-mm thickness (Kieselgel P. F. Merck) eluant diethyl ether (5% vol) in light petroleum ether. The dark brown band of the complex is eluted with dry diethyl ether (peroxide-free).

After evaporation of the diethyl ether, crystals suitable for X-ray analysis are obtained by slowly cooling at $-10\,°C$ a (30–70 in vol) $CHCl_3$–heptane solution, under N_2. Yield: 0.134 g (0.112 mmol) (33.04%).

Anal. Calcd. $(C_{24}H_{15}O_9Ni_3Os_3)$: C, 24.3, H, 1.27, Ni, 14.7, Os, 48.1. Found: C, 23.8, Ni, 13.5, Os, 45.8.

---

*A harmless metal precipitate can be obtained by diluting this solution and adding small zinc pieces.

Properties

The product $Ni_3Os_3(CO)_9(\eta^5-C_5H_5)_3$ is obtained as brown-black crystals and, when solid, is indefinitely air stable. In solution it slowly decomposes under air. The complex is well soluble (with some decomposition) in chloroform and moderately soluble in aliphatic hydrocarbons. IR spectrum (hexane solution), v_{CO}: 2064 (w), 2042 (s), 2010 (sh), 2008 (vs), 1969 (vs), 1956 (sh) cm^{-1}. Mass spectrum: "parent ion" at 822 m/e [the MW is 1194: 822 m/e corresponds to $Os_3(CO)_9$] and loss of 9 CO molecules. The complex is paramagnetic (87 electron count).[2] A discussion of the ^1H NMR is given in Ref. 2.

The structure of $Ni_3Os_3(CO)_9(\eta^5-C_5H_5)_3$ has been established by X-ray diffraction.[3] The complex behaves as an homogeneous catalyst for selective hydrogenation: Hydrogenating catalytic activity is observed also when adsorbed on alumina.[4]

B. DECACARBONYLDIHYDRIDOTRIOSMIUM[5]

$$Os_3(CO)_{12} + H_2(1 \text{ atm}) \longrightarrow Os_3H_2(CO)_{10} + 2CO$$

Procedure

■ **Caution.** *Because of toxicity of metal carbonyls and of evolved CO, and due to highly flammable H_2, this procedure must be carried out in a well-ventilated hood.*

The reaction is run in a three necked 250-mL flat-bottomed flask containing a magnetic stirrer and equipped with a gas inlet tube and a reflux condenser connected to a mercury check valve to ventilate evolved gases; $Os_3(CO)_{12}$ (0.5 g, 0.620 mmol) is suspended into 120 mL of octane (dehydrated as described previously) and the suspension is stirred at room temperature for 10 min under a stream of hydrogen bubbled through the solvent. The solution is then allowed to reflux; a constant stream of hydrogen is bubbled into the suspension and the stirring is continued for all the reaction time. The color, originally light yellow, turns slowly to purple-violet; the reaction time is ~ 2 h and depends on the stream of hydrogen allowed. The disappearance of $Os_3(CO)_{12}$ may be easily followed by IR.[5]

The reaction solution is allowed to cool to room temperature and then filtered on a Whatman paper disc in order to eliminate the small amounts of the sparingly soluble $Os_3(CO)_{12}$ remaining and of the metal powder formed. A nitrogen atmosphere should be preferred, although the stability of the complex allows operations under air.

The filtrate is kept to dryness in a rotating evaporator; the product obtained is pure enough for further reactions. Yield: 0.45 g (0.558 mmol) (90%). A high purity crystalline product is obtained by dissolving the material into a (70–30 in vol) heptane–chloroform solution under nitrogen and allowing it to crystallize at $-20\,°C$ for 12 h. Yield: 0.35 g (0.448 mmol) (72%).

The solutions containing $H_2Os_3(CO)_{10}$ obtained with the previous procedure can be directly used for preparing $NiOs_3(\mu\text{-}H)_3(CO)_9(\eta^5\text{-}C_5H_5)$[2,6,7] and $Ni_3Os_3(CO)_9(\eta^5\text{-}C_5H_5)_3$ via an alternate procedure:

$$3[Ni(CO)(\eta^5\text{-}C_5H_5)]_2 + 4Os_3(\mu\text{-}H)_2(CO)_{10} + \tfrac{1}{2}H_2$$

$$\longrightarrow NiOs_3(\mu\text{-}H)_3(CO)_9(\eta^5\text{-}C_5H_5)\ (70\%)$$

$$+ Ni_3Os_3(CO)_9(\eta^5\text{-}C_5H_5)_3\ (20\%)$$

Anal. Calcd. ($C_{10}H_2O_{10}Os_3$): C, 14.08; H, 0.24; O, 18.76; Os, 66.91. Found: C, 14.12; H, 0.40 (Ref. 7); Os, 67.02.

Properties

The product is obtained as deep purple-violet crystals and, when solid, is air stable for days. It is well soluble in chloroform and carbon tetrachloride, and moderately soluble in aliphatic hydrocarbons; it reacts with acetonitrile. IR spectrum (hexane solution) v_{CO}: 2074 (vs), 2060 (s), 2024 (vs), 2008 (s), 1988 (w) cm^{-1}. (Compare with that reported in Ref. 5). ^{1}H NMR spectrum (CDCl$_3$): $\delta - 11.73$ (s). Mass spectrum: Parent ion at 852 m/e, competitive loss of H and CO groups. A full discussion is given in Ref. 5.

The structure of $Os_3(\mu\text{-}H)_2(CO)_{10}$ has been established by X-ray[8] and neutron diffraction.[9] The 46-electron complex displays a relatively high reactivity under mild conditions, associated with a stable triosmium framework and has been extensively studied as a "model" for the chemisorption of alkenes and alkynes on surfaces and in the catalytic isomerization and hydrogenation of alkenes.[10] When supported onto alumina it is a catalyst for the methanation of CO and CO_2 slightly less efficient than $NiOs_3(\mu\text{-}H)_3(CO)_9(\eta^5\text{-}C_5H_5)$.[4]

References

1. B. F. G. Johnson and J. Lewis, *Inorg. Synth.*, **13**, 93 (1972).
2. M. Castiglioni, E. Sappa, M. Valle, M. Lanfranchi, and A. Tiripicchio, *J. Organomet. Chem.*, **241**, 99 (1983).
3. E. Sappa, M. Lanfranchi, A. Tiripicchio, and M. Tiripicchio Camellini, *J. Chem. Soc. Chem. Commun.*, **1981**, 995.

*See Ref. 7.

4. G. Albanesi, R. Bernardi, P. Moggi, G. Predieri, and E. Sappa, *Gazz. Chim. Ital.*, **116**, 385 (1986).

5. H. D. Kaesz, S. A. R. Knox, J. W. Koepke, and R. B. Saillant, *J. Chem. Soc. Chem. Commun.*, **1971**, 477; *J. Am. Chem. Soc.*, **97**, 3942 (1975).

6. S. G. Shore, W. L. Hsu, C. R. Weisenberger, M. L. Castle, M. R. Churchill, and C. Bueno, *Organometallics*, **1**, 1405 (1982).

7. C. Bergounhou, J.-J. Bonnet, G. Lavigne, and F. Papageorgiou, *Inorg. Synth.*, **26**, 0000 (1989); G. Lavigne, F. Papageorgiou, C. Bergounhou and J. J. Bonnet, *Inorg. Chem.*, **22**, 2485 (1983).

8. M. R. Churchill, F. J. Hollander, and J. P. Hutchinson, *Inorg. Chem.*, **16**, 2697 (1977).

9. A. G. Orpen, A. V. Rivera, E. G. Bryan, D. Pippard, G. M. Sheldrick, and K. D. Rouse, *J. Chem. Soc. Chem. Commun.*, **1978**, 723; R. W. Broach and J. M. William, *Inorg. Chem.*, **18**, 314 (1979).

10. R. D. Adams, in *Comprehensive Organometallic Chemistry*, G. Wilkinson, F. G. A. Stone, and E. W. Abel (eds.) Vol. 4, Pergamon Press New York, 1982, Chapter 33, pp. 967–971.

67. A DICOBALTPLATINUM COMPLEX AND ITS PLATINUM PRECURSOR

$$Na_2[PtCl_4] + (C_6H_5)_2PCH_2CH_2P(C_6H_5)_2$$
$$\longrightarrow Pt[(C_6H_5)_2PCH_2CH_2P(C_6H_5)_2]Cl_2$$
$$Pt[(C_6H_5)_2PCH_2CH_2P(C_6H_5)_2]Cl_2 + 2Na[Co(CO)_4]$$
$$\longrightarrow PtCo_2[(C_6H_5)PCH_2CH_2P(C_6H_5)_2](CO)_7 + CO + 2NaCl$$

Submitted by K. YASUFUKU,* H. NODA,* and H. YAMAZAKI*
Checked by S. MARTINENGO†

Several mixed-metal clusters containing platinum and cobalt are known and some of them have been employed as methanol homologation catalysts.[1] Among them, the title compound[2] was first prepared unambiguously from the reaction of dichloro[1,2-ethanediylbis(diphenylphosphine)]platinum with sodium tetracarbonylcobaltate, $Na[CO(CO)_4]$. The compound also may be prepared by the reaction of [1,2-ethanediylbis(diphenyl-phosphine)]bis(phenylethynyl)platinum with $Co_2(CO)_8$.[1]

The reaction of dichlorobis(triphenylphosphine)platinum(II) with $Na[Co(CO)_4]$ gives only the tetrametal cluster, $Pt_2Co_2(PPh_3)(CO)_8$, in a low yield.[3] The corresponding trimetal cluster, $PtCo_2(PPh_3)(CO)_8$,[4] is obtained

*Riken, Wako-shi, Saitama 351-01, Japan.
†C.N.R. Centro di Studio Sulla Sintesi e La Struttura dei Composti dei Metalli di Transizione Nei Bassi Stati di Ossidazione, 20133 Milano, Italy.

from the reaction of the pentaplatinum cluster, $Pt_5(PPh_3)_4(CO)_6$, with $Co_2(CO)_8$, accompanied by formation of the tetrametal cluster. When dichlorobis(triethylphosphine)platinum(II) is treated with $Na[Co(CO)_4]$, the pentametal cluster, $Pt_3Co_2(PEt_3)_3(CO)_9$, is obtained in a good yield.[5] Structures of these clusters have been determined[3,4] and some redox behavior of the trimetal clusters have been studied.[6]

A. DICHLORO[1, 2-ETHANEDIYLBIS(DIPHENYLPHOSPHINE)]- PLATINUM[7,*]

1,2-Ethanediylbis(diphenylphosphine) (2.6 g) dissolved in CH_2Cl_2 (35 mL) is added dropwise with stirring to the solution of $Na_2[PtCl_4]$ (3.0 g) dissolved in EtOH (50 mL) in a 200-mL flask. A pale pink precipitate formed immediately, is filtered off, washed with water, dried, and placed in a 300-mL flask fitted with a refluxing condenser and a Teflon-coated magnetic stirring bar. A mixture of ethanol (50 mL) and concentrated hydrochloric acid (50 mL) is added to the flask and the flask is warmed to reflux for 5 h at 95 °C. The white powdery percipitate (3.5 g) is filtered, washed with ethanol and recrystallized from hot N,N-dimethylformamide (25 mL) by adding hexane–diethyl ether 1:1 mixture (~ 50 mL). Yield: 2.4 g of the product (mp > 360 °C).

B. HEPTACARBONYL[1, 2-ETHANEDIYLBIS(DIPHENYL- PHOSPHINE)]DICOBALTPLATINUM

■ **Caution.** *Due to possible release of toxic carbon monoxide gas in the reaction of metal carbonyls, and due to the toxic nature of metal carbonyl complexes, this reaction must be carried out in a well-ventilated hood.*

Procedure

Sodium tetracarbonylcobaltate(1 −) is prepared from $Co_2(CO)_8$ (0.6 g) and 3% of Na–Hg (15 g) in tetrahydrofuran (THF).[8] Dichloro[1, 2-ethanediylbis- (diphenylphosphine)]platinum (1.0 g, 1.5 mmol) is suspended in freshly dist- illed, dry, and oxygen-free THF (20 mL) in a 100-mL two-necked flask fitted with a refluxing condenser, a Teflon-coated magnetic stirring bar, and an inlet for nitrogen connected to a mercury or mineral oil bubbler for venting excess gases.

The solution of tetracarbonylcobaltate(1 −) is added dropwise at room temperature with stirring under nitrogen. The flask is then warmed to reflux

*[1, 2-Ethanediylbis(diphenylphosphine)] ≡ [Bis(diphenylphosphino)ethane].

for 1.5 h at 75 °C. The white, powdery suspension turns to a dark brown homogeneous solution to which toluene (20 mL) is added. Solvent is then removed to a residual 20 mL under water pump vacuum.

A silica gel column (100–200 mesh, Wakogel C-200, 3.5 × 13.0 cm) is prepared under inert atmosphere. Before packing the column, the silica gel is placed under a water pump vacuum at 60 °C and then exposed to nitrogen atmosphere three successive times. The packing of the column is done under a nitrogen atmosphere using hexane that has been deaerated by bubbling nitrogen through it.

The condensed reaction mixture is placed on the silica gel column and chromatographed using graduated hexane–toluene mixtures as follows. The column is first eluted by hexane–toluene from 3:1 to 1:1 ratio. A yellow band may elute first after ∼ 200 mL. A dark brown band is eluted next with hexane–toluene ratios from 1:2 to 1:4, requiring ∼ 250 mL. The brown eluate is collected under nitrogen and solvent is removed to a residue of ∼ 40 mL. Degassed hexane (60–160 mL) is carefully layered over the toluene solution leading to the slow crystallization of product. Yield: 0.79–0.9 g (58–66%); mp 200–205 °C (with dec).

Checkers find another 0.1 g of fairly pure product can be recovered from the mother liquor by evaporation to dryness under vacuum, dissolving the residue in the minimum amount of toluene, and carefully layering 4–5 volumes of degassed hexane over the toluene solution.

Anal. Calcd. for $C_{33}H_{24}O_7Co_2Pt$: C, 43.69; H, 2.67. Found: C, 43.79; H, 2.76.

Properties

Heptacarbonyl[1, 2-ethanediylbis(diphenylphosphine)]dicobaltplatinum is a dark brown crystalline solid, which is very slightly air sensitive and can be stored for months at room temperature under nitrogen. IR: ν_{CO} (KBr) = 2049 (s), 2010 (s), 1975 (vs), 1970 (sh), 1729 (s); (CS$_2$) = 2055 (s), 2020 (s), 1995 (s), 1981 (s), and 1752 (m) cm^{-1}. Chemical properties have been described in the introduction.

References

1. M. Hidai, M. Orisaku, M. Ue, Y. Uchida, K. Yasufuku, and H. Yamazaki, *Chem. Lett.*, **1981**, 143.

2. J. Dehand and J. F. Nennig, *Inorg. Nucl. Chem. Lett.*, **10**, 875 (1974).

3. P. Braunstein, J. Dehand, and J. F. Nennig, *J. Organometal. Chem.*, **92**, 117 (1975); J. Fischer, A. Mitschler, R. Weiss, J. Dehand, and J. F. Nennig, *J. Organometal. Chem.*, **91**, C37 (1975).

4. R. Bender, P. Braunstein, J. Fischer, L. Ricard, and A. Mitschler, *Nouv. J. Chem.*, **5**, 81 (1981).

5. J.-P. Barbier, P. Braunstein, J. Fischer, and L. Ricard, *Inorg. Chim. Acta*, **1978**, L361.
6. P. Lemoine, A. Giraudeau, M. Gross, R. Bender, and P. Braunstein, *J. Chem. Soc. Dalton Trans.*, **1981**, 2059.
7. A. D. Westland, *J. Chem. Soc.*, **1965**, 3060.
8. J. K. Ruff and W. J. Schlientz, *Inorg. Synth.*, **15**, 87 (1974).

68. SALTS OF MIXED PLATINUM-TETRARHODIUM CLUSTER COMPLEXES

Submitted by ALESSANDRO FUMAGALLI* and SECONDO MARTINENGO*
Checked by PIERRE BRAUNSTEIN[†] and DOMINIQUE WATT[†]

The anion $[PtRh_4(CO)_{14}]^{2-}$ was first obtained[1] by reduction of $[PtRh_5(CO)_{15}]^-$ and subsequently[2,3] by direct reduction of mixtures of $RhCl_3$ and $Na_2[PtCl_6]$. This latter method, which is described here, makes use of readily available starting materials and gives $[PtRh_4(CO)_{14}]^{2-}$ in good yield (65–70%). The related $[PtRh_4(CO)_{12}]^{2-}$ anion is obtained in nearly quantitative yields from $[PtRh_4(CO)_{14}]^{2-}$ by loss of CO under vacuum. The overall procedure requires ~ 2 days.

General Procedure

■ **Caution.** *The reaction and all the manipulations must be carried out in a well-ventilated hood, owing to the high toxicity of carbon monoxide.*

Both the products described here and the reaction intermediates are highly sensitive to oxygen, thus all the manipulations are carried out using inert atmosphere techniques[4] with standard Schlenk ware. Oxygen-free carbon monoxide (purity 99.95%, $O_2 < 2$ ppm) is from cylinders, and used without further purification.

All the reagent grade solvents are deaerated in vacuum and stored under nitrogen; tetrahydrofuran (THF) is distilled from a solution containing sodium benzophenone ketyl. The salt $Na_2[PtCl_6]\cdot 6H_2O$ is commercially available or prepared according to the literature,[5] and $RhCl_3$ "trihydrate" is commercially available; the Pt and Rh content of these reagents is determined by atomic absorption. The 2 M solution of NaOH in methanol is prepared by dissolving reagent grade NaOH under nitrogen and the titer checked with

*Centro del CNR per la Sintesi e la Struttura dei Composti dei Metalli di Transizione nei Bassi Stati di Ossidazione, and Dipartimento di Chimica Inorganica e Metallorganica, Via G. Venezian 21, 20133 Milano, Italy.
[†]Laboratoire de Chimie de Coordination, Université Louis Pasteur, 4 rue Blaise Pascal, Strasbourg, 67070 France.

standard 1 M HCl. Reagents with a different Pt or Rh content, and NaOH solution with a slightly different titer may be used as well, provided that the same Pt–Rh–NaOH ratio is maintained. The $[(Ph_3P)_2N]Cl$ is available from Strem.

A. BIS[μ-NITRIDO-BIS(TRIPHENYLPHOSPHORUS)((1 +)] [TETRADECACARBONYLPLATINUM-TETRARHODATE(2 −)], $[(Ph_3P)_2N]_2[PtRh_4(CO)_{14}]$

1. $4RhCl_3 + Na_2PtCl_6 + 23CO + 27NaOH$

    $$\xrightarrow[CH_3OH]{CO} Na_2[PtRh_4(CO)_{14}] + 9NaHCO_3 + 18NaCl + 9H_2O$$

2. $Na_2[PtRh_4(CO)_{14}] + 2[(Ph_3P)_2N]Cl$

    $$\longrightarrow [(Ph_3P)_2N]_2[PtRh_4(CO)_{14}] + 2NaCl$$

Procedure A

A 1-L two-necked round-bottomed flask with ground glass joints is equipped with a 4 to 5-cm Teflon-covered magnetic stirring bar, and (on the central neck) a dropping funnel with pressure equalizer tube, and (on the side neck) a three-way stopcock. The flask is charged with $RhCl_3 \cdot 3H_2O$ (2.034 g, Rh content 40%) and $Na_2[PtCl_6] \cdot 6H_2O$ (1.108 g, 1.98 mmol) to give a Rh–Pt ratio of four. Precision in maintaining this ratio is very important.

The three-way stopcock is connected on one side to a vacuum pump and on the other side to a CO line equipped with a mineral oil bubbler and a pressure-release bubbler containing a mercury column (height 50–60 mm). Since the reaction flask will be under slight positive pressure of CO, each of the joints should be fastened by spring attachments. The entire system is evacuated and then filled with CO three times. Methanol (50 mL) is introduced into the flask through the dropping funnel and the stirred mixture is allowed to saturate with CO for 15 min. The dropping funnel is charged with 27 mL of the 2 M solution of NaOH in methanol, in slight excess over the stoichiometric amount, $(OH)^- : Pt = 27.4$. The solution is slowly added over a period of ~ 2 h to the vigorously stirred mixture in the flask, while a slight positive CO pressure is maintained by adding CO at such a rate that some bubbles continuously escape from the pressure-release mercury bubbler. During this period, the color changes first to green, then to brownish-green, and eventually to orange or dark orange, while some white precipitate is formed. At the end of the addition, the mixture is left to stir for three more hours (or overnight) and then filtered, under CO atmosphere, through a medium porosity Schlenk frit, into a 250-mL Schlenk tube. The solid on the filter, prior to discharge, should be washed with 10 mL of methanol. It is important to proceed as rapidly as possible with the treatment of the filtrate, explained as follows.

The combined solution and washings are treated drop by drop, while stirring, with a solution of $[(Ph_3P)_2N]Cl$ (6 g, 10.5 mmol) in methanol (10 mL) to give a yellow finely crystalline precipitate and a pink or slightly red mother liquor. The precipitate is removed by filtration under CO. It is then washed twice with 10 mL of 2-propanol, six times with 10 mL of water, and twice again with 10 mL of 2-propanol. The solid is then vacuum dried, and stored under a CO atmosphere. The product is extracted from the filtering septum into a 100-mL Schlenk tube by dissolution under CO, with ~ 10 mL of THF added in small portions. It is then crystallized by addition of 2-propanol, previously saturated with CO, in one of the two following procedures: (a) 20 mL of 2-propanol is slowly dropped with stirring into the THF solution, then, after 1.5 more hours of stirring, 15 additional mL of 2-propanol is added to complete the precipitation; (b) alternatively, 45 mL of 2-propanol is cautiously layered on the THF solution by means of a syringe. The tube is left standing a few days until the upper layer of 2-propanol has diffused completely into the solution, to give large crystals of the product. In both cases the precipitate obtained is isolated by filtration under CO, washed four times with 2-propanol (10 mL), vacuum dried, and stored under CO. Yields: 2.65–2.80 g (65–70%).

Anal. Calcd. for $C_{86}H_{60}N_2O_{14}P_4PtRh_4$: C, 49.75; H, 2.91; N, 1.35. Found: C, 49.35; H, 2.94; N, 1.33.

Properties

The compound $[(Ph_3P)_2N]_2[PtRh_4(CO)_{14}]$ is a yellow-orange crystalline solid that must be stored under CO. Large crystals are stable in air for a few hours, but its solutions are quickly oxidized. It is soluble in THF, acetone, and acetonitrile, sparingly soluble in methanol, and insoluble in 2-propanol, hexane, and water. The solutions are stable only under CO. Purity can be judged by the color, by analysis, and by the IR spectrum in THF solution. The solution *must* be prepared under CO, and the IR cells must be previously purged with CO, IR bands are observed at 2030 (vw), 1995 (s), 1962 (vs), 1941 (m, sh), 1900 (vw), 1854 (vw), 1807 (ms), 1800 (ms, sh), and 1751 (m) cm^{-1}.

The X-ray structure,[2] and the chemical properties and solution NMR,[3] of $[(Ph_3P)_2N]_2[PtRh_4(CO)_{14}]$ have been reported. The compound reacts with salts of $[Rh_6N(CO)_{15}]^-$ to give salts of $[PtRh_{10}N(CO)_{23}]$.[3–6] Upon thermal treatment, the compound transforms into a mixture containing anions of higher nuclearity. An interesting feature of the anion is its behavior in solution: Under CO it is stable, whereas under nitrogen or vacuum it loses CO to give $[PtRh_4(CO)_{12}]^{2-}$ (see the procedure in Section B), from which it can be restored by simple exposure to CO. This interconversion can be repeated many times without significant losses.

The product has been obtained also as the tetraethylammonium, tetrabutylammonium, and benzyltrimethylammonium salts by metathesis of the sodium salt with salts of the chosen cation. With these cations, however, it is necessary to add water to complete the precipitation.

B. BIS[μ-NITRIDO-BIS(TRIPHENYLPHOSPHORUS)(1 +)] [(DODECACARBONYLPLATINUM-TETRARHODATE(2 −)], $[(Ph_3P)_2N]_2[PtRh_4(CO)_{12}]$

$$[(Ph_3P)_2N]_2[PtRh_4(CO)_{14}]$$

$$\xrightarrow{\text{vacuum}} [(Ph_3P)_2N]_2[PtRh_4(CO)_{12}] + 2CO$$

Procedure B

■ *See cautionary note under General Procedures*

The compound $[(Ph_3P)_2N]_2[PtRh_4(CO)_{14}]$ obtained in the procedure in Section A is dissolved under nitrogen in 20 mL of THF. The solution is then slowly evaporated to dryness by stripping out the solvent under reduced pressure while the solution is rapidly stirred. The residue is redissolved in 15 mL of THF, and the solution is slowly treated, while being stirred, with 10 mL of 2-propanol to give precipitation of a brown-red product, which is completed by concentration under reduced pressure to a final volume of ∼ 20 mL. The product is removed by filtration, washed four times with 10-mL portions of 2-propanol, and vacuum dried. The yields of recrystallized product are in the range 90–95%.

Anal. Calcd. for $C_{84}H_{60}N_2O_{12}P_4PtRh_4$: C, 49.95; H, 2.99; N, 1.39. Found: C, 49.32; H, 3.03; N, 1.37.

Properties

The compound $[(Ph_3P)_2N]_2[PtRh_4(CO)_{12}]$ is a reddish-brown solid that is stable in air for only a few hours and which must be stored under nitrogen. It is soluble in THF, acetone, and acetonitrile, sparingly soluble in methanol, and insoluble in 2-propanol, water, and hexane. The solutions are quickly oxidized in air. Purity can be judged by the IR spectrum, which, in THF solution shows bands at (cm^{-1}): 2015(mw), 1982(s), 1947(s), 1859(vw), 1812(ms), and 1788(m).

The structure and chemical properties have been reported.[2,3] Thermal stability and reactivity with salts of $[Rh_6N(CO)_{15}]^-$ are similar to those described above for the salts of $[PtRh_4(CO)_{14}]^{2-}$.

References

1. A. Fumagalli, S. Martinengo, P. Chini, A. Albinati, S. Bruckner, and B. T. Heaton, *J. Chem. Soc. Chem. Commun.*, **1978**, 195.

2. A. Fumagalli, S. Martinengo, P. Chini, A. Albinati, and S. Bruckner, *XIII Congresso Nazionale di Chimica Inorganica*, Camerino, Italy, 23–26 Sept. 1980, Abstracts, p. 17.

3. A. Fumagalli, S. Martinengo, P. Chini, D. Galli, B. T. Heaton, and R. DellaPergola, *Inorg. Chem.*, **23**, 2947 (1984).

4. D. F. Shriver and M. A. Drezdzon, *The Manipulation of Air-Sensitive Compounds*, 2nd ed., Wiley, New York, 1986.

5. L. E. Cox and D. G. Peters, *Inorg. Synth.*, **13**, 173 (1972).

6. S. Martinengo, G. Ciani, and A. Sironi, *J. Am. Chem. Soc.*, **104**, 328 (1982).

Chapter Six

SOLID STATE

69. SYNTHESIS AND CRYSTAL GROWTH OF $A_3M_2X_9$ (A = Cs, Rb; M = Ti, V, Cr; X = Cl, Br)

Submitted by ANTON STEBLER,* BRUNO LEUENBERGER,* HANS U. GÜDEL*
Checked by BERNARD BRIAT[†]

The spectroscopic, thermal, and magnetic properties of compounds of the type $A_3M_2X_9$ (A = Cs, Rb; M = Ti, V, Cr; X = Cl, Br) are being intensively studied at present.[1-4] Despite their close structural similarities the compounds exhibit a great variability of physical properties. The M^{3+} ions occur in pairs, $M_2X_9^{3-}$, and this dimer character appears to be mainly responsible for their low-temperature magnetic properties. Recent inelastic neutron scattering experiments on $Cs_3Cr_2X_9$ (X = Cl, Br) have revealed that there are interactions also between the dimers, which lead to an energy dispersion of the dimer excitations at low temperatures. They represent a novel class of singlet ground-state magnets.[3]

The crystal structure of these compounds is hexagonal. The detailed structure is known for the compounds $Cs_3Cr_2X_9$ (X = Cl, Br), which belong to the space group $P6_3/mmc$ with two dimers per unit cell. X-ray powder patterns indicate that the other compounds are isostructural to the chromium compounds.

*Institut für anorganische Chemie, Universität Bern, CH-3000 Bern 9, Switzerland.
†Laboratoire d'Optique Physique, ESPCI, 10 rue Vauquelin, 75231 Paris Cédex 05, France.

General Procedure

■   **Caution.**   *Residual traces of water can lead to explosion of the pressure ampules. A protective shield must be placed in front of the furnace during warm-up and reaction periods. Protective glasses and gloves must be worn when handling the ampules. The stated procedures must be closely followed to minimize risk of explosion.*

Preparation from the melt according to the reaction

$$3AX + 2MX_3 \longrightarrow A_3M_2X_9$$

is a straightforward way to obtain the compounds. For $M = Ti$ and V the situation is complicated by the competing equilibrium

$$2MX_3 \leftrightarrow MX_2 + MX_4$$

which is shifted to the right at high temperatures. Thermodynamic properties and phase diagrams are found in Refs. 5–9. Experimental preparative procedures have been described by Saillant and Wentworth.[10] They can be used with a few modifications for the preparation of polycrystalline products. Large single crystals are needed for physical experiments such as magnetooptical, spectroscopic, and neutron scattering measurements. They can be prepared by the Bridgman technique.[11]

Materials	Supplier; Grade
CsCl, CsBr	Merck Suprapur; Cerac 99.9%
RbCl, RbBr	Merck Suprapur; Cerac 99.9%
$TiCl_3$	Fluka pract; Alfa Inorganics 98%
VCl_3	Alfa Inorganics; Cerac 99%
$CrCl_3$	ROC/RIC 99%; Cerac 99.9%
Ti metal powder	Alfa Inorganics 100 mesh 99%; Cerac; Merck
$TiBr_4$	Alfa Inorganics; Great Western Inorganics 99%
VBr_3	Alfa Inorganics; Cerac 99.7%
$CrBr_3$	ROC/RIC 99%; Cerac 99%

The compounds CsCl, CsBr, RbCl, RbBr, $CrBr_3$, and $CrCl_3$ were dried under vacuum at 300 °C for 6 to 12 h. All the other starting materials could be used without further treatment as long as they were kept and handled in a dry atmosphere.

Preparation of Ampules

Silica ($> 99.5\%$ SiO_2) ampules with wall strengths of 1.7 mm are used for both

a   b

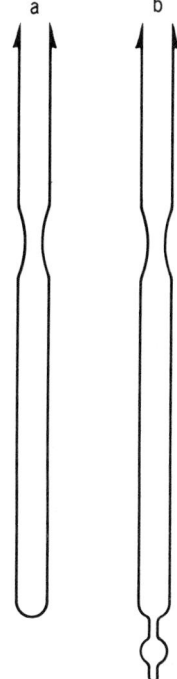

Fig. 1. Ampules for crystal growth.

the synthesis and the crystal growth. They are cleaned with an 8% HF solution, rinsed with distilled water, and dried at 500 °C under vacuum for 3 h. Some ampules for growing crystals are graphitized as follows: After the treatment with HF and water the ampule is rinsed with acetone and then heated with two large burners to produce a thin layer of graphite.

Ampules used for the synthesis of polycrystalline materials are closed by a round seal, whereas a pointed seal (Fig. 1) is necessary for growing single crystals. When large single crystals of several cubic centimeters are needed it is essential to use ampules as shown in Fig. 1(b).

Acetylene–O_2 or propane–O_2 burners are used for sealing the ampules.

Synthesis

All the compounds, both starting materials and products, must be handled in a dry box (humidity $< 0.5\%$) due to their hygroscopic nature. Ampules 20 to 40 cm in length (after sealing) and 2 cm in diameter are used for the preparation of polycrystalline products. A finely powdered mixture of the

TABLE I. Preparative Conditions and Physico-Chemical Properties.

Compound	Melting point (°C)	Reference	Heating Rate (°C h^{-1})	Reaction Temperature (°C)[a]	Reaction Time (days)	Color	Unit Cell (Å) a	c	Ref.	Hygroscopicity
Cs$_3$[Cr$_2$Cl$_9$]	894	8, 12	100	930	3	Pink-violet	7.22	17.93	13	Low
Cs$_3$[Cr$_2$Br$_9$]			100	800	4	Light green	7.51	18.68	14	High
Rb$_3$[Cr$_2$Br$_9$]			100	750	4	Light green	7.33	18.55		Moderate
Cs$_3$[Ti$_2$Cl$_9$]			100	750	4	Dark brown	7.32	17.97		Low
Rb$_3$[Ti$_2$Cl$_9$]			100	700	4	Dark brown	7.50	18.40		Low
Cs$_3$[Ti$_2$Br$_9$]	697	7	100	730	4	Brown	7.58	18.51	15	High
Rb$_3$[Ti$_2$Br$_9$]	647	7	100	670	4	Brown	7.42	18.45		High
Cs$_3$[V$_2$Cl$_9$]	700	8	100	730	5	Dark red	7.24	17.94	13	Low
Rb$_3$[V$_2$Cl$_9$]	688	5, 8	100	720	5	Dark red	7.00	17.64		Low
Cs$_3$[V$_2$Br$_9$]			100	700	5	Brown red	7.52	18.63		High
Rb$_3$[V$_2$Br$_9$]			100	700	5	Brown red	7.38	18.55		High

[a]For crystal growth the temperature is chosen 10 °C lower.

starting materials AX and MX_3 is placed in the ampule. The $AX–MX_3$ ratios, as well as the details of the sealing procedure, are given for the individual compounds in the following sections. For the synthesis the sealed ampule is placed in a slightly inclined ($\sim 15\,°C$) horizontal furnace covering the whole length of the ampule and then slowly heated to the reaction temperature. Heating rates, reaction times, and temperatures are listed in Table I.

Crystal Growth

The Bridgman technique is used,[11] with the ampule (15-cm length, 1-cm diameter) stationary inside a silica or ceramic working tube and the furnace moving on a vertical shaft with variable speed. Kanthal wire AB, 1.0-mm diameter, 1.90-$\Omega\,m^{-1}$ resistance, is used for the furnace coil. The coil has a total length of 30 cm and a resistance of typically 8.6 Ω at room temperature. It is laid out such as to provide two distinct zones, one typically 200 °C cooler than the other. The furnace is placed above the ampule, heated until the hot zone has reached the reaction temperature (Table I) and then lowered at a rate of 1.8 cm h^{-1}. The furnace is kept in this position for 1 to 2 days to ensure proper melting and mixing of the material. The crystal is then grown by raising the furnace at a rate of 1.4 cm 24 h^{-1}. The crystallization point is between the two zones, where the temperature gradient is typically 30 °C cm^{-1}. The crystal is annealed in the cooler zone. The power is switched off after the crystal has emerged from the furnace. The ampule is removed when the whole system has reached room temperature.

A. SINGLE CRYSTALS OF $A_3Cr_2X_9$

$$3AX + 2CrX_3 \longrightarrow A_3Cr_2X_9$$
$$A = Cs; X = Cl, Br \qquad A = Rb; X = Br$$

- *See cautionary notes under the previous General Procedure.*

Crystals of this class of compounds must be grown by using AX and CrX_3 as starting material. Attempts to grow crystals from polycrystalline $A_3Cr_2X_9$ fail due to decomposition of the material possibly accompanied by explosion.

Procedure

The $AX–CrX_3$ ratios are either stoichiometric or with a CrX_3 excess of 10%. The following quantities are suitable for Bridgman ampules of 15-cm length and 1-cm diameter:

$Cs_3Cr_2Cl_9$	5.5 g (33 mmol) CsCl, 3.8 g (24 mmol) $CrCl_3$, ampule not graphitized
$Cs_3Cr_2Br_9$	5.2 g (24 mmol) CsBr, 4.9 g (17 mmol) $CrBr_3$, ampule graphitized
$Rb_3Cr_2Br_9$	5.0 g (30 mmol) RbBr, 5.9 g (20 mmol) $CrBr_3$, ampule not graphitized

The loaded ampules are evacuated at 250 °C to a pressure of $< 10^{-3}$ torr and sealed with an acetylene–O_2 burner. The crystals are grown in a Bridgman apparatus as described in detail in the previous General Procedure.

Properties

The boules are homogeneous and usually consist of one single crystal in the case of $Cs_3Cr_2Cl_9$, one to several crystals in the case of the bromides. Hexagonal crystal faces perpendicular to the *c* axis as well as rectangular faces containing the *c* axis are usually well developed. Some further properties are listed in Table I.

B. SYNTHESIS OF $TiBr_3$

$$3TiBr_4 + Ti \longrightarrow 4TiBr_3$$

■ **Caution.** *Considerable pressure builds up in the ampule during this procedure. Do not exceed the recommended heating rate. A protective shield must be placed in front of the furnace during the warm-up and reaction period. The risk of explosion during this step is considerably higher than in any other procedure described here.*

Synthetic procedures for $TiBr_3$ and its properties are described in Refs. 7, 16, and 17.

Procedure

A sample of 0.625 g (13 mmol) of powdered Ti metal and 14.3 g (39 mmol) of finely powdered $TiBr_4$ are evacuated in a silica ampule (20-cm length, 2-cm diameter) under cooling to -80 °C (CO_2–2-propanol mixture). The lower part of the ampule is cooled while sealing quickly with a very hot flame. This is to reduce the $TiBr_4$ vapor pressure. If the pressure is too high a film of Ti metal is produced near the sealing spot. As a consequence the heat dissipation is so efficient that the ampule is difficult to close. It is important to keep the ampule in a vertical position during the sealing process. The sealed ampule is placed in a slightly inclined ($\sim 15\,°C$) horizontal furnace and heated to $500\,°C$ at a rate of $50\,°C\,h^{-1}$. After a reaction time of 15 days the system is cooled to room temperature. The ampule is opened in a dry box and its contents placed into a sublimation ampule (30-cm length, 2-cm diameter). The excess $TiBr_4$ is sublimed under vacuum at $200\,°C$ in a trap at liquid nitrogen temperature. It is important to remove $TiBr_4$ *quantitatively*. The residue, consisting of a lump of $TiBr_3$ platelets, is used directly for the next step.

C. POLYCRYSTALLINE $A_3[Ti_2X_9]$ and $A_3[V_2X_9]$

$$3AX + 2MX_3 \longrightarrow A_3M_2X_9$$

$$A = Cs, Rb \quad M = Ti, V \quad X = Cl, Br$$

■ *See cautionary notes under the previous General Procedure.*

Due to the high vapor pressure of TiX_4, long ampules are used for the synthesis $3AX + 2TiX_3 \longrightarrow A_3Ti_2X_9$. Polycrystalline $A_3M_2X_9$ serves as starting material for the crystal growth in the next step.

Procedure

A slight excess of TiX_3 and VX_3 is used for the synthesis. The following quantities are suitable for ampules of 30-cm length and 2-cm diameter:

$Cs_3[Ti_2Cl_9]$ 12.3 g (73.1 mmol)CsCl, 7.5 g (49 mmol)$TiCl_3$
$Rb_3[Ti_2Cl_9]$ 10.7 g (88.5 mmol)RbCl, 9.1 g (59 mmol)$TiCl_3$
$Cs_3[Ti_2Br_9]$ 10.5 g (49.3 mmol)CsBr, 10.2 g (35.5 mmol)$TiBr_3$
$Rb_3[Ti_2Br_9]$ 9.3 g (56 mmol)RbBr, 11.6 g (40.3 mmol)$TiBr_3$
$Cs_3[V_2Cl_9]$ 12.3 g (53 mmol)CsCl, 8.3 g (73.1 mmol)VCl_3
$Rb_3[V_2Cl_9]$ 10.7 g (88.5 mmol)RbCl, 10.0 g (63.6 mmol)VCl_3
$Cs_3[V_2Br_9]$ 10.5 g (49.3 mmol)CsBr, 10.3 g (35.4 mmol)VBr_3
$Rb_3[V_2Br_9]$ 9.2 g (56 mmol)RbBr, 11.7 g (40.3 mmol)VBr_3

The filled ampules are evacuated to a pressure of $< 10^{-3}$ torr. As described for the synthesis of $TiBr_3$ (previous section) the lower part of the ampule is cooled while sealing quickly with a very hot flame. The sealed ampule is placed in a slightly inclined ($\sim 15\,°C$) furnace and heated at a rate of $100\,°C\,h^{-1}$ to the reaction temperature (Table I). After the reaction time listed in Table I the heating power is switched off, and the system is left to cool down to room temperature.

Properties

All the products are homogeneous and polycrystalline. Single-crystalline domains of several hundered cubic millimeters are not uncommon for the chlorides. Some further properties are listed in Table I.

D.   SINGLE CRYSTALS of $A_3[Ti_2X_9]$ and $A_3[V_2X_9]$ (A = Cs, Rb; X = Cl, Br)

■ *See cautionary notes under the previous General Procedure.*

Procedure

Polycrystalline $A_3Ti_2X_9$ and $A_3V_2X_9$ are used as starting materials. The Bridgman ampules (15-cm length, 1-cm diameter) are graphitized for the bromides, but not for the chlorides. They are filled to a height of 10-cm with the powdered material and then evacuated at $200\,°C$ to a pressure of $< 10^{-3}$ torr. Immediately before sealing with a very hot flame, the lower part of the ampule is cooled to liquid nitrogen temperature. The crystals are grown in the Bridgman apparatus as described in detail in the General Procedure.

Properties

The boules are homogeneous and consist of one or several crystals. The well-developed crystal faces are the same as in $A_3Cr_2X_9$. Some further properties are listed in Table I.

Comment on $Cs_3MM'Cl_9 (M \neq M')$

■ **Caution.** *One should not attempt to prepare $Cs_3[CrTiCl_9]$ from a mixture of the pure compounds $Cs_3[Ti_2Cl_9]$ and $Cs_3[Cr_2Cl_9]$. Upon a slow heating, the former reduces the latter and liquid $TiCl_4$ is formed with a considerable risk of explosion. Although not checked, this procedure could also be hazardous in the case of Cr—V and Cr—Ru.*

Large single crystals of the type $Cs_3MM'Cl_9$ were grown with M—M' = Cr—Ru and Cr—V.[18] In both cases, a stoichiometric mixture of MCl_3, $M'Cl_3$, and CsCl was used as the starting material. After sealing under vacuum, the ampule (not graphitized) was placed in a vertical Bridgman furnace and slowly heated until the hot zone had reached the reaction temperature ($\simeq 900$ and $650\,°C$ for Cr—Ru and Cr—V, respectively). After 5 days with occasional shaking by rotating the ampule, this one was lowered in the cooler zone at a rate of $1 \, \text{cm} \, 24 \, \text{h}^{-1}$. The crystals were finally annealed around $200 \, ^\circ\text{C}$ after completion of the growth process. The same procedure was used in the case of $M—M' = Cr—Ti$, but it resulted in a powdered material. Optical, magnetooptical and magnetic studies have been conducted on $Cs_3[CrRuCl_9]$,[19] which is isostructural to $Cs_3[Cr_2Cl_9]$ and $Cs_3[Ru_2Cl_9]$.[20] The compound $Cs_3[CrVCl_9]$ has not yet been studied.

References

1. L. Dubicki, J. Ferguson, and B. V. Harrowfield, *Mol. Phys.*, **34**, 1545 (1977); and references 1, 12, and 13 cited therein.

2. B. Briat, O. Kahn, I. Morgenstern Badarau, and J. C. Rivoal, *Inorg. Chem.*, **20**, 4193 (1981); and refs. 2, 3, 7, 8, and 11 cited therein.

3. B. Leuenberger, A Stebler, H. U. Güdel, A. Furrer, R. Feile, and J. K. Kjems, *Phys. Rev. B*, **30**, 6300 (1984); B. Leuenberger, H. U. Güdel, J. K. Kjems, and D. Petitgrand, *Inorg. Chem.*, **24**, 1035 (1985).

4. A. Stebler, Ph.D. thesis, University of Bern (1981).

5. S. A. Shchukarev and I. L. Perfilova, *Russ. J. Inorg. Chem.*, **8**, 1100 (1963).

6. D. V. Korol'kov and G. N. Kudryashova, *Russ. J. Inorg. Chem.*, **13**, 850 (1968).

7. S. A. Shchukarev, I. V. Vasil'kova, and D. V. Korol'kov, *Russ. J. Inorg. Chem.*, **8**, 1006 (1963).

8. I. V. Vasil'kova, A. I. Efimov, and B. Z. Pitirimov, *Russ. J. Inorg. Chem.*, **9**, 493 (1964).

9. N. V. Galitskii, K. P. Minina, and G. M. Borisovskaya, *Russ. J. Inorg. Chem.*, **12**, 1386 (1967).

10. R. Saillant and R. A. D. Wentworth, *Inorg. Chem.*, **7**, 1606 (1968).

11. B. R. Pamplin, *Crystal Growth*, Pergamon Press, Oxford (1975).

12. A. I. Efimov and B. Z. Pitirimov, *Russ. J. Inorg. Chem.*, **8**, 1042 (1963).

13. G. J. Wessel and D. J. W. Ijdo, *Acta Cryst.*, **10**, 466 (1957).

14. R. Saillant, R. B. Jackson, W. E. Streib, K. Folting, and R. A. D. Wentworth, *Inorg. Chem.*, **10**, 1453 (1971).

15. I. Kozhina and D. V. Korol'kov, *J. Struct. Chem.*, **6**, 84 (1965).

16. E. H. Hall and J. M. Blocher, Jr., *J. Phys. Chem.*, **63**, 1525 (1959).

17. E. H. Hall and J. M. Blocher Jr., and I. E. Campbell, *J. Electrochem. Soc.*, **105**, 271 (1958).

18. B. Briat, unpublished results.

19. B. Briat, J. C. Canit, J. Darriet, andj M. Drillon, Proceedings of Int'l Conf. on Coordination Chemistry, Denver, Co., 1984. p. 72.

20. J. Darriet, *Rev. Chem. Min.*, **18**, 27 (1981).

70. HIGHLY CONDUCTING AND SUPERCONDUCTING SYNTHETIC METALS*

Submitted by JACK M. WILLIAMS and COWORKERS[†]

Research in the area of new synthetic metal (synmetal) superconductors continues to be stimulated by rapid developments in the chemistry and physics of these materials.[1] Synmetals are unique in that they have metallic properties even though they contain no metals. However, this field remains very much materials limited because, as yet, it is not possible to predict which new materials will exhibit metallic properties. This situation is improving because structure–property correlations have been developed recently that may be used to predict likely candidates for superconductivity in the 4, 4', 5, 5'-tetramethyl-2, 2'-bi-1-3-diselenolylidene, often called the tetramethyltetraselenafulvalene (TMTSF) class of materials.[2]

The first organic superconductor, $(TMTSF)_2PF_6$, was discovered in 1979 and required pressure (~ 12 k bar $= 12,000$ atm) to induce superconductivity.[3] This was quickly followed in 1981 with the discovery of $(TMTSF)_2ClO_4$ the only *ambient pressure* organic superconductor known in the TMTSF family.[4] Even more recently pressure induced (~ 5 k bar) superconductivity was discovered for the first time in a sulfur-based material, $(BEDT—TTF)_2ReO_4$.[5] This has been a most surprising finding considering that sulfur itself is an electrical insulator. The superconducting transition temperatures (T_c values) for the TMTSF and BEDT–TTF compounds are ~ 1 and 2 K, respectively.

5,5',6,6', tetrahydro-2,2'-Bi-1,3-dithiolo[4,5-*b*][1,4]dithiinylidene

5, 5', 6, 6'-tetrahydro-bis(ethylenedithiolo)tetrathiafulvalene or BEDT–TTF

The synthesis of BEDT–TTF III was first reported in 1978 and was accomplished by phosphite coupling of the key thione intermediate, 1, 3-dithiolo[4, 5-*b*][1, 4]dithiin-2-thione, commonly known as EDT–DTT

*Work performed under the auspices of the Office of Basic Energy Sciences, Division of Material Sciences, of the U.S. Department of Energy under contract W-31-109-ENG-38.
[†]Chemistry and Materials Science Divisions, Argonne National Laboratory, Argonne, IL 60439.

[standing for 4, 5-(ethylenedithio)-1, 3-dithiole-2-thione, **II**].[6] The compound EDT–DTT is synthesized by the addition of 1, 2-dibromoethane to the disodium salt of 4, 5-dimercapto-1, 3-dithiole-2-thione, **I**.

Disodium 4, 5-dimercapto-1, 3-dithiole-2-thione   EDT–DTT

Compound **I** may be prepared by the reduction of carbon disulfide either electrochemically[6] or by use of an alkali metal.[7] In this report we present the synthesis of BEDT–TTF, and derivatives, via the procedure just described beginning with the alkali metal reduction of carbon disulfide. In our hands we find this procedure easier to set up and use than that involving the electrolytic reduction of carbon disulfide.

Materials

The following reagents are distilled prior to use: methanol (Fischer Scientific, acetone-free, absolute, stored over 4A molecular sieve and distilled from CaH_2), and triethyl phosphite (Aldrich, 99%, distilled under Argon). Triethyl phosphite is redistilled under argon *immediately* prior to use. Carbon disulfide (Aldrich, gold label) is purified in the manner described in Ref. 8. *N, N*-Dimethylformamide (DMF) (Aldrich, gold label) is stirred with KOH pellets, distilled over CaO, and dried by passage through Al_2O_3 prior to use. All other chemicals are used as obtained. They are as follows: 1, 2-dibromomethane (Aldrich, gold label), absolute ethanol (Aaper Alcohol and Chemical), diethyl ether (Mallinkrodt, anhydrous), and trichloromethane (Fischer Scientific, reagent grade). The sodium metal is prepared by cutting it into small pieces under hexane in a beaker.

A. SYNTHESIS OF 5,5′,6,6′-TETRAHYDRO-2,2′-BI-1,3-DITHIOLO[4,5-*b*][1,4]DITHIINYLIDENE (BEDT-TTF)

Submitted by PETER E. REED,* JULIE M. BRAAM,* LAUREN M. SOWA,*
ROBERT A. BARKHAU, * GREGORY S. BLACKMAN,* DAVID D. COX,*
GERALD A. BALL,* HAU. H. WANG,† and JACK M. WILLIAMS†
Checked by ALLEN E. UNDERHILL§

(I)   (II) EDT-DTT

(III) BEDT-TTF

Procedure

■ **Caution.**   *Care must be exercised in the handling of metallic sodium and carbon disulfide (inflammable) and a face shield and rubber gloves should be worn when the two are mixed in Section A 1. Gloves should be worn in all steps.*

*Research participants sponsored by the Argonne Division of Educational Programs: Peter E. Reed from the University of Wisconsin-Eau Claire, Eau Claire, WI; Julie M. Braam from St. Mary's College, Winona, MN; Lauren M. Sowa from Grinnell College, Grinnell, IA; Robert A Barkhau from Carthage College, Kenosha, WI; Gregory S. Blackman from Shippensburg State College, Shippensburg, PA; David D. Cox from Bethel College, St. Paul, MN; and Gerald A. Ball from St. Augustine's College, Raleigh, NC.
†Chemistry and Materials Science Divisions, Argonne National Laboratory, Argonne, IL.
‡Correspondent, Chemistry and Materials Science Divisions, Argonne National Laboratory, Argonne, IL.
§Department of Chemistry, University College of North Wales, Bangor, Gwynedd LL57 2UW, Wales, United Kingdom.

The dithiolate, I, is oxygen and moisture sensitive. Carbon disulfide, DMF, and methanol should be purged with argon prior to use.

1. 1,3-Dithiolo[4,5-*b*][1,4]dithiin-2-thione, EDT–DTT, II

To a 100-mL three necked, glass round-bottom flask fitted with a reflux condenser, pressure equalizing dropper funnel, argon sweep, heating mantle, and stirrer is added 18 mL (0.30 mol) of purified[8] carbon disulfide and 2.3 g (0.10 mol) of sodium metal, freshly cut into pieces ~ 3 mm on a side. The carbon disulfide is brought to reflux and 20 mL (0.26 mol) of DMF is added dropwise from the dropper funnel over a 30 min period. The mixture becomes intensely dark red to purple colored over the course of the addition, indicating the presence of **I**. The reflux is continued for an additional 90 min, then the heating mantle is removed and the resulting solution is concentrated under vacuum overnight. It is very important to remove all the DMF by using a good vacuum and warming to a maximum of 50 °C, leaving a very viscous residue.

The remaining residue is dissolved in 65 mL of anhydrous methanol, stirred for 15 min, and suction filtered under argon through a coarse porosity glass frit into a 250-mL round-bottom flask equipped with a magnetic stirring bar and argon sweep. The filtrate is then cooled with stirring in an ice bath, followed by the dropwise addition of 6.5 mL (0.075 mol) of 1,2-dibromoethane from a pressure-equalizing dropper funnel over a 30-min period. The ice bath is then removed and the reaction mixture is allowed to stir for an additional 60 min during which time a yellow precipitate forms. The mixture is then cooled in an ice–acetone bath for 15 min and filtered.

The resulting solid is extracted with 400 mL of boiling absolute ethanol with stirring for 30 min and while hot it is filtered leaving behind a small amount of insoluble brown residue. The filtrate is then cooled in an ice bath to precipitate the crude product, **II**, and filtered (2.95 g, 52% of theoretical yield). Subsequent recrystallization from ethanol (with charcoal treatment), or alternatively, column chromatography on partially deactivated silica gel, gradient eluted with chloroform–hexane, will yield analytically pure product.

Anal. Calcd. for $C_5H_4S_5$: C, 26.76; H, 1.80; S, 71.44. Found:* C, 27.66; H, 1.83; S, 71.45.

Properties

The compound EDT–DTT, **II**, crystallizes as gold needles that melt at 119 to

*Midwest Microlabs, Indianapolis, IN.

120 °C. It is soluble in organic solvents and insoluble in water. Silica gel TLC (thin layer chromatography) gives a spot at $R_f = 0.83$ in 3:1 chloroform–hexane. The ^{1}H NMR spectrum in CDCl$_3$ (TMS, tetramethylsilane) is a singlet at δ 3.41. The IR spectrum (KBr) exhibits bands at 869 (s), 919 (w), 1010 (w), 1035 (w), 1060 (s), 1126 (w), 1255 (w), 1400 (w), 1478 (w), 2925 (w), and 2950 (w). The UV-Vis spectrum shows a λ_{max} at 403 nm, $\varepsilon = 10490 \pm 80$. The compound EDT–DTT is best kept when stored in a cold, dark place.

2.   5,5′,6,6′-Tetrahydro-2,2′-Bi-1,3-dithiolo[4,5-*b*][1,4]dithiinylidene, BEDT–TTF, III

Initially, 20 mL of triethyl phosphite, *always* freshly redistilled over argon prior to use, is added via syringe to a 100-mL three-necked glass round-bottom flask (equipped with a reflux condenser, septum, magnetic stirring bar, heating mantle, and argon sweep) containing 1.00 g of EDT–DTT, **II**. The resulting suspension is stirred and slowly warmed from 110 to 120 °C and held within that temperature range for 45 min. The compound EDT–DTT, **II**, will dissolve upon warming, followed by the precipitation of product, **III**, from the red-orange colored solution.

The reaction mixture is then cooled to -10 °C in an ice–acetone bath, filtered, and the resulting solid is washed with ice-cold anhydrous diethyl ether. The 0.300 g of dark orange solid obtained (36% yield) is recrystallized from 300 mL of chloroform.

Anal. Calcd. for C$_{10}$H$_8$S$_8$: C, 31.22; H, 2.10; S, 66.68. Found: C, 31.20; H, 2.12; S, 66.34.

Properties

The compound BEDT–TTF, **III**, crystallizes as reddish-orange colored needles that decompose at 238 to 240 °C. It is slightly soluble in polar organic solvents and insoluble in water. The ^{1}H NMR spectrum in CDCl$_3$ (TMS) is a singlet at δ 3.29. The IR spectrum (KBr) exhibits bands at 388 (s), 496 (s), 623 (m), 675 (m), 765 (s), 868 (w), 882 (s), 900 (m), 912 (m), 991 (w), 1280 (m), 1407 (m), 2920 (w), and 2955 (w). The UV-Vis spectrum shows a λ_{max} at 319 nm, $\varepsilon = 16310 \pm 590$. The compound BEDT–TTF will slowly oxidize when exposed to light and air, but can be kept for several months if stored under argon in the dark in a freezer.

B. TETRABUTYLAMMONIUM PERRHENATE AND SUPERCONDUCTING BIS(5,5',6,6'-TETRAHYDRO-2,2'-BI-1,3-DITHIOLO[4,5-*b*][1,4]DITHIINYLIDENE) PERRHENATE

Submitted by STEVEN J. COMPTON,* DAVID D. COX,* PETER E. REED,*
and JACK M. WILLIAMS†
Checked by EDWARD M. ENGLER‡

Procedure

Chemical Synthesis

$$[CH_3(CH_2)_3]_4NHSO_4 + HReO_4 \xrightarrow{MeOH} [CH_3(CH_2)_3]_4NReO_4 + H_2SO_4$$

A quantity of tetrabutylammonium hydrogen sulfate 28.7 g (Aldrich), that has been recrystallized from acetone, is added to 300 mL of warm methanol in a 500-mL beaker. To this is added a solution of aqueous perrhenic acid (28.0 g, 80 to 85%, titrated 75.7%) (Alfa) with stirring. The total volume is then reduced by boiling to 125 mL, and the solution is then cooled slowly to 0 °C yielding white needles of $[CH_3(CH_2)_3]_4NReO_4$. The supernatant liquid is further condensed to 50 mL and the resulting needles are combined with the initial crystalline product and twice recrystallized from methanol to give 30.8 g (74% yield) of product (mp 236–237 °C).

Anal. Calcd. for $[CH_3(CH_2)_3]_4NReO_4$: C, 39.00; H, 7.37; N, 2.84; O, 12.99. Found: C, 39.31, 39.12, 39.13; H, 7.56, 7.33, 7.87; N, 2.80, 2.73, 2.88; O, 12.73, 12.82, 13.01.

Electrocrystallization

$$[CH_3(CH_2)_3]_4NReO_4 + 2BEDT-TTF$$

$$\xrightarrow{0.5\,\text{Å}} (BEDT-TTF)_2ReO_4 + [CH_3(CH_2)_3]_4NReO_4$$

The crystal growing apparatus for the electrocrystallization is an H cell

*Research participants sponsored by the Argonne Division of Educational Programs: from Dartmouth College, Hanover, NH; Bethel College, St. Paul, MN; and the University of Wisconsin-Eau Claire, Eau Claire, WI.
†Chemistry and Materials Science Divisions, Argonne National Laboratory, Argonne, IL.
‡IBM Almaden Research Center, 650 Harry Road, San Jose, CA.

(15-mL capacity) and it and the detailed procedures followed for crystal growth have been described previously in *Inorganic Synthesis.*[9] The solutions needed for the cell consist of $[CH_3(CH_2)_3]_4NReO_4$ (0.738 g, 1.5 mmol) in previously distilled, dried, and degassed 1, 1, 2-trichloroethane (TCE) (10 mL) and BEDT–TTF (0.0144 g, 0.0375 mmol) in TCE (5 mL). The crystals are grown at a current of 0.5 μA, and harvested at the desired stage of crystal growth to yield black metallic needle crystals of $(BEDT–TTF)_2ReO_4$. Crystal growth at 0.5 μA can usually be continued for 2 to 3 weeks, until the solution in the anode compartment begins to change in color to brownish-orange, at which point the crystals should be harvested. Crystals grown in this manner using TCE as solvent produce predominantly one crystallographic phase of the superconducting perrhenate compound, whereas three of five different phases (only one superconducting) are grown if the solvent is THF.[5] The crystallographic phase is best confirmed by the unit cell parameters.

Properties

Crystals of $(BEDT–TTF)_2ReO_4$ are lustrous metallic black in color and are superconducting at a temperature of ~ 2 K at a pressure of ~ 5 k bar.[5] In the absence of applied pressure this material shows metallic behavior (increased electrical conductivity as the temperature is decreased) from room temperature to 81 K at which temperature a metal-insulator transition occurs. The crystallographic lattice parameters, for the triclinic unit cell, are[5] $a = 7.78$ Å, $b = 12.59$ Å, $c = 16.97$ Å, $\alpha = 73.0°$, $\beta = 79.89°$, $\gamma = 89.06°$, and unit cell volume, $V_c = 1565$ Å^3.

C. TETRABUTYLAMMONIUM FLUOROSULFATE AND 5,5',6,6'-TETRAHYDRO-2,2'-BI-1,3-DITHIOLO[4,5-*b*][1,4]-DITHIINYLIDENE FLUOROSULFATE

Submitted by DAVID D. COX,* GERALD A. BALL,* ANTHONY S.
ALONSO,* and JACK M. WILLIAMS[†]
Checked by EDWARD M. ENGLER[‡]

Procedure

Chemical Synthesis

$$[CH_3(CH_2)_3]_4NHSO_4 + HFSO_3 \xrightarrow{H_2O} [CH_3(CH_2)_3]_4NFSO_3 + H_2SO_4$$

■ **Caution.** *Fluorosulfuric acid, $HFSO_3$, fumes strongly in air and reacts explosively with water. Proper safety precautions, including wearing a face shield and gloves, should be taken at all time when handling $HFSO_3$.*

A solution containing 105.0 g (31 mmol) of tetrabutylammonium hydrogen sulfate (recrystallized once from acetone) in 150 mL of triply distilled water is placed in a plastic beaker (500 mL) and placed in an ice bath to cool. Then a solution of 95% $HFSO_3$ (18.8 mL, 31 mmol) (Aldrich) is added dropwise with stirring. A white precipitate immediately forms.[§] This precipitate is filtered and washed with copious amounts of ice-cold triply distilled water until the wash solution shows pH near 7.0. The precipitate is then dried and twice recrystallized from ethyl acetate to yield ~9 g (8% of theoretical yield) of colorless, platelike crystals of very pure $[CH_3(CH_2)_3]_4NFSO_3$ (mp = 179–180 °C). A neutral salt of very high purity is essential for any electro-crystallization portions of this synthesis, otherwise spurious reactions occur and crystal growth fails.

Anal. Calcd. for $[CH_3(CH_2)_3]_4NFSO_3$: C, 56.27; H, 10.62; N, 4.10; F, 5.56; S, 9.39. Found: C, 56.38, 56.19; H, 10.56, 10.74; N, 4.09, 4.20; F, 5.68, 5.40; S, 9.38, 9.54.

*Research participants sponsored by the Argonne Division of Educational Programs from Bethel College, St. Paul, MN; St. Augustine's College, Raleigh, NC; and Argo Community High School, Summit, IL.
[†]Chemistry and Materials Science Divisions, Argonne National Laboratory, Argonne, IL.
[‡]IBM Almaden Research Center, 650 Harry Rd., San Jose, CA.
[§]In the literature fluorosulfuric acid is reported to react violently with water, but no difficulties are encountered if the procedure outlined here is followed explicitly.

Electrocrystallization

$$[CH_3(CH_2)_3]_4NFSO_3 + 2BEDT\text{-}TTF$$

$$\xrightarrow[\text{TCE}]{1.0\mu A} (BEDT\text{-}TTF)_2FSO_3(TCE)_{0.5} + [CH_3(CH_2)_3]_4NFSO_3$$

The solutions required consist of BEDT–TTF (0.012 g, 0.03 mmol) in purified (see Section A) TCE (5 mL), and $[CH_3(CH_2)_3]_4NFSO_3$ (0.512 g, 1.5 mmol) in purified TCE (10 mL). The crystals are grown using a current of 1.0 μA and harvested after $\sim 2\frac{1}{2}$ weeks (when color changes may occur) or at any desired stage of crystal growth.

Block-shaped black crystals with a distinct metallic luster are formed which upon crystallographic study are found to crystallize in the triclinic system with lattice constants as follows: $a = 7.786$ (1) Å, $b = 13.033$ (3) Å, $c = 18.590$ (4) Å, $\alpha = 110.09$ (2)°, $\beta = 90.21$ (2)°, $\gamma = 105.27$ (2)°, and $V_c = 1699.8$ (6) Å^3. These lattice parameters indicate that the crystals contain 0.5 molecules of TCE in the formula unit as is the case for isostructural (BEDT–TTF)$_2$ClO$_4$(TCE)$_{0.5}$ in which $V_c = 1689.8$ Å^3.[10] The electrical properties of this material indicate that they are metallic from room temperature to 90 K at which point a metal–insulator transition occurs.

References

1. J. M. Williams, H. H. Wang, T. J. Emge, U. Geiser, M. A. Beno, P. C. W. Leung, K. D. Carlson, R. J. Thorn, and A. J. Schultz, *Prog. Inorg. Chem.*, **35**, 51 (1987).

2. J. M. Williams, M. A. Beno, J. C. Sullivan, L. M. Banovetz, J. M. Braam,G. S. Blackman, C. D. Carlson, D. L. Greer, and D. M. Loesing, *J. Am. Chem. Soc.*, **105**, 643 (1983).

3. K. Bechgaard, C. S. Jacobsen, K. Mortensen, H. J. Pedersen, and N. Thorup, *Solid State Commun.*, **33**, 1119 (1980).

4. K. Bechgaard, K. Carneiro, F. B. Rasmussen, M. Olsen, G. Rindorf, C. S. Jacobsen, H. J. Pedersen, and J. C. Scott, *J. Am. Chem. Soc.*, **103**, 2440 (1981).

5. S. S. P. Parkin, E. M. Engler, R. P. Schumaker, R. Lagier, V. Y. Lee, J. C. Scott, and R. L. Greene, *Phys. Rev. Lett.*, **50**, 270 (1983).

6. M. Mizuno, A. F. Garito, and M. P. Cava, *J. Chem. Soc. Chem. Commun.*, **1978**, 18.

7. G. Steimecke, H.-J. Sieler, R. Kirmse, and E. Hoyer, *Phosphorous Sulfur*, **7**, 49 (1979).

8. A. J. Gordon and R. A. Ford, *The Chemist's Companion*, Wiley, New York, 1972, p. 432.

9. J. M. Williams, *Inorg. Synth.*, **24**, 136 (1986).

10. G. Saito, T. Enoki, K. Toriumi, and H. I. Inokuchi, *Solid State Commun.*, **42**, 557 (1982).

INDEX OF CONTRIBUTORS

SUBJECT INDEX

Names used in this Subject Index for Volumes 26–30 are based upon IUPAC *Nomenclature of Inorganic Chemistry*, Second Edition (1970), Butterworths, London; IUPAC *Nomenclature of Organic Chemistry*, Sections A, B, C, D, E, F, and H (1979), Pergamon Press, Oxford, U.K.; and the Chemical Abstracts Service *Chemical Substance Name Selection Manual* (1978), Columbus, Ohio. For compounds whose nomenclature is not adequately treated in the above references, American Chemical Society journal editorial practices are followed as applicable.

Inverted forms of the chemical names (parent index headings) are used for most entries in the alphabetically ordered index. Organic names are listed at the "parent" based on Rule C-10, *Nomenclature of Organic Chemistry*, 1979 Edition. Coordination compounds, salts and ions are listed once at each metal or central atom "parent" index heading. Simple salts and binary compounds are entered in the usual uninverted way, e.g., *Sulfur oxide* (S_8O), *Uranium(IV) chloride* (UCl_4).

All ligands receive a separate subject entry, e.g., *2,4-Pentanedione*, iron complex. The headings *Ammines, Carbonyl complexes, Hydride complexes,* and *Nitrosyl complexes* are used for the NH_3, CO, H, and NO ligands.

FORMULA INDEX

The Formula Index, as well as the Subject Index, is a Cumulative Index for Volumes 26–30. The Index is organized to allow the most efficient location of specific compounds and groups of compounds related by central metal ion or ligand grouping.

The formulas entered in the Formula Index are for the total composition of the entered compound, e.g., F_6NaU for sodium hexafluorouranate(V). The formulas consist solely of atomic symbols (abbreviations for atomic groupings are not used) and arranged in alphabetical order with carbon and hydrogen always given last, e.g., $Br_3CoN_4C_4H_{16}$. To enhance the utility of the Formula Index, all formulas are permuted on the symbols for all metal atoms, e.g., $FeO_{13}Ru_3C_{13}H_{13}$ is also listed at $Ru_3FeO_{13}C_{13}H_{13}$. Ligand groupings are also listed separately in the same order, e.g., $N_2C_2H_8$, 1,2-Ethanediamine, cobalt complexes. Thus individual compounds are found at their total formula in the alphabetical listing; compounds of any metal may be scanned at the alphabetical position of the metal symbol; and compounds of a specific ligand are listed at the formula of the ligand, e.g., NC for Cyano complexes.

Water of hydration, when so identified, is not added into the formulas of the reported compounds, e.g., $Cl_{0.30}N_4PtRb_2C_4 \cdot 3H_2O$.

$AsFeO_4C_{22}H_{15}$, Iron, tetracarbonyl-
(triphenylarsine)-, 26:61
$As_2C_{10}H_{16}$, Arsine, 1,2-phenylenebis-
(dimethyl)-
gold complex, 26:89
$As_4Au_2F_{10}C_{32}H_{32}$, Gold(I), bis[1,2-
phenylenebis(dimethylarsine)] bis-
(pentafluorophenyl)aurate(I), 26:89
$AuClF_5PC_{31}H_{22}$, Aurate(I), chloro-
(pentafluorophenyl)-
(benzyl)triphenylphosphonium, 26:88
$AuClPC_{18}H_{15}$, Gold, chloro(triphenylphosphine)-, 26:325
$AuClSC_4H_8$, Gold(I), chloro(tetrahydrothiophene)-, 26:86
$AuF_5SC_{10}H_8$, Gold(I), (pentafluorophenyl)-
(tetrahydrothiophene)-, 26:86
$AuF_{15}SC_{22}H_8$, Gold(III), tris(pentafluorophenyl) (tetrahydrothiophene)-,
26:87
$AuMn_2O_8P_2C_{38}H_{25}$, Gold, octacarbonyl-
$1\kappa^4C$,-$2\kappa^4C$-μ-(diphenylphosphino)-
$1:2\kappa P$-(triphenylphosphine)-$3\kappa P$-$trian$-
$galo$-dimanganese-, 26:229

$Au_2As_4F_{10}C_{32}H_{32}$, Gold(I), bis[1,2-
phenylenebis(dimethylarsine)]-
bis(pentafluorophenyl)aurate(I), 26:89
$Au_2F_5NPSC_{25}H_{15}$, Gold(I), (pentafluorophenyl)-μ-thiocyanato-(triphenylphosphine)di-, 26:90
$Au_3BF_4OP_3C_{54}H_{45}$, Gold(1 +), μ_3-oxo-[tris-
[(triphenylphosphine)
tetrafluoroborate(1 –), 26:326
$Au_3CoO_{12}P_3Ru_3C_{66}H_{45}$, Ruthenium, dodeca-
carbonyltris(triphenylphosphine)-
cobalttrigoldtri-, 26:327

$BAu_3F_4OP_3C_{54}H_{45}$, Gold(1 +), μ_3-oxo-[tris-
(triphenylphosphine)]-
tetrafluoroborate(1 –), 26:326
$BClF_4IrN_2P_2C_{36}H_{31}$, Iridium(III), chloro-
(dinitrogen)hydrido[tetrafluoro-
borato(1 –)]bis(triphenylphosphine)-, 26:119
$BClF_4IrOP_2C_{37}H_{31}$, Iridium(III), carbonyl-
chlorohydrido[tetrafluoroborato-
(1 –)]bis(triphenylphosphine)-, 26:117

411